Sustainable Development Goals Series

The **Sustainable Development Goals Series** is Springer Nature's inaugural cross-imprint book series that addresses and supports the United Nations' seventeen Sustainable Development Goals. The series fosters comprehensive research focused on these global targets and endeavours to address some of society's greatest grand challenges. The SDGs are inherently multidisciplinary, and they bring people working across different fields together and working towards a common goal. In this spirit, the Sustainable Development Goals series is the first at Springer Nature to publish books under both the Springer and Palgrave Macmillan imprints, bringing the strengths of our imprints together.

The Sustainable Development Goals Series is organized into eighteen subseries: one subseries based around each of the seventeen respective Sustainable Development Goals, and an eighteenth subseries, "Connecting the Goals", which serves as a home for volumes addressing multiple goals or studying the SDGs as a whole. Each subseries is guided by an expert Subseries Advisor with years or decades of experience studying and addressing core components of their respective Goal.

The SDG Series has a remit as broad as the SDGs themselves, and contributions are welcome from scientists, academics, policymakers, and researchers working in fields related to any of the seventeen goals. If you are interested in contributing a monograph or curated volume to the series, please contact the Publishers: Zachary Romano [Springer; zachary.romano@springer.com] and Rachael Ballard [Palgrave Macmillan; rachael.ballard@palgrave.com].

Susanta Kumar Chakraborty
Lakshmi Sivaramakrishnan
Uday Chatterjee
Ranita Karmakar
Editors

Wetlands and Urban Resilience

A Global Outlook

Springer

Editors
Susanta Kumar Chakraborty
Department of Zoology
Vidyasagar University
Paschim Medinipur, West Bengal, India

Lakshmi Sivaramakrishnan
Department of Geography
Jadavpur University
Kolkata, West Bengal, India

Uday Chatterjee
Department of Geography
Bhatter College Dantan (Autonomous)
Paschim Medinipur, West Bengal, India

Ranita Karmakar
Department of Geography
Jadavpur University
Kolkata, West Bengal, India

ISSN 2523-3084 ISSN 2523-3092 (electronic)
Sustainable Development Goals Series
ISBN 978-3-032-06880-4 ISBN 978-3-032-06881-1 (eBook)
https://doi.org/10.1007/978-3-032-06881-1

This Springer imprint is published by the registered company Springer Nature Switzerland AG
The registered company address is: Gewerbestrasse 11, 6330 Cham, Switzerland

Dedicated to
Young Scholars in the Field of Hydrology, Urban Studies, Environmental Science, Hazard and Disaster Management, Geospatial Science, Biological Science, Geography, and Policymaking

Preface

Writing up a book highlighting some burning problems on two major global environmental issues, "Urban resilience" and "Wetlands," is really challenging to provide the readers proper insights on the interconnections and interrelationships of these two major thrust environmental fields of research areas by bringing together research communities from different disciplines and practitioners at different levels of jurisdictions from the world. This is a book that attempts to deepen understanding of the readers on the ongoing uncontrolled and non-judicious accelerated processes of urbanization, the outcomes of which alongside other forms of environmental perturbations including global climate change have posed real threats to every compartment of global environment, especially on the structure and function of the wetland ecosystem. Wetlands, being the interface between terrestrial and aquatic ecosystems, represent the most valued and precious life supporting water resources and play crucial roles in rendering lot of ecosystem services. They provide food, filter water, offer a unique habitat for many different species of endemic and endangered flora and fauna, act as natural buffer against flood buffers, help recharging groundwater, stabilize and protect shoreline from the devastating effects of natural disasters like catastrophic storms, cyclones, hurricanes, and tsunamis, and generate avenues for different recreational and cultural activities. Despite their vital ecological and socioeconomic roles, wetlands are being considered as the most "Threatened Landscape" of the world mostly due to a number of anthropogenic activities like endangered by non-judicious abstraction of water, eutrophication because of releasing of urban sewage, agricultural drainage, climate-driven changes, but out of all the most important is the destruction of wetlands leading to large shrinkage of wetland area mainly because of accelerated pace of urbanization. In such context, managing wetlands has fostering urban resilience globally is the need of the hour. Both urbanized sectors and wetlands have underwent tremendous ecological transformations during the last century, resulting in eco-degradation of wetlands with drastic depletion of ecosystem services and globalization of urban sectors with profuse changes in the lifestyle patterns of human civilization developing very unique social-ecological systems characterized by a complex web of interactions among "human" and "non-human" components.

Urban resilience takes care of the ability of an urban area to withstand the ongoing man-mediated environmental perturbations and also to adapt to the changing environmental scenario by acquiring resilience in the physical infrastructure, social, ecological, and economic alterations for the survivability of the urban areas. Acknowledging the urgent need to tackle these issues, the present book proposal aims at serving twofold objectives: investigate the intricate relationship between wetlands and global urban resilience, and also to acquaint the readers on innovative research planning and conservation strategies for holistic eco-management of the environment to achieve sustainability. Moreover, the very idea of preparing this book is to pool some salient ideas of governance of water in the inland water bodies, the most precious resource of the world, with the help of ecologically technological applications in view of ongoing urbanization from global perspectives. Each of the 22 different chapters within this book, written by distinguished researchers across the disciplines, has dealt with the in-depth analytical frameworks of the respective subjects adhering to eco-centric and techno-centric approaches.

Drawing on interdisciplinary perspectives from ecology, urban planning, hydrology, economics, and social sciences, the proposed book aims to comprehensively examine and highlight the roles of wetlands in rendering human benefits which in turn, through feedback control, motivate and trigger the interests of environmental planners to promote wetland conservation by promoting and enhancing urban resilience across diverse geographical and socioeconomic contexts of the world. The 22 book chapters, through a series of case study-based empirical research, theoretical analyses, and policy discussions, will help the readers explore multidimensional facets of wetland management, restoration, and conservation in urban settings. Several innovative approaches and best practices are being highlighted for integrating wetland conservation into urban planning and development strategies, fostering synergies between ecological conservation and urban resilience goals. Key themes to be explored include the ecological functions of urban wetlands, such as flood mitigation, water purification, carbon sequestration, and biodiversity conservation; the socioeconomic benefits of wetlands for urban communities, including recreational opportunities, cultural values, and ecosystem services; the challenges and threats facing urban wetlands, including pollution, habitat loss, invasive species, and climate change impacts; the role of governance mechanisms, policy frameworks, and institutional arrangements in promoting wetland conservation and sustainable urban development; the importance of community engagement, stakeholder participation, and indigenous knowledge in wetland management and resilience-building efforts; and the potential for innovative technologies, nature-based solutions, and green infrastructure in enhancing the resilience of urban wetland ecosystems. By synthesizing cutting-edge research, practical insights, and policy recommendations, this book accounts to contribute a deeper understanding of the nexus between wetlands and urban resilience and also to inspire informed decision-making and action toward the sustainable management and conservation of wetlands in urban areas worldwide. With contributions from leading experts and practitioners in the field, the prime envision of this book is to form a repository of valuable research-based academic resources for academics, policymakers, urban planners, environmental professionals, community stakeholders, and students interested in advancing knowledge and practice at the intersection of wetlands conservation and urban resilience. Through collaboration and collective efforts, the authors feel tempted to imagine a future world where wetlands should thrive as integral components of resilient and sustainable urban landscapes, contributing to the well-being of both people and the planet.

The target readers and learners of this volume are undergraduate and postgraduate students, researchers, academics, technicians, administrative planners, decision-makers, and environmental managers who are expected to get the benefits of some new information on a lot of issues with arguments, as contained in this book. The book provides a comprehensive analysis of the intricate relationship between urban environments and the crucial function of wetland ecosystems in enhancing the resilience of cities worldwide. This literary work provides an extensive examination of the various aspects involved in urban wetlands, focusing on their ecological, socioeconomic, and climatic importance within the ever-changing landscape of rapid urban growth. This study thoroughly examines the role of these frequently underestimated natural habitats as crucial contributors to the mitigation of diverse environmental hazards, such as flood control, water purification, and protection against the effects of climate change. Additionally, these habitats provide essential ecosystem services that are vital for the welfare of urban communities. By incorporating a range of viewpoints, examining specific instances, and drawing upon interdisciplinary investigations, the book provides a comprehensive understanding of novel approaches, flexible management methods, and successful policy structures that are essential for the seamless integration of urban wetlands into sustainable urban planning. This integration is vital for promoting resilience in cities situated in various geographical and sociocultural contexts.

This book employs an interdisciplinary methodology, incorporating perspectives from diverse disciplines such as ecology, urban planning, environmental science, and social sciences. Readers can develop a comprehensive understanding of the intricate interplay between urbanization, wetland ecosystems, and resilience. The case studies included in this collection provide valuable insights, exemplary methods, and innovative techniques that may be applied to various urban settings. These findings enhance readers' comprehension and facilitate the implementation of successful strategies in their respective local or regional contexts. In addition to engaging in academic discourse, the book offers practical guidance, actionable insights, and policy relevance. Various stakeholders, including policymakers, urban planners, academics, and environmental practitioners, have the opportunity to utilize these pragmatic observations to develop and execute effective approaches that promote the resilience of cities while safeguarding and optimizing the advantages offered by urban wetland ecosystems.

Paschim Medinipur, West Bengal, India — Susanta Kumar Chakraborty
Kolkata, West Bengal, India — Lakshmi Sivaramakrishnan
Paschim Medinipur, West Bengal, India — Uday Chatterjee
Kolkata, West Bengal, India — Ranita Karmakar

Acknowledgments

This book would not have been possible without the support and guidance of many who have helped in its compilation in various ways. Our gratitude goes out to the contributors, colleagues, researchers, and reviewers who played a dynamic role in the development and publication of the book entitled *Wetlands and Urban Resilience*. We would also like to thank the anonymous reviewers whose valuable comments were a great input. We extend our sincere gratitude to all the contributors whose scholarly insights and dedication have made this edited volume possible. Their commitment to excellence has enriched each chapter, contributing to the depth and diversity of perspectives presented in this work. Lastly, we acknowledge the friends and family who provided encouragement and understanding during the demanding process of compiling and editing this volume. Their unwavering support has been a source of inspiration. We also applaud Springer's publishing staff for their esteemed support, knowledge, and commitment to quality. Their careful editing, design, and production efforts made the book a comprehensive and user-friendly reference. We thank everyone who has played a part in the creation of this book, contributing to its success and impact.

Midnapore (West), West Bengal, India — Susanta Kumar Chakraborty
Kolkata, West Bengal, India — Lakshmi Sivaramakrishnan
Uday Chatterjee
Ranita Karmakar

Disclaimer

The authors of individual chapters are solely responsible for the ideas, views, data, figures, and geographical boundaries presented in the respective chapters of this book, and these have not been endorsed, in any form, by the publisher, the editor, and the authors of forewords, preambles, or other chapters.

Contents

Editors and Contributors

About the Editors

Susanta Kumar Chakraborty is former Professor of Zoology at Vidyasagar University, and he received his MPhil in Environmental Science and PhD in Marine Science from Calcutta University. His research centers on Coastal Ecology, Mangrove and Fresh Water Wetlands, Rivers and Estuaries, emphasizing biodiversity and ecosystem dynamics. He has also worked on the integration of rural bio-technologies for environmental management, eco-restoration, and livelihood generation. He has received a number of awards in recognition of his research commitments. He has taught for over three decades, mostly in the post-graduate level, covering different aspects of Zoology in general and Ecology/Environmental Science in particular, and has successfully supervised 42 PhD students, published around 200 research papers, written books, and has completed more than a dozen research projects, including consultancy work on environmental impact assessment in several countries. He has also served at other universities as a member of selection committees and academic bodies, as well as an expert member in several government departments like UGC and ICAR, and he is a life member of a number of scientific bodies in India and serves on the editorial board of a number of journals.

Lakshmi Sivaramakrishnan is the founder of the Geography Department at Jadavpur University, which was initiated in 2014. She has a long tenure of teaching at different universities of West Bengal, namely, North Bengal University and the

University of Burdwan, as full-time Faculty and as Guest Faculty at the University of Tripura, WBSU, Calcutta University, and Kalyani University. She studied at the University of Calcutta, where she completed her PhD in 1996. She is an Urban Geographer who is proficient in Remote Sensing GIS and therefore combines the strengths of Regional Planning and Human Geography. She has 60 research papers and 3 books to her credit. She has delivered lectures across universities in India and also at the University of Colombo and Dacca and presented papers at national and international conferences across the country. She has taken up projects under UGC and has been supervising several research fellows. She has guided more than 20 scholars who have obtained their doctoral degrees and has also supervised several MPhil scholars. She is on the editorial board of many renowned journals like *Hill Geographer*, *Regional Science Policy and Practice*, and *Indian Journal of Regional Science*, and at present is the Editor of *Geographical Review of India*.

Uday Chatterjee is an Assistant Professor at the Department of Geography, Bhatter College, Dantan, affiliated with Vidyasagar University, West Bengal, India. He earned his doctoral degree in Applied Geography from Ravenshaw University, Cuttack, Odisha. His research interests encompass Urban Planning, Social and Human Geography, Applied Geomorphology, Indigenous Knowledge and Hazards and Disasters, Risk Assessment and Resilience, Community Response, and Environmental Issues. He has been actively involved in editing scholarly works, including the Special Issue on Urbanism, Smart Cities, and Modelling for *GeoJournal*, Springer; special issues in *Global Social Welfare*, Springer, and *Frontiers in Environmental Science* journal, and *Geological Journal* Wiley; Editorial Board Member of *Scientific Reports* (Nature Portfolio) journal, and serving as a Book Series Editor for Developments in Environmental Science, Elsevier (https://www.elsevier.com/books-and-journals/book-series/developments-in-environmental-science) and Sustainability Solutions—

An Interdisciplinary and Transdisciplinary Nexus (https://www.springer.com/series/59999), Springer Nature. His extensive research and editorial contributions highlight his commitment to advancing geographical knowledge and its applications. His research work has been funded by the West Bengal Pollution Control Board (WBPCB), Govt. of West Bengal, India, and Indian Council of Social Science Research (ICSSR), Govt. of India. He has published research papers in reputed journals, edited books with reputed publisher like Springer, Elsevier, CRC Press, and authored books, book chapters, and conference proceedings. Currently, he is working on an international project, in collaboration with Indonesia, Malaysia, and Japan funded by APN (Asia Pacific-Global Change Research, Grant DOI https://doi.org/10.30852/p.22128), and he has visited Saudi Arabia for international collaboration and conference session chair, and has visited Keio University, Tokyo, Japan, as a Visiting Professor.

Ranita Karmakar is a Senior Research Fellow at the Department of Geography, Jadavpur University, Kolkata, India, and is currently working toward her doctoral research. She has completed her graduation and master's in Geography from Lady Brabourne College, Calcutta University, with a specialization in population geography. Currently, she is working as Editor of an Elsevier Book *Named Urban Health: A Global Perspective*. Apart from that she has four more papers that are accepted but yet to be published, many workshops, and seminars. In addition to that, she is currently working on a patent named *Vehicle Pollution Monitoring Device.*

Contributors

Restiyana Agustine National Research and Innovation Agency (BRIN), Jakarta, Indonesia

Azamat Ali School of Liberal and Creative Arts (Journalism and Mass Communication), Lovely Professional University, Phagwara, Punjab, India

Shohrat Ali Department of Civil Engineering, Central University of Jharkhand, Ranchi, India

Prameshwara Anggahegari National Research and Innovation Agency (BRIN), Jakarta, Indonesia

Ulfa Sevia Azni National Research and Innovation Agency (BRIN), Jakarta, Indonesia

Roshan Baa St. Xavier's College, Ranchi, Jharkhand, India

Vibhore Bakshi Department of Urban & Regional Planning, School of Planning and Architecture Bhopal, Bhopal, Madhya Pradesh, India

Sumanta Bhattacharya Asian International University, Imphal, Manipur, India

Uchaimid Biridlo'i Robby National Research and Innovation Agency (BRIN), Jakarta, Indonesia

Arpita Biswas Department of Environmental Sciences, University of Calcutta, Kolkata, West Bengal, India

Shyama Prasad Biswas Department of Life Sciences, Dibrugarh University, Dibrugarh, Assam, India

Abhinab Borah Mahapurusha Srimanta Sankaradeva Vishwavidyalaya, Nagaon, Assam, India

Hrykpb Borah TERI School of Advanced Studies, Delhi, India

Sanchita Boruah Department of Zoology, D.H.S.K. College, Dibrugarh University, Dibrugarh, Assam, India

Susanta Kumar Chakraborty Department of Zoology, Vidyasagar University, Paschim Medinipur, West Bengal, India

Department of Zoology, Vidyasagar University, West Midnapore, West Bengal, India

Department of Zoology, Vidyasagar University, Midnapore, West Bengal, India

Nyasha Chaminuka Faculty of Applied Social Sciences, Department of Development Studies, Zimbabwe Open University (ZOU), Gweru, Zimbabwe

Uday Chatterjee Department of Geography, Bhatter College Dantan (Autonomous), Paschim Medinipur, West Bengal, India

Punarbasu Chaudhuri Department of Environmental Sciences, University of Calcutta, Kolkata, West Bengal, India

Preetisha Choudhury Department of Law, North-Eastern Hill University, Shillong, Meghalaya, India

Anutosh Das Fire and Emergency Management Program, Division of Engineering Technology, Oklahoma State University, Stillwater, OK, USA

Department of Urban & Regional Planning, Rajshahi University of Engineering & Technology, Rajshahi, Bangladesh

Pratik Dash Khejuri College, Vidyasagar University, Midnapore, West Bengal, India

Pradip Debnath Department of Geography and Disaster Management, Tripura University, Agartala, Tripura, India

Parashmoni Deka Gauhati University, Guwahati, Assam, India

Sanjana De Department of Geography, Adamas University, Kolkata, West Bengal, India

Sanjoy Dutta Department of Geography, Pakuahat Degree College, Malda, West Bengal, India

Afreen Fatima Department of Conservation, School of Planning and Architecture, Bhopal, India

Arundhati Ganguly Department of Environmental Science, Asutosh College, Kolkata, West Bengal, India

Trishita Ghosal Department of Geography, Adamas University, Kolkata, West Bengal, India

Anitabha Ghosh Coastal Environmental Studies Research Centre, Egra S.S.B. College affiliated by Vidyasagar University, Kharagpur, West Bengal, India

Borsha Ghosh Department of Urban & Regional Planning, Rajshahi University of Engineering & Technology, Rajshahi, Bangladesh

Srinjana Ghosh Department of Zoology, Bethune College, Kolkata, West Bengal, India

Ruslan Halifu National Research and Innovation Agency (BRIN), Jakarta, Indonesia

Thomas Karakadzai Department of Construction Economics and Management, University of Cape Town, Cape Town, South Africa

Dialogue on Shelter for the Homeless Trust, Harare, Zimbabwe

Ranita Karmakar Department of Geography, Jadavpur University, Kolkata, West Bengal, India

Ajay Khare Department of Conservation, School of Planning and Architecture, Bhopal, India

Kuswinton Politeknik Baubau, Baubau, Indonesia

Terrence Makumbe Department of Community and Social Development, University of Zimbabwe, Harare, Zimbabwe

Snehasish Mandal Department of Biotechnology, Adamas University, Kolkata, West Bengal, India

Isheanesu Daniel Mangwaya Department of Community and Social Development, University of Zimbabwe, Harare, Zimbabwe

Saptarshi Mitra Department of Geography and Disaster Management, Tripura University, Agartala, Tripura, India

Sheila Mitra-Sarkar Department of Urban Studies & Planning, University of California, San Diego, CA, USA

Kasturi Mukherjee Department of Geography, Adamas University, Kolkata, West Bengal, India

La Ode Abdul Munafi University of Dayanu Ikhsanuddin, Baubau, Indonesia

Ephraim Kabunda Munshifwa Lusaka Campus, The Copperbelt University, Lusaka, Zambia

Chilungu Mwiinde Planning Unit, The Copperbelt University, Kitwe, Zambia

Dhanjit Nath Gauhati University, Guwahati, Assam, India

M. C. Nitha Shahul Department of Planning, Environmental Planning Programme, School of Planning and Architecture, Vijayawada, India

G. K. Panda Department of Geography, Kalinga Institute of Social Sciences (Deemed University), Bhubaneswar, Odisha, India

Apurv Bhogibhai Patel Urban Planner-Jamnagar Area Development Authority (JADA), Jamnagar, Gujarat, India

Abhidipta Paul Department of Biotechnology, Adamas University, Kolkata, West Bengal, India

Rajib Pegu Department of Geography, Sankardeva Mahavidyalaya, Pathalipahar, Assam, India

Most. Mousumi Akhter Putul Department of Urban & Regional Planning, Rajshahi University of Engineering & Technology, Rajshahi, Bangladesh

Debasmita Routh Department of Geography, Adamas University, Kolkata, West Bengal, India

Debjani Roy PG Department of Geography, Nirmala College, Ranchi, Jharkhand, India

Bhavneet Kaur Sachdev Suresh Gyan Vihar University, Jaipur, India

Bappa Sarkar Department of Geography, Dinhata College, Dinhata, West Bengal, India

Arkadyuti Seth Sai Nath University, Ranchi, India

Hari Harjanto Setiawan National Research and Innovation Agency (BRIN), Jakarta, Indonesia

Thomas Simbule Planning Department, Livingstone City Council, Livingstone, Zambia

Manoj Kumar Singh Department of Biotechnology, Adamas University, Kolkata, West Bengal, India

Lakshmi Sivaramakrishnan Department of Geography, Jadavpur University, Kolkata, West Bengal, India

Ayon Kumar Tarafdar Department of Planning, School of Planning and Architecture, Vijayawada, India

Robert Wallace Water Protector, Barona Band of Mission Indians, Lakeside, CA, USA

Andy Arya Maulana Wijaya University Muhammadiyah of Buton, Baubau, Indonesia

Part I

Introduction

Understanding Wetlands and Their Role in Urban Resilience

1

G. K. Panda, Susanta Kumar Chakraborty, Lakshmi Sivaramakrishnan, Uday Chatterjee, Ranita Karmakar, and Anitabha Ghosh

Abstract

Wetlands play a significant role in maintaining ecological balance, promoting environmental conservation, and enhancing urban resilience. Rapid sprawling of urban areas across the globe has led to the degradation of the natural environment and ecosystems, especially wetlands. With the current trend of urbanization and climate change impacts, towns and cities are facing threats of flooding, water scarcity, and loss of biodiversity. This chapter explores the critical role of wetlands in ecological, hydrological, environmental, and socioeconomic functions and analyzes how these wetland ecosystems contribute to the sustainability and resilience of urban environments. It aims to examine the importance of wetlands in strengthening urban resilience, particularly against climate-induced stresses like flooding, heatwaves, and water pollution of urban areas through content analysis, case studies on best practices, and current research on conservation and management of the wetlands in major cities and urban areas. This study brings into focus the need for integrating wetland conservation into urban planning, decision-making, and policy interventions for bringing urban resilience and sustainable development of cities and towns.

Keywords

Wetland conservation · Urban resilience · Urban ecology · Ramsar Convention · Sustainable cities

1.1 Introduction: Wetlands—Meaning and Concept

Wetlands are unique ecosystems where land and water meet, forming areas saturated with water either permanently or seasonally. These areas support vegetation and wildlife adapted to wet conditions. According to the Ramsar Convention

G. K. Panda
Department of Geography, Kalinga Institute of Social Sciences (Deemed University), Bhubaneswar, Odisha, India

S. K. Chakraborty
Department of Zoology, Vidyasagar University, Paschim Medinipur, West Bengal, India

L. Sivaramakrishnan
Department of Geography, Jadavpur University, Kolkata, West Bengal, India

U. Chatterjee (✉)
Department of Geography, Bhatter College Dantan (Autonomous), Paschim Medinipur, West Bengal, India

R. Karmakar
Senior Research Fellow, Department of Geography, Jadavpur University, Kolkata, West Bengal, India

A. Ghosh
Senior Research Fellow, Coastal Environmental Studies Research Centre, Egra S.S.B. College affiliated by Vidyasagar University, Kharagpur, West Bengal, India

S. K. Chakraborty et al. (eds.), *Wetlands and Urban Resilience*, Sustainable Development Goals Series,
https://doi.org/10.1007/978-3-032-06881-1_1

(1971), wetlands are defined as "areas of marsh, fen, peatland or water, whether natural or artificial, permanent or temporary, with water that is static or flowing, fresh, brackish, or salt, including areas of marine water with the depth of less than six meters during low tide conditions." This broad definition encompasses a wide variety of habitats, such as swamps, marshes, bogs, estuaries, mangroves, and even rice paddies and reservoirs. Wetlands serve as one of the most productive ecosystems in the world. They are often compared to rainforests and coral reefs for their biodiversity and ecological significance. These ecosystems perform vital functions such as water purification, groundwater recharge, flood control, and shoreline stabilization. They also serve as critical habitats for many species of birds, fish, amphibians, and invertebrates (Mitsch and Gosselink 2015; Chakraborty et al. 2023).

One of the core ecological roles of wetlands is water filtration. As water flows through wetland areas, vegetation and soil trap pollutants, excess nutrients, and sediments, improving water quality. Additionally, wetlands act as carbon sinks, absorbing carbon dioxide and helping mitigate the effects of climate change. According to the Intergovernmental Panel on Climate Change (IPCC 2014), wetland ecosystems store more carbon per unit area than any other ecosystem. Another essential function of wetlands is flood mitigation. During periods of heavy rainfall or snowmelt, wetlands absorb and store excess water, reducing the risk of floods downstream. They act as natural buffers by slowing the movement of water and allowing it to infiltrate into the ground (EPA 2023).

Wetlands also provide economic and social benefits by providing a livelihood to the people who depend on these areas. Many communities rely on wetlands for fishing, agriculture, and tourism. Additionally, wetlands hold cultural and spiritual value in many indigenous and local traditions. Despite their value, wetlands are under severe threat due to urbanization, agriculture, pollution, and climate change. It is estimated that more than 35% of the world's wetlands have been lost since 1970 (Ramsar Convention Secretariat 2018). Therefore, conservation and sustainable management of wetlands are crucial. International efforts like the Ramsar Convention on Wetlands of International Importance and national policies such as India's Wetland Conservation and Management Rules, 2017, aim to protect these ecosystems. Their conservation is not only an environmental concern but also a socio-economic and ethical imperative. Understanding the significance of wetlands and promoting sustainable practices to conserve these habitats are essential steps toward environmental stewardship and global sustainability.

1.2 Wetlands and Urban Resilience

Wetlands are ecosystems that are permanently or seasonally inundated with water, characterized by distinct hydric soils and water-tolerant vegetation. They include marshes, swamps, bogs, and estuaries, and are among the most productive ecosystems on Earth (Ramsar Convention Secretariat 2010). Wetlands provide essential ecosystem services such as water purification, flood regulation, carbon sequestration, and biodiversity support. In the context of increasing urbanization and climate change, wetlands play a critical role in enhancing urban resilience. Urban resilience refers to the ability of cities to absorb, recover, and prepare for future shocks, including climate-related events such as floods, heat waves, and droughts. Wetlands contribute significantly to this resilience in multiple ways. Firstly, they act as natural buffers against flooding. By storing excess rainwater during heavy precipitation events, wetlands help to reduce flood peaks and delay runoff, thereby protecting urban infrastructure and human life (Mitsch and Gosselink 2015). For instance, the East Kolkata Wetlands in India absorb urban wastewater and stormwater, significantly mitigating flood risks for the city (Ghosh 2005). Secondly, wetlands aid in water quality improvement. They act as biofilters, removing pollutants from urban runoff and sewage through processes such as sedimentation, filtration, and microbial activity. This not only protects aquatic ecosystems but also enhances the availability of clean water for urban populations (Kadlec and Wallace 2009).

Another key aspect is temperature regulation. Urban wetlands help to moderate urban heat island effects by cooling the air through evapotranspiration. This becomes particularly important in densely built environments where impervious surfaces trap heat, leading to higher urban temperatures (Gill et al. 2007). Furthermore, wetlands serve as crucial habitats for flora and fauna, contributing to biodiversity conservation within urban settings. They provide nesting, feeding, and breeding grounds for various species, some of which are endangered. This biodiversity has intrinsic ecological value and offers cultural and recreational benefits to urban residents (MEA 2005).

Urban resilience being the ability of an urban area with all of its constituent attributes in terms of economic development, green coverage, biodiversity wealth, cultural activities and technical knowhow to withstand, absorb, reduce and recover by adapting against environmental perturbations. Protecting and conserving ecologically healthy wetlands certainly counterbalance the ecologically deteriorating urban setups because of the multi-fold ecosystem services of the wetlands.

Despite their importance, wetlands in urban areas are under severe threat due to encroachment, pollution, and unplanned development. Urban expansion often leads to the draining and filling of wetlands, resulting in the loss of ecosystem services. Hence, integrating wetland conservation into urban planning is essential. Tools like wetland zoning, eco-sensitive area regulations, and community-based conservation can help protect these vital ecosystems (UN-Habitat 2014). Wetlands are vital allies in building urban resilience. They not only buffer cities against environmental risks but also provide ecological, social, and economic benefits. Protecting and restoring urban wetlands must be prioritized in urban resilience strategies to ensure sustainable and livable cities in the face of a changing climate.

1.3 Ramsar Convention and Urban Resilience

Wetlands are among the most productive ecosystems on Earth, providing critical services such as water purification, flood control, carbon storage, and biodiversity support. As urban areas expand, the role of wetlands in enhancing urban resilience—the capacity of cities to absorb, recover from, and prepare for future shocks (e.g., climate change, flooding, and pollution)—has gained increasing attention. The Ramsar Convention on Wetlands, an international treaty, plays a pivotal role in promoting the conservation and wise use of wetlands worldwide. Adopted in 1971 in Ramsar, Iran, the Ramsar Convention on Wetlands of International Importance is the only global treaty dedicated to wetland conservation. It currently includes over 170 Contracting Parties and covers more than 2400 wetland sites worldwide (Ramsar Convention Secretariat 2022). The convention emphasizes the "wise use" of wetlands, integrating environmental sustainability with human needs. Key objectives of the Convention include (a) designation of wetlands of international importance (Ramsar Sites) and (b) promotion of local and national action for wetland protection, and international cooperation in the management of transboundary wetlands.

Recognizing the mounting challenges posed by urbanization, the Ramsar Convention adopted Resolution XIII.20 on Urban Wetlands in 2018. This resolution encourages cities to integrate wetlands into urban planning and development frameworks to build resilience and reduce disaster risks (Ramsar Secretariat 2018). The convention puts emphasis on the unique benefits of the urban wetlands with regard to flood mitigation by storing stormwater, climate regulation by acting as carbon sinks, wastewater treatment through natural filtration, biodiversity conservation, and creation of green public spaces for recreation and livelihood generation for the urban poor. There are many examples of Ramsar site wetlands contributing to urban resilience in their respective towns and cities. East Kolkata Wetlands (India) naturally treat city sewage and support local livelihoods. The Bañados del Este wetland of Uruguay supports sustainable agriculture and flood control. In Canada, the Vancouver Fraser River estuary has become a habitat for migratory birds and also acts as a buffer zone against sea level rise. Many wetlands contribute directly to urban resilience by reducing vulnerability to climate-related hazards such as floods and droughts and improving water security through groundwater recharge and purification. It has also contributed to the enhancement of socioeconomic resilience by supporting livelihoods (e.g., fishing and farming) and promoting eco-tourism. Wetlands have also contributed to improving public health by mitigating pollution and offering recreational spaces (Meli et al. 2020). Integrating wetlands into urban design aligns with nature-based solutions (NbS) advocated by the United Nations and IPCC for building resilient and sustainable cities (UNEP 2021). The Ramsar Convention serves as a cornerstone of global efforts to conserve wetlands and integrate them into the fabric of urban resilience. It calls for stronger stakeholder engagement, especially involving local communities and urban planners, to safeguard these ecosystems. By recognizing wetlands as critical infrastructure, urban policies can transit toward more sustainable and adaptive systems. Future urban resilience frameworks must prioritize wetland conservation not just as an environmental obligation but as a strategic necessity for community resilience.

1.4 Characteristics and Classification of Wetlands

Wetlands are characterized by showing some criteria like presence of hydric soils, hydrophytes, aquatic fauna and soil must be covered by overlying water at least for couple of months in a year (Chakraborty et al. 2023). Hydrology relates to the presence of water at or near the surface for a significant period of the year. The water may vary from fresh to brackish and saline. Hydrological regimes (inundation, saturation) are primary determinants of wetland type and functioning. This hydrology influences soil oxygen levels and biogeochemical processes operating in the wetlands (Mitsch and Gosselink 2015). Wetlands are characterized by hydric soils, which are saturated long enough to develop anaerobic conditions. These soils often have high organic content and show characteristics such as gleying (grayish coloration) or mottling, which indicates water saturation. Wetlands support hydrophytic vegetation, which is adaptable to saturated soil conditions. Common plant species of the wetlands include reeds, sedges, mangroves, cattails, and mosses. Most of the plant communities are influenced by water depth, flow, and salinity. Wetlands are also known worldwide for their ecological importance as biodiversity hotspots, being the home to numerous species of plants, animals, birds, amphibians, fish, and invertebrates. These are also known for their functions like flood control by storing excess rainwater and releasing it slowly. It also plays a significant role in water purification by filtering pollutants and recharging the groundwater. These are also known for their function as carbon sinks, especially peatlands, which store large amounts of carbon (MEA 2005). Thus, depending on the characteristics, location, origin, and distribution in different parts of the land and sea as well as along the rivers and coasts, wetlands are being classified using various systems. All these characteristics reveal that wetlands are diverse and dynamic ecosystems classified based on their physical and biological characteristics (Chakraborty et al. 2023). The most prominent being the Ramsar classification and the Cowardin classification developed by the US Fish and Wildlife Service. Recognizing and understanding these classifications is essential for wetland conservation, management, and sustainable use (Fig. 1.1).

A. *Ramsar Classification*

   The Ramsar Convention classifies wetlands into three major categories, namely marine and coastal wetlands, inland wetlands, and human-made wetlands.

Fig. 1.1 (**a**) Marine and coastal wetlands, (**b**, **c**) Inland wetlands, and (**d**) human-made wetlands

(i) *Marine and Coastal Wetlands*
These wetlands are associated with the coastal environment and are typically influenced by wave and tidal flows, as well as saline conditions. Mangrove forests found in tropical and subtropical tidal areas such as Sundarbans of India and Bangladesh belong to this category. Estuaries near the river mouths, where rivers meet the sea and freshwater mixes with seawater, also belong to this category. Coral reefs and seagrass beds also form a part of this wetland, which provides one of the most critical habitats for marine biodiversity (Fig. 1.2).

(ii) *Inland Wetlands*
These are located away from coastal areas and include a variety of freshwater and brackish systems. These wetlands include floodplains which are formed as Seasonal Wetlands formed by overflowing rivers. Swamps and marshes found in river basins, deltas, or lake peripheries form a part of this wetland. Peatlands and bogs, which accumulate partially decayed vegetation or organic matter, also belong to this category (Fig. 1.3).

(iii) *Human-Made Wetlands*
Human-made wetlands are artificial wetlands created for various purposes. It includes agriculture and wastewater treatment ponds, village ponds, rice paddies, i.e., Wetland fields used for growing rice, ponds created for treating municipal or industrial wastewater, and salt pans, which are shallow artificial ponds designed to extract salt from sea-water (Ramsar Convention Secretariat 2010) (Fig. 1.4).

B. *Cowardin Classification (US Fish and Wild-life Service)*
This system is more detailed and widely used for mapping and monitoring wetland resources in the United States. In this system, wetlands are classified as marine wetlands, which include open ocean overlying the continental shelf and associated coastlines, rocky intertidal zones, and sandy beaches. Estuarine Wetlands include deep water tidal habitats and adjacent tidal wetlands that are usually semi-enclosed by land but have access to the open

Fig. 1.2 Marine and coastal wetlands

Fig. 1.3 Inland wetlands

ocean. Riverine wetlands comprise deep water habitats within a channel and riverine floodplains, excluding wetlands dominated by trees, shrubs, or emergent vegetation. Lacustrine Wetlands are associated with lakes or reservoirs. Palustrine Wetlands are non-tidal wetlands dominated by trees, shrubs, and emergent vegetation. It includes freshwater marshes and forested swamps (Cowardin et al. 1979).

Fig. 1.4 Human-made wetlands

1.5 Urban Wetlands and Their Ecosystem Services

Urban wetlands are natural, or artificial, permanent or temporary water bodies located in or near urban areas, where water is static or flowing, fresh, brackish, or saline. These ecosystems play a vital role in maintaining the environmental balance of rapidly urbanizing regions. Urban wetlands also offer multiple benefits like recharging of groundwater aquifers, water filtration, flood control, improved air quality, and contribute to biodiversity (Chakraborty et al. 2023). Besides these, urban wetlands are also well recognized for their role in mitigating the urban heat island effect and providing recreational space to the crowded population of the cities. With the current trend of rapidly expanding urbanization, wetlands in urban areas have come under increasing pressure, yet they continue to play a vital role in maintaining ecological and hydrological balance. As emphasized by UN-Habitat and the Ramsar Convention on Wetlands (2020), "wetlands in cities are crucial green infrastructure" that support not only biodiversity but also urban resilience and public well-being. Despite being threatened by anthropogenic activities from the people surrounding it, urban wetlands provide a wide range of ecosystem services. As emphasized by UN-Habitat and the Ramsar Convention on Wetlands (2020), "wetlands in cities are crucial green infrastructure" that support not only biodiversity but also urban resilience and public well-being. Urban wetlands are increasingly recognized as critical components of sustainable urban development. They offer essential ecosystem services that bolster urban resilience against climate-induced challenges such as flooding, heatwaves, and water scarcity

(i) *Water Purification:* Wetlands act as biofilters, removing suspended solids, heavy metals, and nutrients such as nitrogen and phosphorus from wastewater and urban runoff. This occurs through physical sedimentation, chemical absorption, and biological uptake by plants and microbes (Mitsch and Gosselink 2015).

(ii) *Flood Regulation:* By temporarily storing excess rainwater and releasing it gradually, wetlands reduce the risk of urban flooding. This function is particularly critical in flood-prone cities with poor drainage systems

(UN-Habitat & Ramsar Convention 2020). Wetlands function as natural buffers, absorbing excess rainfall and mitigating flood risks in urban areas. In Colombo, Sri Lanka, the degradation of the Colombo Flood Detention Area (CFDA) wetlands due to urbanization has led to increased flood frequency. Hettiarachchi et al. (2014) observed that the transformation of marshy habitats into shrublands and the alteration of peaty soils to semi-mineral soils have significantly reduced the wetlands' water-holding capacity, undermining their flood control services. To address this, Colombo has implemented wetland management strategies that integrate flood protection with urban planning, emphasizing the importance of preserving wetland ecosystems for disaster resilience (GFDRR 2018).

(iii) *Groundwater Recharge:* Wetlands promote infiltration of water into the soil, helping to replenish underlying aquifers. This is especially important in areas facing groundwater depletion (Keddy 2010).

(iv) *Carbon Sequestration and Climate Regulation:* Wetlands store significant amounts of organic carbon in their vegetation and soils. Even small wetlands can serve as carbon sinks, mitigating the urban heat island effect and contributing to climate change mitigation (MEA 2005). Beyond flood mitigation, urban wetlands contribute to climate regulation by moderating temperatures and sequestering carbon. Wetlands International (2020) highlights that urban wetlands can alleviate the Urban Heat Island (UHI) effect by absorbing heat and providing cooling through evapotranspiration. In India, rapid urbanization has led to significant wetland loss, exacerbating UHI effects and increasing vulnerability to heatwaves. Integrating wetlands into urban landscapes as part of blue-green infrastructure can enhance cities' capacity to adapt to climate variability.

(v) *Biodiversity Habitats:* Despite being in urban areas, wetlands support diverse biological communities, including fish, amphibians, reptiles, and migratory birds. Urban wetlands can serve as critical stopovers in bird migratory routes (Davidson 2014).

(vi) *Sociocultural and Recreational Value:* Urban wetlands are important for community recreation, aesthetic enjoyment, spiritual practices, and educational activities. When conserved and integrated into urban green spaces, they enhance the livability of cities (UN-Habitat & Ramsar Convention 2020).

1.6 Contemporary Threats and Challenges for the Urban Wetlands

According to the Ramsar Convention Secretariat (2018), urban wetlands are increasingly under threat due to "encroachment, pollution, and altered hydrological regimes," which diminish their ecological integrity and capacity to function as natural infrastructure to contribute to sustainable development of the surrounding area. Urban wetlands are facing distinct challenges and threats influenced by their location and the anthropogenic activities surrounding them. Some of the major challenges of urban wetlands are given below.

(i) *Location and Anthropogenic Influence:* Urban wetlands are situated in cityscapes and often reflect the impacts of land reclamation, urban expansion, including encroachment, pollution from waste and sewage disposal, and changes in water flow patterns associated with heavy rainfall and city flooding (Ramsar Convention Secretariat 2018).

(ii) *Hydrology:* The hydrological inputs of urban wetlands may include stormwater runoff, treated or untreated sewage, groundwater inflow, and direct precipitation. These sources are often irregular and can contribute to fluctuations in water levels and quality (Mitsch and Gosselink 2015).

(iii) *Vegetation and Biodiversity:* The plant life in these wetlands consists of hardy species such as *Typha*, *Phragmites*, and water hyacinths, which are tolerant to eutrophic and contaminated waters. The presence of such species indicates high nutrient loads, often resulting from untreated wastewater (Keddy 2010).

(iv) *Pollution Load:* Urban wetlands are typically exposed to high levels of nutrients, heavy metals, and other pollutants due to surface runoff and industrial discharge. This leads to degradation in water quality and threatens aquatic life (MEA 2005). Too much of untreated waste disposal into the wetland beyond its carrying capacity turns it into a waste disposal site, creating a foul smell and a filthy environment, leading to the spread of epidemics and contamination of air and water-borne diseases.

(v) *Fragmentation and Size:* Due to land use conversion, urban wetlands tend to be smaller and more fragmented, reducing their capacity to provide ecosystem services (Davidson 2014).

(vi) *Community Engagement and Socio-ecological Resilience*
The involvement of local communities in wetland conservation is pivotal for sustainable urban resilience. In Colombo, community-based initiatives have been instrumental in restoring and managing urban wetlands, leading to enhanced biodiversity and improved livelihoods (UNDP 2022). Similarly, in India, the "Wetland Mitra" program encourages citizen participation in wetland management, fostering a sense of stewardship and promoting sustainable practices (ORF 2023). These examples underscore the importance of integrating social dimensions into wetland conservation strategies

(vii) *Policy Framework and Governance*
Effective governance is crucial for the protection and restoration of urban wetlands. In Latin America, Rojas Quezada (2024) emphasizes the need for comprehensive policies that address the multifaceted roles of wetlands in urban settings. This study advocates for the integration of wetland conservation into urban planning, highlighting successful case studies where policy interventions have led to improved urban resilience. In India, the National Wetland Conservation Programme and the Wetlands (Conservation and Management) Rules (2017), provide a regulatory framework for wetland protection, although challenges in implementation persist (ORF 2023).

1.7 Nature-Based Solutions and Innovative Practices

Nature-based solutions (NbS) offer innovative approaches to enhance urban resilience through wetland restoration and management. Constructed wetlands, for instance, have been employed in Italy for wastewater treatment, demonstrating cost-effectiveness and ecological benefits (Garcia-Herrero et al. 2023). The "Sponge City" concept in China integrates wetlands into urban infrastructure to manage stormwater and reduce flooding, showcasing the potential of NbS in urban resilience strategies (Li et al. 2021). Despite the recognized benefits, urban wetlands face threats from pollution, encroachment, and inadequate policy enforcement. In Bogotá, Colombia, wetlands are under pressure from urban development, leading to biodiversity loss and reduced ecosystem services (The Guardian 2025). Addressing these challenges requires a multidisciplinary approach that combines scientific research, community engagement, and robust policy frameworks. Future research should focus on developing integrated models that quantify the benefits of wetlands in urban resilience and inform evidence-based policymaking. Despite the recognized benefits, urban wetlands face threats from pollution, encroachment, and inadequate policy enforcement. In Bogotá, Colombia, wetlands are under pressure from urban development, leading to biodiversity loss and reduced ecosystem services (The Guardian 2025). Addressing these challenges requires a multidisciplinary approach that combines scientific research, community engagement, and robust policy frameworks. Future research should focus on developing integrated models that quantify the benefits of wetlands in urban resilience and inform evidence-based policymaking.

1.8 Case Studies on Wetland Restoration and Urban Resilience

Urban areas worldwide are increasingly facing environmental and socioeconomic challenges due to rapid urbanization, climate change, and unplanned development. Wetlands, often overlooked in urban planning, play a crucial role in enhancing urban resilience. Realizing the potential of wetlands in bringing urban resilience, many countries across the globe have adopted innovative strategies to manage the wetlands to reap the results of multiple ecological, social, and economic benefits. In many instances, successful cases are marked by strong institutional frameworks, community involvement, and nature-based approaches involving policymakers and planners. The East Kolkata Wetlands (EKW) located on the eastern fringes of Kolkata, India, Bogota of Columbia, Wuhan of China and Colombo of Sri Lanka are a few best practices and innovative models for sustainable urban resilience and ecosystem services, especially in waste management, urban flood control, biodiversity conservation, climate adaptation and livelihood generation (Ghosh 2005; Chakraborty et al. 2023). The case studies presented below are the best examples of wetland-integrated resilience models in diverse urban contexts.

Case Study 1: East Kolkata Wetlands (EKW): An Innovative Model of Urban Resilience

(a) Situational Analysis of the EKW
The East Kolkata Wetlands cover approximately 125 km^2 and comprise a complex network of sewage-fed fish ponds, agricultural land, and salt marshes (Fig. 1.5).

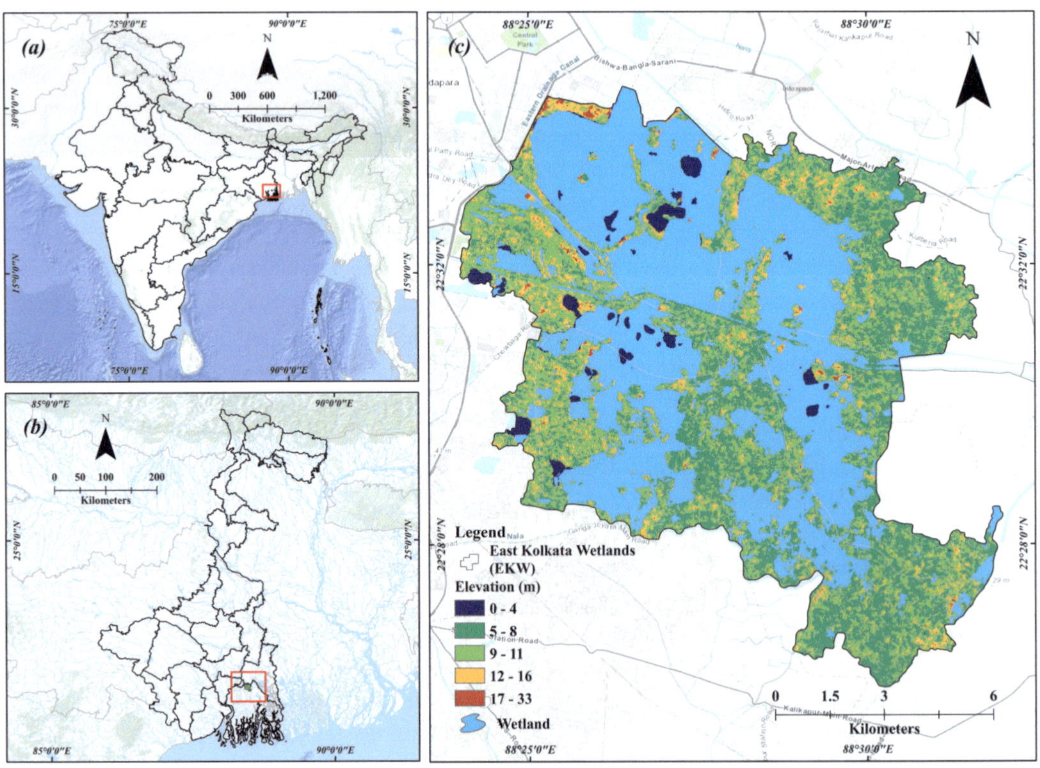

Fig. 1.5 Location map: (**a**) India, (**b**) West Bengal, and (**c**) East Kolkata Wetlands (EKW)

Recognized as a Ramsar site in 2002, the EKW is globally acknowledged for its unique and sustainable practice of using urban wastewater for aquaculture and agriculture (Ramsar Convention Secretariat 2022). These wetlands are home to diverse flora and fauna and support the livelihoods of over 20,000 fisherfolk and farmers (Bunting et al. 2010). The wetland system is a rare example of a symbiotic relationship between urban waste management and livelihood generation (Chakraborty et al. 2023). It stands as a testament to the harmonious coexistence of urban development and ecological preservation. EKW plays a pivotal role in treating around 78% of Kolkata's wastewater through its unique sewage-fed aquaculture system. However, rapid urbanization, especially in areas like Salt Lake and New Town was posing significant threats to this delicate ecosystem. Despite its critical ecological functions, the EKW was facing several challenges and threats such as land encroachment and illegal construction, effects of urban sprawling leading to the shrinkage of wetland areas, and pollution due to disposal of unregulated sewerage, industrial discharge, and plastic waste deteriorating its water quality. There was a lack of community awareness about the significance of the wetland for the city and its people. Although it was legally protected, enforcement was weak due to a lack of coordination among agencies (Dasgupta and Shaw 2021). Despite the enactment of the East Kolkata Wetlands (Conservation and Management) Act in 2006, enforcement remained weak, with instances of illegal land conversion and inadequate stakeholder engagement in conservation efforts. But subsequently proactive measures in its restoration, with the implementation of rennovative strategies and policy guidelines coupled with community participation has brought a rejuvinating change in the wetland ecosystem, as a means toward urban resilience, in a mega city like Kolkata.

(b) Innovative Approaches in Urban Resilience

(i) Wastewater Treatment
One of the most significant contributions of EKW to Kolkata's urban resilience is its capacity to treat nearly 900 million liters of sewage per day through natural bio-filtration processes (Ghosh 2005). This eco-treatment reduces the burden on conventional sewage treatment plants and contributes to maintaining water quality in the region.

(ii) Flood Control and Climate Regulation
As a natural drainage basin, the wetlands absorb excess rainwater and prevent urban flooding during monsoons. They also help in moderating local climate by functioning as carbon sinks and regulating temperatures (Raychaudhuri et al. 2014).

(iii) Livelihood and Food Security
EKW supports urban food systems by producing nearly 13,000 tonnes of fish annually, along with vegetables grown on organically fertilized soil. This contributes to food security and local employment, making the region socioeconomically resilient (Bunting et al. 2010).

(iv) Integrated Management Plans
The East Kolkata Wetlands Management Authority (EKWMA) has developed a Management Action Plan (2021–2026) focusing on pollution abatement, species and habitat conservation, and sustainable resource development (Sharda 2024).

(v) Community Engagement
Empowering local communities through education and participation in conservation efforts are enhancing the effectiveness of wetland management and ensuring sustainable livelihoods.

(vi) Technological Interventions
Employing geospatial technologies and Multi-Criteria Decision Making

(MCDM) methods and advanced technological interventions in monitoring and management of the wetland health, assessing risks, and formulating data-driven management strategies have been found unique in their renovation, contributing to the urban resilience of the city of Kolkata. The EKW utilizes shallow ponds, known as "bheris," where sunlight and algae naturally purify sewage water, supporting fish farming and agriculture. EKW has been a home to over 100 plant species and numerous aquatic organisms. This wetland supports the livelihoods of over 60,000 people engaged in fishing and farming activities. Besides acting as a carbon sink, the EKW has been useful in flood control and groundwater recharge, and it provides a buffer against cyclones and storm surges from the Bay of Bengal (Mondal et al. 2022). Now, East Kolkata Wetlands stand as a global example of nature-based urban infrastructure that enhances resilience against climate change and socioeconomic stresses. Their integration into urban policy and planning is vital not only for Kolkata but for sustainable urban development models worldwide.

Case Study 2: Urban Wetland Restoration and Ecological Resilience in Bogota, Colombia
Bogotá, the capital of Colombia, is a high-altitude city that has undergone rapid urban development, which has significantly reduced its natural wetlands (Fig. 1.6). Once covering approximately 50,000 hectares, the wetland area had shrunk to less than 800 hectares by the late twentieth century due to encroachment, pollution, and unregulated urban sprawl. In response, the city launched a series of initiatives aimed at reversing wetland

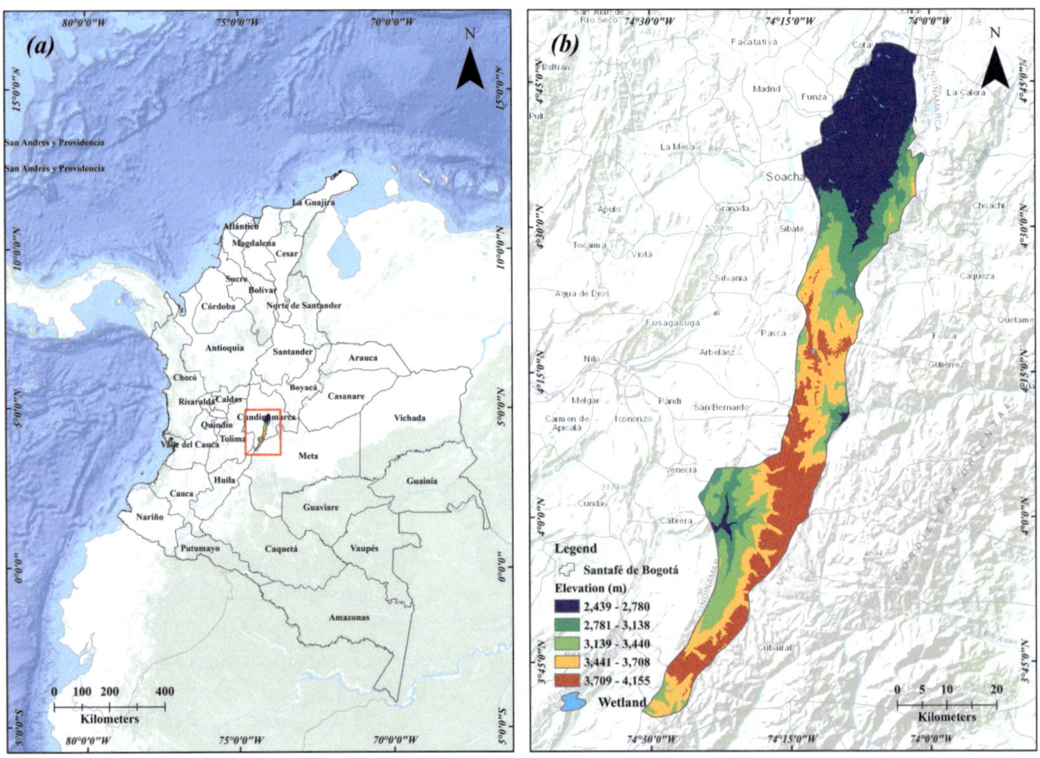

Fig. 1.6 Location map: (**a**) Colombia and (**b**) Santafé de Bogotá

degradation and integrating ecological thinking into urban planning (Jaramillo et al. 2018). Recognizing the ecological and hydrological importance of these urban wetlands, the city government initiated a series of restoration and protection measures. A turning point came with the official designation of 14 wetlands as protected areas under national and municipal environmental regulations (Ramsar Secretariat n.d.; The Guardian 2025). Restoration efforts included the removal of invasive species, reintroduction of native flora, and improvements to hydrological connectivity. These wetlands are now integrated into the city's urban planning strategy, serving as natural infrastructure for stormwater management and flood mitigation. Community involvement has been integral to the success of the project. Environmental education centers located at major wetland sites (e.g., La Conejera and Santa María del Lago) facilitate awareness and participation among local residents and schoolchildren (UNEP 2019). Monitoring programs involving citizens have helped track bird populations and water quality, reinforcing local stewardship. As a result, Bogotá has seen improvements in biodiversity conservation, flood control, and public recreation spaces. The case illustrates the power of community-driven restoration integrated with formal governance and environmental policy. Today, Bogotá's wetlands serve as green infrastructure, contributing not only to flood control and biodiversity conservation but also to improving the quality of urban life through the provision of recreational spaces. This transformation demonstrates how ecological restoration can be embedded into urban resilience planning, with both social and environmental dividends.

Case Study 3: Wetland Conservation in Colombo, Sri Lanka

Colombo, Sri Lanka's commercial capital, has long been vulnerable to seasonal flooding, which has been intensified by urban development and climate variability (Fig. 1.7). It has been historically prone to urban flooding, exacerbated by the loss of wetland ecosystems and poor

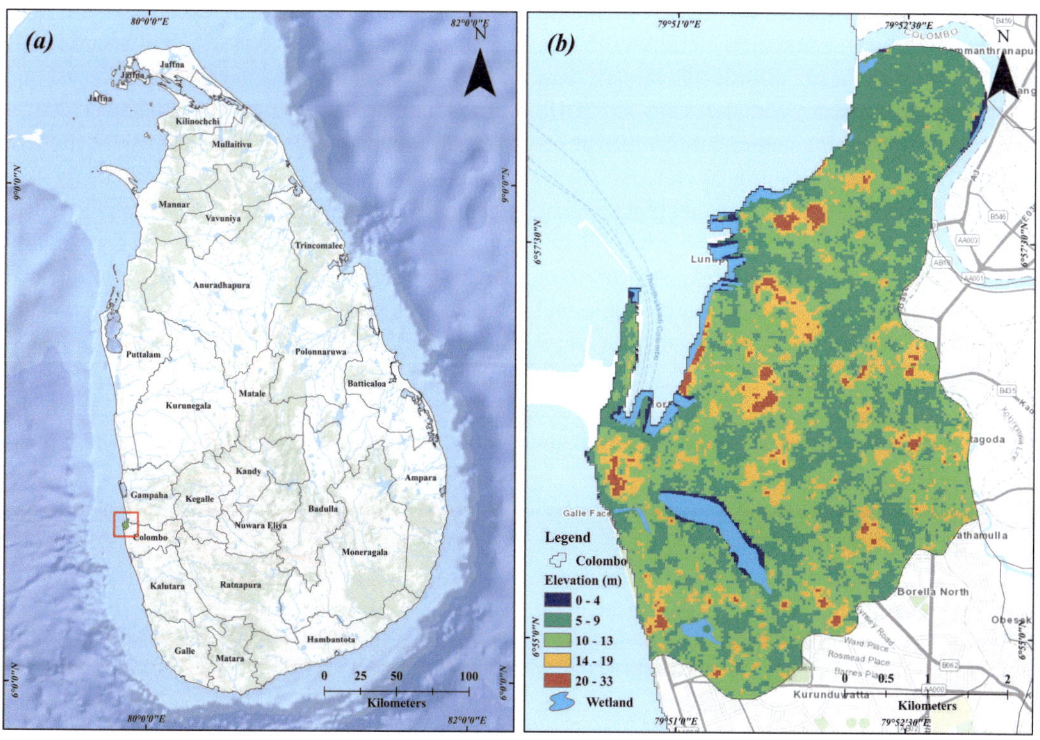

Fig. 1.7 Location map: (**a**) Sri Lanka and (**b**) Colombo

stormwater management. Between 1980 and 2010, more than 60% of Colombo's wetlands were lost due to development (World Bank 2016). In response, the government adopted a novel approach that redefined wetlands as essential urban infrastructure. The Sri Lankan government adopted a wetland-focused strategy for enhancing urban resilience. The Colombo Wetland Management Strategy marked a policy shift toward wetland protection and sustainable urban development. Over 1900 hectares of wetlands were identified and integrated into the urban flood management framework. This was accompanied by the creation of multifunctional wetland parks such as Beddagana Wetland Park and Diyasaru Uyana, which combine biodiversity conservation, flood retention, and public recreation (SLLRDC 2016). Economic valuation studies conducted by the World Bank showed that these wetlands provide ecosystem services worth approximately USD 5 million annually by mitigating flood-related damages (World Bank 2016). Public outreach campaigns and school programs promoted by NGOs and local government agencies further supported community awareness and engagement (UCRSEA 2017). In recognition of these efforts, Colombo was designated a Ramsar Wetland City in 2018, making it a model for nature-based urban resilience in South Asia (Ramsar Secretariat 2018). The integration of wetlands into city planning has helped reduce flood risks, improve urban livability, and safeguard ecological assets. As a result of these efforts, the city became more livable and climate-resilient, setting a benchmark for other South Asian urban centers.

Case Study 4: Integrated Wetland Management in Wuhan, China

Wuhan often referred to as the "City of Lakes," lies at the intersection of the Yangtze and Han rivers and is highly susceptible to seasonal flooding (Fig. 1.8). With rapid urbanization increasing the volume of impervious surfaces, the city struggled to manage stormwater and urban heat. Recognizing these challenges, Wuhan was selected as one of China's first pilot cities for the ambitious "Sponge City" initiative launched in 2015. It was a national program promoting nature-based solutions to urban water management (ADB 2021). The Sponge City strategy in Wuhan combined gray and green infrastructure to create an urban environment capable of absorbing, storing, and reusing rainwater. Key interventions include permeable pavements, bioswales, rain gardens, and green roofs. These were coupled with extensive wetland restoration efforts, particularly around East Lake and Liangzi Lake, where buffer zones were rehabilitated, and eco-tourism facilities were developed (Liu and Jensen 2020). Technology has played a central role in this initiative. Wuhan deployed IoT-based water monitoring systems and GIS tools to manage flood risks and track environmental indicators in real time (World Bank 2019). The city also implemented early warning systems to better respond to extreme weather events. The program was coordinated across municipal, provincial, and national agencies, with additional support from international institutions such as the World Bank and Asian Development Bank. As a result, pilot areas of Wuhan achieved 70% rainwater absorption, significantly reducing flood impacts and enhancing urban aesthetics (MOHURD 2018). Wuhan's experience demonstrates how modern urban centers can blend traditional ecosystem restoration with cutting-edge technology to create climate-resilient cities. Wuhan's Sponge City model has led to a significant reduction in flood incidents in pilot areas. It has also improved water quality, enhanced urban aesthetics, and provided important lessons for scaling similar efforts in more than 30 other Chinese cities. Wuhan stands today as a global example of how wetland management in combination with ecological designs, community engagement, and smart technologies can build urban resilience. Here is a clear and concise comparative table summarizing the key features, approaches, and impacts of the case studies on the wetlands of Bogotá (Colombia), Colombo (Sri Lanka), and Wuhan (China) (Table 1.1).

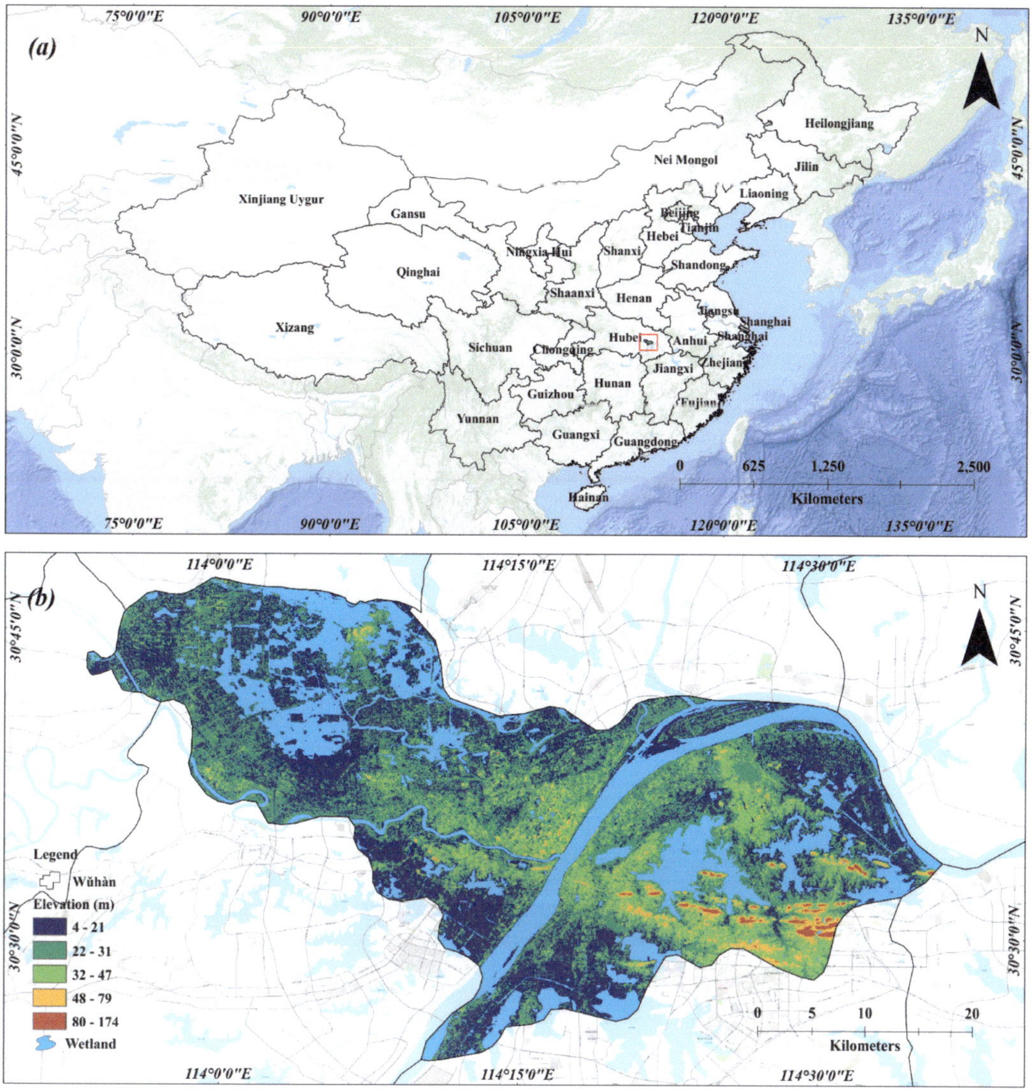

Fig. 1.8 Location map: (**a**) China and (**b**) Wuhan

1.9 India's National Policy on Wetland Conservation and Urban Resilience

The National Wetland Conservation Programme (NWCP) of the Government of India was launched in 1985 by the Ministry of Environment, Forest and Climate Change (MoEFCC), to protect and conserve wetlands across the country. This was further strengthened by the Wetlands Conservation and Management Rules, 2017, which replaced the earlier 2010 rules (MoEFCC 2017). These rules provide a regulatory framework for the conservation of wetlands, emphasizing community participation and the integration of wetland management into development planning. In the face of rapid urbanization and climate change, the conservation of wetlands has emerged as a strategic component of environmental governance and urban resilience in India. India is also a signatory to the Ramsar Convention (1971), an international treaty for the

Table 1.1 Key features, approaches, and impacts of the case studies on the wetlands of Bogotá (Colombia), Colombo (Sri Lanka), and Wuhan (China)

Aspect	Bogotá, Colombia	Colombo, Sri Lanka	Wuhan, China
Context	Rapid urbanization drastically reduced wetlands from 50,000 to ~800 hectares	Urban flooding worsened by wetland loss and climate change	Located at Yangtze-Han rivers confluence, faces seasonal flooding exacerbated by urbanization
Primary focus	Wetland restoration and integration into urban planning	Wetlands as functional urban infrastructure for flood mitigation and biodiversity	"Sponge City" concept: natural water absorption and reuse combined with wetland restoration
Key strategies	Legal protection, invasive species removal, native vegetation restoration, community engagement, environmental education	Wetland protection via policy, development of urban wetland parks, ecosystem services valuation, public engagement	Permeable pavements, green roofs, bioswales, urban rain gardens, wetland restoration, real-time monitoring, multi-level governance
Community involvement	Active involvement in biodiversity monitoring, wetland maintenance, education centers	Citizen participation through awareness campaigns, school programs, NGO partnerships	Limited direct community role; focus on multi-level governance and technology integration
Technology use	Moderate (monitoring and education)	Moderate (ecosystem valuation, education campaigns)	High (IoT sensors, GIS, early warning systems)
Ecosystem services	Flood control, biodiversity conservation, recreational spaces	Flood regulation, climate mitigation, recreation, economic savings estimated	Stormwater management, flood reduction, water quality improvement, urban cooling
Governance model	Local government coordination, legal frameworks, community partnerships	Government-NGO collaboration, policy enforcement, Ramsar City designation	Multi-level governance involving national, provincial, municipal authorities, international partners
Impact/outcomes	Increased wetland area and biodiversity, improved flood management, enhanced urban green space	Reduced flooding risk, biodiversity conservation, increased public awareness, improved urban livability	70% rainwater absorption in pilot areas, reduced floods, improved aesthetics, replicable model for other cities
Unique strength	Strong community and education focus, combined with legal protection	Economic valuation of wetlands, strengthening conservation, urban wetland parks as multi-use spaces	Integration of advanced technology with nature-based solutions in a large-scale urban context

conservation and wise use of wetlands. As of 2025, India has over 75 Ramsar sites, highlighting its global commitment. It is needless to mention that in urban areas, wetlands play a key role in enhancing urban resilience by mitigating flood risks, controlling pollution, and acting as heat and carbon sinks (Ramsar Convention 2025). With the increase in frequency of extreme weather events due to climate change, preserving urban wetlands is critical for reducing disaster risks. The National Mission on Sustainable Habitat under the National Action Plan on Climate Change (NAPCC) recognizes the importance of ecological infrastructure, such as wetlands, for building climate-resilient cities. Some of the important Ramsar sites of urban wetlands working for their conservation in India are Depor Beel, a flood plain of the river Brahmaputra near Guwahati, Pallikarnai freshwater marsh of Chennai, Mithi River and wetlands of Mumbai, and the Yamuna floodplains of Delhi. Located near Guwahati, Deepor Beel is a Ramsar site under pressure from urban encroachment and pollution. It is a key flood buffer for the city and serves as a habitat for the migratory birds. The Assam government has begun wetland reclamation initiatives to address these threats (Wetlands International South Asia 2020). Pallikaranai Marsh in Chennai, one of India's last surviving freshwater swamps, is vital for stormwater drainage and biodiversity. Despite threats from waste dumping and construction, it has been notified as a protected wetland under the 2017 rules (NIUA 2022). Mumbai has lost over 70% of its wetlands in the last three decades. The 2005 Mithi River floods demonstrated the catastrophic consequences of wetland encroachment. Citizen groups and NGOs now play a major role in wetland restoration (Wetlands International South Asia 2020). Delhi has more than 600 identified water bodies, many of which are degraded. The Yamuna floodplains, acting as seasonal wetlands, are essential for urban cooling and aquifer recharge. Delhi's Master Plan 2041 includes steps for wetland rejuvenation (MoEFCC 2022). Integrated urban planning that incorporates wetland conservation ensures not only ecological sustainability but also supports livelihoods, biodiversity, and long-term climate adaptation. Efforts like the Atal Mission for Rejuvenation and Urban Transformation (AMRUT) and the Smart Cities Mission increasingly acknowledge the role of green-blue infrastructure in enhancing urban resilience. India's wetlands are under immense pressure from urban expansion, pollution, and climate change. However, strong national policies combined with region-specific strategies, stakeholder participation, and integration with urban development schemes can transform wetlands into assets for sustainable and resilient urban ecosystems. India's wetlands are under immense pressure from urban expansion, pollution, and climate change. However, strong national policies combined with region-specific strategies, stakeholder participation, and integration with urban development schemes are transforming wetlands into assets for sustainable and resilient urban ecosystems.

1.10 Conclusions: Summary and Way Forward

Wetlands are indispensable for ensuring ecological sustainability and enhancing urban resilience, especially in the context of rapid urbanization and climate change. Their conservation and restoration, integrating into urban planning in order to generate and offer pathways and avenues for chalking out proper eco-management strategies to ensure climate adaptation, disaster risk reduction, and enhanced quality of urban life. These ecosystems act as natural sponges that absorb excess rainfall, thereby mitigating floods, recharging groundwater, and regulating urban microclimates. In densely populated cities, where concrete surfaces dominate, wetlands provide critical relief by supporting biodiversity, purifying water, recharging groundwater aquifers, and offering green spaces that improve the quality of life. In urban areas dominated by impervious surfaces, wetlands offer essential ecological services, including water filtration, habitat support, and recreational spaces that enhance urban livability. However, increasing encroachment, pollution, and land-use change have led to the alarming degradation of wetlands, more specifically in the developing countries to excessive utilization of natural resources. The loss of these natural buffers has exacerbated urban flooding, water scarcity, and heat stress in many metropolitan regions across the globe. Recognizing these challenges, efforts are being made at the global and national level for integrating wetland conservation into urban planning. Furthermore, community participation and decentralized governance models are also evolving as effective strategies for sustainable wetland management. Increasing realization is also visible in building truly resilient and climate-adaptive cities by treating wetlands as vital urban infrastructure rather than wastelands. Their conservation is not just an environmental imperative but a socioeconomic necessity that supports sustainable development, safeguards livelihoods, and builds climate-ready urban futures. In the long run, there is a need to build up political will, inclusive governance practices, and community participation to sustain these wetlands for future posterity. Moving forward, it is essential to recognize wetlands as critical components of urban infrastructure. Their protection is the key not only to environmental health but also to sustainable urban development. Prioritizing wetland conservation is, ultimately, a strategic step toward creating cities that are resilient, equitable, and prepared for future climate challenges. There is a need to foster multi-stakeholder partnerships and community engagement in wetland stewardship as well as to strengthen policy frameworks and legal protections for urban wetlands at local, national, and international levels. Investing in wetlands of urban areas is, therefore, an investment in urban resilience and the development of sustainable cities.

References

Asian Development Bank (2021) Sponge cities in China: addressing water challenges in the face of climate change. https://www.adb.org/publications/sponge-cities-china

Bunting SW, Pretty J, Edwards P (2010) Wastewater-fed aquaculture in the East Kolkata Wetlands, India: anachronism or archetype for resilient ecocultures? Rev Aquac 2(3):138–153. https://doi.org/10.1111/j.1753-5131.2010.01031.x

Chakraborty SK, Sanyal P, Ray R (2023) Eco-biological uniqueness of East Kolkata Wetland, a Ramsar site in India. In: Springer eBooks, pp 257–301. https://doi.org/10.1007/978-3-031-09253-4_5

Cowardin LM, Carter V, Golet FC, LaRoe ET (1979) Classification of wetlands and deepwater habitats of the United States. U.S. Fish and Wildlife Service

Dasgupta R, Shaw R (2021) Urban resilience and ecological restoration: a case study of East Kolkata Wetlands. Environ Manag 67(3):347–359

Davidson NC (2014) How much wetland has the world lost? Mar Freshw Res 65(10):934–941. https://doi.org/10.1071/MF14173

EPA (2023) Wetlands protection and restoration. United States Environmental Protection Agency

Garcia-Herrero L, Lavrnic S, Guerrieri V, Toscano A, Milani M, Cirelli GL, Vittuari M (2023) Cost-benefit of green infrastructures for water management: a sustainability assessment of full-scale constructed wetlands in Northern and Southern Italy. arXiv preprint arXiv:2305.06284

GFDRR (2018) Urban wetlands management in Colombo: a new model for urban resilience. Global Facility for Disaster Reduction and Recovery. Retrieved from https://www.gfdrr.org/en/publication/results-resilience-urban-wetlands-management-colombourban-response.org+2GFDRR+2GFDRR+2

Ghosh D (2005) Ecohydrology of the East Kolkata Wetlands. World Water Congress

Gill SE et al (2007) Adapting cities for climate change. Built Environ 33(1):115–133

Hettiarachchi M, Athukorale K, Wijekoon S, de Alwis A (2014) Urban wetlands and disaster resilience of Colombo, Sri Lanka. Int J Disaster Resil Built Environ 5(1):79–89

IPCC (2014) Climate change 2014: mitigation of climate change. Intergovernmental Panel on Climate Change

Jaramillo L, Rodríguez F, Morales P (2018) The role of urban wetlands in biodiversity and water management in Bogotá. J Urban Ecol 4(1):juy005. https://doi.org/10.1093/jue/juy005

Kadlec RH, Wallace SD (2009) Treatment wetlands, 2nd edn. CRC Press

Keddy PA (2010) Wetland ecology: principles and conservation, 2nd edn. Cambridge University Press

Li Q, Wang F, Yu Y, Huang Z, Li M, Guan Y (2021) Comprehensive performance evaluation of LID practices for the sponge city construction: a case study in Guangxi, China. arXiv preprint arXiv:2112.10347

Meli P, Herrera B, Rey Benayas JM, Martínez Ramos M (2020) The role of wetland restoration in urban resilience: a global review. Environ Res Lett 15(7):073002

Millennium Ecosystem Assessment (MEA) (2005) Ecosystems and human well-being: wetlands and water synthesis. World Resources Institute

Ministry of Environment, Forest and Climate Change (MoEFCC) (2017) Wetlands (conservation and management) rules, 2017

Ministry of Environment, Forest and Climate Change (MoEFCC) (2022) Master plan Delhi 2041—draft document

Ministry of Housing and Urban-Rural Development of China (MOHURD) (2018) Sponge city pilot program progress reports. Beijing, China

Mitsch WJ, Gosselink JG (2015) Wetlands, 5th edn. Wiley

Mondal BK, Kumari S, Ghosh A, Mishra PK (2022) Transformation and risk assessment of the East Kolkata Wetlands (India) using fuzzy MCDM method and geospatial technology. Geogr Sustain 3(3):191–203

National Institute of Urban Affairs (NIUA) (2022) Urban resilience strategies in Indian cities

ORF (2023) Governing urban wetlands in India: a pathway to sustainable urbanisation. Observer Research Foundation. Retrieved from https://www.orfonline.org/english/research/governing-urban-wetlands-in-india-a-pathway-to-sustainable-urbanisation

Ramsar Convention (2025) List of wetlands of international importance. Retrieved from https://rsis.ramsar.org

Ramsar Convention Secretariat (2010) Wise use of wetlands: a conceptual framework. Ramsar Convention

Ramsar Convention Secretariat (2018) Global wetland outlook: state of the world's wetlands and their services to people. Ramsar Convention

Ramsar Convention Secretariat (2022) Ramsar Sites Information Service (RSIS): East Kolkata Wetlands. Retrieved from https://rsis.ramsar.org

Ramsar Convention Secretariat (n.d.) Urban wetlands in Bogotá, Colombia. Retrieved from https://www.ramsar.org

Raychaudhuri S, Dasgupta R, Bhattacharya S (2014) Wetlands and urban resilience: the case of Kolkata. Urban Ecol 18(2):235–250

Rojas Quezada CA (ed) (2024) Urban wetlands in Latin America: protection, conservation, innovation, restoration, and community for sustainable and water sensitive cities. Springer

Sharda D (2024) East Kolkata Wetlands: analysing the socio-legal protections in the face of rapid urban sprawl. Centre for Development Policy and Practice. cdpp.co.in

Sri Lanka Land Reclamation and Development Corporation (SLLRDC) (2016) Colombo wetland management strategy. Colombo, Sri Lanka

The Guardian (2025) 'Even the sound of the water has changed': can Bogotá bring its wetlands back from the brink? Retrieved from https://www.theguardian.com/global-development/2025/feb/13/water-colombia-bogota-bring-its-wetlands-back-from-the-brink-urban-development

UNDP (2022) Enhancing socio-ecological resilience in the urban wetlands of Colombo. United Nations Development Programme. Retrieved from https://www.undp.org/srilanka/press-releases/enhancing-socio-ecological-resilience-urban-wetlands-colombo

UNEP (United Nations Environment Programme) (2021) Making peace with nature: a scientific blueprint to tackle the climate, biodiversity and pollution emergencies. Nairobi

UN-Habitat (2014) Planning for ecosystem services in cities. United Nations Human Settlements Programme

UN-Habitat & Ramsar Convention on Wetlands (2020) The value of wetlands in urban and peri-urban areas

United Nations Environment Programme (UNEP) (2019) Nature-based solutions for climate adaptation in urban areas: lessons from Bogotá. GRID-Arendal. Retrieved from https://www.grida.no

Urban Climate Resilience in South Asia (UCRSEA) (2017) Urban wetlands as a tool for flood risk reduction: Colombo case study. Retrieved from https://www.iucn.org

Wetlands (Conservation and Management) Rules (2017) Ministry of Environment, Forest and Climate Change, Government of India

Wetlands International South Asia (2020) Urban wetlands of India: a review

World Bank (2016) Valuing ecosystem services of Colombo wetlands: a cost-saving framework. Retrieved from https://www.worldbank.org

World Bank (2019) Wuhan urban flood management project. Project Information Document. Retrieved from https://projects.worldbank.org

2 Decolonizing the Research Narrative: From "Drain the Swamps" to "Water Back" Movement

Sheila Mitra-Sarkar and Robert Wallace

Abstract

In colonized North America, the destruction of swamps from the 1600s to the 1900s was labeled as "civilizing" the ecosystem, with wetlands seen as unproductive and disease-ridden. This mindset contributed to the displacement of Indigenous peoples and ecosystem degradation for profit. Today, wetlands are disappearing three times faster than forests, primarily due to changes in agricultural land use. The Ramsar Convention of 1971 has seen over half of the Wetlands of International Importance damaged, and a 2009–2019 US report noted a 50% increase in wetland loss. To address this, it's crucial to acknowledge the historical ecocide and genocide in the name of progress. Urban planners can aid in preservation by adopting decolonized approaches and incorporating Indigenous Ecological Knowledge Systems. This chapter seeks to educate planning students and the community about past decisions and the ongoing loss of wetlands due to neglect of Indigenous practices.

S. Mitra-Sarkar (✉)
Department of Urban Studies & Planning, University of California, San Diego, CA, USA
e-mail: shsarkar@ucsd.edu

R. Wallace
Water Protector, Barona Band of Mission Indians, Lakeside, CA, USA

Keywords

Wetlands · Draining swamps · Water back · Decolonize wetlands · Civilized swamps · Indigenous ecological knowledge systems · Producing knowledge · Standing rock

2.1 Introduction

Decolonized scholarly writing is crucial in the humanities and social sciences, particularly for addressing environmental issues. It allows researchers to critically explore the links between their work, imperialism, and racial hierarchies affecting original stewards of wetlands (Thambinathan and Kinsella 2021). We recognize the term "research" carries weight for those with colonized ancestors and challenge the notion of "objective research," which has often led to predatory behaviors toward Indigenous communities (Thambinathan and Kinsella 2021). This oppressive scholarship can perpetuate harmful stereotypes and marginalize Indigenous voices, disrupting their deep connections to the land and water (Bishop 1998; Datta 2017). Consequently, Indigenous communities are wary of participating in research that claims neutrality, rooted in past exploitation and misrepresentation (Rigney 1999).

In the Western approach to knowledge acquisition, colonizers and travelers often observed Indigenous methods of nature preservation and

S. K. Chakraborty et al. (eds.), *Wetlands and Urban Resilience*, Sustainable Development Goals Series,
https://doi.org/10.1007/978-3-032-06881-1_2

cultural traditions through a lens claiming to be "objective" and "neutral." However, these observations reflected biases that downplayed Indigenous wisdom (Masta 2022; Datta 2018; Cajete 1994; Said 1979). Tuhiwai Smith (2012) draws on Edward Said's insights from *Orientalism* (1979) to show how portrayals of Indigenous peoples have created dominant narratives that misrepresent their complexity (Masta 2022; Cajete 1994). This "Othered" framing has influenced public perceptions, often overlooking Indigenous knowledge of environmental stewardship (Masta 2022; Rose 2024; Wilson 2001).

Despite the barriers, there has been a notable increase in scholars focused on decolonizing qualitative research methodologies. These researchers prioritize Indigenous and non-Eurocentric narratives to provide more authentic historical accounts of ecological destruction in wetland areas (Datta 2018; Brayboy 2005; Cajete 1994; Said 1979). This shift aims to enrich the understanding of environmental issues by incorporating diverse perspectives and experiences (Omodan 2025; Hayward et al. 2021; Thambinathan and Kinsella 2021; Tuhiwai Smith 2012; Rigney 1999).

The authors of this chapter recognize the harms of Eurocentric perspectives in environmental research, particularly regarding Indigenous values and wisdom. The ongoing degradation of wetlands for settler ecologies[1] is often overlooked in mainstream scholarship. We aim to share decolonized truths about wetlands, highlighting the importance of incorporating Indigenous knowledge and the historical context of settler colonialism in discussions about wetland restoration. Decolonized scholarship focuses on two primary goals (Omodan 2025; Thambinathan and Kinsella 2021):

(a) Expanding the academic discourse to include anticolonial and non-Western perspectives while amplifying critical Indigenous voices that challenge traditional academic frameworks.
(b) This emerging scholarship examines the connection between colonialism and ecological theory, highlighting how colonial practices have shaped landscapes and modern environmental concepts. By normalizing resource extraction and imposing management on ecosystems and Indigenous communities, these practices necessitate reevaluating ecological thought. Integrating counterstorytelling can clarify issues of environmental exploitation and genocide (Fig. 2.1).

[1] Repopulating nature to create inclusive wild spaces, rescuing endangered species to support settler ecologies, and expanding wild areas to extend settler ecologies.

2.2 Counterstorytelling as a Research Methodology

Tuhiwai Smith (2012) underscores the need for both Eurocentric and non-Eurocentric researchers to critically reflect on the assumptions and values that shape their research practices. This examination is crucial for understanding how colonial histories impact research methodologies and outcomes. Counterstorytelling serves as a decolonizing method that integrates Tribal Critical Race Theory (Brayboy 2005), emphasizing the importance of incorporating knowledge from marginalized Indigenous communities (Tuhiwai Smith 2018; Datta 2018; Brayboy 2001). Decolonizing research means centering the perspectives of non-Western individuals and engaging with previously marginalized theories (Datta 2018; Tuhiwai Smith 2012; Brayboy 2005; Crenshaw 1991). This chapter promotes an anti-oppressive qualitative research approach rooted in decolonial critical race feminist theories, highlighting respect, responsibility, and humility in addressing intersectional struggles (Crenshaw 1991). There is no standardized model for decolonizing research methodologies, making it essential to find ways to integrate decolonial learning into qualitative research and elevate critical storytelling, often overlooked in Western contexts (Thambinathan and Kinsella 2021).

Fig. 2.1 Examples of settler ecologies. (Top left) The Kumeyaay were dispossessed of their connections with the bay by the 1700s. The bay inlet shown here is mainly used for recreation. Harbor Island, an artificial peninsula created in 1961 from harbor dredgings, is located in San Diego Bay in San Diego, California. (Top right) Currently, San Elijo Lagoon reflects a strong focus on human interests and activities. (Center left) Vegetated San Elijo Lagoon has numerous alterations (railroad berm) affecting tidal circulation. The Lagoon has been modified for human recreational and housing needs. (Center right) Mission Bay, California, was once home to the Kumeyaay. From July 1952 to December 1959, San Diego operated a landfill in Mission Bay Park, accepting millions of gallons of industrial waste from the aerospace industry. Toxic substances were often buried in steel drums or dumped in unlined holes below the groundwater level. Environmental studies show that the San Diego River and Mission Bay aquatic organisms are the main at-risk species from landfill contamination. (Bottom left) Minnehaha Creek, Minneapolis, Minnesota, is fragmented by the parking lot of a billboard, buildings, a big box store, and Highway 7, making it perilous for the wildlife to raise their young. (Bottom right) Interpretive sign on how the European settlers changed the habitat surrounding what they renamed as Emigration Creek in Salt Lake City, Utah

Counterstorytelling is a significant research method that emphasizes marginalized communities affected by colonization and imperialism, emerging from Critical Race Theory in the late 1980s (Solórzano and Yosso 2002; Crenshaw 1991). This methodology combines various data-gathering techniques influenced by Critical Race Theory and seeks to humanize quantitative data while amplifying silenced voices in qualitative research. It contrasts sharply with dominant Western research practices, often shaped by those with privilege. Decolonized qualitative research through counterstorytelling offers meaningful alternative perspectives and uncovers historical truths ignored in mainstream narratives about the impact of colonization on Indigenous peoples and Nature (Mitra-Sarkar 2024; Deloria Jr 2023).

In this chapter, written as counterstorytelling, we use four key practices of decolonized writing (Thambinathan and Kinsella 2021):

1. *Critical Reflexivity*
   Decolonized academic researchers have shifted their perspective from interpreting community knowledge to becoming learners coauthoring research with Indigenous activist scholars (Cajete 1994; Datta 2018).
2. *Reciprocity and Respect for Self-determination*
   Decolonizing research methodology addresses the fundamental issues underlying conflicts by employing the principle of reciprocity. This approach fosters collaboration with Indigenous scholars from the beginning to the end of a study. It aims to establish collective ownership over the entire research process and counterstorytelling (Cajete 1994; Tujiwai Smith 2012; Datta 2018; Solórzano and Yosso 2002).
3. *Embracing "Other(ed)" Ways of Research*
   Unlearning and reimagining the value of knowledge are essential for decolonizing research. Aimé Césaire, in his 1955 essay "Discourse on Colonialism," critiques Western civilization's historical violence and exploitation linked to colonialism. He labels certain academics as "goitrous academics" for perpetuating capitalism and undermining Indigenous ecological knowledge. Césaire's compelling message inspires a vision for knowledge that promotes justice and equity, recognizing the transformative power of accessible expertise in empowering communities (Césaire 1955; Cajete 1994; Tujiwai Smith 2012; Datta 2018; Thambinathan and Kinsella 2021).
4. *Transformative Praxis*
   The concept of "margin" is a vital analytical tool for understanding oppression, inequality, and power dynamics in various social contexts (Drame et al. 2011). Many researchers work within these margins, focusing on social justice and collaborating with marginalized communities to highlight their experiences and amplify their voices (Tujiwai Smith 2012). Emancipatory research engages these communities to represent their historically silenced narratives authentically. The aim of decolonized researchers is to convey these stories with integrity and nuance (Hayward et al. 2021; Rigney 1999; Tujiwai Smith 2012; Brayboy 2001).

Counterstorytelling serves as our research method to highlight the uncomfortable truths regarding the destruction of wetlands due to colonial practices and the absence of Indigenous wisdom in their protection (Pidgeon and Riley 2021; Kovach 2021). To protect our wetlands during the climate crisis, we need to understand how they were initially destroyed and recognize that these destructive practices continue today. We also want to amplify the work of Indigenous scholars who have always advocated for a biocentric approach to restore our wetlands by counterstorytelling.[2]

2.3 Cultural History of Wetland Losses in the United States

Wetlands are unique ecosystems where the soil is water-saturated for part of the year. Vegetated wetlands enhance resilience to storms and sea level rise, improve water quality, capture carbon dioxide, and provide crucial habitats for both endangered and commercially valuable species (US Fish & Wildlife Service 2024). For First Nations (Indigenous) peoples, these wetlands are integral to the environment, serving as the veins of Mother Earth, with trees symbolizing her lungs (Hill 2016). Wetlands also filter sediment, contributing to a thriving ecosystem (Wetland Info 2022). Indigenous peoples of Turtle Island view wetlands as sacred spaces rich in spiritual significance, reflected in their art, dances, and

[2] One such site is the Windover Pond near Cape Canaveral in Florida, which was used for 300 burials about 8000 years ago.

stories. They have traditionally used wetlands for burials, as evidenced by archaeological findings (Giblett 2024; The Ramsar Bureau 1971).

2.4 Naming It: Studying Wetland Loss Through the Lens of Ecocide and Ethnocide due to Settler Colonialism

In December 2019, Vanuatu's EU ambassador proposed making ecocide a crime at the International Criminal Court. Bartolomé Clavero (2008) highlighted that ethnocide harms cultures and individual souls. Indigenous peoples' spiritual practices, like constructing sweat lodges near water for purification, were disrupted by settlers, resulting in both ethnocide and genocide (Teixeira 1991; Dillon 2021; Gilo-Whitaker 2019). In colonized North America, the destruction of swamps from the 1600s to the 1900s was often referred to as "making it civilized" (Dahl and Allord 1996). This term was used to justify the dispossession of Indigenous peoples and to render the land more suitable for white settlers moving from the East to the West Coast (Carlson 2010). We must recognize the intentional destruction of wetlands and their habitats as ecocide; the area of wetlands decreased from 221 million acres in the 1600s to just 103 million acres remaining by the mid-1980s (Dahl and Johnson 1991). Six States (California, Missouri, Iowa, Illinois, Indiana, and Ohio) lost 85% or more of their wetland acreage, and 22 states lost 50% or more of their wetlands (Dahl and Allord 1996; Dahl and Johnson 1991). By 2019, there were an estimated 116.4 million acres of wetlands, accounting for less than 6% of land area within the conterminous United States[3] (US Fish & Wildlife Service 2024).

The destruction of wetlands coincided with the deception, murder, and disenfranchisement of Indigenous peoples from their ancestral lands. The loss of wetlands and wildlife is closely linked to the genocide of these tribes in Turtle Island, yet it's seldom addressed together. Survivors were often denied access to sacred sites vital for their spiritual ceremonies, which are closely tied to locations like wetlands and oceans. It is essential to recognize how settler colonialism impacted Native spirituality and well-being when tribes were relocated away from these crucial areas, hindering their ability to perform essential rituals[4] (Trope 2010) (Table 2.1).

2.5 Colonizer Ecologies

Colonial capital formation often involved appropriating resources without clear ownership, gaining legitimacy in capitalist systems. Mukherjee and Chakraborty (2016) note that concepts like res nullius (unassigned resources belong to the sovereign) and terra nullius (no man's land) provided colonial states with legal justification to seize community resources, labeling them as 'waste' for profit (p. 78). In their quest for capital, colonizing countries aimed to drain swamps before establishing settler ecologies in the New World. Di Palma (2014) explores the seventeenth-century efforts to drain England's eastern fens through the General Drainage Act of 1600, which transformed wetlands into profitable land—this labeling of wetlands as wastelands dispossessed peasants who relied on the landscape's resources for their economies.

Wetland drainage in Colonial America remains underexplored historically, yet addressing its significance and destruction is crucial. When European colonizers settled in the early 1600s, they brutally displaced the tribes. The area that would become the contiguous United States, referred to as Turtle Island, had approximately 221 million acres of wetlands, marshes, swamps, and robust ecological diversity. During the 1600s and 1700s, European

[3] There was a net gain in non-vegetated wetlands of 488 K acres, and a related increase in pond area of over 7% (U.S. Fish & Wildlife Service 2024).

[4] Even now, access is often restricted to certain individuals at locations regarded as particularly sacred, and specific protocols must be followed regarding permissible and prohibited behaviors at these sites.

Table 2.1 Wetland ecological systems and settler colonialism

Types	Features of wetlands	The outcome of settler colonialism
Marsh	Marshes are wetlands characterized by water-saturated soil and soft-stemmed vegetation. Types include prairie potholes and the Everglades. They can be coastal or inland, freshwater or saltwater. Rich in nutrients, marshes primarily receive water from surface runoff.	Like many other wetland ecosystems, freshwater marshes have faced significant acreage losses due to colonial expansion and extensive draining. The loss of habitat and the displacement of Indigenous communities have not been thoroughly examined.
Non-tidal marsh	Non-tidal marshes are the most common wetlands in North America. They mainly consist of freshwater, though they can also be brackish or alkaline. They typically develop along streams, in poorly drained areas, and near lakes and rivers. Water levels range from a few inches to three feet, and some, like prairie potholes, may dry out completely. Freshwater marshes are highly productive ecosystems rich in nutrients. They support diverse plant communities, which provide habitat for various wildlife.	
Tidal marsh	Tidal marshes are mainly located along the eastern US coast, from Maine to Florida and into Louisiana and Texas along the Gulf of Mexico. They can be freshwater, brackish, or saline, all influenced by ocean tides. Tidal marshes play vital roles by buffering stormy seas, reducing shoreline erosion, and absorbing excess nutrients before they enter oceans and estuaries. They also provide food and habitat for clams, crabs, juvenile fish, and nesting sites for migratory waterfowl.	Since the 1600s, pressure to fill in these wetlands for coastal development has led to significant and continuing losses of tidal marshes along the Atlantic coast and the Gulf of Mexico.
Swamps	A swamp is a wetland dominated by woody plants, such as Red Maple swamps in the Northeast and bottomland hardwood forests in the Southeast. They have saturated soils and standing water, creating a nutrient-rich habitat for water-tolerant trees like Cypress. Common shrubs include Buttonbush and Smooth Alder. Swamps support diverse wildlife, including rare species like the endangered American Crocodile, and can be classified into shrub and forested swamps based on vegetation type.	Historically, European colonizers portrayed swamps as frightening wastelands. Their perception, combined with the nutrient-rich soils found in swamps, led to the draining and clearing of many fertile woodlands for agriculture and other developments. This viewpoint contributed to the significant destruction of vast tracts of swampland over the past 200 years, including the loss of more than half of the Great Dismal Swamp in southeastern Virginia.
Bogs	Bogs are unique North American wetlands with spongy peat deposits, acidic waters, and a layer of sphagnum moss. They receive water mainly from precipitation, resulting in low nutrient levels for plant growth. Bogs play a vital role in flood prevention by absorbing excess precipitation and supporting unique species like the carnivorous Sundew. They also provide habitat for animals threatened by human activities.	Bogs in the United States are mainly located in the glaciated Northeast and Great Lakes regions. Colonial expansion led to a significant loss of bog acreage due to conversion to cropland and peat extraction for fuel and soil conditioning. Scientists recognize bogs' importance in regulating the global climate by storing carbon in peat. These unique ecosystems can be destroyed in days but take centuries to form.
Pocosins	"Pocosin" comes from an Algonquin word meaning "swamp on a hill." These evergreen landscapes, primarily found on the Atlantic Coastal Plain in North Carolina, have saturated soil from a shallow water table but typically lack standing water. They vary in size, can be isolated from streams, and provide crucial habitats for wildlife as they remain largely undisturbed despite development pressures in the Eastern United States.	From the 1600s to the 1900s, agriculture primarily threatened pocosins. However, timber harvesting, peat mining, and phosphate mining have joined agriculture as the biggest threats to undisturbed pocosins.

Source: US EPA (2015), Dahl & Johnson (1991)

monarchs promoted the establishment of settler-colonized environments to ensure a continuous source of resource extraction for their countries. This involved creating a strong, manipulated presence in the ecosystems, which we call "colonizer ecologics." The impacts of this colonization on the landscape became evident in the early to mid-1700s (Lightfoot et al. 2013). Initially, settlements consisted primarily of shelters and subsistence farms on small tracts of land.

Wetlands on small tracts were drained using hand-dug ditches to maximize land use. In the mid-to-late 1700s, as settlers arrived, land clearing transformed larger areas, turning coastal wetlands into farmland. This conversion increased agricultural productivity but devastated local ecosystems and wildlife (Di Palma 2014). The English Enclosure movement and European expansion spurred this aggressive settlement, driven by individualism and commercial interests. Enclosure involved appropriating "waste" lands, denying commoners access and rights to the land. This narrative of progress often marginalized those who depended on these natural resources for their livelihoods (Dillon 2021; Di Palma 2014).

Widespread wetland destruction occurred in the southern colonies during the 1600s and 1700s. In 1754, South Carolina permitted the destruction of Cacaw Swamp for agriculture, while Virginia and North Carolina surveyed the Great Dismal Swamp in 1763 for water transportation routes. Plantation farming in the South often required altering or destroying wetlands. By the 1780s (Fig. 2.2), settlers took over fertile river valleys in present-day Georgia, severely impacting wetlands, wildlife habitats, and Indigenous livelihoods. Small towns and farms emerged along rivers in states like Massachusetts and Pennsylvania, with settlement expanding into Virginia and the Carolinas by the 1800s (Dahl and Allord 1996).[5]

Capitalism and the pursuit of profit have transformed ecologically diverse wetlands, often regarded as wastelands, into private property for agricultural or urban development. This transformation has fueled policies leading to land alienation. The phenomenon known as "wetlands grab" has dispossessed and displaced Indigenous inhabitants, disrupting their sustainable ways of living (Dillon 2021). Extractive practices aimed at draining swamps to make wetlands economically productive prioritized the health of Anglo-European settlers while expanding the settler nation (Dillon 2021; Di Palma 2014; Vileisis 1997). The acquisition of wetlands for capital gain and generational wealth was further cemented by the Indian Removal Act of 1830, which was signed into law by President Andrew Jackson on May 28, 1830.[6] This law, as outlined by Congress, facilitated "an (inequitable) exchange of lands" with the tribes residing in any of the states or territories. In the mid-1800s, the US Congress also passed the Swamp Land Acts, initially aimed at assisting Louisiana in draining its swamplands and constructing levees. Other states soon adopted these laws. The Acts allowed individuals to claim wetlands or marshlands as their own if they drained the land and used it for agriculture. This resulted in a significant increase in levee construction and wetland drainage across the country (Parrish 2024; Knox et al. 2022).

The period from 1800 to 1860 saw significant growth in the United States, marked by major land acquisitions like the Louisiana Purchase (1803), the cession of Florida (1819), the annexation of Texas (1845), the Oregon Compromise (1846), and the lands ceded by Mexico

[5] Repopulating nature to create inclusive wild spaces, rescuing endangered species to support settler ecologies, and expanding wild areas to extend settler ecologies.

[6] The progress of removal would "perhaps cause them gradually…to cast off their savage habits and become an interesting, civilized, and Christian community." The legislation was the culmination of multiple parties' advocacy for the seizure of tribal lands and the relinquishment of tribal sovereignty (Tracey 2024).

Fig. 2.2 Top Left: The Lakota phrase "Mní wičhóni" ("Water is life") was the protest anthem from Standing Rock heard around the world. It also has a spiritual meaning rooted in Indigenous respect for Mother Earth by living sustainably. (Top Right) Ohlone rowing their tule boat in San Francisco Bay. ((Center) Artist's recreation of Cahokia is the site of an ancient Native city along Mississippi river valley near modern day St. Louis which existed pre-contact. (Bottom Left). About 3,000 people including Anishanabee and Dakota tribes marched in 2017 during President Donald Trump's first inauguration to protect ground water from Dakota Access Pipeline. (Bottom right) Thousands have parked along Missouri River at Cannon Ball in the Standing Rock Sioux Reservation to stop Dakota Access Pipeline (DAPL). Native led groups like Barona Band of Mission Indians have united in mass gatherings of Water Protectors at Camp Oceti Sakowin, Red Warrior Camp, and Sacred Stone Camp.Photo: Bobby Wallace)

(1848). The population surged from 7.2 million in 1810 to 12.8 million in 1830. This expansion fueled land speculation and a perception of available resources, often at the expense of Native tribes, who faced prohibitions on their religious practices and suffered crimes committed by the government to benefit European settlers (Fig. 2.3). The Indian Removal Act of 1830 (Trail of Tears)[7] and the General Allotment Act

[7] Some historians classify Trail of Tears as ethnic cleansing (Gilio-Whitaker 2019); others refer to it as genocide (Ostler 2015). Ruth Muskrat Bronson, a Cherokee scholar and poet, aptly describes the removal of her Cherokee ancestors in the poem Trail of Tears:

> From the homes their fathers made// From the graves the tall trees shade// For the sake of greed and gold//The Cherokees were forced to go// To a land they did not know; // And Father Time or wisdom old// Cannot be erased, through endless years//The memory of the trail of tears.

Fig. 2.3 States with notable wetland loss from 1780 to 1980 (Dahl and Allord 1996)

of 1887[8] forced hundreds of tribes to relocate from their ancestral lands (Trail of Tears), notably displacing them to the Central Plains, away from the tidal wetlands and the Great Lakes (Fig. 2.4) (Crepelle 2021; Fenelon and Trafzer 2014) (Fig. 2.5).

During significant westward expansion, settlers moved rapidly into the wetland-rich regions of the Ohio and Mississippi River Valleys (see Fig. 2.2 in Dahl and Allord 1996). This large-scale conversion of wetlands into farmland began to impact wetland destruction across the United States, notably. Figure 2.6 in Dahl and Allord (1996, p. 20) illustrates the areas where significant wetland ecocide occurred between 1800 and 1860 (Dahl and Allord 1996).

The California Gold Rush of 1849 significantly harmed the environment and Indigenous tribes. Before the Gold Rush, over 150,000 Indigenous Peoples lived in California, and their cultures were sustainable, based on hunting, gathering, and fishing. By 1870, their population had plummeted to about 31,000, primarily due to diseases introduced by Gold Rush settlers, as well as violence, forced relocations, and massacres (International Indian Treaty Council 1973).[9] Around 300,000 settlers migrated to California, leading to societal upheaval and environmental damage (Clarke 2016). Before Spanish settlement in 1769, the region was home to many Indigenous communities, including up to 70,000 Yokuts near Tulare Lake and numerous Chumash and Tongva villages along the coast. The Bay Area was inhabited by tens of thousands of Ohlone, Coast Miwok, Sierra Miwok, Patwin, and Wappo people (Clarke 2016). Placer (miners direct the water and sediment mixture through sluice boxes to separate the gold) and hydraulic mining significantly altered these ecosystems, causing habitat

8 The Dawes Act of 1887, also known as the General Allotment Act, regulated land rights on tribal territories in the United States. It converted communal tribal land into private allotments for Native American families and individuals, imposing a capitalist property system that was foreign to their traditional land tenure practices. (Blansett, 2015)

9 The 1870 federal census tallied 7241 remaining California Natives (Given the state of the federal 1870 census, some natives may have been missed.) (International Indian Treaty Council 2020).

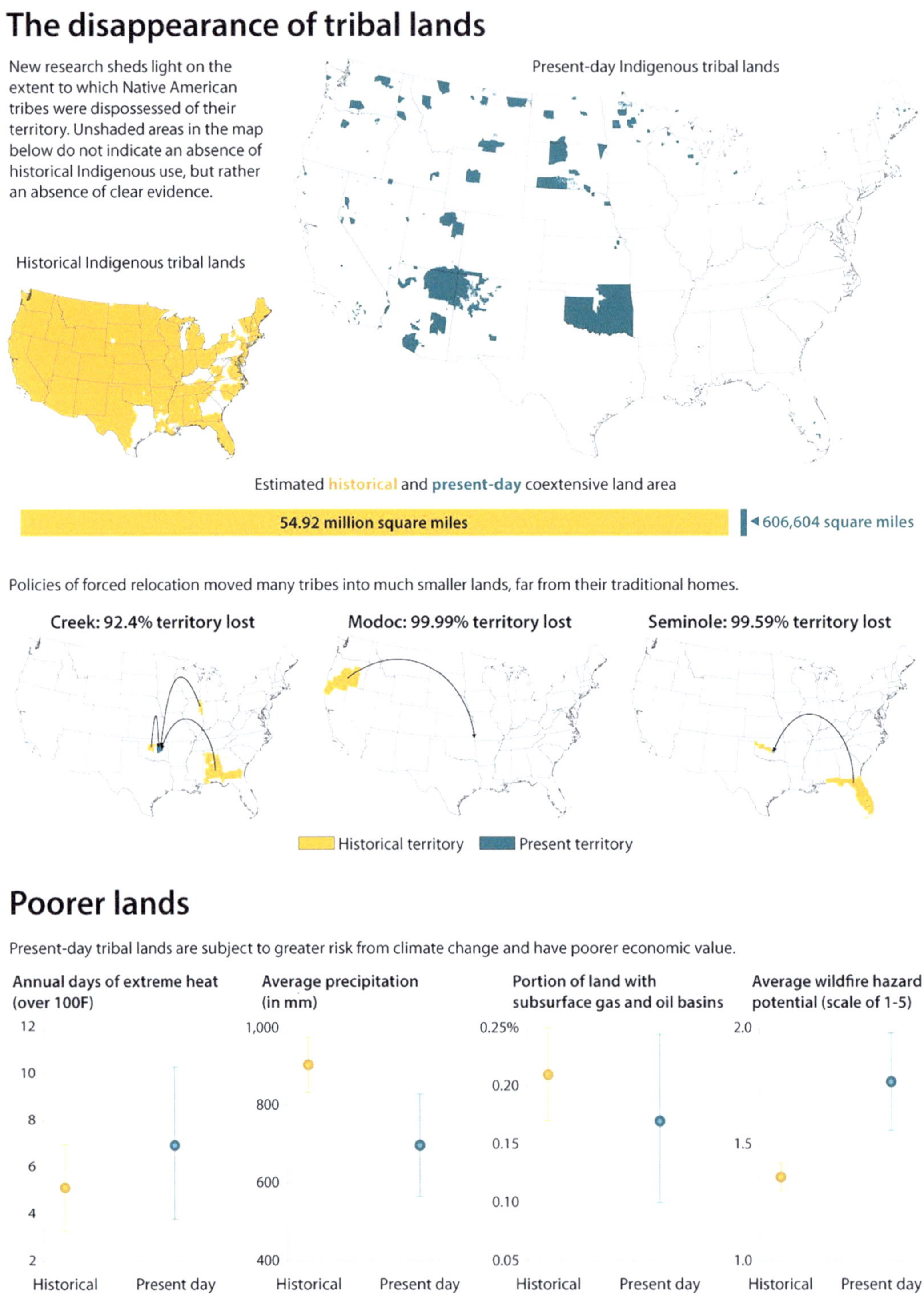

Fig. 2.4 Proportion reduction for estimating coextensive land area considers multiple tribes in one area. Historical record limitations may lead to an underestimation of total historical land. Creek, Modoc, and Seminole tribes lost over 90% of their land and wetlands. Plots illustrate changes in tribal land conditions (mean and 95% confidence interval) for selected variables. (Source: Farrell et al. 2021)

Fig. 2.5 (Top left) An artist's rendition of the Trail of Tears, National Park Service, Public Domain, https://www.fws.gov/media/artists-rendition-trail-tears. https://www.goodfreephotos.com/united-states/wisconsin/horicon-national-wildlife-refuge/farms-and-marsh-grasses.jpg.php

destruction and wildlife loss, and disrupting Indigenous practices due to increased flooding and erosion. During the Gold Rush, miners excavated 12 billion tons of earth and released approximately 7600 tons of mercury into Northern California's ecosystems (Clarke 2016; Nuwer 2013).[10]

The early twentieth century witnessed important engineering and drainage projects spurred by the World Wars, population growth, and industrial expansion. Key projects impacting wetlands include California's Central Valley Project and the Mississippi River's lock and dam system (Dahl and Allord 1996). During the 1900s, urban and agricultural expansion projects resulted in the drainage of many wetlands (Fig. 2.4). Notable efforts included the attempted full drainage of Horicon Marsh in Wisconsin in 1904, commercial timber harvesting in southern Georgia

[10] Today, just one gram of mercury in a small lake would be enough to violate federal health standards. Mercury is found throughout the waterways and ecosystems of the Sierra Nevada. This contamination is a lasting legacy of the California Gold Rush, which occurred in the 1850s and significantly influenced California's history. Even though the Gold Rush ended more than a century ago, thousands of abandoned mine sites and the sediment from these mines continue contributing to the problem (Clarke 2016).

starting in 1908 before draining the Okefenokee Swamp through the Suwanee Canal, and the draining of Lake Mattamuskeet, North Carolina's largest natural lake, in 1914 to create farmland (Dahl and Allord 1996; Dahl and Johnson 1991) Fig. 2.5.

In the early 1900s, land developers created drainage ditches to dry peatlands north of Red Lake, Minnesota (Glaser 1987). On July 29, 1917, the Minneapolis Sunday Tribune advertised the area as "unsettled, uncut timberland in the middle of the country" (Wright 1984, cited from Dahl and Allord 1996). By 1930, most of the prairie wetlands in Iowa, southern Minnesota, and the Red River Valley had been drained (Schrader 1955). In the 1930s, the US government offered free engineering services to farmers for wetland drainage, and by the 1940s, it started sharing the costs of these projects (Burwell and Sugden 1964). Organized drainage districts nationwide coordinated efforts to remove surface water from wetlands (Wooten and Jones 1955). Settlers altered much of the Everglades in southern Florida by building over 400 miles of drainage canals by the 1930s, which reduced sustainable living conditions for the Seminoles. The Seminoles had been relocated to the area by the US military during the Seminole Wars from 1835 to 1842 (Seminole Nation Museum 1973).[11]

By the 1960s, significant political and financial incentives for draining wetlands were in place. The Federal Government supported this through various policies, including the Watershed Protection and Flood Prevention Act of 1954, which promoted wetland drainage around flood-control projects. Subsidies and public works projects by the US Department of Agriculture also facilitated wetland losses (Erickson and Raymond 2018). Not surprisingly, tile and open-ditch drainage were deemed conservation practices under the Agriculture Conservation Program, leading to annual wetland losses of approximately 550,000 acres from the mid-1950s to the mid-1970s (Office of Technology Assessment 1984), with agriculture responsible for over 80% of these losses (Frayer et al. 1983).

From 1987 to now, federal policies have taken a 360 turn, from destroying to protecting wetlands, at least on paper (Dahl and Allord 1996).[12] After destroying more than 50% of the vegetated wetlands and habitat, Congress enacted the Emergency Wetlands Resources Act (Public Law 99-645) in 1986. Under the provisions of this Act, the US Fish and Wildlife Service must produce publicly available maps of US wetlands and report to Congress every 10 years on the status and trends of the Nation's wetlands. Despite the Army Corps of Engineers' focus on the monetary value of ecological services provided by wetlands, the sixth report, covering 2009–2019 (Fig. 2.7), found that the US experienced a net wetland loss of 221,000 acres, with losses surpassing gains (Fig. 2.6) (Office of the Technology Assessment 1984). Figure 2.5 shows a change in wetland types: vegetated wetlands have declined by 670,000 acres, whereas non-vegetated wetlands and deepwater areas have increased by 488,000 and 182,000 acres, respectively (US Fish & Wildlife Service 2024). Eurocentric approaches to wetland ecosystems prioritize financial gain, overlooking their spiritual, cultural, and ecological significance. The commodification of wetlands has not countered the loss of area due to ongoing exploitation (Fig. 2.2) ("Valuing Wetlands" 2021).

The US Supreme Court's May 2023 decision in Sackett v. Environmental Protection Agency weakened the Clean Water Act of 1972, restricting protections mainly to large waterways and allowing development in inland wetlands. An estimated 7–36 million hectares of non-tidal wetlands may have lost federal protections (Fitzerald Guth 2024). The US Army Corps of Engineers is expediting over 600 permit applications linked to President Trump's "Declaring a National Energy Emergency" Executive Order, raising concerns from environmental groups. These projects include a crude oil pipeline in Louisiana affecting 234 acres of wetlands

[11] Attempts were underway to restore some of The Everglades (Lord, 1993)

[12] See Dahl and Allord (1996) p. 23/

Fig. 2.6 (Top 1 (Library of Congress's Prints and Photographs division under the digital ID cph.3c00734 (Library of Congress Prints and Photographs Division, Washington, D.C. 20540 USA http://hdl.loc.gov/loc.pnp/pp.print)old. https://www.loc.gov/pictures/item/ca1040.photos.019042p. Public domain in the US as it was published before January 1, 1930

and another pipeline beneath Lake Michigan, which supplies over 10 million people with drinking water. Some proposed projects, like the Stibnite Gold Mine in Idaho, appear unrelated to energy and would affect 145 acres of wetlands and 21 miles of streams (Green 2025).

2.6 Promoting Radical Transformative Research to Save Wetlands: Indigenous Ecological Knowledge Systems

The climate crisis is intensifying, marked by the alarming loss of wetland acreage in the United States and globally. This deterioration is primarily attributed to our ongoing practice of monetizing and commodifying these vital ecosystems. Through the lens of counterstorytelling presented in this chapter, it becomes evident that traditionally biased, Eurocentric cost-benefit analyses (USDA 2015) tend to prioritize the extraction and destruction of wetlands. This approach neglects the intricate ecological and cultural values these areas provide, perpetuating a cycle of degradation that threatens our environment and communities. We urge scholars to cite and encourage research that centers on biocentrism. This ethical viewpoint broadens the concept of morality to encompass

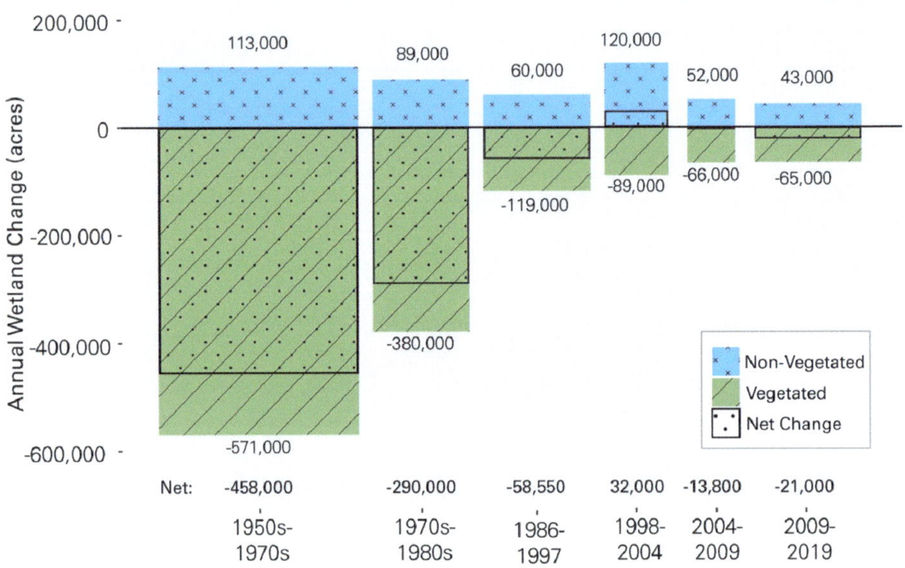

Fig. 2.7 The vegetated wetland loss and non-vegetated wetland gain have been ongoing for at least ~70 years (US Fish & Wildlife 2024)

all living beings, granting them intrinsic value and moral consideration while promoting urban resilience during a climate catastrophe. It highlights the interconnectedness of all life and underscores the significance of biodiversity.

Indigenous tribes have long generated and applied knowledge of biocentrism through direct experience with ecosystems and species. Over millennia, they have developed complex knowledge systems that have contributed significantly to ecology and evolution, but these contributions have not yet been comprehensively synthesized in Western academic work (Jessen et al. 2021). We must acknowledge that Indigenous peoples were the first ecologists to practice sustainable lifestyles. Their place-based knowledge incorporated customary laws and guidelines rooted in cultural and spiritual understandings of holistic practices (Fig. 2.1). The foundational principles reveal four Indigenous Ecological Knowledge Systems (IEKS) types that have shaped numerous Indigenous-led ecological movements (Cameron 2022):

1. **Traditional Knowledge**: A precious heritage passed down through generations, rooted in ancestral lore and wisdom
2. **Current Knowledge**: A dynamic journey of experimentation and adaptation, embracing the ever-changing nature of our world
3. **Revealed Knowledge**: Deep insights gained through intuitive spiritual experiences, connecting us to the sacred
4. **Embodied Knowledge**: A profound understanding forged through sensory connections with human and non-human realms

We strongly recommend decolonizing our research methods and incorporating Indigenous Ecological Knowledge Systems (IEKS) to revitalize and protect our vegetated wetlands and wildlife. The insights provided by Indigenous scholars, who actively engage in social movements using IEKS, contribute to ecological restoration in ways that differ from those of their Western counterparts (Estes 2020). Their ancestral wisdom, which is deeply connected to Nature, as outlined in previously described foundational principles, is vital for addressing the blind spots present in the perspectives of scholars from the Eurocentric countries regarding environmental conservation (Ferreira et al. 2023; Cameron 2022; Jensen et al. 2021). Worldwide, activists and scholars from non-Eurocentric countries increasingly acknowledge the significance of IEKS for science, policy, environmental conservation, and the cultural heritage of Indigenous peoples (Fabiano et al. 2021).

The 2010 People's Accords of Cochabamba is a key Indigenous environmental document highlighting one of the inspiring non-Eurocentric movements. Bolivia's most advanced ecological movement emerged under Indigenous leadership, particularly through the Movement Toward Socialism (MAS in Spanish) and Evo Morales. The Indigenous Andean concept of Vivir Bien, or "Living Well," and Pachamama, or "Mother Earth," are vital for understanding the People's Accords and IEKS (Solon 2018). Vivir Bien focused on an earth-centric approach that prioritizes the well-being of the entire ecosystem. This perspective emphasizes that capitalist exploitation of nature stems from patriarchal colonial systems, with overconsumption, primarily fueled by Eurocentric countries, being part of the issue rather than the solution (Estes 2020). For example, if the global population consumed resources at the same rate as the average American, we would need four Earths to sustain that level of consumption. This underscores the unsustainable nature of such patterns, contributing to the loss of wetlands. As decolonized scholars in the United States, we must recognize this Eurocentric capitalist burden and guide research in decolonized environmentalism (Estes 2020).

There is also a more extended history of Indigenous internationalism that is explicitly anti-imperialist and anti-neocolonialist (the use of economic, political, cultural, or other pressures to control or influence Indigenous territories), fighting to protect the environment that is underexplored (Estes 2020). These movements have shown that academic exploration that dismisses the Indigenous approaches to preserve the ecosystems will not stop wetland loss and climate catastrophe (Solon 2018; Estes 2020). The resistance at Standing Rock against the Dakota Access Pipeline (2016–2017) highlights how Indigenous Ecological Knowledge Systems informed revolutionary environmental theory, playing a key role in the movement led by the Standing Rock Sioux Tribe. The Indigenous tribes opposed the project because it threatened their spiritual and cultural values about "sacred land" and their water supply. The authentic nature of the peaceful protest attracted thousands of "Water Protectors" from across the United States and around the world over time (Estes 2020; Water Protectors 2019). The Standing Rock water protectors forged stronger connections with non-human communities and educated non-Indigenous people on how to decolonize their anthropocentric views of nature. The Standing Rock Movement aimed to reach a broad audience to share its message and succeeded despite facing brutal military suppression. Most importantly, Americans witnessed the Indigenous struggle, which includes violence, resource theft, and the collusion between corporations and governments—issues that are inherently connected to the fight for environmental protection (Guha 2023; Monet 2018).

2.7 Conclusions

We cannot fully understand the long-term climate impacts of land dispossession, ecocide, forced migration, and genocide against Indigenous peoples in North America. However, it's crucial to acknowledge these issues in scholarly work regarding Indigenous rights, as we have attempted through counterstorytelling. Farrell et al. (2021) analyzed a new dataset on Indigenous land dispossession, revealing a significant decline in the density and distribution of Indigenous lands. Many Indigenous peoples have had to move to areas more susceptible to climate change risks, often lacking valuable wetlands. Tragically, land dispossession came with "water dispossession," loss of wetlands, and rights to the river systems (Fig. 2.3) (Frayer et al. 1983).

As decolonized researchers, we have shown why we must become activist scholars to truly make a difference in climate change. Our research methods will always be part of Indigenous movements because they have contributed significantly to "producing ecological knowledge" and revolutionary theories (Estes 2020). We align with scholar Nick Estes, a Kul Wicasa citizen of the Lower Brule Sioux Tribe. Estes writes,

> I situate myself as a movement intellectual, actively participating in social movements (like Standing Rock), learning from them, and producing knowledge for and about them. There is something profoundly different than producing knowledge and writing for academic spaces. The scholarship retains its sharpness, as Ruthie Gilmore (Center for Place Culture and Politics) points out. There's a key difference between analytical sharpness and narrowness, the latter often constrained by the academy's limited function rather than being liberatory. (Estes 2020)

To protect our wetlands through academic scholarship, we must embrace Indigenous Ecological Knowledge Systems and actively engage in actions that resist the destruction of ecological systems by governments and corporations. We must be movement scholars, aiming to produce knowledge and decolonize research methods that prioritize Eurocentric approaches. We encourage scholars from diverse backgrounds to create a new framework that emphasizes actions, incorporates the voices and perspectives of Indigenous communities, integrates traditional ecological knowledge, and challenges existing power dynamics in environmental research. We hope that groundbreaking research practices will cultivate collaborative partnerships that honor and celebrate Indigenous spiritual and cultural heritage while advancing the sustainable stewardship of these essential ecosystems.

References

AP Archive (2024, May 23) Trucks carrying aid roll into Wadi Gaza as desperate Palestinians line up to snatch food boxes. Retrieved May 30, 2024, from YouTube website https://www.youtube.com/watch?v=nd8K3byKBv0

Baka J (2013) The political construction of wasteland: governmentality, land acquisition and social inequality in South India. Dev Chang 44(2):409–428

Bell DA (1993) Faces at the bottom of the well: the permanence of racism. Basic Books, A Member of the Perseus Books Group

Bhandar B (2018) Colonial lives of property: law, land, and racial regimes of ownership. Duke University Press, Durham

Bishop R (1998) Freeing ourselves from neocolonial domination in research: a Maori approach to creating knowledge. Int J Qual Stud Educ 11(2):199–219

Bogel-Burroughs N (2024, May 6) Police raid U.C. San Diego encampment and arrest 64 protesters. The New York Times. Retrieved from https://www.nytimes.com/2024/05/06/us/ucsd-campus-protests-arrests.html

Brayboy B (2001, November) Toward a tribal critical theory in higher education. Paper presented at the Association for the Study of Higher Education, Richmond, VA

Brayboy BMJ (2005) Toward a tribal critical race theory in education. Urban Rev 37(5):425–446. https://doi.org/10.1007/s11256-005-0018-y

Burwell RW, Sugden LG, (1964) Potholes–Going, going..., in Linduska, J.P., ed., Waterfowl tomorrow: Washington, D.C., U.S. Fish and Wildlife Service, p.369–380

Cajete G (1994) Look to the mountain: an ecology of indigenous education, 1st edn. ERIC. Kivaki Press, 585 E. https://eric.ed.gov/?id=eD375993

Cameron L (2022) Indigenous ecological knowledge systems—exploring sensory narratives. Ecol Manag Restor 23(S1):27–32. https://doi.org/10.1111/emr.12534

Carlson AE (2010) The Other Kind of Reclamation: Wetlands Drainage and National Water Policy, 1902–1912. Agricultural History, 84(4):451–478. http://www.jstor.org/stable/27869012

Césaire A (1955) Discourse on colonialism. Monthly Review Press

Clavero B (2008) Genocide or ethnocide, 1933-2007. Retrieved October 11, 2025, from Open Library website: https://openlibrary.org/books/OL44889285M/Genocide_or_ethnocide_1933-2007

Clarke C (2016, September 26) Untold history: the survival of California's Indians. Retrieved from PBS SoCal website: https://www.pbssocal.org/shows/tending-the-wild/untold-history-the-survival-of-californias-indians

Collin S, Collin Y, Koskey M (2018) Protecting the right to exist as a people: intellectual property as a means to protect traditional knowledge and indigenous culture. Scholarworks.alaska.edu. https://scholarworks.alaska.edu/handle/11122/10584

Cowan J, Chen MK (2024, June 16) In California, tribal members are reclaiming the "land of the flowing water." The New York Times. Retrieved from https://

www.nytimes.com/2024/06/16/us/california-native-american-tribes.html

Crenshaw K (1991) Mapping the margins: intersectionality, identity politics, and violence against women of color. Stanford Law Rev 43(6):1241–1299. https://doi.org/10.2307/1229039

Crepelle A (2021) Lies, damn lies, and Federal Indian Law: the ethics of citing racist precedent in contemporary Federal Indian Law. N.y.u. Rev Law Soc Chang 44:565. Archived from the original (PDF) on August 21, 2024

Dahl T, Allord G (1996) Technical aspects of wetlands history of wetlands in the conterminous United States. Retrieved from https://biotech.law.lsu.edu/blog/History-of-Wetlands-in-the-Conterminous-United-States.pdf

Dahl TE, Johnson CE (1991) Wetlands: status and trends in the conterminous United States mid-1970s to mid-1980s, report. Washington, DC. https://digital.library.unt.edu/ark:/67531/metadc948734/. Accessed 8 Dec 2024

Datta R (2018) Decolonizing both researcher and research and its effectiveness in indigenous research. Res Ethics 14(2):1–24. https://doi.org/10.1177/1747016117733296

Davidson J, da Silva FC (2024) Decolonising the earth: anticolonial environmentalism and the soil of empire. Theor Cult Soc. https://doi.org/10.1177/02632764241242380

Deloria V Custer Died For Your Sins: An Indian Manifesto, New York: Macmillan, 1969. ISBN 0-8061-2129-7; later edition with new preface: Norman, University of Oklahoma Press, 1988. ISBN 978-08061-2129-1

Di Palma V (2014) Wasteland : A History. New Haven: Yale University Press

Dillon L (2021) Civilizing swamps in California: Formations of race, nature, and property in the nineteenth century U.S. West. Environment and Planning D: Society and Space, 40(2), 258–275. https://doi.org/10.1177/02637758211026317 (Original work published 2022)

Drame ER, Toro Martell S, Mueller J, Oxford R, Wisneski DB, Xu Y (2011) Engaged scholarship in the academy: reflections from the margins. Equity Excell Educ 44(4):551–565. https://doi.org/10.1080/10665684.2011.614874

Ely Museum (2006) How were the Fens drained? Retrieved from https://www.elymuseum.org.uk/wp-content/uploads/2020/06/Drainage-who-drained-the-fens-1.pdf

Erickson R, Raymond L (2018) Stream Channelization (P.L. 83-566) INCREASED Wetland Losses in Dakotas! Retrieved December 11, 2024, from https://www.semanticscholar.org/paper/STREAMCHANNELIZATION-(P.L.-83-566)

Estes N (2020) Water protectors gave us a vision of the future. www.societyandspace.org. https://www.societyandspace.org/articles/water-protectors-gave-us-a-vision-of-the-future

Estes N (2024) Our history is the future. Haymarket Books

Fabiano E, Schulz C, Martín Brañas M (2021) Wetland spirits and indigenous knowledge: implications for the conservation of wetlands in the Peruvian Amazon. Curr Res Environ Sustain 3:100107. https://doi.org/10.1016/j.crsust.2021.100107

Farrell J, Burow PB, McConnell K, Bayham J, Whyte K, Koss G (2021) Effects of land dispossession and forced migration on indigenous peoples in North America. Science 374(6567). https://doi.org/10.1126/science.abe4943

Fenelon JV, Trafzer CE (2014) From colonialism to denial of California genocide to misrepresentations: special issue on indigenous struggles in the Americas. Am Behav Sci 58(3):3–29. https://doi.org/10.1177/0002764213495045

Ferreira CSS, Kašanin-Grubin M, Solomun MK, Sushkova S, Minkina T, Zhao W, Kalantari Z (2023) Wetlands as nature-based solutions for water management in different environments. Curr Opin Environ Sci Health 33:100476. https://doi.org/10.1016/j.coesh.2023.100476

Fitzerald Guth A (2024, December 2) Controversial U.-S. Supreme Court decision threatens the country's inland wetlands with new development, study says. Retrieved from Mongabay Environmental News website: https://news.mongabay.com/2024/12/controversial-u-s-supreme-court-decision-threatens-the-countrys-inland-wetlands-with-new-development-study-says/

Fournier C, Stewart S, Adams J, Shirt C, Mahabir E (2023) Systemic disruptions: decolonizing indigenous research ethics using indigenous knowledges. Res Ethics Rev 19(3): 174701612311692–174701612311692. https://doi.org/10.1177/17470161231169205

Frayer WE, Monahan TJ, Bowden DC, Graybill FA (1983) Status and trends of wetlands and deepwater habitat in the conterminous United States, 1950s to 1970s. Department of Forest and Wood Sciences, Colorado State University, Fort Collins

Gibbons A (2012) Ancient American skeletons safe from reburial, but only for the moment. Retrieved from https://www.science.org/content/article/ancient-american-skeletons-safe-reburial-only-moment

Giblett R (2024) Introduction to Wetland Cultures, Past and Present. Literatures, Cultures, and the Environment, 1–33. https://doi.org/10.1007/978-3-031-57365-1_1

Gilio-Whitaker D (2019) As the grass grows, the indigenous fight for environmental justice, from colonization to standing rock. Beacon Press

Glaser PH (1987) The ecology of patterned boreal peatlands of northern Minnesota-A communi ty profi le: U.S. Fish and Wildlife Service, Report 85 (7 .14), 98 p

Gramling C (2018, August 20) The Everglades has endured decades of human meddling. Now it faces a more relentless foe: rising seas. Retrieved from Science News website: https://www.sciencenews.org/article/florida-everglades-freshwater-saltwater-sea-level-rise

Green A (2025, February 19) Trump order fast-tracks projects that would damage wetlands, environmental groups say—Inside Climate News. Inside Climate News. https://insideclimatenews.org/news/19022025/trump-fast-tracks-projects-that-would-damage-wetlands/

Guha N (2023, October 9) 6 Years after standing rock, native tribes still fight Dakota access pipeline. Truthout. https://truthout.org/articles/5-years-after-standing-rock-native-tribes-still-fight-dakota-access-pipeline/

Hargrove D (2024) What about Mission Bay's toxins, sea world? Retrieved November 1, 2024, from San Diego Reader website: https://www.sandiegoreader.com/news/2016/jan/05/ticker-what-about-mission-bays-toxins-seaworld/

Hayward A, Wodtke L, Craft A, Robin T, Smylie J, McConkey S, Nychuk A, Healy C, Star L, Cidro J (2021) Addressing the need for indigenous and decolonized quantitative research methods in Canada. SSM Popul Health 15:100899. https://doi.org/10.1016/j.ssmph.2021.100899

Hill D (2016) Forgotten Narrative: the First Nations Reality, [YouTube Video]. Retrieved from https://www.youtube.com/watch?v=HOAYbxJGgDQ

History.com (2009, December 15) Westward expansion. Retrieved from HISTORY website: https://www.history.com/topics/19th-century/westward-expansion

IITC (1973) International Indian Treaty Council | IITC. Retrieved from International Indian Treaty Council website: https://www.iitc.org/

Jessen TD, Ban NC, Claxton NX, Darimont CT (2021) Contributions of indigenous knowledge to ecological and evolutionary understanding. Front Ecol Environ 20(2):93–101. https://doi.org/10.1002/fee.2435

Justo (2023, June 22) The power of water in native American spirituality—native tribe info. Retrieved September 8, 2024, from Native Tribe Info website: https://nativetribe.info/the-power-of-water-in-native-american-spirituality/#:~:text=Water%20is%20a%20powerful%20element%20that%20has%20been

Justo (2024, April 13) Uncover the sacred traditions: native American burial rights—native tribe info. Retrieved September 8, 2024, from Native Tribe Info website: https://nativetribe.info/uncover-the-sacred-traditions-native-american-burial-rights/

Knox RL, Wohl EE, Morrison RR (2022) Levees Don't Protect, They Disconnect: A Critical Review of How Artificial Levees Impact Floodplain Functions. Science of the Total Environment, 837, Article ID: 155773. https://doi.org/10.1016/j.scitotenv.2022.155773

Kovach M (2021) Indigenous Methodologies Characteristics, Conversations, and Contexts, Second. University of Toronto Press. https://utppublishing.com/doi/book/10.3138/9781487525644

Leonard K, David-Chavez D, Smiles D, Jennings L, ʻanolani Alegado R, Tsinnajinnie L, … Begay R (2023) Water back: a review centering rematriation and indigenous water research sovereignty. Retrieved from https://www.water-alternatives.org/index.php/alldoc/articles/vol16/v16issue2/707-a16-2-10/file

Lightfoot KG, Panich LM, Schneider TD, Gonzalez SL (2013) European colonialism and the Anthropocene: A view from the Pacific Coast of North America. Anthropocene, 4, 101–115. https://doi.org/10.1016/j.ancene.2013.09.002

Lord LA (1993) Guide to Florida environmental issues and information: Winter Park, Fl a., Florida Conservation Foundation, 364 p

Long H, Zhang Y, Ma L, Tu S (2021) Land use transitions: progress, challenges and prospects. Land 10(9):903. https://doi.org/10.3390/land10090903

Masta S (2022) Theory-to-practice: researching indigenous education in the United States. Int J Multicult Educ 24(1):1–15. https://doi.org/10.18251/ijme.v24i1.1937

Monet J (2018, March 18) What standing rock gave the world. Common Dreams. https://www.commondreams.org/views/2018/03/18/what-standing-rock-gave-world

Mukherjee J, Chakraborty G (2016) Commons vs commodity: urban environmentalisms and the transforming tale of the East Kolkata Wetlands. Urbanities 6(2). Retrieved from https://www.anthrojournal-urbanities.com/wp-content/uploads/2016/11/8-Mukherjee-Chakraborty.pdf

Nuwer R (2013) The Gold Rush Left Behind Mercury That's Still Contaminating California. Smithsonian Magazine. https://www.smithsonianmag.com/smart-news/the-gold-rush-left-behind-mercury-thats-stillcontaminating-california-7581172/

Office of the Technology Assessmenty Assessment (OTA) (1984) Wetlands: their use and regulation. U.S. Congress, Washington, DC. Retrieved from https://govinfo.library.unt.edu/ota/Ota_4/DATA/1984/8433.PDF

Olmsted N (2007, November 26) Walking on water—a history of mission bay. Retrieved December 4, 2024, from Walking on Water—A History of Mission Bay | UC San Francisco website: http://www.ucsf.edu/news/2007/11/7374/walking-water-history-mission-bay2

Omodan BI (2025) Building reciprocal relationships through decolonial practices in academic research. Cogent Soc Sci 11(1). https://doi.org/10.1080/23311886.2024.2443558

Ostler J (2015, March 02) Genocide and American Indian History. In: Oxford research encyclopedia of American history. Retrieved November 27, 2024, from https://oxfordre.com/americanhistory/view/10.1093/acrefore/9780199329175.001.0001/acrefore-9780199329175-e-3

Parrish A (2024, February 5) Draining America. Worlds revealed. Retrieved from The Library of Congress website: https://blogs.loc.gov/maps/2024/02/draining-america/

Pidgeon M, Riley T (2021) Understanding the application and use of Indigenous research methodologies in the Social Sciences by Indigenous and non-Indigenous scholars. International Journal of Education Policy and Leadership 17 (8), 1–17. https://doi.org/10.22230/ijepl.2021v17n8a1065

Prucha FP (1984) The great father: The United States Government and the American Indians, vol I. The University of Nebraska Press, Lincoln, p 206

Reservation News (2016) News. Retrieved September 8, 2024, from Kumeyaay.com website: https://www.kumeyaay.com/news/215-park-employees-get-away with-breaking-laws-at-sacred-monument.html

ReWild (2024, March 8) History of ReWild Mission Bay. Retrieved October 31, 2024, from ReWild Mission Bay website: https://rewildmissionbay.org/history-of-rewild-mission-bay/

Rigney L-I (1999) Internationalization of an indigenous anticolonial cultural critique of research methodologies: a guide to indigenist research methodology and its principles. Wicazo Sa Rev 14(2): 109–121. https://doi.org/10.2307/1409555

Rose N (2024, July 5) Wetland cities: how challenging colonial distinctions of land and water promote urban resilience. Yale Environment Review. https://environment-review.yale.edu/wetland-cities-how-challenging-colonial-distinctions-land-and-water-promotes-urban-resilience

Said EW, Internet Archive (1979) Orientalism. In Internet Archive. New York: Vintage Books. Retrieved from https://archive.org/details/orientalism0000said

Schrader TA (1955) Waterfow l and the potholes of the north central states, in The yea rbook of agriculture: Washington, D.C., U.S. Department of Agriculture, 84th Congress, I st Session, House Docu ment no. 32, p. 59–04

Seminole Nation Museum (1973) The Seminole wars. Seminole Nation Museum. Retrieved from seminolenationmuseum.org website: https://seminolenationmuseum.org/history-seminole-nation-the-seminole-wars/

Solon P (2018, March 26) Vivir Bien: Old Cosmovisions and New Paradigms . Retrieved from Great Transition Initiative website: https://greattransition.org/publication/vivir-bien

Solórzano DG, Yosso TJ (2002) Critical race methodology: counter-storytelling as an analytical framework for education research. Qual Inq 8(1):23–44. https://doi.org/10.1177/107780040200800103

Staff (2024) Enclosure (Inclosure). Practical Law UK Glossary w-022-3305. Thomson Reuters Practical Law. Retrieved 20 March 2024

Straw R (2021, March 16) KQED story on opposition to plans for gas drilling in our Suisun Marsh. Retrieved December 12, 2024, from The Benicia Independent ~ Eyes on the Environment/Benicia news & views website: https://beniciaindependent.com/kqed-story-on-opposition-to-plans-for-gas-drilling-in-our-suisun-marsh/

Teixiera L (1991) Access to information on the Costanoan/Ohlone Indians of the San Francisco and Monterey Bay area: a descriptive guide to research. Master's thesis

Thambinathan V, Kinsella EA (2021) Decolonizing methodologies in qualitative research: creating spaces for transformative praxis. Int J Qual Methods 20(2): 1–9. https://doi.org/10.1177/16094069211014766

The Ramsar Bureau (1971) The Convention on Wetlands text, as originally adopted in 1971 Final Act Annex I Convention on Wetlands of International Importance Especially as Waterfowl Habitat Final Text adopted by the International Conference on the Wetlands and Waterfowl at Ramsar, Iran, 2 February 1971. Retrieved from https://www.ramsar.org/sites/default/files/documents/library/original_1971_convention_e.pdf

Tracey L (2024, September 27) Andrew Jackson's speech on the Indian Removal Act: annotated—JSTOR Daily. Retrieved from JSTOR Daily website: https://daily.jstor.org/andrew-jacksons-speech-on-the-indian-removal-act-annotated/

Trope JF (2010) Existing federal law and the protection of sacred sites: possibilities and limitations. Retrieved from www.culturalsurvival.org website: https://www.culturalsurvival.org/publications/cultural-survival-quarterly/existing-federal-law-and-protection-sacred-sites

Tuhiwai Smith L (2012) Decolonizing methodologies: research and indigenous peoples. Zed Books

U.S. Fish & Wildlife Service (2024, April 1) 2019 Wetlands status and trends report. Retrieved from www.fws.gov website: https://www.fws.gov/project/2019-wetlands-status-and-trends-report

UC Hastings Scholarship Repository (1960) Assessment of golf courses. Retrieved from https://repository.uclawsf.edu/cgi/viewcontent.cgi?article=1613&context=ca_ballot_props

UNDP (2022) Wadi Gaza master plan and photo album—natural park and greenway. Retrieved May 31, 2024, from UNDP website: https://www.undp.org/papp/publications/wadi-gaza-master-plan-and-photo-album-natural-park-and-greenway

UNDP (2023) Wadi Gaza. Retrieved from UNDP website: https://www.undp.org/sites/g/files/zskgke326/files/2023-06/wadi_gaza_pamphlet.pdf

UNDP Palestinian (2023, July 17) Wadi Gaza cleaning—final results. Retrieved May 31, 2024, from YouTube website: https://www.youtube.com/watch?v=HvMwi3xIqoc

United States Fish and Wildlife Service California Department of Fish and Game (2010) Kinder Morgan Suisun Marsh diesel fuel oil spill. Retrieved from https://nrm.dfg.ca.gov/FileHandler.ashx?DocumentID=22852&inline

United States. Army. Corps of Engineers (1959) The United States Department of Commerce. Office of Area Development. Future San Francisco Bay area development, 1960–2020. [Washington]. https://catalog.hathitrust.org/Record/009645476/Home

US EPA (2015, April 9) Classification and types of wetlands. Retrieved from www.epa.gov website: https://www.epa.gov/wetlands/classification-and-types-wetlands#undefined

USDA (2015) Wetlands benefits and costs vary with location. Economic Research Service. Usda.gov. https://www.ers.usda.gov/amber-waves/2015/may/wetlands-benefits-and-costs-vary-with-location

Valuing wetlands (2021). Nature Geoscience 14(3):111–111. https://doi.org/10.1038/s41561-021-00713-4

Vileisis A (1997) Discovering the unknown landscape: a history of America's wetlands. Island Press, Washington, DC

Water Protecters (2019) The oral stories from standing rock. The Water Protectors Community. https://waterprotectorscommunity.org/

Wetland Info (2022) First Nations' values. Retrieved from wetlandinfo.des.qld.gov.au website: https://wetlandinfo.des.qld.gov.au/wetlands/management/wetland-values/first-nations/

Wilson C (2001) Decolonizing methodologies: research and indigenous peoples. Soc Policy J N Zeal (17). https://www.msd.govt.nz/documents/about-msd-and-our-work/publications-resources/journals-and-magazines/social-policy-journal/spj17/17-pages214-217.pdf

Wooten HH, and l ones, LA (1955) The history of our drainage enterprises, in The yearbook of agriculture, 1955: Washington, D.C., U.S. Department of Agriculture, 84th Congress, I st Session, House Document no. 32, p. 478–498

Wright HE Jr (1984) Red Lake peatland- Its past and patterns: Minneapolis, Minn., University of Minnesota, James Ford Bell Museum of Natural History, v. 1, 7 p

Yates SR, Cowart LB (2019) The impact of shuttered golf courses on property values. J Sustain Real Estate 11(1): 2–15. https://doi.org/10.22300/1949-8276.11.1.2

Yeo S (2020, November 6) Ecocide: should killing nature be a crime? Retrieved from www.bbc.com website: https://www.bbc.com/future/article/20201105-what-is-ecocide

Zedler JB (1992) Read "Environmental science in the coastal zone: issues for further research" at NAP.edu. In: nap.nationalacademies.org. National Academies of Sciences. Retrieved from https://nap.nationalacademies.org/read/2249/chapter/9#114

Part II

Managing the Wetlands: Strategies for Conservation

3 Toward Sustainable Wetland Management: Addressing Policy and Implementation Challenges in Urban India

Apurv Bhogibhai Patel and Vibhore Bakshi

Abstract

The proliferation of issues of preserving wetlands has become a cumbersome process for Indian cities. However, the sensitivity of the issue of wetland conversation is humongous. This chapter examines the multifaceted process of wetland degradation in urban India, scrutinizing the intersection of various contributory factors. A comprehensive review of the global discourse on wetland conservation reveals diverse strategies employed across different contexts, highlighting both successes and gaps. This research article underpins the criticality of the emerging factors for wetland conversation including ineffective policy implementation, administrative inertia, political indifference, and lack of public awareness. Despite these various policies and administrative structures, the efficacy of wetland preservation in Indian cities remains questionable. Urban expansion, industrialization, and agricultural encroachment further exacerbate the depletion of wetlands, often due to a disconnect between policy and ground realities. The study underscores the need for a paradigm shift from reactive to proactive conservation strategies, emphasizing adaptive management practices that can accommodate the dynamic nature of urban wetlands. By addressing the systemic issues and leveraging the strengths of existing policies, it is posited that we can make significant strides in safeguarding urban wetlands for future generations.

Keywords

Urban wetlands · Wetland conservation · Wetland degradation · Sustainable preservation · Conservation strategies · Ecological restoration

3.1 Introduction

Wetlands, often referred to as the "kidneys of the Earth," perform critical ecosystem services that are indispensable for ecological balance and human well-being. They purify water by filtering pollutants, support an astonishing array of biodiversity, regulate local microclimates, and act as natural buffers against urban flooding (Hettiarachchi et al. 2014a). Their role in mitigating climate change impacts, such as absorbing carbon and stabilizing water cycles, underscores their ecological and socioeconomic value (Datri et al. 2024). Despite their irreplaceable contributions, wetlands in urban India face

A. B. Patel (✉)
Urban Planner-Jamnagar Area Development Authority (JADA), Jamnagar, Gujarat, India

V. Bakshi
Department of Urban & Regional Planning, School of Planning and Architecture Bhopal, Bhopal, Madhya Pradesh, India
e-mail: vibhore.bakshi@spabhopal.ac.in

S. K. Chakraborty et al. (eds.), *Wetlands and Urban Resilience*, Sustainable Development Goals Series,
https://doi.org/10.1007/978-3-032-06881-1_3

alarming rates of degradation, driven by rapid urbanization, industrial expansion, and neglect in policymaking (Hettiarachchi et al. 2014a; Matthews 1993).

The degradation of these ecosystems has profound implications for urban sustainability. As cities expand, wetlands are often the first victims of land reclamation for infrastructure development (Mishra et al. 2021). Encroachments, pollution from industrial effluents and untreated sewage, and the dumping of solid waste exacerbate the problem, leading to the destruction of these fragile ecosystems (Zhang et al. 2022). This not only undermines their ecological integrity but also heightens the vulnerability of urban populations to water scarcity, flooding, and the loss of biodiversity (Dooley and Stelk 2021; Lyons et al. 2016). Urban wetlands hold cultural and aesthetic significance, providing recreational spaces and contributing to the quality of urban life (Ahn and Schmidt 2019). Their disappearance marks an ecological crisis and a cultural and social loss. This underscores the urgency of addressing the factors contributing to their decline (Mitchell 2013; Lewis 1995). However, efforts to conserve wetlands in India often falter due to systemic challenges such as ineffective policy implementation, administrative inertia, and public awareness (Díaz-Pinzón et al. 2022; Hettiarachchi et al. 2014b).

A critical analysis of the current state of urban wetlands reveals a pressing need for integrated, multi-stakeholder approaches to conservation (Alikhani et al. 2021). Moving beyond piecemeal efforts, there is a need for a comprehensive framework that aligns policy, technology, community engagement, and urban planning (Jörg 2013; Matthews 1993). Such a structure must be based upon the nuanced understanding of the institutional and sociopolitical dimensions of wetland degradation, and how these occur through a lens of challenges in governance and implementation in urban areas. Addressing these challenges requires not only political will but also an acknowledgment of the dynamic and adaptive nature of wetland ecosystems (Amaranath et al. n.d.).

The chapter highlights a significant gap in the discourse on wetland conservation, by investing the nexus between policy design and the ground realities of the management of urban wetlands. Through policy analysis with policy implementation, the study illustrates the contradictions and inefficiencies associated with wetland conservation and the degree of complexity in stakeholder relations and values that intersects the process of safeguarding urban wetlands.

This research is driven by three objectives:

- To critically assess the significant policy gaps that have implications for the conservation of urban wetlands
- To investigate the sociopolitical and administrative constraints to the effective implementation of wetland-related laws and policies
- To develop an integrated, context-specific governance framework that appropriately weighs conservation priorities alongside urban development priorities

These objectives will be pursued through a qualitative and policy-oriented lens that synthesizes relevant interdisciplinary literature and selected illustrative case studies to produce recommendations for effective governance of urban wetlands in India.

3.2 Understanding Urban Wetlands

Urban wetlands, including lakes, marshes, ponds, and swamps, are dynamic systems that are part of urban environments. They are multifunctional and have major ecological, hydrological, socio-economic, and cultural benefits and play a key role in urban sustainability and urban resilience development (Hettiarachchi et al. 2015; Jia et al. 2011). Ecologically, the wetlands are biodiversity hotspots and the habitats for numerous species of animals, most of which are rare, endangered, or migratory species (Erwin 2009; Hughey et al. 2004; Li et al. 2010).

Their role in sustaining ecological linkages is crucial in densely populated urban zones (Alberto et al. 2019). Wetlands are hydrologically dynamic environments that act like wide-ranging natural sponges: they can absorb stormwater and prevent urban flooding, support aquifer recharge, and decrease urban pollution (Li et al. 2010). These hydrological functions are critically important for climate adaptation in rapidly urbanizing areas and provide climate mitigation benefits by acting as carbon sinks and microclimate regulators (Amaranath et al. n.d.), where they prevent or reduce the impacts of heat islands and weather extremes (Huang et al. 2011; Hughey et al. 2004). Urban wetlands are not just about ecology; they are also about providing community identity, beauty, and public green space, often associated with local history, sacred traditions, and informal economic practices (Dhir et al. 2009; Niemeijer and de Groot 2008). Similarly, wetlands provide ways to connect urban communities to nature and serve as an avenue for environmental learning and civic cultural engagement (Matthews 1993). Unfortunately, government policies and planning often overlook their multifunctional values, cementing wetlands as neglected remnants of urbanization that suffer both ecological service and public loss (Hettiarachchi et al. 2015; Jia et al. 2011; Lehner and Döll 2004). Accepting them as part of urban infrastructure could revive their conservation from nominal preservation to actual transformative integration of wetlands into urban planning and governance (Hettiarachchi et al. 2014a). The missing piece is an adequate understanding of the connections of dependencies between wetlands and urban systems that improve inclusive, resilient, and sustainable communities (Jörg 2013; Díaz-Pinzón et al. 2022; Lolu et al. 2020; Datri et al. 2024).

3.3 Current State of Wetlands in Urban India

The state of wetlands in urban India is on an alarming trajectory of environmental degradation because of rapid urbanization, industrialization, and neglect by institutions (Author et al. 2023). In the last 30 years, urban areas in India have lost over 30% of wetland area, and these losses reflect failures in governance, administration, and planning (Díaz-Pinzón et al. 2022; Lima et al. 2017). Case studies from cities in southern Vietnamese megacities (Bengaluru, Chennai, and Kolkata) show systematic degradation. In Bengaluru, wetland losses are severe, as the city once contained many lakes that were interlinked (Alikhani et al. 2021). At present, there is a visible loss of wetlands exemplified by the failure of the entire Bellandur Lake under the weight of industrial effluents and dumping of sewage and waste (Al Khaddar et al. 2020; Chakraborty et al. 2021a). In Chennai, the degradation of Pallikaranai Marshland—once a flood buffer—has intensified the city's flood vulnerability, evidenced by the 2015 deluge (Manzoor et al. 2018). Encroachments and solid waste dumping have critically altered the wetland's ecological function (Hails 1996; Manzoor et al. 2018). Although Kolkata's East Kolkata Wetlands have Ramsar designation and are critical urban wastewater treatment wetlands, they are still vulnerable because of ineffective enforcement and unchecked land conversion (Das Sarkar et al. 2020; Mondal et al. 2017). Wetlands are degrading, which has implications for not only their ecology but also their socioeconomics. And degrading wetland ecosystems will negatively affect flood regulation, groundwater recharge, and biodiversity while making urban heat islands worse and urban quality of life worse (Lima et al. 2017; Soanes et al. 2020; Jigar et al. 2016).

Addressing this crisis requires integrated action: stringent enforcement, community participation, and embedding wetlands into urban resilience and climate adaptation strategies (Ghosh and Das 2019; Sengupta and Dalwani 2008). Without sustained and coordinated efforts, India risks irreversibly compromising its natural infrastructure (Alikhani et al. 2021; Author et al. 2023; Díaz-Pinzón et al. 2022; Soanes et al. 2020).

3.4 Factors Contributing to Wetland Degradation

The degradation of wetlands in urban India is a multidimensional factor due to a variety of anthropogenic activities (Zedler and Fuller 2017). These factors are mainly caused by rapid economic growth and increased population, upsetting the fragile equilibrium of wetland ecosystems. This paper critically analyses three key contributors to wetland degradation: urbanization and infrastructure development, industrialization and pollution, and encroachment for agriculture and settlements.

3.4.1 Urbanization and Infrastructure Development

The unexpected growth and encroachment of cities and towns have caused extensive development on wetlands, which has irreparably affected the wetlands' important ecological and hydrological functions (Banerjee and Dey 2017; Rajaram and Das 2008). Wetland morphology has altered which has also reduced the groundwater recharge, stormwater regulation, and biodiversity of wildlife species (Pavithrapriya et al. 2015). In Mumbai, the destruction of wetlands has resulted from not only construction waste but also the introduction of land-filling practices to create artificial land (Rode 2019). The combination of these factors, coupled with the increasing floods and ecological destruction in already degraded areas, which is evident in the Mithi River basin, is concerning (Ghosh et al. 2018). Thus, it seems that there is a paramount imperative on the part of urban policy to integrate wetland protection into urban planning.

3.4.2 Industrialization and Pollution

Industrial effluents, especially in the form of raw effluents, pose a significant threat to wetland ecology (Phukan and Saikia 2014). Heavy metals, toxic chemicals, and solid wastes pollute water bodies and disturb biodiversity (Ghosh et al. 2018; Bandyopadhyay et al. 2020). In addition, such pollution is also a threat to human health through the process of bioaccumulation in waterborne organisms (Bandyopadhyay et al. 2020; Patil 2020). Delhi Yamuna floodplains, as well as wetland areas of Kolkata, depict how weak regulatory control and enforcement have turned wetlands into dumping sites for industrial wastage (Pavithrapriya et al. 2015; Rode 2019). The absence of stringent regulatory policies expands the process of ecological degradation, requiring policy reforms aimed at pollution control in areas around wetlands (Wright et al. 2006).

3.4.3 Encroachment for Agriculture and Settlements

The fringes of wetlands are often occupied for agricultural purposes or unauthorized settlements by economically disadvantaged populations, which results in habitat destruction and ecological disturbance (EnviStats India 2020; Gagan et al. 2016). Fertilizer runoffs and the application of pesticides result in eutrophication and thus worsen water quality and compromise aquatic life (Prakash Panwar et al. 2017). Moreover, unauthorized settlements do not have proper sanitation facilities, and thus, the waste is discharged directly into wetlands (Bandyopadhyay et al. 2020; Phukan and Saikia 2014). East Kolkata Wetlands, being a Ramsar site, also suffer from intensive pressures of illegal land use, thus demoralizing their role as natural wastewater treatment systems (Ghosh et al. 2018; Pavithrapriya et al. 2015).

3.5 Policy Landscape for Wetland Management

India has a strong policy framework that is mainly governed by central legislation that supports state- and globally mandated regulations, providing a backbone for implementing conservation policies concerning wetland protection (Barbier and Knowler 1997). Nonetheless, effective protection of such wetland degradation largely remains ineffective at implementation and enforcement, as well as in public engagement. The overview of policymaking would expose current gains that could have been taken, but it persists in the problems of maintaining such a vital ecosystem (Fig. 3.1).

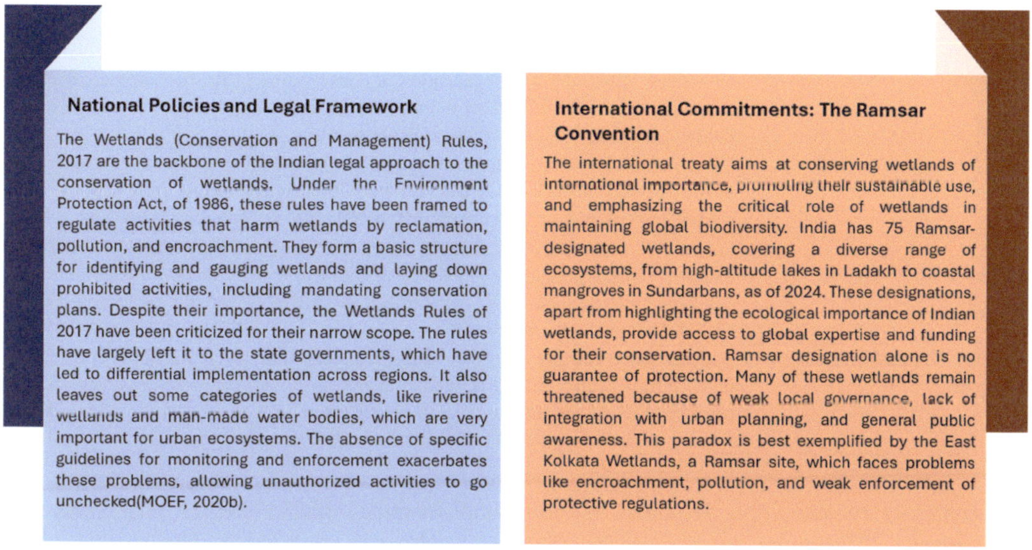

Fig. 3.1 National and international policy landscape for wetland management

3.5.1 Disconnect Between Policy and Implementation

A critical gap in India's policy landscape lies in the disconnect between well-intentioned regulations and ground-level implementation. This dissonance is rooted in several systemic issues (Table 3.1):

3.5.2 Comparison of Urban Wetland Policies in Indian Cities and Governments

This analysis highlights a fragmented implementation landscape across Indian cities. The national Wetlands (Conservation and Management) Rules, 2017, provide a framework for the conservation of wetlands in India, but because of poor coordination with state and municipal agencies, enforcement happens at the city level with varying degrees of effectiveness. Bengaluru and Chennai rely on enforcement challenges and over-reliance on litigation that often impedes timely intervention. Conversely, the East Kolkata Wetlands (EKW) model demonstrates a more structured approach, particularly within the framework of GIS, but remains a localized project without a scaled approach to the needs of the city (Li et al. (2016); Mondal et al. 2017). The extent to which technology has been adopted also varies. While both central and state policies endorse GIS and remote sensing. The traditional community engagement mechanisms that are regarded as valuable for promoting sustainability remain weak in practice (Das Sarkar et al. 2020). While there have been some limited efforts in cities such as Bengaluru and Chennai, they often lacked a formalized forum or model for engaging community members to sustain their participation. Financial limitations and lack of coordinated funding from central government sources have led to a fragmented approach to implementing wetland projects. Furthermore, institutional inertia and poor coordination across agencies impede wetland conservation efforts, which have further eroded accountability and commitment to environmental health (Das et al. 2021; Hettiarachchi et al. 2014a; Shivakrishna et al. 2021) (Table 3.2).

3.5.3 Recommendations for Improvement

Central, state, and city policies should be aligned to ensure a cohesive strategy for wetland conservation. A centralized task force for the management of urban wetlands can help to bridge gaps

Table 3.1 Disconnect between policy and implementation

Administrative inertia	Bureaucratic delays and overlapping jurisdictions among local, state, and central authorities hinder the timely execution of conservation plans. For instance, many wetlands identified for protection under the Wetlands Rules are yet to be officially notified, leaving them vulnerable to exploitation
Inadequate funding and resources	Conservation efforts are often constrained by limited financial and technical resources. State agencies tasked with wetland management frequently lack the capacity to conduct scientific assessments or enforce regulations effectively
Lack of public awareness and engagement	Wetland conservation policies rarely incorporate community participation, leading to a lack of ownership and accountability at the local level. Public awareness campaigns, when undertaken, are sporadic and fail to convey the ecological and socioeconomic value of wetlands
Conflict with development priorities	Policies aimed at wetland conservation often clash with development projects such as urban infrastructure expansion, industrial zones, and agricultural intensification. The absence of integrated land-use planning exacerbates this conflict, with wetlands being viewed as obstacles rather than assets in the development process

and enhance accountability. Standardizing the adoption of GIS mapping, remote sensing, and real-time monitoring systems across cities and states will provide accurate and actionable data. A centralized digital platform could facilitate the sharing of data and collaboration among stakeholders. Develop structured frameworks for community involvement. Public awareness campaigns, citizen science initiatives, and participatory planning can empower local stakeholders to take ownership of wetland conservation. Dedicated financial mechanisms, such as a national urban wetland restoration fund, can address resource constraints. Incentivizing public-private partnerships and CSR investments can also build in more financial support. Training programs for local authorities and improved inter-agency coordination can hasten decision-making and implementation. Integrating wetland management into curricula for professionals in urban planning can foster a long-term vision for conservation.

3.6 Implementation Challenges

Despite a thorough legal framework, India's wetlands conservation efforts are hindered by inadequate enforcement, political indifference, and poor management, which all contribute to the continuous degradation of urban wetlands (Bassi and Kumar 2012). A significant impediment to effective implementation is conceptualizing policies, rather than implementing them. Legal protections for wetlands exist, but the lack of technical capacity, monitoring capacity, and accountability renders those protections ineffective in practice. Encroachment, dumping of wastes, illegal development, etc., continue unabated and unregulated in places such as Mumbai and Bengaluru, despite the presence of legal protections (Singh and Biswas 2021, 2022). A delay in imposing penalties for offenses removes respect for the law and erodes public trust, showcasing normalized and accepted violations as it relates to the conservation of wetlands (Chakraborty et al. 2021b; Mathevet et al. 2020). Fragmentation is also another vital barrier to the implementation of wetland conservation projects. A historic overlap in mandates among central, state, and local agencies results in jurisdictional ambiguity and delays in enforcement and other administrative action (MOEF 2020; Singh and Biswas 2022).

The absence of standardized operating procedures delays critical actions like wetland notification, boundary delineation, and conservation planning. Political will is often lacking (Al Khaddar et al. 2020; Molur et al. 2020). Competing needs for development, compounded by limited electoral dividends, lead to wetlands being low priorities for funding and urban planning (Manzoor et al. 2018). Local decision-makers may be pressured by powerful economic interests, such as real estate and industry associations, to support development over wetland conservation (Chakraborty et al. 2021a;

Table 3.2 Comparison of urban wetland policies in Indian cities and government

Aspect	Bengaluru	Chennai	Kolkata	Central government policies	State government policies
Key legislation	Bengaluru Development Authority (BDA) laws and Master Plan provisions for wetland protection	Tamil Nadu Protection of Tanks and Eviction of Encroachment Act, 2007	Kolkata Metropolitan Development Authority (KMDA) regulations for wetlands	Wetlands (Conservation and Management) Rules, 2017; compliance with Ramsar Convention	Varies by state; often includes state-specific environment or irrigation acts
Focus	Flood mitigation, biodiversity conservation	Restoration of water bodies for flood control and drinking water	Emphasis on the East Kolkata Wetlands (EKW) as a Ramsar site for wastewater management	Broad conservation, sustainable use of wetlands, and Ramsar site preservation	State-specific; often prioritizes irrigation and agricultural concerns
Implementation challenges	Limited enforcement of urban planning guidelines	Weak coordination between agencies and rampant encroachments	Rising urbanization pressure despite EKW protection	Inadequate monitoring mechanisms and limited funding	State bureaucracies often face coordination and funding issues
Community involvement	Minimal public participation	Local-level advocacy exists but lacks scale	Active stakeholder involvement in EKW management	Encourages community roles in Ramsar sites but lacks structured engagement mechanisms	Rarely engages local communities actively in wetland management
Monitoring mechanisms	Sporadic and heavily reliant on public-interest litigation	Limited GIS usage; relies on physical surveys	Uses basic GIS and participatory monitoring methods	Promotes GIS and remote sensing but suffers from fragmented applications	Implementation varies widely; it is not standardized
Financial support	Reliance on municipal budgets and ad hoc grants	State funding with occasional central support	State and KMDA funding with EKW-specific allocations	Allocates funds under central environmental programs	Dependent on central grants and state budgets, leading to financial constraints
Technological adoption	Limited GIS adoption	Minimal technological interventions	GIS and traditional methods blend	Advocates for GIS mapping and smart monitoring systems	Inconsistent across states; lacks nationwide uniformity

Hazbavi et al. 2020). Politicians unwilling to confront powerful interests are in effect undermining conservation interests (Mathevet et al. 2020). In the absence of strong political resolve and institutional leadership, wetlands are left as a marginal contribution to urban development, resulting in ongoing degradation and policy neglect (Kumar and Singh Maan 2019; Singh and Biswas 2022).

3.7 Public Awareness and Community Participation

Public awareness and community involvement play a pivotal role in sustaining wetlands (Shivakrishna et al. 2021). Wetlands are tremendously important from the ecological, social, and cultural point of view. But over the years, they have been viewed as wastelands and neglected or lost (Singh and Biswas 2021, 2022). Educating and creating awareness around wetlands will close this gap by creating a sense of collective responsibility to sustainably utilize and conserve wetlands (Peetabas and Panda 2014). Many urban residents do not know that wetlands offer ecosystem services to people, such as water purification, flood abatement and biodiversity support. Not knowing something about wetlands leads to littering, encroachments and excessive use (Shivakrishna et al. 2021; Kumar and Singh 2020). Public education through schools and colleges and local workshops, and experiential learning can help foster awareness and eventually change behaviors (Kumar Sharma et al. 2020; Rani et al. 2011). Wetland activities can be done with citizens, like cleaning a wetland or conducting biodiversity monitoring, that take people from observing a wetland to active participation in stewardship (Kumar and Singh 2020). Community based conservation is often the most successful method of achieving conservation objectives when residents who know the landscape are provided with tools and training and are also included in the decision-making process (Manju et al. 2005). An example is the Loktak Lake project in Manipur, where local fisherfolk helped to restore and monitor the changes in the lake in this project (Mohapatra et al. 2007; Mondal et al. 2017; Das Sarkar et al. 2020).

Optimal community engagement is frequently based on multi-stakeholder collaborations. NGOs, academic institutions, and private-sector partners can contribute to the community by providing technical support, training, and infrastructure (Alikhani et al. 2021; Li et al. 2010). PPP would improve the facilities they manage, such as a waste treatment plant, by involving the local community (Alikhani et al. 2021; Liu and Diamon 2005; Lal 2008).

3.7.1 Challenges to Community Engagement

Despite the obvious advantages of community involvement, several obstacles still exist. These include (Fig. 3.2):

To overcome these obstacles, inclusive strategies that prioritize fair involvement and guarantee that all parties gain from conservation efforts are needed.

Fig. 3.2 Challenges to community engagement

3.8 Case Studies of Successful Wetland Conservation

Analyzing the best-case studies would help highlight the exemplary models of good wetland conservation both globally and locally. It has been experienced that, with such comprehensive planning, community engagement, and adaptive management, urban wetlands can also be protected and restored. These cases serve not merely as success stories but as empirical evidence supporting the feasibility of integrated wetland governance.

3.8.1 Colombo, Sri Lanka: Integrating Wetlands into Urban Planning

Wetland conservation is a key element of urban development, and the city of Colombo provides a convincing example. The Colombo Wetland Management Plan has always been an exemplary model for maintaining ecological balance with high-speed urban development (Mohri et al. 2013; Wang 2018). Central to Colombo's approach is the recognition of wetlands as integral infrastructure—providing ecosystem services such as flood mitigation, water purification, and biodiversity support—rather than as peripheral environmental spaces (Wattage and Mardle 2005; Rathnayake 2024). Buffer zones developed around wetlands form an important element of Colombo's strategy, which mitigates the impact of urban sprawl and industrial encroachment (McInnes and Everard 2017). Through public awareness campaigns and educational programs, Colombo has created information among the urban populace concerning wetland ecosystems (Rathnayake 2024). Strengthened community participation and the development of stewardship at the local level have made the entire efforts of conservation more robust and sustainable (Cobracollactive 2024; Slykan Trust 2023). According to the UNDP (2022), the integration of over 2000 hectares of wetlands into the city's green infrastructure network has contributed to a 40% reduction in flood-related losses in key urban zones. The successful integration of wetlands into urban planning in Colombo is also supplemented by a policy framework that remains in line with the Ramsar Convention on Wetlands. The synergy between ecological planning and development agendas has enabled the city to preserve ecological health while advancing green infrastructure goals (Hettiarachchi et al. 2014b; UNDP 2022).

3.8.2 Singapore: A Model of Urban Wetland Preservation

Singapore stands out as the leader in urban sustainability with its approach to wetland conservation serving as a model for cities around the world (Chan and Liang Jim 2023; Tan et al. 2013). The commitment to the preservation and restoration of wetlands is well illustrated through initiatives such as the Kranji Marshes, once a neglected area, which now thrives as a habitat supporting over 170 species of birds and an impressive diversity of flora (Nguyen Thi Hong et al. 2023; Teo 2023; Woon 2023). Singapore's approach is characterized by technological innovation and the integration of ecological functions into city infrastructure. Integrated water management systems combining stormwater drainage, flood control, and wetland preservation (Chan n.d.; Er et al. 2023). The Active, Beautiful, Clean Waters (ABC Waters) Programme, launched in 2006, now incorporates more than 100 wetland-related projects across the city-state, with data showing a 30% improvement in water quality across participating catchments (Lena Chan and Liang Jim (2023)). This comprehensive approach ensures not only the protection but also the utilization of wetlands to boost urban resilience in fighting climate change impacts like higher rainfall amounts and increasing sea levels (Liang Jim and Chan n.d.). The policies of Singapore focus on the collaboration between government agencies, private-sector partners, and community groups (Chuan et al. 2023; Tan et al. 2013). This enables joint research, environmental education, and sustainable tourism programs that contribute to high levels of public engagement and funding (Leng 2021; Dooley and Stelk 2021; Chan and Liang Jim 2023). A strong legislative framework, such as the Biodiversity Conservation Act, institutionalizes adaptive management and ecological monitoring in the city's wetland strategy (Ismail and Yixiong 2023; Tan et al. 2013; Yee 2023).

3.8.3 Chilika Lake, India: Lessons in Adaptive Management

In India, the case of Chilika Lake has been considered a success story about the feasibility of adaptive management practices for reviving and maintaining vital wetland ecosystems (Mohapatra et al. 2007). Chilika Lake, the largest coastal lagoon in India and a Ramsar site, has long faced pressures from overfishing, habitat degradation, and pollution (Ghosh et al. 2006). The restoration of Chilika was enabled by a multi-level, collaborative governance strategy that included the active participation of local fisherfolk, NGOs, state agencies, and international partners (Dujovny 2009). Active participation was required from local fisherfolk, which empowered communities to practice sustainable fishing and manage habitats. Such a people-centered approach ensured culturally appropriate conservation and contributed to the ownership and time span (Peetabas and Panda 2014). An adaptive management framework was followed, that allowed for constant monitoring, and changes could be accommodated as needed according to the ecological status of the lake. Notably, fish catch increased from 1600 metric tonnes in the late 1990s to over 14,000 metric tonnes in 2018, with a concurrent 60% rise in migratory bird species recorded by the Chilika Development Authority. Scientific research, which includes water quality analysis and biodiversity assessments, helped to direct policy and operational changes and ensure that conservation strategies kept pace with environmental changes (Ghosh et al. 2006; Jena 2007). Contributions involved the State Government of Odisha, central government agencies, environmental NGOs, and international organizations such as WWF. This collective effort allowed for resource-sharing and funding and the transfer of expertise, thus making a stronger model for managing wetlands (Basu et al. 2021). The revitalization of Chilika Lake has been accompanied by great benefits—an increase in fish populations and water quality, and increased species of birds, particularly the ones that migrate (Pattnaik 2022). Economically, the restoration enhanced ecotourism and diversified local incomes, demonstrating the viability of conservation as a development strategy (Pattnaik 2022; Basu et al. 2021; Ghosh et al. 2006; Peetabas and Panda 2014).

These global and local case studies provide empirical validation of how well-planned, community-oriented, and adaptive governance mechanisms can reverse wetland degradation trends. Colombo and Singapore exemplify strategic urban integration and policy innovation, while Chilika Lake demonstrates the effectiveness of grassroots engagement and ecological adaptability. Together, they underscore the need for a multi-scalar, collaborative, and dynamic approach to wetland governance. They also reinforce the core argument of this chapter: that wetland conservation must be embedded within broader urban planning processes and supported by participatory governance and continuous policy evolution.

Comparison table of the case studies on successful wetland conservation in *Colombo*, *Singapore*, and *Chilika Lake* (Table 3.3):

3.9 Need for a Paradigm Shift

The current state of wetland conservation in urban India underlines the need for a paradigm shift in the management of these vital ecosystems. Historically, conservation efforts have been largely reactive, responding to environmental degradation only after significant damage has occurred. This approach has proven inadequate in the face of the rapid and often unpredictable changes posed by urbanization, climate change, and socio-economic pressures. Therefore, a transition has to be made to a proactive model of conservation that

Table 3.3 Comparison of the case studies

Aspect	Colombo, Sri Lanka	Singapore	Chilika Lake, India
Location	Colombo, capital of Sri Lanka	Singapore, city-state in Southeast Asia	Odisha, East Coast of India
Type of wetland	Urban wetlands, including marshes and lakes	Coastal wetlands, including marshes and lagoons	Coastal lagoon (the largest in India)
Conservation objective	Integrate wetlands into urban planning	Preserve and restore wetland biodiversity and ecosystem services	Revitalize and restore a threatened ecosystem
Key strategies	Buffer zones around wetlands	Integrated water management systems	Community involvement and adaptive management
	Public awareness campaigns	Use of technology and innovative planning	Multi-stakeholder partnerships
	Educational programs	Strong policy framework	Continuous ecological monitoring
Community engagement	High, with public participation in education and stewardship	High, with collaboration among agencies, NGOs, and communities	Very high, with active local fishing communities
Government involvement	Strong, with policies supporting conservation	Very strong, with legal and policy support	Strong, involving both state and central governments
Challenges addressed	Urban encroachment, pollution	Urban expansion, climate change	Pollution, overfishing, and habitat degradation
Outcomes achieved	Improved urban green infrastructure, flood mitigation	Thriving biodiversity, enhanced climate resilience	Increased fish populations, improved water quality, and bird biodiversity

would predict and proactively mitigate the challenges of change before ecosystems reach tipping points.

3.9.1 The Limitations of Reactive Conservation

Reactive strategies—most often employed after decline—do not consider systemic drivers, such as the rapid expansion of urban areas and pollution (Hettiarachchi et al. 2015; Li et al. 2010). Restoration involves considerable expense and time, and rarely results in full recovery of the original ecological function (Alberto et al. 2019; Huang et al. 2011; Alikhani et al. 2021).

3.9.2 Proactive Conservation: Embracing Adaptive Management

Achieving sustainable wetland conservation requires a paradigm shift toward proactive management. Proactive management involves anticipating threats, using measures to prevent degradation, and instilling long-term resilience in wetland ecosystems. Proactive conservation fundamentally revolves around adaptive management: the dynamic, flexible approach to assessing ecosystem state and adjusting strategies under the state of the system. This is very relevant to urban wetlands, which are usually characterized by changing conditions, so it requires a quick response (Dooley and Stelk 2021). Adaptive management merges monitoring, research, and stakeholder input into a feedback loop informing decision-making (Lyons et al. 2016). For example, continued data collection on water quality, biodiversity, and land use may indicate trends or potential threats before they occur. Through early detection of such issues, urban planners and conservationists can take preventive action, for instance, by introducing buffer zones or strengthening wetland regulations to reduce harm. Rather than reacting to a crisis, this approach proactively prevents one and saves conservation efforts in the long run.

3.9.3 Benefits of a Proactive Approach

Proactive management improves ecosystem resilience, lowers long-term restoration expenditures, and maintains principal wetland functions for flood prevention, water filtration, and carbon capture (Mishra et al. 2021; Zhang et al. 2022). Proactive management further encourages inclusive conservation, allowing for community observation and guard duties to create local stewardship, improve compliance, and add indigenous knowledge for adaptive management (Pablo Torres Lima et al. 2017; Soanes et al. 2020).

3.9.4 Challenges to Implementing Proactive Management

The implementation of proactive conservation is impeded by weak policy frameworks, fragmented governance, and insufficient financial resources (Author et al. 2023; Jigar et al. 2016; Soanes et al. 2020). Institutional coordination and resource continuity are required for conservation projects, including cooperative partnerships across government, civil society, and private sectors, including access to international grants and interventions from CSR financial support (Pablo Torres Lima et al. 2017; Díaz-Pinzón et al. 2022).

3.10 Technological Interventions in Wetland Management

Technology has become an essential and growing tool in the promotion of wetland conservation in urban India, given its capabilities for real-time data collection, predictive modeling, and evidence-based decision-making (Shivakrishna et al. 2021; Das et al. 2021). With the integration of modern technologies such as GIS and smart monitoring systems, authorities can gain data that is precise, supports informed decision-making, and has rapid response to emerging threats (Garg 2015; Kumar and Singh 2020). GIS have transformed spatial analysis, offering accurate mapping of wetland boundaries, tracking land-use transitions, and delineating areas of high vulnerability to encroachment and pollution (Rani et al. 2011; Hettiarachchi et al. 2014a). The spatial layering in GIS is advantageous for urban planners and conservationists because it can reveal emerging patterns that were not obvious in traditional surveys and be used to simulate possible future scenarios using scenarios of projected urban growth and climate change datasets (Kumar and Singh 2020; Ma and Yan 2024). Advanced monitoring systems utilize IoT sensors, drones, and remote sensing to enable ongoing monitoring of ecological variables like water levels, pollutant loads, and habitat changes (Kumar Sharma et al. 2020; Shivakrishna et al. 2021). Such systems allow for near-immediate recognition of threats and responses to and prevent irreversible damage (Kumar and Singh 2020; Ma and Yan 2024).

Nevertheless, the implementation of technology is limited by real-world impediments, particularly in municipalities with limited funds. High upfront costs for software and hardware, the lack of technical capacity, and the challenges associated with sharing confidential data continue to limit full implementation (Fu et al. 2014; Groot et al. 2006). Innovations in technology, especially AI and Machine Learning, will likely support increased predictive capacity—allowing for dynamic modeling of degradation trends, monitoring of species, and resource optimization (Barbier and Knowler 1997; Lathuillière et al. 2016). Community engagement, through mobile applications and digital feedback, complements high-tech solutions, adding social intelligence to ecological monitoring (Kumar Sharma et al. 2020; Shivakrishna et al. 2021).

To reconcile the competing demands of ecological considerations and urban growth, technological interventions have considerable potential for change (Kumar and Singh 2020; Ma and Yan 2024). While there are obstacles relating to finances and ability, the long-term benefits of technology—from improved responsiveness to a wider range of planning rationales—are an important factor for effective wetland governance in India (Das et al. 2021; Shivakrishna et al. 2021). By integrating smart tools with strategic policy and community stewardship, India can be an early example of imaginative, tech-enabled, forward-looking wetland conservation (Kumar Sharma et al. 2020; Shivakrishna et al. 2021).

3.11 Economic Perspectives

Urban wetlands provide an important economic value from ecosystem services including flood control, water purification, carbon sequestration and supporting biodiversity; all of which could be restored artificially using significant amounts of money (Bystrom et al. 2000; Venkatachalam 2023). Wetlands are considered 'natural infrastructure' and this idea reframes the idea of preservation of ecological integrity, to the value of urban planning and urban environments, which represent measurable savings to city governments (Barbier and Knowler 1997; Lathuillière et al. 2016). Urban wetlands' capacity to reduce flooding, develop stormwater infrastructure, and improve water quality offers a measurable cost saving to city governments (Le et al. 2024). Given long-term water quality benefits in urban planning, the water cleaning benefits of wetlands reduce the need for a more expensive treatment plant, while north of a whole range of economic benefits from wetland benefits, urban wetlands can offer longer-term cost-savings (Fu et al. 2014; Groot et al. 2006; Lathuillière et al. 2016). Including the cost of ecosystem services into valuations in the urban plans allows wetland benefits to be visible in monetary terms to policymakers and investors.

3.11.1 Incentivizing Conservation

Economic inducements, including payments for ecosystem services (PES), tax incentives, and targeted subsidies, can encourage stakeholder participation in wetland conservation, especially for local communities, industries, and private landowners (Barbier and Knowler 1997; Fu et al. 2014). In PES systems, service beneficiaries of wetland ecosystem services—such as municipalities or industries—water purification or flood protection (Blundo et al. 2018; Le et al. 2024). Tax incentives can incentivize industries to switch to sustainable practices, such as water recycling measures and pollution reductions. CSR frameworks can also fund restoration and conservation projects (Datri et al. 2024). The application of cost-benefit analysis to urban planning underscores the economic case for long-term wetland protection compared to short-term gains from land conversion (Harrington and McInnes 2009); Hettiarachchi et al. 2015). As means of quantifying avoided disaster costs—as with flood-related damages—and ecosystem services like carbon sequestration and temperature regulation, CBA provides an additional economic rationale for conservation (Jia et al. 2011; Li et al. 2010) (Fig. 3.3).

3.11.2 Employment and Livelihood Opportunities

Wetlands contribute to local economies through ecotourism, sustainable fisheries, and agriculture, which are compatible with the wetlands, creating jobs and, at the same time, enhancing community stewardship (Erwin 2009; Lal 2008; Alikhani et al. 2021). Although wetlands have considerable economic potential, barriers exist that include limited awareness by decision-makers, the challenge of valuing intangible services, and little guidance by institutions regarding the application of economic tools in policy (Huang et al. 2011). Potential solutions include capacity building in urban planners, and a framework for standard values could support incorporating economic data into policy processes (Alberto et al. 2019; Hettiarachchi et al. 2015). Mobilizing resources from the state, private sectors, and civil society could provide funding for long-term conservation (Datri et al. 2024).

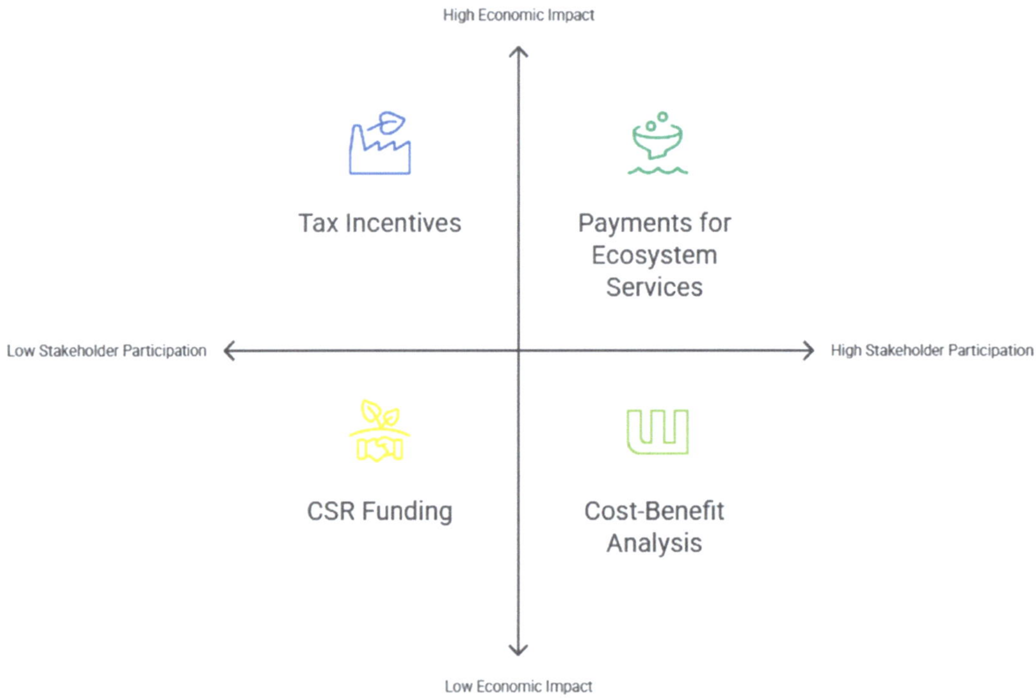

Fig. 3.3 Economic incentives for wetland conservation

3.12 Role of Urban Planning

Urban planning is pivotal in shaping the course of wetland conservation in rapidly expanding cities. As centers of population, commerce, and infrastructure grow, the deliberate incorporation of wetlands into urban development blueprints becomes critical. Cities can preserve these vital ecosystems while meeting the demands of urbanization by adopting innovative and sustainable planning approaches (Fig. 3.4).

3.12.1 Integrating Wetlands into Urban Development Frameworks

The first step in utilizing urban planning for wetland conservation is the recognition of wetlands as critical natural assets. Planners must treat wetlands as integral components of urban infrastructure rather than surplus or encroachable land (Amaranath et al. n.d.; Datri et al. 2024). This requires detailed mapping of existing wetlands and their ecological functions to identify areas that must be prioritized for protection (Jörg 2013). This way, through the inclusion of wetlands in master plans, they are protected from competing land-use pressures like residential or commercial development. The planners can provide ecological reserves or buffer zones in city layouts that will be afforded legal and administrative protection (Alikhani et al. 2021; Díaz-Pinzón et al. 2022). This also involves planning setbacks and safe zones around the wetlands to prevent pollution and encroachment by urban activities.

3.12.2 Promoting Green Infrastructure Solutions

Green infrastructure is a new alternative to the conventional gray infrastructures, which include concrete drains and flood barriers (Díaz-Pinzón et al. 2022). Constructed wetlands, for example, are created to mimic natural wetland processes

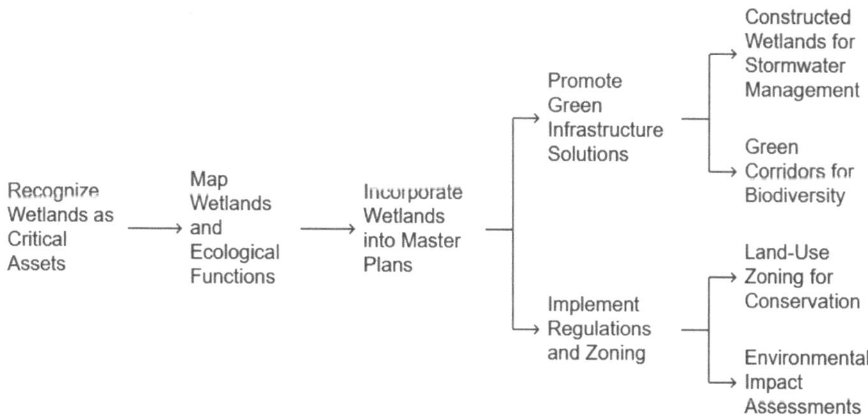

Fig. 3.4 Planning incentive for wetland conservation

but provide the same level of service, including stormwater management, water filtration, and habitat creation. Constructed wetlands slow down runoff, trap sediments, and filter pollutants to handle urban runoff (Hettiarachchi et al. 2014b; Mitchell 2013). This reduces the burden on the traditional stormwater drainage systems, besides mitigating urban flooding (Díaz-Pinzón et al. 2022). For instance, Singapore and Melbourne have implemented constructed wetlands in cities to great success and thus demonstrate their potential for achieving a balance between ecological preservation and utility in an urban context. Beyond constructed wetlands, green corridors linking urban wetlands to other natural ecosystems improve biodiversity and support species movement in the general health of an ecosystem of cities (Liang Jim and Chan n.d.; Tan et al. 2013). Interconnected systems provide aesthetic and recreational values that make them attractive to community engagement.

3.12.3 Regulations and Zoning

Urban planning frameworks should have strict regulations and zoning laws that safeguard wetland ecosystems (Chuan et al. 2023; Dooley and Stelk 2021). Land-use zoning will ensure that wetland areas are only used for conservation purposes or other compatible activities such as ecotourism or sustainable agriculture. The proper demarcation of wetland boundaries, implemented through legal instruments, can prevent encroachment and unapproved development (Leng 2021; Chan and Liang Jim 2023). SEAs and EIAs should be required for any proposed urban project adjacent to wetlands. These tools allow planners to assess the possible ecological impacts of development and apply mitigation measures before the start of construction (Chan n.d.; Er et al. 2023).

3.12.4 Challenges in Urban Planning for Wetland Conservation

Despite its potential, planning for wetland conservation poses many challenges. Rapid urbanization often places immediate economic considerations above long-term ecological benefits, leading to wetland degradation (Author et al. 2023). Poor coordination between different government departments and agencies often leads to fragmented decision-making, undermining comprehensive efforts at conservation (Alikhani et al. 2021). The lack of technical expertise among urban planners is another significant barrier. Some training programs that will equip the planners with skills in ecological restoration, wetland hydrology, and green infrastructure design are necessary to bridge the gap (Juwarkar et al. (1995); Jigar et al. 2016). The general public is not aware of the benefits of wetlands and is often less responsive to community mobilization toward conservation-friendly urban planning measures (Díaz-Pinzón et al. 2022).

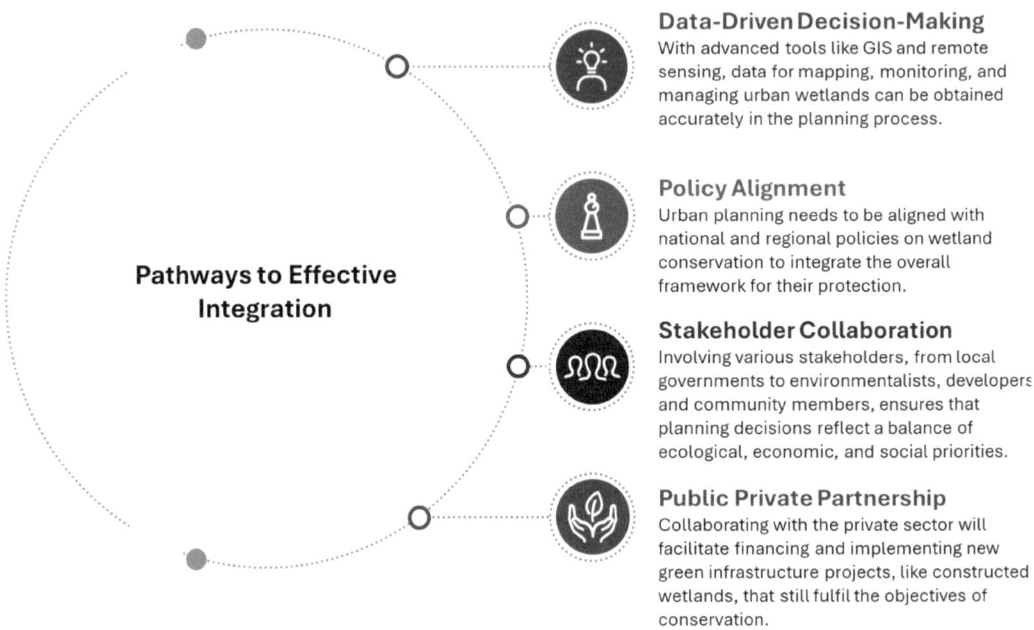

Fig. 3.5 Pathways to effective integration

3.12.5 Pathways to Effective Integration

To overcome these challenges, urban planning must adopt multidisciplinary and participatory approaches (Fig. 3.5).

3.13 Policy Recommendations

Effective wetland conservation in India's urban spaces requires an effective policy framework that covers legislative, financial, and collaborative strategies (Bassi and Kumar 2012; Hughey et al. 2004). Targeted interventions to address the systemic factors that fuel wetland degradation must also improve enforcement but build capacities and partnerships (MOEF I (2020); Bassi and Kumar 2012; *Reiteration-OM_Protection-of-Wetlands_08032022* n.d.). The following represent some of the key policy recommendations to strengthen the conservation and sustainable management of urban wetlands.

3.13.1 Strengthening Legislative Frameworks to Deter Violations

The legal mechanisms in place in India, such as the Wetlands (Conservation and Management) Rules, 2017, and the Environmental Protection Act, are already there. But the problem is that their enforcement is very weak because of a weak deterrent and loopholes in regulations (Bera et al. 2021; MOEF 2020). The curbing of violations like encroachment, pollution, and illegal dumping would require stronger legislative frameworks (Al Khaddar et al. 2020) (Fig. 3.6).

Comparative evidence from regions such as Southeast Asia and Latin America shows that decentralized, data-driven governance is effective in wetland management. In Thailand and Costa Rica, for example, giving jurisdictional authority to municipal units resulted in greater efficiency in enforcement in wetland areas and improved biodiversity protection. India's policy reform can benefit from these models by localizing enforcement through tiered legal frameworks.

Fig. 3.6 Strengthening legislative frameworks to deter violations

3.13.2 Fostering Collaboration Among Government, Private Sectors, and NGOs

A multi-stakeholder approach to wetland conservation is required to leverage the expertise and resources of government agencies, private enterprises, and NGOs (Molur et al. 2020). This partnership can result in innovative and sustainable conservation strategies. PPPs can help mobilize resources for projects related to wetland restoration, like funding for green infrastructure or constructed wetlands. The study reveals that the lack of inter-departmental coordination is one of the principal drivers behind ineffective wetland governance. Overlapping jurisdictions and siloed responsibilities delay responses and dilute accountability. The involvement of the private sector, as mandated by CSR, can provide financial backing while creating accountability (Billore et al. 1999; Hettiarachchi et al. 2015). The role of NGOs is more prominent in spreading awareness, carrying out research, and galvanizing local communities. Policies must promote cooperation with NGOs actively to ensure the work reaches the grassroots (Jia et al. 2011). Wetland management requires collaboration between sectors such as urban development, water resources, and environmental protection. The policy must make provisions for the effective working of departments across various lines of departments in the interest of reducing overlapping jurisdictions and jurisdictional conflict (Ghosh et al. 2006; Hettiarachchi et al. 2014a; Li et al. 2010).

3.13.3 Ensuring Adequate Financial and Technical Support for Wetland Restoration Projects

This also comes at a considerable cost of financing and technical know-how; proper support in these two sectors is paramount to the sustainability of these conservation endeavors

(Fu et al. 2014). Dedicated funds toward conserving wetlands should be created through budgetary appropriation, foreign aid, and corporate social responsibility contributions (Bassi et al. 2014). It would then be easy for some allocations of the city development budget to be provided periodically to ensure sustainable financial support (Kumar and Singh Maan 2019). A recurring theme in wetland degradation is the absence of institutionalized financing mechanisms. Case studies in India suggested that a typical pattern of wetland conservation in India is delayed fund disbursal, possibly coupled with weak capacity to absorb funds, causing alleged permanent loss of wetlands. This pattern does not occur in cities like Singapore and New York which have taken a lead and learning from the environmental sustainability movement, established climate resilience bonds and incentives in the form of tax-based funding instruments and other funding models to ensure continued year-on-year funding of blue-green infrastructure. Policies must insist that training for wetland management be conducted on these professionals: urban planners, engineers, and local authorities; some of the wetland management techniques that must be emphasized are adaptive management, GIS-based monitoring, and ecological restoration (Kumar and Singh 2020). R&D in wetland management technologies will help foster new inventions. Grants or tax breaks for green infrastructure inventions such as constructed wetlands and real-time monitoring systems help popularize it quickly (Ma & Yan 2024; Rani et al. 2011; Singh et al. 2022).

3.14 Building a Sustainable Future

The study emphasizes the necessity of concurrent changes in legal frameworks and financing methodologies and levels of community engagement for isolated conservation initiatives to provide sustainable impact. By augmenting institutional analysis through exploration of local features, the present research provides a multidimensional appreciation of wetland degradation that can be reproduced for policy reform in comparable contexts of the Global South. The analysis makes a clear case from a policy standpoint for moving beyond a reactive and compliance-based governance model to a proactive, adaptive, and inclusionary governance model. In practical implication, wetlands must be re-envisioned as key assets in climate resilience and sustainable urbanization. These insights are consistent with global lines of research, while also contributing to the growing literature on environmental governance for providing context-specific solutions in rapidly urbanizing areas. This not only enhances India's jurisdiction on sustainable wetland governance practices but also provides a model that other countries undergoing a similar ecological transition can draw upon (Fig. 3.7).

3.15 Conclusion

Urban wetlands are essential for the ecological, social, and economic vitality of urban areas. They are natural shields against climate change impacts and manage hydrological systems that promote biodiversity, making them an important component of sustainable urban ecosystems. Despite their considerable value, the degradation of urban wetlands in India is alarming at a rate triggered by urban expansion, industrialization, and weak policy implementations. These are now a compulsion rather than an option. Wetland conservation in the urban context calls for an approach that understands its fluid and integrative nature. It starts with dealing with the core issues of weak enforcement of regulations, bureaucratic logjam, and political lethargy. There has to be a strong structure, which not just buttresses the policies in place but also enforces their effective implementation at the grassroots level.

Community involvement is the foundation of a successful wetland management practice. Education of the local population for the proper care of these systems can turn the conservation approach into the collective goal of society. Loktak Lake is an initiative that shows how, with knowledgeable citizens and active stakeholders, significant outcomes can be achieved through collaborative

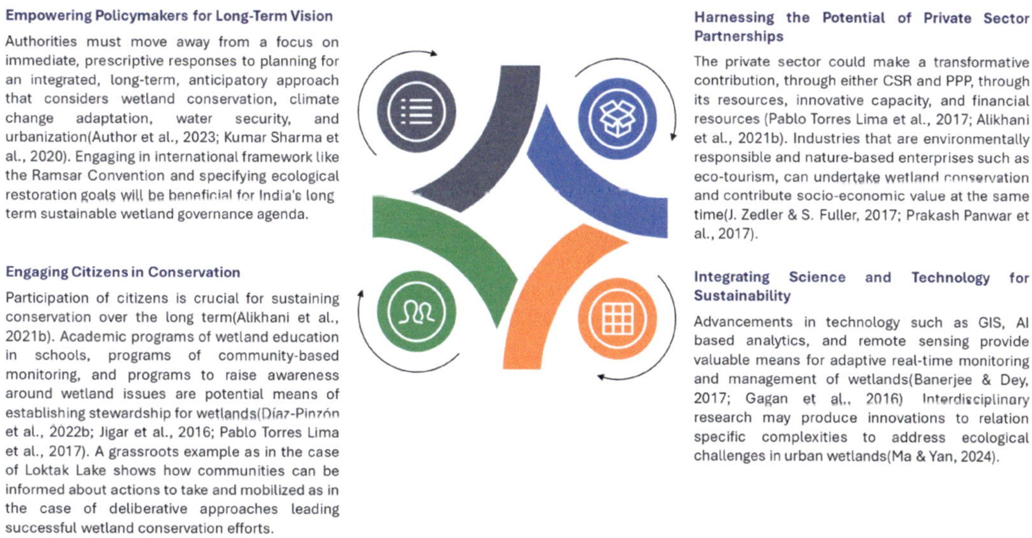

Fig. 3.7 Initiatives to provide sustainable impact

efforts. Advances in technology, such as GIS mapping and real-time monitoring systems, have become significant in modern wetland management. These tools offer information that is accurate and implementable, thus allowing governing bodies to act in advance on risks such as encroachments and environmental degradation. Similarly, incorporating wetlands into urban planning is necessary for the creation of urban spaces that harmonize growth with ecological sustainability. Techniques like green infrastructure and artificial wetlands not only protect natural ecosystems but also improve the quality of life in cities.

This chapter demonstrates the importance of creating urban planning frameworks that include wetland ecosystems within city master plans, rather than considering them as leftover pieces of urban fabric. A significant policy implication is to create a decentralized governance system that provides local bodies and communities a voice in policy decisions. In addition, bridging wetland conservation with climate adaptation and sustainable development goal policies can elevate the importance of wetland efforts at various governance scales. Globally, there are many parallels to urban settlements, where wetlands are integrated into global cities' resiliency planning. In the Indian context, this compares to the global emphasis on nature-based solutions but is underscored by a need for more attention to contextualized and equity-based models in countries experiencing rapid urbanization in India. The results of this research advance the collective discourse concerned with environmental governance, demonstrating how institutional gap, stakeholder fragmentation, and unyielding regulations impede conservation of wetlands. By framing a mechanism that includes policy analysis as well as implementation understanding, this research prescriptively adheres to integrating ecological conservation principles with urban development demands.

It will be an opportunity to redefine its relationship with urban wetlands by moving from merely reactive conservation efforts to proactive adaptive management practices. Such a paradigmatic shift, in close association with the collaborative policymaking practices between policymakers, private enterprise entities, researchers, and the community, can lead the way into sustainability. By quantifying the economic values of their services and incentives for their preservation, governments and various stakeholders can make ecological priorities fall in line with development goals.

Urban wetlands are not just water bodies but are the lifelines of sustainable cities. Their conservation is essential for building resilience in cities in the context of rapid environmental and demographic changes. Urgent, collective action is required to overcome the existing challenges and capitalize on emerging opportunities to secure these ecosystems for the future. Research aims to add to the global body of knowledge on urban ecological systems, specifically by presenting a scalable model that may be applied across developing economies. The results distinctively contribute to environmental policy research by bridging the legal framework with the ground practices. The chapter could also be extended in future research with longitudinal studies on metrics of wetland health or comparative policy studies across urban areas in South Asia, Africa, and Latin America, which are undergoing a similar transition of urbanization with ecological conservation.

References

Ahn C, Schmidt S (2019) Designing wetlands as an essential infrastructural element for urban development in the era of climate change. Sustainability (Switzerland) 11(7). https://doi.org/10.3390/su11071920

Al Khaddar R, Kaushika ND, Singh SK, Tomar Editors RK (2020) Advances in energy and environment select proceedings of TRACE 2020. http://www.springer.com/series/15087

Alberto RT, Biagtan AR, Isip MF, Tagaca RC (2019) Hot spot area analysis of onion armyworm outbreak in Nueva Ecija using geographic information system. Spat Inf Res 27(6):673–680. https://doi.org/10.1007/s41324-019-00266-0

Alikhani S, Nummi P, Ojala A (2021) Urban wetlands: a review on ecological and cultural values. Water (Switzerland) 13(22) MDPI. https://doi.org/10.3390/w13223301

Amaranath AT, Rama A, Subrahmanian R (n.d.) Water urbanism-integrating urban wetlands into urban landscape for sustainable development

Author P, Author C, Pathania S, Scholar R (2023) Sustainable urban development in India: assessing policies, challenges and opportunities for achieving sustainable urbanization (2001–2023). Int J Creat Res Thoughts 11(6):e252–www.ijcrt.org

Bandyopadhyay S, Magsi H, Sen S, Ponce T, Editors D (2020) Water management in South Asia socioeconomic, infrastructural, environmental and institutional aspects. Springer. http://www.springer.com/series/15344

Banerjee S, Dey D (2017) Eco-system complementarities and urban encroachment: a SWOT analysis of the East Kolkata Wetlands, India. Cities Environ 10. https://digitalcommons.lmu.edu/cate/vol10/iss1/2

Barbier EB, Knowler D (1997) Economic valuation of wetlands a guide for policy makers and planners. Ramsar Convention Bureau, Gland

Bassi N, Kumar MD (2012) Addressing the civic challenges: perspective on institutional change for sustainable urban water management in India. Environ Urban ASIA 3(1):165–183. https://doi.org/10.1177/097542531200300109

Bassi N, Kumar MD, Sharma A, Pardha-Saradhi P (2014) Status of wetlands in India: a review of extent, ecosystem benefits, threats and management strategies. J Hydrol Reg Stud 2:1–19. Elsevier B.V. https://doi.org/10.1016/j.ejrh.2014.07.001

Basu T, Das A, Pham QB, Al-Ansari N, Linh NTT, Lagerwall G (2021) Development of an integrated peri-urban wetland degradation assessment approach for the Chatra Wetland in Eastern India. Sci Rep 11(1). https://doi.org/10.1038/s41598-021-83512-6

Bera B, Bhattacharjee S, Shit PK, Sengupta N, Saha S (2021) Anthropogenic stress on a Ramsar site, India: study towards rapid transformation of the health of aquatic environment. Environ Chall 4. https://doi.org/10.1016/j.envc.2021.100158

Billore SK, Singh N, Sharma JK, Dass P, Nelson RM (1999) Horizontal subsurface flow gravel bed constructed wetland with Phragmites Karka in Central India. Water Sci Technol 40(3):163–171. https://doi.org/10.1016/S0273-1223(99)00461-8

Blundo G, Bax V, Marcela RA, Lissat (2018) The different dimensions of livelihood impacts of payments for environmental services (PES) schemes: a systematic review. Ecol Econ 149:160–183

Bystrom O, Andersson H, Gren I-M (2000) Economic criteria for using wetlands as nitrogen sinks under uncertainty. Ecol Econ 35:35–45

Chakraborty B, Roy S, Bera A, Adhikary PP, Bera B, Sengupta D, Bhunia GS, Shit PK (2021a) Cleaning the river Damodar (India): impact of COVID-19 lockdown on water quality and future rejuvenation strategies. Environ Dev Sustain 23(8):11975–11989. https://doi.org/10.1007/s10668-020-01152-8

Chakraborty B, Roy S, Bera A, Pratim Adhikary P, Bera B, Sengupta D, Sankar Bhunia G, Kumar Shit P (2021b) Eco-restoration of river water quality during COVID-19 lockdown in the industrial belt of eastern India. Environ Sci Pollut Res. https://doi.org/10.1007/s11356-021-12461-4/Published

Chan L (n.d.) City in a garden-Singapore's urban biodiversity conservation efforts

Chan L, Liang Jim L (2023) Habitat restoration and enhancement as a thrust in the nature conservation masterplan and city in nature vision. http://nparks.gov.sg/biodiversity/our-national-plan-for-conservation/nature-

Chuan TY, Yang NM, Ng A, Ng D, Hern C (2023) Restoration of a diversity of ecosystems on Coney Island Park. In: Chan L, Ng D, Lim LJ (eds) handbook on habitat restoration: general principles and case studies in Singapore. National Parks Board, Singapore

Cobracollactive (2024) Colombo wetland management framework-draft. Cobra Collactive

Das Sarkar S, Sarkar UK, Lianthuamluaia L, Ghosh BD, Roy K, Mishal P, Das BK (2020) Pattern of the state of eutrophication in the floodplain wetlands of eastern India in context of climate change: a comparative evaluation of 27 wetlands. Environ Monit Assess 192(3). https://doi.org/10.1007/s10661-020-8114-8

Das S, Bhunia GS, Bera B, Shit PK (2021) Evaluation wetlands ecosystem health using geospatial technology: evidence from Lower Gangetic Flood Plain in India. Environ Sci Pollut Res Int 29(2):1858–1874. https://doi.org/10.21203/rs.3.rs-522708/v1

Datri LA, Lopez M, Buchter S, Pazcel EM, Gandini M (2024) Functional urban wetlands in dysfunctional cities. Curr Landsc Ecol Rep 9(2):21–30. https://doi.org/10.1007/s40823-024-00097-2

Dhir B, Sharmila P, Saradhi PP (2009) Potential of aquatic macrophytes for removing contaminants from the environment. Crit Rev Environ Sci Technol 39(9):754–781. https://doi.org/10.1080/10643380801977776

Díaz-Pinzón L, Sierra L, Trillas F (2022) The economic value of wetlands in urban areas: the benefits in a developing country. Sustainability (Switzerland) 14(14). https://doi.org/10.3390/su14148302

Dooley W, Stelk M (2021) Urban wetlands protection and restoration guide—a resource to support efforts to restore and protect urban wetlands for economic, ecological, and social benefits. Association of State Wetland Managers. Windham

Dujovny E (2009) The deepest cut: political ecology in the dredging of a new sea mouth in Chilika Lake, Orissa, India. Conserv Soc 7(3):192–204. https://doi.org/10.4103/0972-4923.64736

EnviStats India (2020) EnviStats India 2020: Vol. II—environment accounts. www.ramsar.org

Er K, Zhi Qiang S, Wright J, Velautham E, Sia A (2023) The learning forest: a habitat restoration project that ecologically connects with the Singapore botanic gardens rainforest. In: Chan L, Ng D, Lim LJ (eds) handbook on habitat restoration: general principles and case studies in Singapore. National Parks Board, Singapore

Erwin KL (2009) Wetlands and global climate change: the role of wetland restoration in a changing world. Wetl Ecol Manag 17(1):71–84. https://doi.org/10.1007/s11273-008-9119-1

Fu B, Wang YK, Xu P, Yan K, Li M (2014) Value of ecosystem hydropower service and its impact on the payment for ecosystem services. Sci Total Environ 472:338–346. https://doi.org/10.1016/j.scitotenv.2013.11.015

Gagan M, Amit C, Avinash K, Ajendra K (2016) Impact of industrial effluent on ground water and surface water quality—a case study of Dhampur region (U.P.), India. J Chem Pharm Sci 9(2) www.jchps.com

Garg JK (2015) Wetland assessment, monitoring and management in India using geospatial techniques. J Environ Manag 148:112–123. https://doi.org/10.1016/j.jenvman.2013.12.018

Ghosh S, Das A (2019) Urban expansion induced vulnerability assessment of East Kolkata Wetland using Fuzzy MCDM method. Remote Sens Appl Soc Environ 13:191–203. https://doi.org/10.1016/j.rsase.2018.10.014

Ghosh AK, Pattnaik AK, Ballatore TJ (2006) Chilika Lagoon: restoring ecological balance and livelihoods through re-salinization. Lakes Reserv Res Manag 11(4):239–255. https://doi.org/10.1111/j.1440-1770.2006.00306.x

Ghosh S, Dinda S, Chatterjee ND, Das K (2018) Analyzing risk factors for shrinkage and transformation of East Kolkata Wetland, India. Spat Inf Res 26(6):661–677. https://doi.org/10.1007/s41324-018-0212-0

Groot D, Stuip M, Finlayson M, Davidson N (2006) Valuing wetlands Guidance for valuing the benefits derived from wetland ecosystem services. Ramsar Convention Secretariat

Hails AJ (1996) Wetlands, biodiversity and the Ramsar Convention: the role of the convention on wetlands in the conservation and wise use of biodiversity. Ramsar Convention Bureau

Harrington R, McInnes R (2009) Integrated Constructed Wetlands (ICW) for livestock wastewater management. Bioresour Technol 100(22):5498–5505. https://doi.org/10.1016/j.biortech.2009.06.007

Hazbavi Z, Sadeghi SH, Gholamalifard M, Davudirad AA (2020) Watershed health assessment using the pressure–state–response (PSR) framework. Land Degrad Dev 31(1):3–19. https://doi.org/10.1002/ldr.3420

Hettiarachchi M, McAlpine C, Morrison TH (2014a) Governing the urban wetlands: a multiple case-study of policy, institutions and reference points. Environ Conserv 41(3):276–289. https://doi.org/10.1017/S0376892913000519

Hettiarachchi M, Morrison TH, Wickramsinghe D, Mapa R, De Alwis A, McAlpine CA (2014b) The eco-social transformation of urban wetlands: a case study of Colombo, Sri Lanka. Landsc Urban Plan 132:55–68. https://doi.org/10.1016/j.landurbplan.2014.08.006

Hettiarachchi M, Morrison TH, McAlpine C (2015) Forty-three years of Ramsar and urban wetlands. Glob Environ Chang 32:57–66. https://doi.org/10.1016/j.gloenvcha.2015.02.009

Huang HF, Kuo J, Lo SL (2011) Review of PSR framework and development of a DPSIR model to assess greenhouse effect in Taiwan. Environ Monit Assess

177(1–4):623–635. https://doi.org/10.1007/s10661-010-1661-7
Hughey KFD, Cullen R, Kerr GN, Cook AJ (2004) Application of the pressure-state-response framework to perceptions reporting of the state of the New Zealand environment. J Environ Manag 70(1):85–93. https://doi.org/10.1016/j.jenvman.2003.09.020
Ismail E, Yixiong C (2023) Lessons learnt from habitat restoration at Marsiling Park. In: Chan L, Ng D, Lim LJ (eds) handbook on habitat restoration: general principles and case studies in Singapore. National Parks Board, Singapore
Jena J (2007) Impact of Mahandi basin development on eco-hydrology of Chilika. https://www.researchgate.net/publication/267233186
Jia H, Ma H, Wei M (2011) Urban wetland planning: a case study in the Beijing central region. Ecol Complex 8(2):213–221. https://doi.org/10.1016/j.ecocom.2011.03.002
Jigar E, Gebru S, Ayalew A (2016) Socio-economic values, threats and legal protection aspects of wetland ecosystem in Afar region, Ethiopia. www.ijedr.org
Jörg J (2013) Strategic approaches to urban wetlands: reconciling nature conservation, engineering and landscape architecture. www.ecofibraperu.com
Juwarkar AS, Oke B, Juwarkar A, Patnaik SM (1995) Domestic wastewater treatment through constructed wetland in India. Water Sci Technol 32(3):291–294. https://doi.org/10.1016/0273-1223(95)00637-0
Kumar Sharma N, Jaya Kullu N, Anshula K (2020) Spatiotemporal analysis of wetlands in Ranchi Urban Area, Jharkhand: proposed national level urban wetland census. J Water Resour Eng Manag:16–22. www.stmjournals.com
Kumar G, Singh KK (2020) Mapping and monitoring the selected wetlands of Punjab, India, using geospatial techniques. J Indian Soc Remote Sens 48(4):615–625. https://doi.org/10.1007/s12524-020-01104-9
Kumar G, Singh Maan G (2019) study on status and health of wetlands in India: a geographical analysis. https://doi.org/10.6084/m9.figshare.11684394
Lal R (2008) Carbon sequestration. Philos Trans R Soc Lond Ser B Biol Sci 363(1492):815–830). Royal Society. https://doi.org/10.1098/rstb.2007.2185
Lathuillière MJ, Coe MT, Johnson MS (2016) A review of green-and blue-water resources and their trade-offs for future agricultural production in the Amazon Basin: what could irrigated agriculture mean for Amazonia? Hydrol Earth Syst Sci 20(6):2179–2194. Copernicus GmbH. https://doi.org/10.5194/hess-20-2179-2016
Le T-AT, Vodden K, Jianghua W, Sabau G, Bullock R (2024) Payments for ecosystem services programs: a global review of contribution towards sustainability. Heliyon 10(1):e22361
Lehner B, Döll P (2004) Development and validation of a global database of lakes, reservoirs and wetlands. J Hydrol 296(1–4):1–22. https://doi.org/10.1016/j.jhydrol.2004.03.028
Lena Chan E, Liang Jim L (2023) Handbook on habitat restoration general principles and case studies in Singapore parks let's make Singapore our city in nature. www.nparks.gov.sg
Leng AS (2021) Nature conservation in Singapore. Biblioasia
Lewis WM (1995) Wetlands: characteristics and boundaries. National Academy of Sciences
Li Y, Zhu X, Sun X, Wang F (2010) Landscape effects of environmental impact on bay-area wetlands under rapid urban expansion and development policy: a case study of Lianyungang, China. Landsc Urban Plan 94(3–4):218–227. https://doi.org/10.1016/j.landurbplan.2009.10.006
Li X, Mitra C, Marzen L, Yang Q (2016) Spatial and temporal patterns of wetland cover changes in East Kolkata Wetlands, India from 1972 to 2011. Int J Appl Geospatial Res 7(2):1–13. https://doi.org/10.4018/ijagr.2016040101
Liang Jim L, Chan L (n.d.) Part I: Chapter 2—Principles and approaches to habitat restoration and enhancement with particular reference to the Singapore context. In: Chan L, Ng D, Lim LJ (eds) handbook on habitat restoration: general principles and case studies in Singapore. National Parks Board, Singapore
Lima PT, Gómez KC, Sánchez RB (2017) Socio-environmental perception of an urban wetland and sustainability scenarios: a case study in Mexico City. Society of Wetland Scientists
Liu J, Diamon J (2005) China's environment in a globalizing world. Nature 435:1179–1186
Lolu AJ, Ahluwalia AS, Sidhu MC, Reshi ZA, Mandotra SK (2020) Carbon sequestration and storage by wetlands: implications in the climate change scenario. In: Restoration of wetland ecosystem: a trajectory towards a sustainable environment. Springer Singapore, pp 45–58. https://doi.org/10.1007/978-981-13-7665-8_4
Lyons JE, Collazo JA, Herring G (2016) Testing assumptions for conservation of migratory shorebirds and coastal managed wetlands. Wetl Ecol Manag 24(5):507–520. https://doi.org/10.1007/s11273-015-9477-4
Ma J, Yan B (2024) Numerical simulation of mercury conversion during pulverized coal combustion. Processes 12(12):2910. https://doi.org/10.3390/pr12122910
Manju G, Chowdary VM, Srivastava YK, Selvamani S, Jeyaram A, Adiga S (2005) Mapping and characterization of inland wetlands using remote sensing and GIS. J Indian Soc Remote Sens 33(I):51–61
Manzoor J, Sharma M, Sofi IR, Fayaz M, Bhat MH (2018) Status of Indian wetlands with special reference to pesticides and their impact, pp 64–83. https://doi.org/10.4018/978-1-5225-6111-8.ch004
Mathevet R, Targowla S, Munisamy A, Govindan V, Narayanan A, Bautès N (2020) Wetlands for a sustainable urban future: insights from Pondicherry, South

India. Grassroots J Nat Resour 3(4):74–93. https://doi.org/10.33002/nr2581.6853.03045
Matthews GVT (1993) The Ramsar Convention on Wetlands: its history and development. Ramsar Convention Bureau
McInnes RJ, Everard M (2017) Rapid assessment of wetland ecosystem services (RAWES): an example from Colombo, Sri Lanka. Ecosyst Serv 25:89–105. https://doi.org/10.1016/j.ecoser.2017.03.024
Mishra HS, Bell S, Wilczyńska A, Balicka J (2021) Urban wetlands and storm water management. In: Urban blue spaces: planning and design for water, health and well-being. Taylor and Francis, pp 326–371. https://doi.org/10.4324/9780429056161-17
Mitchell SA (2013) The status of wetlands, threats and the predicted effect of global climate change: the situation in Sub-Saharan Africa. Aquat Sci 75(1):95–112. https://doi.org/10.1007/s00027-012-0259-2
MOEF (2020) Assessment and monitoring framework for Indian Wetlands. MOEF
MOEF I (2020) Guidelines for implementing wetlands (Conservation and Management) Rules, 2017 M. MOEF I
Mohapatra A, Mohanty RK, Mohanty SK, Bhatta KS, Das NR (2007) Fisheries enhancement and biodiversity assessment of fish, prawn and mud crab in Chilika lagoon through hydrological intervention. Wetl Ecol Manag 15(3):229–251. https://doi.org/10.1007/s11273-006-9025-3
Mohri H, Lahoti S, Saito O, Mahalingam A, Gunatilleke N, Irham, Hoang VT, Hitinayake G, Takeuchi K, Herath S (2013) Assessment of ecosystem services in homegarden systems in Indonesia, Sri Lanka, and Vietnam. In: Ecosystem services, vol 5. Elsevier B.V, pp 124–136. https://doi.org/10.1016/j.ecoser.2013.07.006
Molur S, Smith KG, Daniel BA, Darwall WRT (2020) The status and distribution of freshwater biodiversity in the Western Ghats, India. www.iucn.org/species
Mondal B, Dolui G, Pramanik M, Maity S, Biswas SS, Pal R (2017) Urban expansion and wetland shrinkage estimation using a GIS-based model in the East Kolkata Wetland, India. Ecol Indic 83:62–73. https://doi.org/10.1016/j.ecolind.2017.07.037
Nguyen Thi Hong N, Tun K, Yit Chuan T, Chan L (2023) Artificial tidal pools: habitat enhancement of built-up shorelines of Singapore. In: Chan L, Ng D, Lim LJ (eds) handbook on habitat restoration: general principles and case studies in Singapore. National Parks Board, Singapore
Niemeijer D, de Groot RS (2008) A conceptual framework for selecting environmental indicator sets. Ecol Indic 8(1):14–25. https://doi.org/10.1016/j.ecolind.2006.11.012
Patil MS (2020) Impact of industrialization on environment in India. Vivek Res J 5(2):2581–8848. http://environmentinsider.com/wp-content/uploads/2014/06/ase17.jpg
Pattnaik AK (2022) Chilika lake: restoring ecological balance and livelihoods through re-salinization. In: The wetland book. https://doi.org/10.1007/978-94-007-6173-5_177-2
Pavithrapriya S, Mahimairaja S, Sivasubramanian K (2015) Pollution due to heavy metals in Coimbatore Wetlands, India. Res J Agric For Sci 3(6). www.isca.me
Peetabas N, Panda RP (2014) Conservation and management of bioresources of Chilika Lake, Odisha, India. Int J Sci Res Publ 5(7). www.ijsrp.org
Phukan P, Saikia R (2014) Wetland degradation and its conservation: a case study of some selected wetlands of Golaghat district, Assam, India. Res J Rec Sci 3. www.isca.me
Prakash Panwar S, Sharma M, Solanki N (2017) Urbanization, population and environment in India: a review. Int J Emerg Technol 8(1):103–108. www.researchtrend.net
Rajaram T, Das A (2008) Water pollution by industrial effluents in India: discharge scenarios and case for participatory ecosystem specific local regulation. Futures 40(1):56–69. https://doi.org/10.1016/j.futures.2007.06.002
Rani M, Kumar P, Yadav M, Hooda RS (2011) Wetland assessment and monitoring using image processing techniques: a case study of Ranchi, India. J Geogr Inf Syst 03(04):345–350. https://doi.org/10.4236/jgis.2011.34032
Rathnayake Z (2024) From wastelands to wetlands the fight to save Sri Lanka's natural flood buffers. BBC
Reiteration-OM_Protection-of-Wetlands_08032022 (n.d.)
Rode S (2019) Wetlands degradation and conversion due to urbanisation in Mumbai metropolitan region: acute problems with long term solutions. Bus Excell Manag 9(1):33
Sengupta M, Dalwani R (2008) East Kolkata wetlands: a resource recovery system through productive activities. https://www.researchgate.net/publication/267547694
Shivakrishna A, Ramteke KK, Kesavan S, Prasad P, Naidu BC, Dhanya M, Abidi ZJ (2021) Monitoring of current land use pattern of Ramsar designated Kolleru Wetland, India using geospatial technologies. J Environ Biol 42(1):106–111. Triveni Enterprises. https://doi.org/10.22438/JEB/42/1/MRN-1404
Singh S, Biswas R (2021) Critical review of existing policies, plans, schemes, programs and guidelines related to management and rejuvenation of urban water bodies in India and Delhi. ResearchGate. https://www.researchgate.net/publication/377242372
Singh S, Biswas R (2022) Analysis of land use change effects/impacts on surface water resources in Delhi. Urban Sci 6(4). https://doi.org/10.3390/urbansci6040092
Singh G, Chandra R, Shukla S (2022) India 2 Scientist-SE & Head, Geospatial Data Division, Remote Sensing Applications Centre. www.irjet.net

Slykan Trust (2023) On becoming and staying a Ramsar Wetland City. Colombo's Urban Wetlands

Soanes K, Threlfall CG, Taylor L, Kirk H, Ramalho CE, Cumpston Z, Parris KM (2020) Recognising the conservation and cultural value of urban wetlands. www.nespurban.edu.au

Tan PY, Wang J, Sia A (2013) Perspectives on five decades of the urban greening of Singapore. Cities 32:24–32. https://doi.org/10.1016/j.cities.2013.02.001

Teo R (2023) Mangrove restoration in Pulau Ubin. In: Chan L, Ng D, Lim LJ (eds) handbook on habitat restoration: general principles and case studies in Singapore. National Parks Board, Singapore

UNDP (2022) Enhancing socio-ecological. UNDP

Venkatachalam L (2023) Economic valuation of wetland ecosystem services—current status and the way forward (242). Madras Institute of Development Studies

Wang S (2018) Urban wetlands management in Colombo a new model for urban resilience. www.gfdrr.org

Wattage P, Mardle S (2005) Stakeholder preferences towards conservation versus development for a wetland in Sri Lanka. J Environ Manag 77(2):122–132. https://doi.org/10.1016/j.jenvman.2005.03.006

Woon J (2023) Singapore index on cities' biodiversity—a monitoring tool for biodiversity conservation efforts. In: Chan L, Ng D, Lim LJ (eds) handbook on habitat restoration: general principles and case studies in Singapore. National Parks Board, Singapore, pp 272–280

Wright T, Tomlinson J, Schueler T, Cappiella K, Kitchell A, Hirschman D (2006) Direct and indirect impacts of urbanization on wetland quality. Wetlands & watersheds article #1. www.stormwatercenter.net

Yee A (2023) Habitat enhancement aided by geographic information systems (GIS). In: Chan L, Ng D, Lim LJ (eds) handbook on habitat restoration: general principles and case studies in Singapore. National Parks Board, Singapore

Zedler J, Fuller S (2017) Key findings for impacts of urbanization on wetlands. Capital Area Regional Planning Commission

Zhang Y, You X, Huang S, Wang M, Dong J (2022) Knowledge atlas on the relationship between water management and constructed wetlands—a bibliometric analysis based on CiteSpace. Sustainability (Switzerland) 14(14). https://doi.org/10.3390/su14148288

Conservation Challenges of Wetlands in Upper Assam

4

Shyama Prasad Biswas and Sanchita Boruah

Abstract

The wetlands of upper Assam are unique ecosystems with significant ecological and economic values. Four major types of wetlands are found in the region: (a) Oxbow lakes (open *beel*); (b) tectonic lakes (closed *beel*); (c) seasonal water-logged areas; and (d) man-made tanks. The floodplain lakes, mainly oxbow and tectonic in origin, are the dominant types of wetlands. The huge man-made tanks of upper Assam have aesthetic beauty and are the center of attraction for tourists. These wetlands support a wide variety of aquatic plants and animals, including many rare and endemic ornamental species like *Badis assamensis, Channa aurantimaculata, C. barca, C. bleheri*, and *Mystus dibrugarensis*. Migratory birds during the winter months are an attraction for thousands of tourists, and even the critically endangered Gangetic dolphin (*Platanista gangetia*) occasionally enters certain wetlands during high flood. These wetlands provide important ecological services such as preventing waterlogging and flood mitigation. Additionally, these wetlands are vital for fisheries and livelihood options. However, many prized species have become rare due to anthropogenic disturbances, rapid urbanization, high siltation rates, embankment construction, high fishing pressure, improper disposal of solid waste and nonbiodegradable materials, and contamination of wetlands by agricultural pesticides. Although aquatic biota have a natural resilience capacity to withstand unfavorable changes to a certain limit, it is crucial to adopt a holistic approach involving all stakeholders to restore wetland habitats and also to protect urban and peri-urban areas from waterlogging conditions, thereby reducing the chances of waterborne diseases.

Keywords

Floodplain lakes · Aquatic fauna · Gangetic dolphin · Restoration strategy

4.1 Introduction

Upper Assam has ten administrative districts: Biswanath, Golaghat, Jorhat, Majuli, Sivasagar, Charaideo, Dibrugarh, Tinsukia, Lakhimpur, and Dhemaji. This region is rich in natural resources, particularly oil, coal, and tea plantations. However, one of the most important resources in upper Assam is its wetlands, which are scattered throughout the state. These beautiful waterbodies

S. P. Biswas (✉)
Department of Life Sciences, Dibrugarh University, Dibrugarh, Assam, India
e-mail: spbsdu@dibru.ac.in

S. Boruah
Department of Zoology, D.H.S.K. College, Dibrugarh University, Dibrugarh, Assam, India

S. K. Chakraborty et al. (eds.), *Wetlands and Urban Resilience*, Sustainable Development Goals Series,
https://doi.org/10.1007/978-3-032-06881-1_4

cover approximately 0.1 million hectares and is home to over 300 species of fish and other aquatic megafauna (Boruah and Biswas 2002); however, seasonal species composition and diversity vary widely. Most of the fish species are either temporary or permanent "residents" of over 3500 floodplain lakes (*beels*), which are the feeding and breeding grounds of most of the riverine fish species in Assam (Biswas and Michael 1992; Biswas 1996; Boruah and Biswas 2002). However, the Brahmaputra river system has been experiencing a gradual decline in habitat quality due to changes in its water regime, particularly in the seasonality and hydrograph of the river (Biswas and Boruah 2000; Boruah and Biswas 2002). This impacted the linked ecosystem also, as evident from the wetland habitats, as alteration and shrinkage of *beel* habitats have become increasingly common in recent years (Abujam et al. 2014; Nayak and Biswas 2020). The *beels* of this region are also subjected to the adverse effects arising from the pesticides and other agricultural runoff. The tea gardens use a huge amount of different types of pesticides, insecticides, herbicides, and acaricides like paraquat, endosulfan, dicofol, etc. annually. Various types of pesticides, agricultural fertilizers, and other harmful chemicals used in the tea gardens and also in paddy fields over the years ultimately find their way into the nearby wetlands (Biswas 1996, 2014). Moreover, disposal of outdated stock, containers, and packets of insecticides and fertilizers, and wastewater from industries are serving as major sources of pesticide pollution in such aquatic systems. This is responsible for the mass mortality of fish as well as other aquatic biota in the *beels*. Such practices should be stopped entirely to save the wetland habitats and biota.

Urbanization in the developing world is a matter of great concern for policymakers. The northeastern states of India are also not the exception; every prominent township of the region is more or less affected by rapid urbanization, which has badly impacted the wetlands in the periphery of urban areas. These wetlands are the natural sinks for waste generated and excess rainwater during the monsoon. The floodplain lakes are connective links and an integral part of the riverine system. However, in the last four decades or so, rapid urbanization has greatly contributed to the shrinkage of wetlands, resulting in a phenomenal increase in waterlogging days in cities of Assam. During the 1960s and 1970s, waterlogging in cities and towns of Assam was unheard of, as there were lots of seasonal water bodies (*beels*) which retained excess rainwater, besides acting as breeding and nursery grounds of small and medium-sized fish species. Encroachment of these water bodies not only affected urban life but also disrupted the crucial ecological balance. Converting a wetland into a stadium in the 1980s has increased the waterlogging incidents many times during rainy months. In this chapter, an attempt has been made to quantify the changes occurring in a specific wetland system within the upper Brahmaputra basin, assessing the cumulative impact on biodiversity and human communities, as well as the shifting trends in dependency on the wetland.

4.2 Study Area

The Upper Brahmaputra Valley has 15,196.0 ha area of wetland, which is 15.01% of the total 101,231.60 ha area of wetlands in the entire state of Assam. Seven prominent wetlands of upper Assam have been selected (Fig. 4.1) for the present study:

(a) Maguri *Beel:* Maguri *beel* is one of the largest wetlands in the upper Brahmaputra basin and lies between 27°33′79.5″ and 27°34′ 35.7″ N and 95°22′8.1″ and 95°22′19″ E within Tinsukia district, which is adjacent to the Dibru-Saikhowa National Park.
(b) Mer *Beel:* This wetland (27°18′57.8″ N and 95°12′08.8″ E, 71 msl) is about 10 km from Duliajan in Dibrugarh District. The *beel* is mostly a swampy area, and in between there are grass, fallow land, and human settlements.
(c) Borsola *Beel:* Borsola (26°50′22.3″ N and E 94°13′12″ E) of Jorhat district is an "open"

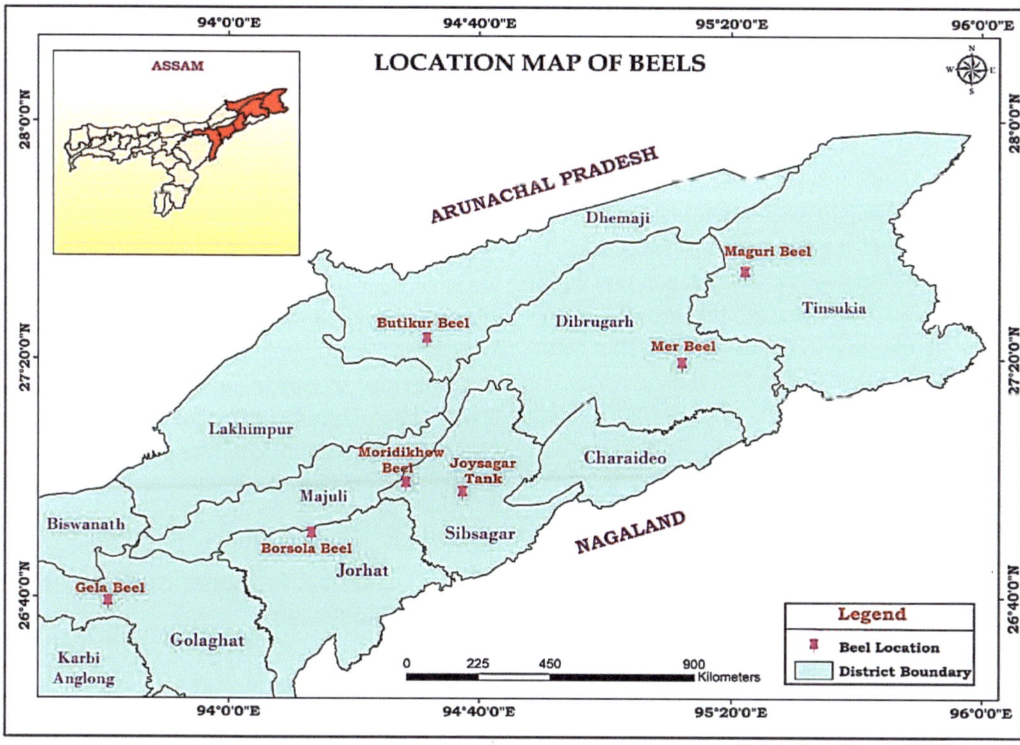

Fig. 4.1 Map of the study area

type of floodplain lake *(beel)* which is situated about 10 km from Jorhat town.

(d) Butikur *Beel*: This floodplain lake is in Dhemaji District (26°75′ N & 94°56′ E, 104 msl) and is an irregularly shaped *beel*.

(e) Moridikhow: Moridikhow is an oxbow lake which was originally a part of the river Dikhow, a tributary of the Brahmaputra. This lake (26°98′ N and 94°46′ E) is about 25 km away from Sivasagar town in upper Assam.

(f) Gelabeel: Gelabeel is an oxbow lake, situated about 16 km north of Golaghat town, between 93°40′ & 94°5′ E and 26°4′ & 26°45′ N).

(g) Joysagar tank: Joysagar tank (26°58′ N and 94°41′ E) is in the Sivasagar District and about 5 km from the Sivasagar town. It was excavated during the Ahom King Rudrasingha in 1707 AD.

4.3 Materials and Methods

Wetlands were selected based on their potential fishery resources. Based on the fish landings, waterbodies were identified and classified, and water and fish samples were collected through regular field surveys. Each wetland is again divided into three segments, of which one sampling station from each segment is selected for monitoring water quality (Trivedy et al. 1987; APHA 2005) and recording fish species. Seven physicochemical parameters of the selected wetlands were recorded seasonally (pre-monsoon, monsoon, post-monsoon and winter). Fish species were identified following Talwar and Jhingran (1991) and FishBase. The conservation status (CAMP 1998) of ichthyofauna and their abundance frequency (Gupta Choudhury 2011) in the studied wetlands were also recorded. Condition factor (K) is used to compare the condition or the health status of the fish and is based on the hypothesis that fish with a heavier weight for a given length exhibit a better physiological condition (Bagenal and Tesch 1978). The relationship between the total length and the total weight was determined using the formula $W = aL^b$ (Le Cren 1951). The coefficient of growth (b) obtained from the regression equation is a good indicator of the growth of the species in question (Biswas 1993). The entire plan of work is depicted in Fig. 4.2.

K-value	Remark
>1.10	Excellent condition
1.00	Good health, well-proportioned fish
0.90	Moderate health
0.80	Poor health, long and thin body
<0.70	Extremely poor health, having a big head, a narrow and thin body

At least two species of cyprinids, catfishes, loaches, and murrels which are commonly found in wetlands, were randomly selected from the surveyed wetlands for health status assessment. The health status of fish species was evaluated based on the average K-value for each species.

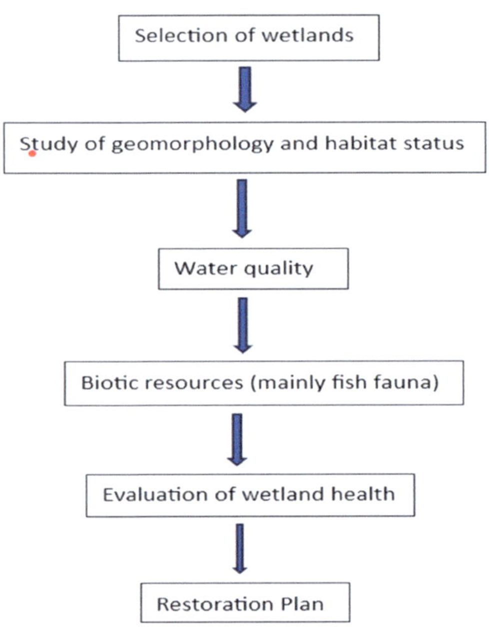

Fig. 4.2 Methodological flowchart

4.4 Results and Discussion

Wetlands in the upper Brahmaputra basin may be broadly categorized into (a) Oxbow lakes (open beel); (b) Tectonic lakes (close beel); (c) Seasonal water-logged areas; (d) Man-made tanks (Biswas and Abujam 2022). Floodplain lakes, mainly of oxbow and tectonic origin, are common. These wetlands sustain a variety of aquatic plants and animals, including many rare and unique species. Wetlands are important in preventing waterlogging in urban areas and mitigating floods. Economically, they are the reservoir of fisheries, providing livelihoods for thousands of people in upper Assam.

4.4.1 General Features of Wetland Ecosystem

Among all hydrological conditions, the "flood pulse" is likely the most critical factor influencing other abiotic factors and the faunal distribution in the wetlands of the Brahmaputra basin. Typically, three to four major floods occur between May and October, and fish migration is closely linked to this flood regime (Biswas and Boruah 2000). The upper Brahmaputra basin usually experiences heavy rainfall in June, July, and August. However, the rainfall pattern over the past few decades has changed significantly, even though the total annual rainfall in the region has remained almost the same (Biswas 2014). The aquatic fauna can adapt to normal seasonal fluctuations in water and adjust their physiological and biological activities.

Over 100 species from the wetlands have so far been identified—Indian major carps (IMC), minor carps, barbs, catfishes, feather backs, air-breathing fishes, loaches, and eels (Table 4.1). On average, minor carps, barbs, and small catfishes together contribute about 60% of the landings. The air-breathing group shares about 30% of the catch, whereas giant catfishes and IMCs contribute about 10% of the catch between November and April. Physical features and the biota of the studied *beels* (wetlands) have been summarized below.

Table 4.1 Fish species recorded from the selected wetlands of upper Assam

Family	Scientific name	IUCN status	Population trend	Economic importance
Notopteridae	*Notopterus notopterus* (Pallas, 1769)	LC	S	F
	Chitala chitala (Hamilton, 1822)	NT	D	F
Clupeidae	*Gudusia chapra* (Hamilton, 1822)	LC	D	F
Cyprinidae	*Amblypharyngodon mola* (Hamilton, 1822)	LC	S	F, OR
	Barilius barila (Hamilton, 1822)	LC	*	F, OR
	Cabdio jaya (Hamilton, 1822)	LC	D	F, OR
	C. morar (Hamilton, 1822)	LC	*	F, OR
	Cirrhinus mrigala (Hamilton, 1822)	LC	S	F
	C. reba (Hamilton, 1822)	LC	*	F
	Devario devario (Hamilton, 1822)	LC	*	OR
	D. aequipinnatus (McClelland, 1839)	LC	*	OR
	Esomus dandrica (Hamilton, 1822)	LC	*	F, OR
	Labeo bata (Hamilton, 1822)	LC	*	F
	L. calbasu (Hamilton, 1822)	LC	*	F
	L. (Catla) catla (Hamilton, 1822)	LC	*	F
	L. gonius (Hamilton, 1822)	LC	*	F
	L. rohita (Hamilton, 1822)	LC	*	F
	Tor putitora (Hamilton, 1822)	EN	D	F
	Osteobrama cotio (Hamilton, 1822)	LC	*	F
	Pethia gelius (Hamilton, 1822)	LC	*	OR
	P. ticto (Hamilton, 1822)	LC	*	F, OR
	P. conchonius (Hamilton, 1822)	LC	*	F, OR
	P.chola (Hamilton, 1822)	LC	*	F, OR
	P. Sophore (Hamilton, 1822)	LC	*	F, OR
	Rasbora daniconius (Hamilton, 1822) *R. rasbora* (Hamilton, 1822)	LC LC	* *	F, OR F, OR
	Systoma sarana (Hamilton, 1822)	VU		F
Nemacheilidae	*Acanthobitis botia* (Hamilton, 1822)	LC	*	F, OR
Cobitidae	*Botia dario* (Hamilton, 1822)	LC	*	F, OR
	B. rostrata	LC	*	F, OR
	Lepidocephalichthyes guntea (Hamilton, 1822)	LC	*	F, OR
	L. thermalis	LC	*	F, OR
Bagridae	*Hemibagrus menoda* (Hamilton, 1822)	VU		F
	Mystus bleekeri (Hamilton, 1822)	LC	D	F, OR
	M. cavasius (Hamilton, 1822)	LC	D	F, OR
	M. dibrugarensis (Chaudhuri, 1913)	LC	*	F, OR
	M. tengara (Hamilton, 1822)	LC	*	F, OR
	M. gulio (Hamilton, 1822)	LC		F, OR
	M. vittatus (Bloch, 1794)	LC	D	F, OR
	Rita rita (Hamilton, 1822)	LC	D	F
	Sperata aor (Hamilton, 1822)	LC	S	F
	S. seenghala (Sykes, 1839)	LC	*	F
Siluridae	*Ompok bimaculatus* (Bloch, 1794)	NT	*	F
	O. pabo (Hamilton, 1822)	NT	D	F
	Wallago attu (Bloch and Schneider, 1801)	VU	D	F
Schilbeidae	*Ailia coila* (Hamilton, 1822)	NT	D	F
	Clupisoma garua (Hamilton, 1822)	LC	D	F
	Eutropiichthys vacha (Hamilton, 1822)	LC	D	F

(continued)

Table 4.1 (continued)

Family	Scientific name	IUCN status	Population trend	Economic importance
Sisoridae	*Bagarius bagarius* (Hamilton, 1822)	VU	D	F
	Gagata cenia (Hamilton, 1822)	LC	*	F, OR
	G. gagata (Hamilton, 1822)	LC	D	F, OR
Heteropneustestidae	*Heteropneustes fossilis* (Bloch, 1794)	LC	*	F
Clariidae	*Clarias magur* (Linnaeus, 1758)	EN	D	F
Chacidae	*Chaca chaca* (Hamilton, 1822)	LC	D	OR
Belonidae	*Xenentodon cancila* (Hamilton, 1822)	LC	*	F, OR
Synbranchidae	*Ophichthys cuchia* (Hamilton, 1822)	LC	*	F
Mastacembelidae	*Macrognathus aral* (Lacepede, 1800)	LC	S	F
	M. pancalus (Hamilton, 1822)	LC	*	F, OR
	Mastacembelus armatus (Bloch & Schneider, 1801)	LC	S	F, OR
Ambassidae	*Chanda nama* (Hamilton, 1822)	LC	D	F, OR
	Parambassis ranga (Hamilton, 1822)	LC	S	F, OR
Anabantidae	*Anabus testudineus* (Bloch, 1792)	LC	*	F, OR
Badidae	*Badis badis* (Hamilton, 1822)	LC	*	F, OR
	B. assamensis			
Osphronemindae	*Trichogaster fasciata* (Bloch & Schneider, 1801)	LC	*	F, OR
	T. Labiosa (Day, 1877)	LC	*	F, OR
	T. lalius (Hamilton, 1822)	LC	*	F, OR
Nandidae	*Nandus nandus* (Hamilton, 1822)	LC	*	F, OR
Gobiidae	*Glossogobius giuris* (Hamilton, 1822)	LC	*	F
	Channa aurantimaculata (Musikasinthorn, 2000)	DD	*	F, OR
	C. barca (Hamilton, 1822)	VU	*	F, OR
	C. bleheri (Vierke, 1991)	VU	*	F, OR
	C. gachua (Hamilton, 1822)	LC	*	F, OR
	C. marulius (Hamilton, 1822)	LC	*	F, OR
	C. punctata (Bloch, 1793)	LC	S	F, OR
	C. stewartii (Playfair, 1867)	LC	*	F, OR
	C. striata (Bloch, 1793)	LC	S	F, OR
Tetraodontidae	*Leiodon cutcutia* (Hamilton, 1822)	LC	*	OR

#IUCN status: LC least concerned, *NT* near threatened, *VU* vulnerable, *EN* endangered, *DD* data deficient
Population trend: S stable, *D* declining, * unknown
Economic value: F food fish, *OR* ornamental fish

4.4.1.1 Maguri Beel

At the Full Storage Level (FSL), 167.40 ha of area is covered by Maguri beel, and 117.18 ha at the Dead Storage Level (DSL). The length of the wetland ranges from 1850 to 2800 m, and its breadth ranges from 640 to 800 m. The maximum depth of Maguri was recorded as 4.3 m, and the depth at the dead storage level is ~0.95 m. It is an open *beel* with a diverse ecosystem and is blessed with an affluent ichthyofaunal diversity, connected to the River Dibru, and in fact, the river flows from east to west direction through the wetland. The temperature of water showed the variation from 15.2 to 30.7 °C, pH from 6.35 to 8.27, and DO from 4.4 to 8.7 mg/l. The *beel* can be called moderately to highly productive, which can be inferred by looking at the total alkalinity, which ranges from 40 to 80 mg/l. This wetland is an important and significant destination for tourists in upper Assam and is surrounded by swamp forests, semi-evergreen forests, deciduous forests, and wet evergreen forests. When it comes to endemic and rare ornamental fish, Maguri serves as a crucial source of fishery. It is also a major source of fisheries, especially endemic and rare ornamental fish species like *Badis assamensis* and *Channa aurantimaculata. C. barca, C. bleheri, C. stewartii* (Biswas et al. 2007). This *beel* is highly infested by aquatic vegetation, predominantly *Eichhornia crassipes* and *Nelumbo nucifera.* In the past, this wetland system was hydrologically connected to the Lohit River through its tributary called Dangori, as well as the R. Dibru. During the 1990s, R. Dibru started flowing through this *beel* carrying water from the combined Dangori-Lohit due to the avulsion of R. Lohit. Maguri is now separated from the Dibru-Saikhowa National Park by the Dibru River and a small sand bar. As a result, Maguri *beel* now acts as both a lentic and a lotic system throughout the year. Over 12,000 people inhabit 10 adjoining villages and are directly dependent on their bioresources. Out of 2297 total families in the 10 villages, 70.3% directly depend on fishing for their livelihood. A handful of youth are also involved in aqua-tourism. The dairy business is limited to a particular community. The oil and gas spillage of the Baghjan oil field in 2020, for more than 4 months, caused havoc on the livelihood of the people since they are completely dependent on the ecosystem services provided by this wetland (Nayak and Biswas 2023). Maguri *beel* deserves special attention under the Man and Biosphere Programme (MAB) of UNESCO, as it is one of the major sources of fisheries and also a prominent tourist spot of Assam.

4.4.1.2 Mer *Beel*

This wetland has a water spread area of 207.51 ha. A major part of the beel is a swampy area, having diverse types of macrophytes and medicinal plants, and in between there are grass and fallow land (vegetated areas) and human settlements. It has now become a popular tourist destination. Wild animals, including leopards and pythons, are common in the vicinity of the *beel.* Mer *Beel* is a hotspot of migratory birds during winter. This wetland is a repository of a variety of small and air-breathing fishes as well as giant catfish. Cypriniformes (carps and barbs) and Siluriformes (catfishes) dominate in this *beel.* Air breathing fishes also contribute significantly to the fish species of the beel. Apart from a variety of food fishes, 36 ornamental fish species were recorded under 23 genera and 17 families were recorded. The K-value indicated that 30% of the studied fish species have poor health. This ultimately resulted in a decline in fish production, impacting the livelihood of the neighborhood.

4.4.1.3 Borsola *Beel*

The *beel* is about 9 km north of Jorhat town, having an area of 90 ha. This *beel* is directly connected through a connecting channel (open type) and, as such, there is a seasonal change in the water volume as well as the biota of the wetland. Chlorophyceae was the most dominant class (50.04%) among phytoplankton (Saikia and Biswas 2023), followed by Cyanophyceae (27.78%) and Bacillariophyceae (22.22%). Thick growth of water hyacinth and other macrophytes also alters the characteristics of the *beel* during non-rainy months. Borsola *beel* is an important fishery resource of Jorhat District and a breeding ground for riverine fishes including major Indian carps. However, a gradual shrinkage of the beel due to the encroachment and siltation has been observed during the study period. This might eventually lead to a shortage of fish cover and, eventually, affect Jorhat city and its suburbs to a large extent.

4.4.1.4 Butikur *Beel*

This floodplain lake is in Dhemaji District, spreads over an area of about 15 ha, and its depth ranges between 0.7 and 13.5 m depending on the season. This *beel* is irregularly shaped and it has a maximum length of 940 m and a breadth of 810 m at the live storage level (LSL), while the length and breadth were recorded as 680 m and 510 m at the dead storage level (DSL). It is covered with a thick growth of hydrophytes dominated by *Eichhornia crassipes, Jussia* sp., *Najas* sp., and *Otellia* sp. (Hussain 2011). Major physicochemical parameters of the beel were recorded as follows: Water temperature ranged from 17.6 to 30 °C; pH 6.9–7.2; transparency from 72 to 177 cm; DO ranged from 4.8 to 11.2 mg/l; free CO_2 from 4.2 to 7.6 ml/l; total alkalinity fluctuated between 25 and 110 mg/l. The *beel* depicted a hard water character, and chloride content varied from 8.7 to 21.8 mg/l (Hussain 2011), indicating the unpolluted nature of the *beel*. Dhemaji town in upper Assam depends on this *beel* for fish supply and flood mitigation. Unplanned fishing and siltation of the bed have affected fish production, as has been experienced in recent years.

4.4.1.5 Moridikhow

Moridikhow is an oxbow lake that was originally a part of the river Dikhow, a tributary of the Brahmaputra. This lake (26°98′ N and 94°46′ E) is about 25 km away from Sivasagar town in upper Assam. It is a perennial water body, mostly choked with aquatic macrophytes, especially during dry months (Borah 2017). The value of DO was highly variable (2.63 ± 1.58 to 5.98 ± 2.82 mg/l). This might be due to profuse vegetation growth in the water body. High aquatic vegetation may also be the reason for a high CO_2 value (1.75–5.81 mg/l). pH of water showed that the water of the studied lake is near neutral (6.9) to slightly alkaline (7.3). Total alkalinity ranged from 46 to 106 mg/l. The oxbow lake is considered highly productive. About 40 fish species were recorded from the water body (Borah 2017). This oxbow lake is vital for the people in and around Gaurisagar town in Sivasagar District. Fish diversity, however, has declined in this lake due to anthropogenic stress, especially fishing nets with small mesh sizes is also responsible for the degradation of fish production. As a pilot project, cage culture of major carps was successfully demonstrated at Moridikhow (Borah 2017).

4.4.1.6 Gelabeel

This is another oxbow lake, well-known for its fish production and catering to the needs of Dergaon town in Golaghat District for a long time. As in other *beels* of Assam, cyprinids were the most dominant group represented by 21 species (Sarma 2017). The water quality was generally in the safe range for aquatic life. Although the DO and free CO_2 content showed significant seasonal variations, the values remained within the normal range and never crossed the critical values during the study period. The water of Gelabeel showed medium hardness, and the total alkalinity value indicated the wetland is fairly productive. 62 fish species, including some vulnerable and rare fish, were recorded from the beel, indicating its rich biodiversity. Minor carps, featherbacks, and medium-sized catfishes were mostly encountered in the catch. This water body was once an important sighting place of the Gangetic dolphin. However, the average annual fish production from Gelabeel has been showing a downward trend (Sarma 2017).

4.4.1.7 Joysagar Tank

Of the many excavated huge man-made tanks during the reign of the Ahom dynasty, Joysagar Tank is of great importance due to historical reasons as well as its scenic beauty. Thus, it has developed as a heritage site of Assam. It covers approximately 1.3 km^2 water spread area with a peripheral length of about 48,000 m^2. The maximum depth of the tank is about 9 m. Apart from being enriched with the diverse fish and benthic fauna (Sarma 2015), it is also a crucial habitat for migratory birds. Aquatic macrophytes infest the peripheral area, but the tank water is mostly crystal clear. The tank is a repository of fish species, including major carps and giant catfish (Sarma and Biswas 2012). Moreover, it is the prime source of potable water for a vast area in the suburbs of Sivasagar town. Like other historical tanks of the region, the water level remains nearly constant throughout the year.

4.4.2 Wetland Fish Health

The condition factor (K) is an important indicator of a fish's overall well-being. Values of "K" > 1 suggest that the fish are in good health (Table 4.2). In the monsoon season, 51.02% of the fish were found to be in excellent health,

Table 4.2 Pooled condition factor (*K*) of the selected wetland fish species

Sl. no.	Fish species	*a*	*B*	*r*	*K*	Health status
1.	*Lepidocephalichthys guntea* (Hamilton, 1822)	0.35	2.32	0.87	0.76	Poor
2.	*Botia dario* (Hamilton, 1822)	0.05	2.77	0.92	1.02	Good
3.	*Trichogaster fasciata* (Bloch & Schneider)	0.02	3.17	0.84	1.16	Excellent
4.	*Anabas testudineus* (Bloch, 1792)	0.67	1.85	0.87	0.78	Poor
5.	*Mystus tengara* (Hamilton, 1822)	0.05	3.15	0.97	1.04	Good
6.	*M. vittatus* (Bloch, 1794)	0.04	2.32	0.86	1.03	Good
7.	*M. cavasius* (Hamilton, 1822)	0.34	1.44	0.82	0.99	Moderate
8.	*Chanda nama* (Hamilton, 1822)	0.02	2.36	0.91	1.04	Good
9.	*Parambasis ranga* (Hamilton, 1822)	0.03	2.42	0.87	1.02	Good
10.	*Channa punctata* (Bloch, 1793)	0.05	2.53	0.89	1.19	Excellent
11.	*C. gachua* (Hamilton, 1822)	0.04	2.33	0.90	0.97	Moderate
12.	*Gagata cenia* (Hamilton, 1822)	0.03	2.56	0.93	1.01	Good
13.	*Puntius sophore* (Hamilton, 1822)	0.06	2.99	0.89	1.07	Good
14.	*Parambassis lala* (Hamilton, 1822)	0.03	2.41	0.91	0.93	Moderate
15.	*Glossogobius giuris* (Hamilton, 1822)	0.03	3.24	0.92	1.19	Excellent
16.	*Macrognathus aral* (Bloch & Schneider)	0.06	3.28	0.91	1.12	Excellent
17.	*M. pancalus* (Hamilton, 1822)	0.06	2.82	0.87	1.02	Healthy
18.	*Mastacembelus armatus* (Lacepede, 1800)	0.02	3.11	0.93	1.16	Excellent
19.	*Nandus nandus* (Hamilton, 1822)	0.02	2.76	0.90	1.17	Excellent
20.	*Ompok pabda* (Hamilton, 1822)	0.07	2.13	0.91	0.96	Moderate
21.	*Pseudotropius atherinoides* (Bloch, 1794)	0.02	23.23	0.95	1.12	Excellent
22.	*Xenentodon cancila* (Hamilton, 1822)	0.04	3.13	0.94	1.19	Excellent

28.57% were classified as good, 16.33% were fairly healthy, and 4.08% were in poor health. During the pre-monsoon period, only 10.20% of the fish were in excellent condition, while 28.57% were in good health, 46.94% were moderately healthy, 10.20% were in poor health, and 4.08% were classified as having extremely poor health. In the post-monsoon season, the health of the fish declined further, with 46.94% exhibiting poor health, 32.66% in moderate health, 8.16% in good health, and only 12.25% in excellent health. In winter, the situation was even worse: 30.61% of fish were in weak health, 56.18% were in poor health, 8.16% were in fairly good health, and just 2.04% were assessed as being in good health. This study indicated that the highest number of healthy fishes was found in the monsoon, while the lowest was documented in winter, potentially due to food scarcity during that season. Our findings align with those of Haque and Biswas (2014), who studied the condition factor of certain ornamental fishes in the Brahmaputra River.

A "*K*" value greater than 1 indicates better condition in fish (Le Cren 1951). Among the specimens studied, the highest "*K*" value was recorded during the monsoon for *Chanda nama*, at 1.29 ± 0.45, while the lowest value was observed in winter for *Lepidocephalichthys guntea*, at 0.74 ± 0.17. Most fish had "*K*" values close to or exceeding 1, suggesting they were in favorable condition. The hypothesis states that fish with a heavier weight for a given length show better physiological conditions (Bagenal and Tesch 1978). The "*b*" and "*K*" values can be used to estimate a species' growth and overall well-being. However, the "*K*" value is influenced by several factors, including the age and sex of the fish, the season, stage of maturation, gut fullness, and type of food consumed, and seasonal variation in *K*-values in fish is quite natural (Biswas 1993).

4.4.3 Land Use and Land Cover in the Selected Area

Identified three potential responses of fish to climate change and alterations in land use. First, fish may adapt to the new environmental conditions; second, their distribution patterns could change;

and lastly, they may face extinction if environmental changes exceed the tolerance levels of their species. While the specific mechanisms are still uncertain, evidence suggests that climate-related changes in recruitment success play a crucial role in marine fish populations (Rijnsdorp et al. 2009), a concept that may also apply to freshwater systems. Freshwater ecosystems, particularly wetlands, are vital as they provide numerous ecosystem services and harbor rich biodiversity. Assam is fortunate to have many wetlands and abundant water sources due to significant rainfall, making its water bodies a valuable heritage of biodiversity (Boruah et al. 2008). However, Assam's wetlands, which are essential for biodiversity and ecosystem services, are threatened by changes in river course, bank erosion, and siltation that ultimately impact wetland habitats (Nayak and Biswas 2020). As viewed by Parra et al. (2023), the fishery potential of any aquatic body is fundamentally determined by the qualitative and quantitative abundance of its fish fauna. Land use and land cover (LULC) conversion refers to the changes in various land-use categories and represents the complex interrelationship between people and the environment (Wubie et al. 2016). Notably, the distribution of fish species can vary significantly due to the geographical and geological characteristics of their habitat. In this study, we observed considerable variation in the occurrence of fish species, influenced by the geographical features of the areas examined. The land use and land cover (LULC) of the study areas were classified into seven categories: agricultural land/cropland, pasture lands (forest plantations), water bodies, built-up land (including rural village settlements and urban areas), barren land/unculturable land, grass/grazing land, and wetlands. Changes in LULC are significant contributors to global change, which can drastically affect ecosystem functioning, biological cycles, and biodiversity (Basommi et al. 2016). The data collected from this study will serve as a foundation for developing better management strategies for wetlands in Assam. Similar LULC studies were conducted by Vivekananda et al. (2021).

4.4.4 Major Issues Related to Wetlands in Upper Assam

4.4.4.1 "Flood Pulse" and Its Impact on Aquatic Biota

Beels are seen to receive a good amount of water after the monsoon from the adjacent huge catchment areas connecting the river as backflow. The influx of water leads to the autostocking of the open beels. During this time, many riverine fish enter the open *beels,* including IMC, for spawning (Biswas and Michael 1992). Though perceived negatively, the annual flood is very essential for the wetlands as they not only recharge and rejuvenate the *beels* with fresh water and nutrients, stocking them with many fishes but also delay the process of eutrophication by cleaning out the floating weeds and other materials which otherwise would have choked the *beels.*

4.4.4.2 Aquatic Pollution

The extensive use of pesticides in tea cultivation and agriculture adversely affects aquatic life. These chemicals often mix with seasonal waterbodies through runoff, posing significant dangers to fish and the entire food chain, ultimately impacting wetland habitats. Additionally, the dumping of municipal waste is another serious threat to the survival of wetlands (Biswas 2014).

4.4.4.3 River Embankments

One reason for the decline in the natural population of Indian major carp in the Brahmaputra system is the construction of ecologically harmful river embankments (Biswas et al. 2018). Similarly, damming rivers or tributaries negatively affects the environmental flow, which in turn severely hinders the autostocking and breeding of fish and other aquatic organisms in the wetlands.

4.4.4.4 Protection of Ornamental Fish Species

Along with a high nutritive value, certain small indigenous fish (SIF) also have ornamental

(Mahapatra et al. 2004; Biswas et al. 2007, 2020) and larvicidal properties. Out of the 217 fish species recorded from Assam, 150 fish are stated to have ornamental values (Bhattacharyya et al. 2003). Further, Choudhury and Biswas (2005) reported 70 ornamental fish species from upper Assam having commercial value in overseas and domestic markets. Phukon and Biswas (2013) reported that native small fish species in a seasonal water body near Dibrugarh that consume mosquito larvae serve as both bio-control and bio-monitoring agents. With the rich piscine biodiversity, outstanding sub tropical climate, vast aquatic resources, and good transport facilities to the neighboring states for exportation, Assam can conquer a good position in the global fish market and can provide employment to thousands of people adding growth in the economy of the state in near future and these suggest a lot of prospects be considered regarding the small indigenous fish culture in Assam.

4.4.4.5 Role of Women in Wetland Fisheries

Women actively take part in community fishing in the *beels* of Assam. There is also potential to earn income through the preparation of feed or through integrating ornamental fish culture into existing farming practices. The transfer of technologies for a wide range of aquaculture techniques will be helpful in entrepreneurship development and women's empowerment. Formation of self-help groups and cooperative societies will open up new dimensions that will attract funding support from various government schemes. Successful implementation will enable women to increase household income and purchasing power, elevate their societal status, reduce poverty, and better utilize available resources.

4.4.4.6 Legislation Related to Wetland Conservation

There are adequate rules and regulations for the protection of the environment, habitat, and biodiversity in India. Among these, the Indian Fisheries Act (1897), Assam Fisheries Act (1953), Fish Farmers and Piscicultural Society Rules (1971), and Indian Biodiversity Act (2002) are notable. The River Board Act (1956) was enacted to advise the state governments in matters related to the regulation or development of an interstate river or river valley or any specified part thereof. Recently, a new draft bill, The River Basin Management Bill (2018), envisages among other things, the integration of (a) the hydrological and ecological features; (b) land use appropriate to the relevant area; (c) relationship between surface and groundwater; (d) inter-linkages between water quality and broader environmental approaches; and (e) a holistic view of the relationship of all such activities with one another and with the basin/sub-basin as a whole.

4.4.4.7 Impact of Climate Change

The impact of climate change on the environment and economic activity is reflected in changes in the frequency, duration, and intensity of extreme weather events (Banholzer et al. 2014). Climate change is one of the most important global issues in the twenty-first century. Variations in temperature and rainfall have a profound impact on both water availability and demand which serve as major indicators of climate change on water resources (Soundharajan et al. 2016). A systematic literature review reveals rising average global temperatures (1.5 °C) and altered precipitation patterns, increased flood and drought frequency, shifts in water quality and vegetation composition, decreased waterbird populations, and negative impacts on livelihoods, agriculture, and human health. Adaptive management strategies are necessary to mitigate these impacts, including integrated wetland management plans, climate-resilient agriculture practices, community-based conservation initiatives, and enhanced monitoring and research efforts (Macusi et al. 2015).

Human activities have been accelerated by significant population growth, migration, and increased socioeconomic activity (Mahmoud and Gan 2018). The Intergovernmental Panel on Climate Change (IPCC) in its fifth assessment (AR5) reported that global temperatures increased by 0.85 °C from 1880 to 2012 and the temperature would rise by 0.3–0.7 °C in 30 years. This is an alarming situation that will affect, among others, the freshwater biota, as many of the wetland habitats will be under severe threat. It is also important to understand how aquatic systems and wider landscapes will respond to climate change. As a variety of forces continue to change the water landscape in the Himalayan region, adaptation to the changing environment needs to be focused.

4.4.4.8 Culture-Based Fishery in Wetlands

It is widely reported that the fish resources from the rivers and *beels* have declined drastically over the years. The decline has greatly impacted the socioeconomic conditions of the fisherfolk, which thrive exclusively on capture fisheries. The rapidly developing road networks and various construction activities, especially the embankments blocking the path of feeder channels of these wetlands, have also prevented autostocking and flushing out of floating hydrophytes. These changing dynamics of the habitat and their significant alterations result in the shallow channels remaining dry during the non rainy months. Thus, it can be said that all these natural and anthropogenic factors have detrimentally impacted the production potential of the natural water bodies. Culture-based fishery in natural waterbodies has gained importance worldwide. Encouragement of such communities to adopt cage culture was and is still considered a viable alternative means of subsistence, and it has also been adopted abroad as a result of the relocation of rural, agricultural communities caused by reservoir impoundment (Abery et al. 2005). Ross et al. (2008) have established the conditions for introducing species into aquaculture in Latin America by improving the closed reproductive cycle, feeding, and growing techniques. Improved water quality, fish nutrition, antibacterial properties, and increased fish production are some of the key advantages technology brings to culture practices that are true in Assam conditions (Biswas 2014; Gogoi et al. 2015). Improved technology, such as automatic fish feeding devices, and captive breeding of culturable native fish species (Gogoi et al. 2015; Nayak et al. 2022; Dohutia et al. 2023) could increase the efficacy and profitability of aquaculture practices (El Shal et al. 2021).

Anthropogenic perturbations and physical disturbances from the adjacent areas and local community, like illegal fishing, deliberate use of pesticides, and run-offs from adjoining agricultural lands, have various negative impacts on the wetland bio resources, especially fishes (Nayak and Biswas 2020). The modifications and changes made by man on wetlands to cater to increasing agricultural, housing, and developmental needs were not seriously considered in the initial years. However, the effects of these losses and modifications are cumulative and subtle. Another paradoxical view is that "nature destroys nature more than humans." If we base our present study on this, this could probably be validated well in the Maguri-Motapung wetland complex. One particular effect of natural events that occur regularly in the Brahmaputra basin is the frequent change in river course, avulsion, and excessive erosion and aggradation (Boruah and Biswas 2007). Natural changes have a bearing on land use, including land cover, habitat alterations, and the pattern of livelihood of the dependent residents and local communities close to the wetland (Boruah and Biswas 2010).

4.4.4.9 Restoration Plan

The present level of productivity in all categories of seasonal and perennial wetlands is much below the potential of the resources. One of the ways widely suggested by the experts is to explore the possibilities of rearing commercially important carps in suitable low-cost cages in *beels* and to popularize culture-based fisheries among the fishermen. However, the entire *beel* should not be converted into an aquaculture firm; instead, a maximum of 10–15% of the periphery of the beel area may be utilized for culture-based fishery, leaving the rest of the beel area for the growth and propagation of wild fish species.

There are adequate rules and regulations for the protection of the environment, habitat, and biodiversity in India. Among these, the Indian Fisheries Act (1897), Assam Fisheries Act

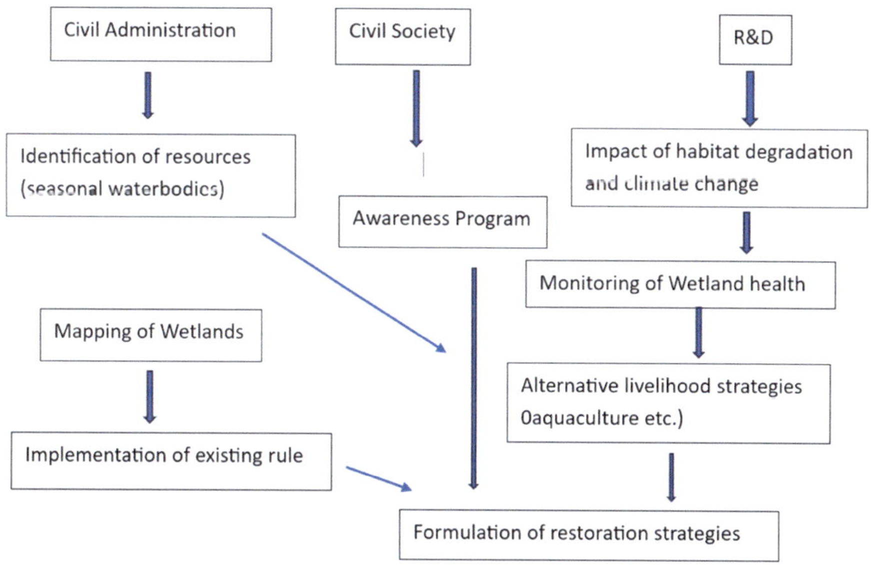

Fig. 4.3 Proposed restoration plan

(1951), Fish farmers and Piscicultural Society Rules (1971), and Indian Biodiversity Act (2002) are notable. The River Board Act (1956) was enacted to advise the state governments in matters related to the regulation or development of an interstate river or river valley or any specified part thereof. The River Basin Management Bill (2018) also emphasized on (a) integration of the hydrological and ecological features; (b) appropriate land-use pattern to the relevant area; and (c) a holistic view of the relationship of all such activities with one another and with the basin/sub-basin as a whole. Restoring the degraded wetlands and mitigating floodwater in urban areas requires strictly enforcing existing laws and the involvement of civil society in the restoration program (Fig. 4.3).

4.5 Conclusion

It is widely accepted that the fish resources from the rivers and *beels* have declined drastically over the years. The decline has greatly impacted the socioeconomic conditions of the fisherfolk, which thrive exclusively on capture fisheries. The construction activities, river embankments, improper waste disposal, etc., are some prominent threats. These factors contribute to the deterioration of the production potential of the *beels*. Habitat alteration poses a serious threat, as many connecting channels are drying up. These anthropogenic perturbations and other factors like illegal fishing, chemical accumulation due to agricultural wash-offs, and accumulation of untreated wastes are negatively impacting the fish resources of these wetlands (Nayak and Biswas 2020). The modifications and changes made by man on wetlands to cater to increasing agricultural, housing, and developmental needs were not seriously considered in the initial years. However, the effects of these losses and modifications are cumulative and subtle. One particular effect of natural events that occur regularly in the Brahmaputra basin is the frequent change in river course, avulsion, and excessive erosion and aggradation (Boruah and Biswas 2007).

Wetlands act as natural buffers against flood hazards. Natural changes have impacted land use, including land cover, habitat alterations (both faunal and human), and livelihood patterns of the residents (Boruah and Biswas 2010; Gogoi et al. 2015). These challenges and the growing pressure on the wetlands must be addressed immediately to achieve the long-term goal of conserving the wetlands and their associated biota of the Upper Brahmaputra basin. For this, a solution-oriented, holistic approach is a must, both at the level of policymakers and the stakeholders. Mass awareness and community engagement can play a pivotal role in dealing with the challenges. Proper management and disposal of waste should be considered as a priority to save these wetlands from further deterioration. Alternate sources of livelihood among the stakeholders should also be promoted so that these water bodies can boost urban resilience and improve civic life during the monsoon season. In concluding remarks this chapter, highlighted how wetland is important on biodiversity and human communities across the world.

Acknowledgments The authors acknowledge the help rendered by Ms. Swarupa Bhattacharjee, Dr. Shatabdi Deb, Dr. A. Santoshkumar, Dr. H. Saikia, Dr. Md. Hussain, Dr. Utpal Sarmah, Dr. Anamika Sarmah, and Dr. Seemashri Borah for their help in the preparation of this chapter.

Potential Conflicts of Interest The authors declare that they have no known competing financial interests or any other competing issues that could have appeared to influence the work reported in this chapter.

Informed Consent Informed consent was obtained from all individual participants included in the study.

Authors Contribution Both authors have contributed equally to this work.

Data Availability Statement The datasets during and/or analyzed during the current study may be made available from the corresponding author on reasonable request.

Funding The research was supported partially by ICAR-NATP and NFDB, Government of India.

References

Abery NW, Sukadi F, Budhiman AA, Kartamihardja ES, Koeshendrajana S, de Silva SS (2005) Fisheries and cage culture of three reservoirs in West Java, Indonesia: a case study of ambitious development and resulting interactions. Fish Manag Ecol. https://doi.org/10.1111/j.1365-2400.2005.00455.x

Abujam SS, Shah RK, Deori DJ, Biswas SP (2014) Diversity of macrophytes in Maijan Wetland, Upper Assam, India. South Asian J Exp Biol 4(4):164–171

APHA (2005). Standard methods for the examination of water and wastewater, 21st edn. APHA, AWWA, WPCF, Washington, DC

Bagenal TB, Tesch FW (1978) Methods of assessment of fish production in fresh waters, 3rd edn. IBP handbook no. 3. Blackwell Scientific Publications, Oxford/London, pp 101–136

Banholzer S, Kossin J, Donner S (2014) The impact of climate change on natural disasters. In: Singh A, Zommers Z (eds) Reducing disaster: early warning systems for climate change. Springer, Dordrecht, pp 21–49. https://doi.org/10.1007/978-94-017-8598-3_2

Basommi LP, Qing-Feng G, Dan-dan C, Singh SK (2016) Dynamics of land use change in a mining area: a case study of Nadowli District. J Mt Sci 13(4):633–642

Bhattacharjya BK, Choudhury M, Sugunan VV (2003) Ichthyofaunistic resources of Assam with a note on their sustainable utilization. In: Proceedings of the workshop on participatory approach for fish biodiversity conservation in Northeast India. National Bureau of Fish Genetics, Lucknow, pp 85–105

Biswal A, Paul T, Debroy S, Debnath S (2020) Indigenous ornamental fish of North-East India: present status, threats and conservation strategies. Biotica Res Today 2(6):445–447

Biswas SP (1993) Manuals and methods in fish biology. South Asian Publishers, Delhi, 157 pp

Biswas SP (1996) Global water scarcity: issues and implications with special reference to India. Verhandlungen Internationale Vereinigung für Theoretische und Angewandte Limnologie 26:115–121

Biswas SP (2014) Conservation plan for aquatic fauna in the Brahmaputra Basin. In: Sinha RK, Ahmed B (eds) Rivers for life—proceedings of the international symposium on river biodiversity: Ganges-Brahmaputra-Meghna river system, ecosystems for life. A Bangladesh-India initiative. IUCN, International Union for Conservation of Nature, pp 45–53

Biswas SP, Abujam SK (2022) Ecosystem services of riverine wetlands with special reference to the Upper Brahmaputra Basin. Indian J Agric Econ 77(3): 521–529

Biswas SP, Boruah S (2000) Fisheries ecology of the north-eastern Himalaya with special reference to the Brahmaputra River. Ecol Eng 16:39–50

Biswas SP, Michael RG (1992) Fishery characteristics and the present status of fisheries of the River Brahmaputra. In: Proceedings of the seminar on the conservation of river dolphin in the Indian Sub-continent, August 18–19, 1992, New Delhi

Biswas SP, Das JN, Sarkar UK, Lakra WS (2007) Ornamental fishes of North East India—an atlas. NBFGR (ICAR) Publication, Lucknow, 111 p

Biswas SP, Boruah S, Sharma A (2018) Environmental protection of the Brahmaputra river from environmental, ecological and legal perspectives. Soochow Law J Taipei 15(2):135–157

Borah S (2017) Studies on habitat ecology and fisheries of Moridikhow in Sivasagar District, Assam. Unpublished PhD thesis, Dibrugarh University

Boruah S, Biswas SP (2002) Ecohydrology and fisheries of the Upper Brahmaputra Basin. Environmentalist 22(2):119–131. https://doi.org/10.1023/A:1015369313873

Boruah S, Biswas SP (2007) Natural and anthropogenic hazards on the hydrobiology of the Brahmaputra river system. In: Goswami UC (ed) Natural and anthropogenic hazards on fish and fisheries. Narendra Publishing House, Delhi, pp 233–244

Boruah S, Biswas SP (2010) An ecohydrological approach to habitat restoration of the Brahmaputra river: bank stabilization by phytoremediation. In: 4th Aqua conference, 8–10 December 2010, India Habitat Centre, Delhi, pp 79–87

Boruah S, Ganguly P, Biswas SP, Sharma A (2008) Land use land cover changes: its impact on the wetland Ecosystem of Maguri beel, Assam, India. In: Choubey VK, Singh RD, Singh O, Sharma MK (eds) Conservation and restoration. National Institute of Hydrology, Roorkee, pp 605–616

C.A.M.P (1998) Report of the Workshop on "Conservation assessment and management plan for freshwater fishes of India" (Sanjay M, Sally W (eds)). Technical report. Zoo-Outreach Organization, Coimbatore

Choudhury M, Biswas SP (2005) Habitat degradation of flood plain lakes of Assam—a great threat to fishery industry in the state. Indian Farming 55(6):14–21

Dohutia S, Nayak N, Sonowal J, Chutia P, Biswas SP (2023) Investigation on the captive propagation of the endemic murrel *Channa stewartii* (Playfair, 1867). Egypt J Aquat Biol Fish 27(2):209–219

El Shal AM, El Sheikh FM, Elsbaay AM (2021) Design and fabrication of an automatic fish feeder prototype suits tilapia tanks. Fishes 6:74. https://doi.org/10.3390/fishes6040074

Gogoi B, Kachari A, Dutta R, Darshan A, Narayan D (2015) Fishery based livelihood approaches and management of fishery resources in Assam, India. Int J Fish Aquat Stud 2(4):327–330

Gupta Choudhury A (2011) Ichthyofaunal diversity of Noa Dihing drainage system in Namdapha National Park, Arunachal Pradesh. PhD thesis, Assam University, Silchar

Haque S, Biswas SP (2014) Length-weight relationship and condition factor of *Botia dario* (Hamilton-Buchanan) from Sivasagar district, Assam. Int J Fish Aquat Stud 2(1):244–247

Hussain M (2011) Studies on different limnological aspects of certain beels of Dhemaji, Assam and their bearing in fish production. Unpublished PhD thesis, Dibrugarh University

Le Cren ED (1951) The length-weight relationships and seasonal cycle in gonad weight and condition in perch (*Perca fluviatilis*). J Anim Ecol 20:201–219. https://doi.org/10.2307/1540

Macusi ED, Abreo NAS, Cuenca GC, Ranara CTB, Cardona LT, Andam MB, Guanzon GC, Katikiro RE, Ashoka Deepananda KHM (2015) The potential impacts of climate change on freshwater fish, fish culture and fishing communities. J Nat Stud 14(2): 14–31

Mahapatra BK, Vinod K, Mandal BK (2004) Ornamental fish of North Eastern India—its distribution and conservation status. J Environ Ecol 22(3):674–683

Mahmoud SH, Gan TY (2018) Impact of anthropogenic climate change and human activities on environment and ecosystem services in arid regions. Sci Toal Environ 633(15):1329–1344

Nayak N, Biswas SP (2020) Wetland shrinkage: a threat to the indigenous fish population of Assam. NeBIO 11(1):7–8

Nayak N, Biswas SP (2023) Impact of oil spillage on biodiversity in and around Maguri-Motapung Beel of Assam. Environ Pollut Clim Change 7(5):1–5. https://doi.org/10.4172/2573-458X.1000343

Nayak N, Dohutia S, Sonowal J, Dutta J, Biswas SP (2022) Captive breeding of the recently discovered snakehead species *Channa andrao* Britz, 2013 by habitat manipulation. Uttar Pradesh J Zool 43(24): 448–451. https://doi.org/10.56557/UPJOZ/2022/v43i243343

Parra H, Christopher KP, Miguel M, Santos M, Karen AB, Vandeperre F (2023) The Portuguese industrial pelagic longline fishery in the Northeast Atlantic: catch composition, spatio-temporal dynamics of fishing effort, and target species catch rates. Fish Res 264. https://doi.org/10.1016/j.fishres.2023.106730

Phukon HK, Biswas SP (2013) An investigation on larvicidal efficacy of some indigenous fish species of Assam, India. Adv Biores 4(3):22–25

Rijnsdorp AD, Peck MA, Engelhard GH, Möllmann C, Pinnegar JK (2009) Resolving the effect of climate change on fish populations. ICES J Mar Sci 66:1570–1583

Ross LG, Martinez Palacios CA, Morales EJ (2008) Developing native fish species for aquaculture: the interacting demands of biodiversity, sustainable aquaculture and livelihoods. Aquac Res 39:675–683. https://doi.org/10.1111/j.1365-2109.2008.01920.xr

Saikia H, Biswas SP (2023) An investigation on certain water quality of Borsola Beel (Wetland) from Upper Assam. Int J Ecol Environ Sci 49. ISSN: 2320-5199. https://doi.org/10.55863/ijees.2023.3082

Sarma U (2015) Investigation on benthic macrofauna of the Joysagar tank in relation to its bottom water quality. Unpublished PhD thesis, Dibrugarh University

Sarma A (2017) Studies on limnology with special reference to fish and fisheries of Gelabeel in Golaghat District, Assam. Unpublished PhD thesis, Dibrugarh University

Sarma U, Biswas SP (2012) Studies on bottom water quality and Macrobenthic Community as bioindicators in the Joysagar Tank, Sivasagar, Assam (India). J Frontline Res Arts Sci 2:93–100

Soundharajan BS, Adebayo JA, Remesan R (2016) Evaluating the variability in surface water reservoir planning characteristics during climate change impacts assessment. J Hydrol 538:625–639

Talwar PK, Jhingran AG (1991) Inland fisheries of India and adjacent countries (vol. 1 & 2). Oxford IBH Publication, New Delhi, 1158 pp

Trivedy RK, Goel PK, Trisal CL (1987) Practical methods in ecology and environmental science. Enviro Media Publication, Karad, 340 pp

Vivekananda GN, Swathi R, Sujith AV (2021) Multi-temporal image analysis for LULC classification and change detection. Eur J Remote Sens 54:189–199. https://doi.org/10.1080/22797254.2020.1771215

Wubie MA, Assen M, Nicolau MD (2016) Patterns, causes and consequences of land use/cover dynamics in the Gumara watershed of Lake Tana basin, Northwestern Ethiopia. Environ Syst Res. https://doi.org/10.1186/s40068-016-0058

5 Safeguarding Deepor Beel: Analyzing Legal Tools for Urban Resilience and Wetland Preservation

Preetisha Choudhury

Abstract

In today's developing India, rapid urbanization presents formidable obstacles to the preservation of wetlands, which are vital ecosystems that offer indispensable ecological services. Deepor Beel, a prominent wetland situated in the southwest of Guwahati city, Assam, India, is particularly vulnerable to the impacts of urbanization, environmental degradation, and climate change. This chapter studies the diverse range of laws, regulations, and policies governing land utilization, environmental preservation, and conservation endeavors with the aim of providing insights into fostering urban development practices that prioritize wetland conservation. Through a doctrinal legal methodology, including an analysis of international conventions, national legislations, and judicial decisions, this study elucidates the legal tools available for maintaining a delicate balance between urban development and ecological conservation in the context of Deepor Beel. Despite being an important Ramsar site, the Beel faces threats like encroachment, waste dumping, and other urban developments, all of which have tremendously diminished its ecological value in the present times. The Wetlands (Conservation and Management) Rules, 2017, and other legal safeguards remain largely ineffective in practice. Ultimately, the chapter identifies major regulatory and implementation gaps and also suggests innovative approaches such as zoning regulations, environmental impact assessments, and participatory governance models to strengthen the resilience of urban areas while protecting the ecological integrity of Deepor Beel.

Keywords

Wetlands · Deepor Beel · International · National frameworks · Conservation strategies

5.1 Introduction

Wetlands are ecologically distinct areas where land is saturated with water, either seasonally or permanently, forming a unique interface between terrestrial and aquatic ecosystems (Ramsar Convention 1971). They include inland marshes like "permanent and seasonal rivers, inland deltas and floodplains, permanent and seasonal lakes and ponds, marshes, freshwater swamps, and peatlands, as well as human-made wetlands such as reservoirs, barrages and dams, aquaculture ponds, excavations and burrow pits, wastewater treatment ponds, irrigation canals, ditches, irrigation ponds, and rice fields (EnviStats India 2022)." Globally, wetlands are critical for biodiversity conservation, water regulation, and

P. Choudhury (✉)
Department of Law, North-Eastern Hill University, Shillong, Meghalaya, India

S. K. Chakraborty et al. (eds.), *Wetlands and Urban Resilience*, Sustainable Development Goals Series,
https://doi.org/10.1007/978-3-032-06881-1_5

maintaining ecological balance. They are natural water purifiers, filtering pollutants and recharging groundwater. Their ability to store carbon makes them crucial for mitigating climate change. Wetlands also act as natural infrastructure, protecting communities from floods, droughts, and other disasters. They are vital for sustaining agriculture, fisheries, and livelihoods, contributing to poverty alleviation and food security. Furthermore, wetlands support global biodiversity, providing critical habitats for an extensive array of flora and fauna, many of which are imperiled (Saikia 2019). A number of Sustainable Development Goals (SDGs), such as clean water and sanitation (SDG 6), climate action (SDG 13), and life on land (SDG 15), are directly impacted by wetlands. Wetlands contribute to biodiversity, energy, transport, and tourism, addressing SDGs by supporting sustainable agriculture, water management, economic growth, disaster resilience, climate action, and biodiversity conservation. Despite their importance, an estimated 50% of wetlands worldwide have been lost, with Asia witnessing significant wetland degradation (Davidson 2014).

India, with its diverse geography and climate, is home to a wide range of wetlands that cover approximately 4.7% of the nation's land area, as detailed in the National Wetland Atlas (Ministry of Environment, Forest and Climate Change 2018). These wetlands offer numerous advantages to human culture and are among the planet's most productive ecosystems (Meena and Sharma 2019). Assam, located in Northeast India, is known for its wetlands, locally called *beels*. These beels are freshwater bodies of varying sizes, often formed by cut-off river channels or tectonic activity, connected to the Brahmaputra and Barak rivers. There are over 1392 officially designated floodplain wetlands in Assam. Of these, 423 are registered, meaning they have been officially recognized and supervised, whereas the other 969 wetlands are still unregistered (Chandralekha et al. 2019). These beels collectively cover more than one lakh hectares, making up 61% of the state's total lentic water bodies (Das et al. 2009). Among these, *Deepor Beel* stands as the most expansive and sole Ramsar site within the Brahmaputra Valley, having been recognized in 2002 for its significant ecological value. Deepor Beel, a former waterway of the majestic Brahmaputra River, stands as a remarkable wetland characterized by its central core and lush outlying zones. This beautiful ecosystem supports diverse wildlife, including endangered species like leopards, tigers, fishing cats, and Asiatic elephants, as well as numerous migratory birds that visit its shores. Among its avian diversity, 102 species have been recorded, including 55 resident, 27 locally migratory, and 20 migratory species. The wetland has a natural ability to purify itself. However, human activities are harming its ecosystem and affecting its environmental quality and ecosystem services.

The irony lies in the fact that, although recognized as an important Ramsar site, Deepor Beel continues to suffer from increasing man-made elements such as urbanization, pollution, and unchecked urban development. This ongoing degradation highlights a serious gap between the legal protections that exist on paper and their actual implementation on the ground. While several studies have explored the ecological and social aspects of the wetland, there is limited work that critically examines the legal and policy frameworks and the gaps that lie within them. This chapter seeks to highlight all those gaps by exploring significant legal tools meant to protect the wetland, identifying where they fall short, and at last suggesting some measures that might strengthen wetland governance in the face of urban challenges. The study follows a doctrinal legal approach, making an in-depth analysis of international conventions such as the Ramsar Convention of 1971 and national legislation such as the Environment (Protection) Act, 1986, Wetlands (Conservation and Management) Rules, 2017, and other key judicial decisions that have shaped wetland protection in India. By having key objectives such as (i) to assess the key legal frameworks that apply to Deepor Beel, (ii) to highlight major gaps and challenges in its effective implementation, and (iii) to suggest measures that could better support the wetland's ecology, this chapter becomes significant as it looks at Deepor Beel not just as an environmental issue, but as a part of a larger effort to make the city of Guwahati more sustainable, flood-resilient and environmentally friendly zone.

5.2 Deepor Beel as Guardian of Guwahati's Ecology

Deepor Beel, also spelled *Dipor Beel*, is an exquisite natural freshwater lake positioned to the southwest of the vibrant city of Guwahati in Assam's Kamrup district. The name "Deepor Beel" derives from the Sanskrit expression "*dipa*" (meaning elephant) and the Assamese word "*beel*" (referring to a wetland), reflecting its historical association with elephants. Once part of the Brahmaputra River's channel, it is officially recognized as spanning 6.89 sq. km as per the Guwahati Water Bodies (Preservation and Conservation) Act, 2008, although recent evaluations indicate that its actual area may range from 13 to 15 square kilometers (National Wetland Atlas: Assam 2010; National Wetland Atlas: North-Eastern States 2011; Government of Assam 2024). The beel serves as a vital reservoir for Guwahati and provides a locale for plentiful resident and migrant bird species, such as the Lesser and Greater Adjutant Stork (*Leptoptilos javanicus* and *dubius*), the globally threatened Spotbilled Pelican (*Pelecanus philippensis*), and Baer's Pochard (*Aythya baeri*) (Saikia 2005). The waters are abundant with 50 fish species, supporting the livelihoods of many nearby villages. Furthermore, *nymphaea nuts and flowers*, along with ornamental fish, seeds of the giant water lily *Euryale ferox*, and other medicinal plants, contribute significantly to local market incomes. The surrounding forest is also home to commercially valuable orchids, enhancing the region's natural wealth (Ramsar Sites Information Service 2022). Additionally, it provides habitat for wild Asian elephants from the nearby Rani and Garbhanga forests (Kalita and Das 2020). Furthermore, the beel's connectivity with the Brahmaputra River enhances its hydrological importance, while its biodiversity and ecosystem services make it invaluable for both nature and humanity. The beel also plays a central ecological part in upholding regional environmental balance by regulating floods naturally and absorbing excess water during monsoons, thereby preventing urban flooding in Guwahati. Its rich biodiversity supports numerous species, while its aquatic plants sequester carbon, aiding in climate change mitigation. The beel also sustains local livelihoods through its thriving fish population and ecosystem services (International Collective in Support of Fishworks 2014).

Deepor Beel also holds significant value across multiple dimensions. The environmental values include its role in biodiversity preservation and water purification. Further, the social values include its important role played in the life of local communities by providing resources like fish and serving as a cultural hub. Economic values are evident in the wetland's contributions to agriculture through nutrient cycling and its potential for ecotourism. Educational and research opportunities also make it the most significant site for scientific study and environmental awareness. Importantly, the beel serves as a natural buffer in times of environmental hazards. In this sense, the beel acts as a "green infrastructure," offering essential ecosystem services that strengthen the city's ability to adapt to climate-related vulnerabilities and ecological stress. However, despite of incredible worth, Deepor Beel in Assam faces threats from urbanization, encroachments, and pollution caused by industrial and domestic waste, which degrade its water quality and ecosystem health (Das 2021).

5.3 Urbanization and Other Emerging Threats to Deepor Beel

Despite being a distinguished Ramsar site and protected under national and state laws, Deepor Beel is threatened by multiple threats (Mozumder and Tripathi 2014). These challenges not only affect the ecology of the wetland but also expose the inadequacy of the legal enforcement and institutional management. Major threats include:

(i) *Encroachment and Urbanization:* Over the years, Deepor Beel has faced severe encroachment, reducing its original expanse from over 4000 hectares to a mere 500 hectares. Unauthorized constructions and land reclamation for agriculture and urban settlements have been significant contributors to this decline. A striking example is the encroachment by a school, which had erected approximately 20 structures and a wall within the wetland area. Although the Kamrup (Metro) district administration demolished these encroachments in 2015, such instances underscore the persistent threat of encroachment and highlight the urgent need for stronger protective measures (Choudhury 2024a, b).

(ii) *Garbage Dumping and Pollution:* The disposal of wastes by the Guwahati Municipal Corporation (hereafter, GMC) near the beel heavily pollutes the wetland and harms its biodiversity. The Pamohi River exacerbates the issue by carrying waste into the beel. Despite efforts to shift the dumping site, the pollution persists, endangering water quality, aquatic life, and nearby communities. Farmers, residents, and environmentalists have voiced concerns over the declining biodiversity and ecological health of the area (The Times of India 2024). Moreover, untreated effluents from Guwahati flow into Deepor Beel via the Bahini and Bharalu rivers, further compromising the ecosystem (Das 2022a, b).

(iii) *Climate Change and Biodiversity Loss:* Pollution, including untreated sewage and waste dumping, has enhanced the growth of invasive species like water hyacinths, which choke native aquatic plants, reduce oxygen levels, and devastate fish populations and other aquatic organisms (Sur and Mandal 2019). This degradation severely impacts the "bread and butter" of the nearby communities that rely on the wetland's resources. Furthermore, rising temperatures have further compromised the survival of aquatic species, leading to a notable 26% decline in fish varieties (Hazarika 2023).

(iv) *Pressure from Agricultural and Industrial Practices:* The runoff from adjacent agricultural fields introduces pesticides and fertilizers into the wetland, leading to water pollution and jeopardizing the delicate balance of the wetland's ecosystem (Ramsar Convention on Wetlands 2013). Moreover, the construction of nearby industries contributes to air and water pollution, further accelerating the environmental degradation of the beel.

(v) *Human-Wildlife Conflict:* Encroachment into the buffer zones of the wetland disrupts its vital function as a sanctuary for birds and other wildlife, thereby endangering its ecological balance. Additionally, the establishment of a wide-gauge railway line that cut through the Deepor Beel bird sanctuary presents grave dangers to the local fauna. Over the years, the line, which traverses close to the Garbhanga Hills and Rani Reserve Forest, has caused tragic incidents by upsetting the Asiatic elephants' habitat (2021).

These ongoing threats directly violate key environmental laws, which aim to protect the natural ecosystem of India. Additionally, the Guwahati Water Bodies (Preservation and Conservation) Act, 2008, which provides for the protection and conservation of waterbodies in the state of Assam, fails to truly protect the beel. The continued deterioration of the beel due to urban pressure highlights serious enforcement failures as well as weak institutional accountability. This raises an important question: Are existing legal frameworks and penalties imposed strong enough to create a deterrent effect and protect critical ecosystems such as the Deepor Beel? In contrast to other countries, where public awareness and environmental ethics remain ingrained within the people, India continues to remain a country where people are more worried about their rights, including the "*right to a clean environment,*" while neglecting their most important fundamental duty, "*to protect and preserve the environment.*" These pressing issues call for a deeper legal inquiry, which is explained in the next chapter through an analysis of the existing regulatory frameworks and their role in preserving Deepor Beel in India.

5.4 Legal Frameworks for Wetland Protection and Analysis

Wetlands stand as some of the most essential ecosystems on our planet, providing services such as water filtration, flood control, carbon sequestration, and biodiversity support. However, these ecosystems are under immense threat due to urbanization, pollution, and climate change. In response to these challenges, various legal frameworks have been established on both national and international fronts. The Ramsar Convention on Wetlands of International Importance Especially as Waterfowl Habitat (1971) is the first current international legal document that claims to specifically protect wetlands and establish national obligations on a global scale for the prudent use of wetlands. This landmark treaty, signed by 18 nations, holds countries accountable for preserving wetlands by prohibiting irresponsible exploitation and requiring national policies to recognize, guard, and preserve wetlands of both national and international significance (Ramsar Convention 1971). India, having ratified the Convention, embraces its obligations to protect and nurture its wetlands. As of August 2024, the nation boasts 85 Ramsar sites (Hindustan Times 2024), including Deepor Beel, and has implemented the Wetlands (Conservation and Management) Rules, 2017, to support compliance (Balwan and Kour 2021; Baruah and Goswami 2023). Further, supplementing the Ramsar Convention, 1971, the Convention on Biological Diversity (hereafter, CBD 1992) emphasizes the critical role of wetlands in its broader goal of conserving biodiversity, while the Sustainable Development Goals (SDGs 2015), particularly, SDG 15 (emphasizing Life on Land) talk about "protect, restore and promote sustainable use of terrestrial ecosystems, sustainably manage forests, combat desertification, and halt and reverse land degradation and halt biodiversity loss." The Paris Agreement of 2015, on the other hand, also implicitly supports the conservation of wetlands through its emphasis on natural carbon sinks and climate change adaptations (The World Economic Forum 2023).

However, although India has aligned her national laws with these global commitments, the continuing deterioration of the beel highlights the gap between India's ratification of these conventions and the real implementation. Despite being of international importance, Deepor Beel continues to suffer from numerous threats. One key challenge lies in the nature of the Ramsar Convention, as the treaty "is not a regulatory regime and has no punitive sanctions for violations of or defaulting upon treaty commitment" (The Ramsar Convention Manual, 6th edition, p. 14). As a result, the success of the Convention depends entirely on how seriously each nation enforces it at the national and local levels, something that remains significantly weak in the case of Deepor Beel.

5.4.1 Wetland Protection Laws in Neighboring Countries

Wetland protection laws in neighboring countries of India, such as Bangladesh, Pakistan, Sri Lanka, and Nepal, are critical for preserving these ecosystems, but each country faces unique challenges in their implementation. In Bangladesh, the current management of wetlands is primarily governed by the National Water Policy, 1999 (Byomkesh et al. 2009). The country also follows the guidelines set by the Ramsar Convention and also its own laws and policies, such as the National Environment Management Action Plan, 1995; National Environment Policy, 1992; and the Bangladesh Environment Conservation Act, 1995, which enforces protection to wetlands, though illegal land reclamation and pollution continue to pose significant threats. Further, Pakistan, a Ramsar Convention signatory, has established the Pakistan Wetland Programme

(PWP), which is a national initiative designed to support the conservation of wetlands across Pakistan. It is funded by multiple organizations, including the Global Environment Facility (GEF) and the United Nations Development Programme (UNDP), with additional contributions from the WWF Network, the Royal Netherlands Embassy (RNE), and the Pakistan Poverty Alleviation Fund (Rao 2009). Moreover, for the preservation and maintenance of wetlands in Pakistan and for environmental education, WWF Pakistan has established a wetland center at Sandspit, west of Karachi. WWF Pakistan wetland projects are aimed at conserving the wetlands of Pakistan, identifying the wetland complexes, enhancing biodiversity, creating awareness among the people, preparing, and carrying out national wetland conservation plans into action for wetlands' sustained growth (Khan and Arshad 2014).

Looking at Sri Lanka, wetlands stand as a cornerstone of the nation's exceptional ecological and biological richness, serving as essential sanctuaries for a diverse array of wildlife and plant life (IUCN Sri Lanka 2004). Sri Lanka made a significant advancement in wetland conservation with the establishment of a National Policy on Wetlands in 2004, following the National Wetlands Steering Committee in 2003 (IUCN Sri Lanka and the Central Environmental Authority 2006). Additionally, the framework for wetland protection is boosted by the National Environmental Act, 1980, the Flora and Fauna Protection Ordinance, 1937, and the Central Environmental Authority. These national regulations, complemented by international obligations such as the Ramsar Convention, are instrumental in safeguarding the wetlands of Sri Lanka (Kariyawasam 2024). In Nepal, the National Wetland Policy, 2059 (2003), is designed to engage local communities in the supervision of wetlands while promoting the prudent use of their resources in order to protect the rich biodiversity found within these ecosystems. Furthermore, the policy directives outlined in the Nepal Policy and Action Plan and the Nepal Biodiversity Strategy explicitly focus on wetland biodiversity and provide clear strategies for its management. Although there is no singular, comprehensive legislation governing wetlands in Nepal, the conservation of these vital areas is addressed through various existing laws (Poudel 2009).

While these countries have frameworks in place for wetland protection, effective implementation often faces obstacles such as limited enforcement, urban pressure, and a lack of knowledge among the public and policymakers. However, each country also leaves some lessons that can be learned by India. For instance, Bangladesh relies on community initiatives such as co-management plans. Further, the Bangladesh Haor and Wetland Development Board (BHWDB) is working on a Haor Master Plan to protect its wetlands (Thompson 2025). Again, Pakistan's Wetland Programme is funded by international organizations such as the UNDP or the WWF; however, the lack of a clear, centralized national policy and the interference of too many ministerial activities in India make it harder to protect wetlands efficiently. Similarly, Nepal's policy highlights the importance of community leadership and the sustainable use of wetland resources, which is lacking in India. Sri Lanka, on the other hand, has a dedicated National Policy on Wetlands, permitting larger institutional precision, whereas India's wetland policies suffer from weak institutional management and uneven jurisdiction, a big issue that affects enforcement in Deepor Beel too.

5.4.2 Constitutional Vision for Wetland Protection in India

In India, wetland protection is governed by a combination of national and state-level laws, regulations, and policies. The Constitution of India, while not specifically mentioning wetlands, provides a strong legal foundation for their protection through several provisions that support environmental conservation. Article 48A, inserted through the Constitution (42nd Amendment) Act of 1976, mandates "the state shall endeavour to protect and improve the environment and to safeguard the forests and wild life of the country." This is supported by Article 51-A (g), which establishes a fundamental duty for citizens to actively engage in the "protection and improvement of the natural environment including forests, lakes, rivers, and wild life, and to have compassion for living creatures." Furthermore, Article 253 authorizes the Parliament to frame legislation concerning international agreements, thereby facilitating the conservation of global environmental resources, including wetlands (Constitution of India 1950).

5.4.3 National Laws Protecting Wetlands in India

The Environment (Protection) Act, 1986, stands as a pivotal legislative measure dedicated to the preservation of India's natural surroundings, including wetlands. Its broad definition of "environment" encompasses air, water, land, and their interrelationships, thereby allowing it to serve as the legal framework for wetland protection [Section 2(a)]. The Act authorizes the Central Government under Section 3 to take necessary actions for environmental protection, pollution control, and conservation. This includes prohibiting certain industrial activities in environmentally sensitive areas [Section 3(2)(v)] and setting standards for environmental pollutants [Section 3(2)(iv)]. This Act has been foundational for the development of rules and regulations that protect wetlands and Ramsar sites of international significance. Additionally, the Wetlands (Conservation and Management) Rules, 2017, aim to regulate and protect wetlands in India by providing guidelines for their conservation, management, and preservation without restricting their wise use. These rules emphasize institutional conservation and the prevention of wetland degradation. Section 4 of these Rules places specific prohibitions on certain harmful activities within protected wetland sites. Section 4(iii) enforces a prohibition on the storage, disposal, manufacture, or handling of hazardous substances, adhering to the guidelines set forth by the Manufacture, Storage, and Import of Hazardous Chemicals Rules of 1989; Section 4(iv) explicitly forbids the disposal of municipal solid waste in proximity to wetlands. Furthermore, Section 4(v) establishes stringent regulations against the release of untreated waste and effluents from residential, urban, and industrial sources into wetland regions.

Further, the Municipal Waste (Management and Handling) Rules, 2000, outline comprehensive guidelines for the treatment and elimination of toxic municipal trash in India. These guidelines were the first significant legal framework to address municipal waste management, aiming to ensure environmentally sound practices while reducing public health risks. Furthermore, legislations such as the Indian Fisheries Act, 1897, which penalizes the poisoning of waters causing fish deaths; the Water (Prevention and Control of Pollution) Act, 1974, and related Rules, 1975, which address water pollution and restrict sewage or pollutant discharge into water bodies under Section 25; the Wildlife (Protection) Act, 1972; the Forest Conservation Act, 1980; the Air (Prevention and Control of Pollution) Act, 1981; and the Biological Diversity Act, 2002, all in whole have been playing a crucial part in protecting marine bodies, including wetlands, in the country.

However, wetland protection laws in India suffer from fragmented frameworks, limited coverage, and poor enforcement, leaving many wetlands unprotected. Ambiguities in definitions and weak integration with land use policies further hinder conservation. Exclusion of smaller and agricultural wetlands, lack of community involvement, and insufficient data exacerbate the challenges. These issues demand stronger laws, better enforcement, and coordinated efforts.

5.4.4 State Laws Protecting Deepor Beel in India

In Assam, Deepor Beel exemplifies nature's beauty, protected by the Guwahati Water Bodies (Preservation and Conservation) Act of 2008. This law aims to conserve and regulate water bodies under the Guwahati Metropolitan Development Authority, promoting ecotourism while preventing encroachment and degradation. As per Schedule IV of the Act, Deepor Beel has been declared as a "waterbody," and waste dumping in areas listed in this Schedule is strictly prohibited (Section 4), the penalty for which is imprisonment for a term that may extend up to 3 years or with a fine which may extend to 50,000 rupees or both (Section 6). Apart from this, the beel is also entitled to protection under the Wetlands (Conservation and Management) Rules, 2017, which in particular prohibits hazardous activities within protected wetlands *(applicable in case of Deepor beel)*. The guidelines for implementing Wetlands (Conservation and Management) Rules, 2017, state lucidly that certain activities such as solid waste dumping, encroachment of any kind, and discharge of untreated wastes are severely prohibited in a notified wetland. The guidelines further state that "any activity that falls under the list of prohibited actions in Rule 4(2) of the Wetlands (Conservation and Management) Rules, 2017, cannot be carried out in a notified wetland unless special permission is granted by the Ministry of Environment, Forest and Climate Change (MoEF&CC). And to get this permission, the State Government must submit a formal request to the Ministry, but only after receiving a recommendation from the State Wetland Authority." However, the harsh reality shows that Deepor Beel, being a designated Ramsar site, fails to enjoy this utmost legal protection.

In addition to the aforementioned, various Indian states have enacted specific laws addressing wetland issues. Notable examples include the East Kolkata Wetlands (Conservation and Management) Act, 2006; West Bengal Wetlands and Water Bodies Conservation Policy, 2012; Kerala Conservation of Paddy Land and Wetland Act, 2008; the Jammu and Kashmir Wildlife (Protection) (Amendment) Act, 2002; and the Andhra Pradesh Water, Land and Trees Act, 2002.

5.4.5 Critical Analysis of the Laws Governing Deepor Beel in India

According to the 2011 census, the Guwahati Metropolitan Area had a population of around 1 million, which resulted in the unregulated dumping of hazardous wastes near the beel. From 2006 to 2021, the GMC dumped "approximately 550 tons of solid waste daily, or about 62 truckloads," at a 24-hectare site just 500 meters east of the wetland. Following strong agitations and a National Green Tribunal order (2019) this dumping area was closed. However, the dumping continued, but this time, at an area named "*Belortol in Paschim Boragaon*," only a kilometer away from the beel. This relocation rather than removal reflects the ignorance of treating the beel as ecologically important. Further, as per the GMC estimates, around 1.5 million metric tons of old waste have accumulated at the original site. While nearly 600,000 metric tons have been processed, the remaining is expected to be settled by 2026. However, until then, the environmental threat remains. A study conducted by the Pollution Control Board, Assam (PCBA) from January to December 2023 revealed severe contamination of the beel's water and a significant loss in the aquatic biodiversity (The Times of India 2024). Another major challenge is the failure to demarcate the wetland's official boundaries, making it more vulnerable to urbanization. In March 2024, the Assam Cabinet decided to "denotify the beel as a wildlife sanctuary," a step that would do no good rather than further destroying its ecological value (Bharadwaj 2024).

Despite legal prohibitions and penalties, threats continue to affect the beel. A significant failure of the Guwahati Water Bodies (Preservation and Conservation) Act, 2008, is that it does not integrate the beel into the city's urban development plans, making it difficult to recognize Deepor Beel as part of the city's climate and effective flood resilience infrastructure. The role of the State Wetland Authority also remains limited due to institutional overlapping coming from various state bodies. Additionally, the absence of any organized structure at the grassroots level to address concerns about wetland management and protection is further weakening the enforcement and public accountability. Thus, reforms are required, but they should go beyond the law to include community involvement, transparency, and accountability.

5.5 Indian Judiciary and Its Role in Protecting Wetlands

The Indian judiciary has been a strong advocate for environmental protection, playing a key role in safeguarding the environment and ensuring sustainable growth. Courts in India have interpreted the right to a clean and healthy environment as a fundamental right enshrined in Article 21 of the Constitution, recognizing that a sound environment is integral to the right to life.

Over the years, the judiciary has used this interpretation to tackle a myriad of environmental challenges, such as contamination, deforestation, and the degradation of natural habitats like wetlands. One of the most significant contributions of the Indian Judiciary's commitment to environmental protection has been the application of the "Polluter Pays Principle," which embodies the elements of "responsibility" and "accountability." This principle holds polluters accountable for expenses associated with environmental remediation and pollution. In the landmark *Vellore Citizen's case*, the court construed the "Polluter Pay Principle" as an absolute responsibility, requiring polluters to compensate victims and restore degraded environments. Similarly, in *Indian Council for Enviro-Legal Action case*, the court established that individuals engaged in activities deemed inherently hazardous or perilous are liable for any resultant harm, regardless of the precautions taken. Similarly, in the case of *M.C Mehta v Kamal Nath & Ors* (1997), pollution was classified as a civil wrong and a tort against the public at large.

Furthermore, the National Green Tribunal (hereafter, NGT) has also actively addressed issues related to solid waste disposal, sewage management, sewage discharge, and the contamination of wetland water bodies with hazardous substances. In the case of *Subhas Datta v State of West Bengal & Others* (2024), the NGT invoked the East Kolkata Wetlands (Conservation and Management) Act, 2006, mandating the state government to cease all detrimental activities in proximity to the wetland. The tribunal explicitly banned pollution from plastic, leather, and rubber, as well as the operation of tannery waste processing facilities in the vicinity. Again, in another significant case, *Court on its own Motion v State of Karnataka* (2024), the tribunal recognized the issue of municipal solid waste pollution affecting the water bodies surrounding Bangalore. The tribunal focused on remedial actions for restoring Varthur, Agara, and Bellandur lakes while also addressing the issue of solid waste dumping and effluent discharge into these lakes. In Jammu and Kashmir, in the case of *Raja Muzaffar Bhat v Union Territory Of Jammu and Kashmir* (2023), the tribunal ordered that encroachment, sewage channeling, and the clearance of municipal solid trash in Wular Lake, Hokersar, and Kreento-Chandhara wetlands be avoided. Similarly, in Bombay, the High Court has intervened to put an end to the indiscriminate and widespread practice of municipal solid waste dumping on mangrove plots and wetland beds in the Mulund-Thane belt area (Thomas 2012).

Since 2007, several PILs have highlighted the dumping of municipal waste and encroachment into Deepor Beel, emphasizing the abrogation of environmental rights. In 2007, residents filed a PIL against the GMC for dumping contaminated trash in the area (Barman 2019). During a petition hearing submitted by RTI activist Rohit Choudhury, the NGT underscored that the earth belongs to all creatures and not humans alone (*Rohit Choudhury v Union of India* 2014). In 2018, the NGT ordered the removal of the municipal waste disposal site from Deepor Beel by 2019 (Bahn 2019). Despite delays, the tribunal persisted, directing the GMC to cease waste dumping at West Boragaon, near Deepor Beel. After prolonged efforts and multiple hearings, the corporation finally halted waste dumping in the area on June 28, 2021, providing renewed hope to activists and locals who had been fighting for this cause since 2014 (Das 2022a, b). Further, in 2023, in Pramod Kalita & 2 Ors v Union of India & Ors (2023), the Division Seat of the Gauhati High Court, comprising Chief Justice Sandeep Mehta and Justice Suman Shyam, addressed the significant environmental challenges faced by Deepor Beel and instructed the Assam Government to undertake initiatives aimed at safeguarding, enhancing the aesthetic appeal, and promoting the biodiversity enhancement of the beel (Aaranyak 2003; Gupta 2023).

Additionally, the Supreme Court has also consistently emphasized the strict enforcement of wetland conservation laws and the accountability of governments in ensuring their protection. In *M K Balakrishnan and Others v Union of India and Others* (2017), the Apex Court instructed both the central and state governments to notify wetlands in accordance with the Wetlands (Conservation and Management) Rules, 2010. The judgment emphasized the ecological importance of wetlands and mandated their protection through proper management and conservation policies. Thus, this underscores the judiciary's critical role in safeguarding wetlands through proactive interventions, which uphold ecological integrity while promoting sustainable development.

5.6 Conservation Strategies for Strengthening the Ecosystem of Deepor Beel

Deepor Beel, a vital wetland in Assam, is in jeopardy due to urbanization, pollution, encroachment, and invasive species that are affecting its rich biodiversity and crucial ecological and cultural significance. And so, to guarantee its long-term preservation, a thorough conservation approach is needed that involves the active role of local communities, environmental organizations, and government authorities. Some of the conservation strategies that can be advanced are:

(i) *Strict Enforcement of Wetland Protection Laws:* An important strategy is to strictly implement and enforce regulations such as the Environment Protection Act, 1986, the Wetlands (Conservation and Management) Rules, 2017, and the Guwahati Water Bodies (Preservation and Conservation) Act, 2008, for the purpose of combating illegal activities such as encroachment and pollution. Moreover, it is also necessary that the penalties for polluters are tightened and the role of the enforcement agencies, such as the State Wetland Authority is strengthened from time to time.

(ii) *Pollution Control and Waste Management:* It is essential to set up industrial wastewater management systems in order to stop the influx of untreated sewage and industrial runoff from entering the area (Agarwal et al. 2015; Singh et al. 2023). In this regard, establishing sewage treatment plants and encouraging sustainable agricultural practices to reduce chemical pollution have become necessary. Further, in light of the ongoing dumping near the beel, it has become of the utmost necessity to set up decentralized waste treatment units and transfer the hazardous wastes away from the wetland.

(iii) *Community Involvement and Awareness:* The conservation of the beel can be greatly aided by encouraging the local communities to take part in conservation efforts. This can be achieved through awareness campaigns that demonstrate the ecological importance of the wetland, as well as through the inclusion of local stakeholders in decision-making processes affecting the wetland (Conservation of Deepor Beel Wetland Through Community Participation & Human Resources Development 2023; Pollution Control Board, Assam 2023). Lessons can also be drawn from community-led initiatives undertaken in the Loktak Lake, Manipur, and East Kolkata Wetlands that have proved to be successful (Mehrotra et al. 2024).

(iv) *Management of Invasive Species:* Another strategy is addressing the increasing challenges posed by invasive species such as water hyacinths and the intrusion of non-native plants (Ramsar Convention on Wetlands 1999). Regular removal of invasive species by institutional bodies can help restore the wetland's natural ecosystem and maintain its biodiversity.

(v) *Restoration of Buffer Zones and Habitat Management:* Zoning regulations serve as a vital instrument in safeguarding Deepor Beel by clearly demarcating areas for conservation, limiting urban development, and controlling activities that could harm the wetland. These regulations can create buffer zones around the wetland to prevent encroachment and reduce pollution (Ramachandra 2001). They can also designate areas for ecotourism, promoting sustainable use of resources. By restricting industrial and agricultural expansion near the Beel, zoning can help maintain its ecological integrity.

(vi) *Monitoring and Research of Deepor Beel:* Regular monitoring of wetland health, water quality, and biodiversity is crucial for timely interventions. Scientific research and collaboration with institutions can improve monitoring (Krishna et al. 2023; Centre of Excellence for Sustainable Development 2023). Further, using AI technologies can enhance conservation by enabling real-time monitoring, detecting degradation, predicting threats, and optimizing resource management for effective decision-making (Alliance for Innovation and Infrastructure 2024). For instance, the successful use of drones in Chilika Lake to keep a watch on the poachers can offer a promising model that could be adopted for Deepor Beel to give early warning systems and enable immediate restoration.

(vii) *Strengthening Regional Cooperation and Funding:* Cooperation between local, state, and national governments, as well as NGOs, is necessary to ensure the effective implementation of conservation policies. Securing funding from both governmental and international sources (e.g., Ramsar and UNDP) will provide the resources needed for long-term management plans.

(viii) *Promoting Ecotourism:* Developing ecotourism initiatives, such as in Kaziranga National Park and Majuli Island, can create a sustainable revenue stream for conservation efforts while raising awareness about the wetland's ecological value. This can include guided bird-watching tours, boat rides, and educational programs that highlight Deepor Beel's biodiversity and ecosystem services (Dutta and Archana 2020; Deori and Das 2013).

(ix) *Climate Change Adaptation and Resilience Building:* Wetlands, such as Deepor Beel, serve an essential function in alleviating the effects of climate change, including rising water levels and shifting weather patterns. Conservation strategies must emphasize climate resilience to enhance their ecological functions. Again, as wetlands help secure water availability during droughts, effective flood management can protect surrounding areas from frequent floods, particularly those with lower return periods (Hoque and Sharma 2020; Ferreira et al. 2023).

The following *Threats and Response Strategies for Conserving Deepor Beel* highlight the interconnectedness of legal protection, community involvement, and restoration efforts in addressing the challenges faced by wetlands like Deepor Beel. By integrating these strategies, a holistic approach can be adopted that ensures the long-term protection and resilience of these vital ecosystems (Table 5.1).

And, together, these strategies highlight the fact that the restoration and protection of Deepor Beel depend not only on stronger laws and policies but also on locally rooted initiatives that involve communities and long-term ecological planning.

5.7 Conclusion

Deepor Beel, an internationally significant Ramsar site and a lifeline for Guwahati's atmosphere and the locals that survive on it, is one of Assam's most vital wetland ecosystems. However, despite its ecological and cultural value, the beel has suffered and still suffers from

Table 5.1 Threats and response strategies for conserving Deepor Beel

Threats	Legal protection	Community role	Restoration efforts
Pollution	Enforcing laws such as the Environment (Protection) Act, 1986, the Wetlands (Conservation and Management) Rules, 2017	Organizing local clean-up campaigns and monitoring dumping of waste materials	Setting up sewage treatment plants and reviving wetland ecosystem
Encroachment	Strengthening land use and zoning regulations to prevent illegal construction	Engaging local communities in boundary monitoring and awareness programs	Creation of buffer zones and implementation of smart and eco-friendly urban planning
Invasive species	Strengthening regulations on invasive species, through India's National Biodiversity Action Plan (Target 6)	Supporting local group to report and remove invasive plants	Restoration of native habitats and regular cleaning
Human-wildlife conflict	Strict implementation of the Wildlife Protection Act, 1972	Community guarding and creating awareness on harmonious coexistence	Reducing conflict zones and creating safe corridors
Wetland degradation	Recognizing Deepor Beel under national and international conservation frameworks	Promoting more such community-led-initiatives like "*Simang*," "*Kumbhi Kagaz*", where the locals are taking actions to revive Deepor Beel	Reforestation, wetland restoration, and implementing eco-sensitive plans

Source: Author's own compilation

modern-day stresses. While laws exist for its protection, their implementation on the ground remains poor and inconsistent. And so, for safeguarding Deepor Beel, a multifaceted approach is needed that combines legal tools with effective community involvement and environmental planning. Strengthening the present legislation with stricter penalties for encroachments and pollution, alongside enhanced monitoring, will improve compliance. Furthermore, wetlands require topmost priority in urban planning since they are the natural assets that can help control floods, purify the air, and offer a beautiful home for other creatures to live in. And so, measures such as the creation of eco-sensitive zones, buffer areas, and sewage treatment plants and using drones and satellites can significantly enhance wetland protection.

As we look to the future, we need to question ourselves whether the available laws, policies, and city planning are adequate to answer the growing challenges of urbanization and climate change. Along with the introduction of stricter penalties, a lot else can be done, like using nature-based solutions like wetland restoration and sustainable stormwater management in urban resilience strategies, and most importantly, encouraging and listening to the local communities who have never left any stone unturned in protecting nature. Community involvement through awareness campaigns and participatory conservation programs is thus required, as it will foster local ownership and responsibility among the people who have always cared for and lived close to the Deepor Beel.

Lastly, this chapter concludes by emphasizing that the future of Deepor Beel depends on the choices we make as responsible citizens of India. Its fate lies in our hands, where we need to act urgently and with long-term goals in mind and treat Deepor Beel not as a "*dumping ground or leftover space*" but as a vital part of Guwahati's ecological balance and urban resilience.

While this chapter offers a doctrinal perspective, an empirical study on the legal effectiveness and community perceptions surrounding Deepor Beel can further enrich future work.

Potential Conflicts of Interest The author declares no conflict of interest.

Informed Consent Not applicable.

Author's Contribution The author has sole contribution.

Data Availability Statement Not applicable.

AI Disclosure The author is solely responsible for the research, legal analysis, and original writing presented in this chapter. However, AI-assisted tools were used only for language refinement and formatting improvements. Rather than that, the content, interpretations, and conclusions are the author's own work.

Funding No funding received for this study.

References

Aaranyak (2003) Conservation of Deepor Beel wetland through community participation and human resources development. Final technical report, Project No. SWP 2020910AS36, Aaranyak, Guwahati, India

Agarwal R, Chaudhary M, Singh J (2015) Waste management initiatives in India for human well being. Eur Sci J 11(10):105–127

Alliance for Innovation and Infrastructure (2024) Advancing conservation: harnessing AI to preserve vital wetlands. https://www.aii.org/advancing-conservation-harnessing-ai-to-preserve-vitalwetlands/. Accessed 2 Dec 2024

Bahn D (2019) NGT directs Assam government to declare 'Deepor Beel' as an eco-sensitive zone. The New Indian Express. https://www.newindianexpress.com/nation/2019/Aug/22/ngt-directs-assam-government-to-declare-deepor-beel-as-an-eco-sensitive-zone-2022734.html. Accessed 5 Dec 2024

Balwan WK, Kour S (2021) Wetland-an ecological boon for the environment. East Afr Scholars J Agric Life Sci 4(3):38–48. https://doi.org/10.13140/RG.2.2.15728.79368

Barman U (2019) A socio-legal study on impact of solid waste on wetland environment: a case study from 'Deepor Beel' wetland, Assam. Nat Law School J. https://repository.nls.ac.in/nlsj/vol15/iss1/7

Baruah D, Goswami K (2023) Efficacy of legal institutions in protecting Assam's Ramsar site: assessing waste pollution in Deepor Beel through the legal lens. NLUA Law Policy Rev 7:55–89. https://nluassam.ac.in/docs/Journals/NLUALPR/Volume-7/Article-3.pdf

Bharadwaj S (2024) As Assam's only wetland protected under global treaty deteriorates, local fishing community faces loss of livelihood. Article-14. https://article-14.com/post/as-assam-s-only-wetland-protected-under-global-treaty-deteriorates-localfishing-community-faces-loss-of-livelihood-66bc2476cfc17. Accessed 29 Sept 2025

Byomkesh T, Nakagoshi N, Md Shahedur R (2009) State and management of wetlands in Bangladesh. Landsc Ecol Eng 5:81–90. https://doi.org/10.1007/s11355-008-0052-5

Chandralekha D, Bhabesh N, Das B (2019) A study on plankton diversity along with conservation status of Jhagra Beel, Dhubri, Assam, India. J Adv Zool 40(1): 118–122. https://doi.org/10.53555/jaz.v40i01.4136

Choudhury T (2024a) Deepor Beel under threat due to encroachment. Guwahati Plus. https://www.guwahatiplus.com/exclusive-news/deepor-beel-under-threat-due-toencroachment-urban-expansion. Accessed 3 Dec 2024

Choudhury T (2024b) Dumping ground pollution threatens Assam's Deepor Beel, local residents demand action. Guwahati Plus. https://www.guwahatiplus.com/guwahati/dumping-ground-pollution-threatens-assams-deepor-beel-local-residents-demand-action. Accessed 16 Mar 2025

CII-ITC Centre of Excellence for Sustainable Development (2023) Wetland monitoring framework a toolkit for corporates. https://sustainabledevelopment.in/wp-content/uploads/2023/08/Wetlands-Monitoring-Framework.pdf. Accessed 2 Dec 2024

Conservation of Deepor Beel Wetland Through Community Participation & Human Resources Development (2023) Final technical report, Project No: SWP 2020 910AS36. https://www.ndsu.edu/pubweb/~bezbarua/em/Documents/ANCDeepor%20Final%20report-iucn-1.pdf. Accessed 2 Dec 2024

Convention on Biological Diversity (adopted 5 Jun 1992 de 29 Dec 1993) 1760 UNTS 79 (CBD 1992)

Court on its own Motion v. State of Karnataka, Original Application No. 125/2017 (I.A. No.105/2024 AND I.A. No.106/2024)

Das S (2021) Changing trends in urbanization of Deepor Beel wetland. Int J Innov Sci Res Technol 6(11): 664–669

Das B (2022a) NGT orders GMC to speed up scientific solid waste disposal management in Guwahati. The News Mill. https://thenewsmill.com/2022/05/ngt-orders-gmc-to-speed-up-scientific-solid-waste-disposal-management-in-guwahati/. Accessed 5 Dec 2024

Das B (2022b) Untreated legacy waste is polluting the sensitive wetland ecosystem of Deepor Beel. https://india.mongabay.com/2022/08/untreated-legacy-waste-is-polluting-the-sensitive-wetland-ecosystem-of-deepor-beel/. Accessed 30 Nov 2024

Das P, Behera BK, Ngachan SV, Kosygin L (2009) Beels of Assam: an overview. In: Ecology, aquatic bio-resources and conservation of wetlands of North east India. Akansha Publishing House, New Delhi, pp 29–40

Davidson NC (2014) How much wetland has the world lost? Long-term and recent trends in global wetland area. Mar Freshw Res 65(10):934–941. https://doi.org/10.1071/MF14173

Deori S, Das N (2013) Ecotourism in wetland environment: a case study of Deepor Beel Wildlife Sanctuary (Ramsar Site) of Assam (India). Revista Brasileira de Ecotourism 6(2). https://doi.org/10.34024/rbecotur.2013.v6.6313

Dutta J, Archana S (2020) Valuing fishing activity of the Deepor Beel. Space Cult India 7(4):122–132. https://doi.org/10.20896/saci.v7i4.607

Ferreira CS, Kašanin-Grubin M, Solomun MK, Sushkova S, Minkina T, Zhao W, Kalantari Z (2023) Wetlands as nature-based solutions for water management in different environments. Curr Opin Environ Sci Health 33:100476. https://doi.org/10.1016/j.coesh.2023.100476

Government of Assam, Assam Science Technology and Environment Council, Development and Management of Deepor Beel. https://astec.assam.gov.in/portlet-sub-innerpage/development-and-management-of-deepor-beel. Accessed 3 Dec 2024

Gupta V (2023) High Court suggests remedial measures to conserve Deepor Beel Lake, a Ramsar Site. Latest laws. https://www.latestlaws.com/latest-news/high-court-suggests-remedial-measures-to-conserve-deepor-beel-lake-a-ramsar-site-203993. Accessed 5 Dec 2024

Guwahati Smart City Limited, Deepor Beel (Lake/Wetland). https://gscl.assam.gov.in/portlet-innerpage/deeporbeel-lake-wetland. Accessed 29 Nov 2024

Hazarika R (2023) Locational suitability of Paschim Boragaon landfill adjacent to Deepor Beel Wetland, Assam. Curr World Environ 18(3):1022–1031

Hoque NA, Sharma A (2020) Sustainable management and farmer's willingness to participate: a case study of Deepor Beel. Ecol Environ Conserv 26(4):1806–1815

HT Correspondent (2024) India's Ramsar wetland tally at 85 with 3 new sites. Hindustan Times. https://www.hindustantimes.com/india-news/three-new-ramsar-wetlands-added-101723658977711.html. Accessed 30 Nov 2024

Indian Council for Enviro-Legal Action v Union of India & Ors, AIR 1996 SC 1446

International Collective in Support of Fishworks (2014) Assam: how fishermen of Deepor Beel are fighting to save the wetland and their livelihood. https://icsf.net/newss/assam-how-fishermen-of-deepor-beel-are-fighting-to-save-the-wetland-and-their-livelihood/. Accessed 3 Dec 2024

IUCN Sri Lanka (2004) Wetland conservation in Sri Lanka. In: Proceedings of the national symposium on wetland conservation and management: Sri Lanka. iii +75pp. https://portals.iucn.org/library/efiles/documents/2004-026.pdf. Accessed 15 Dec 2024

IUCN Sri Lanka and the Central Environmental Authority (2006) National wetland directory of Sri Lanka, Colombo, Sri Lanka. http://www.cea.lk/web/images/pdf/7-1.Book-National-Wetland-Directory-Low%20res%281%29.pdf. Accessed 15 Dec 2024

Kalita A, Das AN (2020) Status of elephant corridors in Garbhanga-Rani reserve forest in an around Deepor beel. Int J Emerg Technol Innov Res 7(3):329–337. https://doi.org/10.13140/RG.2.2.31848.01281

Kariyawasam S (2024) Sustainable practices and legal pathways to Bellanwila-Attadiaya Wetland conservation project report. https://doi.org/10.13140/RG.2.2.25374.75841

Khan AA, Arshad SANA (2014) Wetlands of Pakistan: distribution, degradation and management. Pak Geog Rev 69(1):28–45

Krishna S, Chakraborty S, Chhavi Tayal GIZ (eds) (2023) Cultural significance of Indian wetlands. Ministry of Environment, Forest and Climate Change of India & Deutsche Gesellschaft für Internationale Zusammenarbeit (GIZ) GmbH

M K Balakrishnan and Others v Union of India and Others, (2017) 7 SCC 808

M.C Mehta v Kamal Nath & Ors, (1997) 1SCC 388

Meena AK, Sharma C (2019) Wetlands in India: ecosystem benefits, threats and management strategies. In: Jayakumar R (ed) Research trends in multidisciplinary research, vol 10. AkiNik Publications, pp 115–130

Mehrotra S, Singha L B, Garg, YK, Borikar P (2024) Enhancing Climate Resilience through Community-led Localised Strategies for Loktak Wetland. Urban India, 44(I.Sup): 28–43.

Ministry of Environment, Forest and Climate Change (2018) World Wetlands Day–2018. Government of India. https://moef.gov.in/uploads/2019/08/Final-Report-WWD-2018.pdf. Accessed 14 Dec 2024

Ministry of Statistics and Programme Implementation (2022) Wetlands. Government of India. https://mospi.gov.in/sites/default/files/reports_and_publication/statistical_publication/EnviStats/Chap4Wetlands_envst22.pdf. Accessed 3 Dec 2024

Mozumder C, Tripathi NK (2014) Geospatial scenario based modelling of urban and agricultural intrusions in Ramsar wetland Deepor Beel in Northeast India using a multi-layer perceptron neural network. Int J Appl Earth Obs Geoinf 32:92–104. https://doi.org/10.1016/j.jag.2014.03.002

National Wetland Atlas: Assam, SAC/RESA/AFEG/NWIA/ATLAS/18/2010. Space Applications Centre (ISRO), Ahmedabad, India, 174p

National Wetland Atlas: North-Eastern States, SAC/EPSA/ABHG/NWIA/ATLAS/35/2011. Space Applications Centre (ISRO), Ahmedabad, India, 318p

Pollution Control Board, Assam (2023) Action plan for Deepor Beel, priority III. https://www.pcbassam.org/RRC%20Action%20Plan%20Final/Action%20Plan%20P34/Action_Plan_Deepor_Beel.pdf. Accessed 2 Dec 2024

Poudel BS (2009) Wetland conservation in Nepal: policies, practices, problems and possibilities. Banko Janakari:5–9. https://doi.org/10.3126/banko.v19i3.2205v

Pramod Kalita & 2 Ors v Union of India & Ors, PIL/18/2023

Raja Muzaffar Bhat v Union Territory Of Jammu and Kashmir, Original Application No. 351/2023, (I.A. No. 519/2023)

Ramachandra TV (2001) Restoration and management strategies of wetlands in developing countries. Electron Green J 1(15) https://doi.org/10.5070/G311510435

Ramsar Convention on Wetlands (1999) Ramsar COP7-DOC: COP7 DOC. 24—report of the Secretary General on the implementation of the Convention. https://www.ramsar.org/sites/default/files/documents/cop7docs/NON%20RESRECS%20FINAL/COP7%2024E.pdf. Accessed 2 Dec 2024

Ramsar Convention on Wetlands (2013) Wetlands and agriculture: Impacts of farming practices and pathways to sustainability (Briefing note no. 13). https://www.ramsar.org/sites/default/files/documents/library/bn13_agriculture_e.pdf. Accessed 3 Dec 2024

Ramsar Sites Information Service (2022) Deepor Beel. https://rsis.ramsar.org/ris/1207. Accessed 29 Nov 2024

Rao LA (2009) Protection and management of Pakistan wetlands. United Nations Development Programme (UNDP). https://www.undp.org/sites/g/files/zskgke326/files/migration/pk/Protection-andManagement-of-Pakistan-Wetlands.pdf. Accessed 15 Dec 2024

Rohit Choudhury v Union of India through the Secretary, Original Application No. 19/2014/EZ with I.A. No. 71/2020/EZ

Saikia PK (2005) Qualitative and quantitative study of lower and higher organisms and their functional role in the Deepor Beel ecosystem. Project report submitted to North Eastern Space Applications Centre (NESAC), Department of Space, Government of India, Umium, Meghalaya, Shillong, 97 pp

Saikia JL (2019) Deepor Beel wetland: threats to ecosystem services, their importance to dependent communities and possible management measures. Nat Resour Conserv 7(2):9–24. https://doi.org/10.13189/nrc.2019.070201

Singh G (2021) How Assam's Deepor Beel is perishing from concretisation, waste dumping. Down To Earth. https://www.downtoearth.org.in/water/how-assam-s-deepor-beel-is-perishing-from-concretisation-waste-dumping-78870. Accessed 3 Dec 2024

Singh BJ, Chakraborty A, Sehgal R (2023) A systematic review of industrial wastewater management: evaluating challenges and enablers. J Environ Manag 348:119230. https://doi.org/10.1016/j.jenvman.2023.119230

Space Applications Centre (NESAC). Department of Space, Government of India. Umium, Meghalaya, Shillong (97 pp). https://rsis.ramsar.org/ris/1207. Accessed 29 Nov 2024

Subhas Datta v State of West Bengal & Others, Original Application No. 85/2024/EZ

Sur S, Mandal J (2019) Deepor Beel: a weeping and blemishing Ramsar site. Jalaplavit 9(3):48–58

The Constitution of India, 1950

The Guwahati Waterbodies (Preservation and Conservation) Act, 2008

The Municipal Solid Wastes (Management and Handling) Rules, 2000

The Ramsar Convention on Wetlands of International Importance especially as Waterfowl Habitat, 1971 (2 Feb 1971 de 21 Dec 1975) 996 UNTS 245 (the Ramsar Convention 1971)

The Times of India (2024) Smoke, pollutants from dumping ground pose threat to Deepor Beel. https://timesofindia.indiatimes.com/city/guwahati/smoke-and-pollutants-from-dumping-ground-pose-threat-to-deepor-beel/articleshow/107899252.cms. Accessed 3 Dec 2024

The Wetlands (Conservation and Management) Rules, 2017

The World Economic Forum (2023) Why wetlands are key to mitigating impact of climate change. https://www.weforum.org/stories/2023/12/wetlands-carbon-sink-climate-change-mitigation/. Accessed 1 Dec 2024

Thomas S (2012) Halt dumping of waste on Thane mangrove plot: HC. The Times of India. https://timesofindia.indiatimes.com/city/mumbai/halt-dumping-of-waste-on-thane-mangrove-plot-hc/articleshow/11382244.cms. Accessed 4 Dec 2024

Thompson PM (2025) Bangladesh Haor and Wetland Development Board (BHWDB) is working on a Haor Master Plan. Convention on Biological Diversity. https://www.cbd.int/financial/values/bangladesh-hailhaorwetlands.pdf. Accessed 15 April 2025

United Nations, Goal 15: Life on land. https://sdgs.un.org/goals/goal15. Accessed 1 Dec 2024

Vellore Citizens Welfare Forum v Union of India & Ors, AIR 1996 SC 2715

Part III

The Resilient City: Exploring Urban Wetlands and Climate Resilience

6 Urban Wetlands as Green Infrastructure for Sustainable Cities

Afreen Fatima and Ajay Khare

Abstract

Urban wetlands serve as critical green infrastructure, providing essential services such as water filtration, flood control, and groundwater recharge. This chapter explores the interconnected roles of wetlands and aquifers in sustaining urban ecosystems and advancing sustainable development. Through case studies from India, particularly Telangana, it examines the challenges, opportunities, and innovative approaches for integrating these systems into urban planning. Emphasizing decentralized, landscape-based solutions, the discussion highlights how working with water's natural behavior—slowing, sinking, and sharing—can restore ecological function and social trust. The socioeconomic and cultural significance of traditional systems is underscored, especially in the semiarid state of Telangana. The chapter delves into the science of aquifers in riverbeds and floodplains, exploring their role in water harvesting and recharge. It also addresses watershed governance, the reconnection of surface and groundwater, and projected 2050 water stress, proposing strategies rooted in local geology, ecology, and cultural values.

A. Fatima (✉) · A. Khare
Department of Conservation, School of Planning and Architecture, Bhopal, India
e-mail: ajaykhare@spabhopal.ac.in

Keywords

Urban wetlands · Green infrastructure · Aquifers · Sustainable cities · Water management · India

6.1 Introduction

Urban wetlands and aquifers have long supported the lifelines of cities, offering vital services such as flood control, water purification, biodiversity support, and groundwater recharge. As cities expand and climates shift, these systems—often overlooked or fragmented—emerge as essential foundations for urban resilience. They bridge the past and future, the natural and built, the visible and invisible. Yet many urban plans treat them as afterthoughts rather than as living infrastructures rooted in place, memory, and ecology.

Across India and elsewhere, there remains a gap in how wetlands and aquifers are understood—not just as hydrological systems, but as cultural and social frameworks. Conventional water policies often rely on centralized infrastructures—pipelines, treatment plants, and diversion systems—that are energy-intensive and ecologically disruptive. These approaches tend to serve dominant urban populations, leaving behind the knowledge and needs of rural, peri-urban, and indigenous communities. The deeper interdependence between land, water, and people often goes unrecognized. In contrast,

S. K. Chakraborty et al. (eds.), *Wetlands and Urban Resilience*, Sustainable Development Goals Series,
https://doi.org/10.1007/978-3-032-06881-1_6

traditional water systems—such as tanks, stepwells, and restored wetlands—reflect generations of environmental knowledge. They were designed to respond to seasonal rhythms, support biodiversity, and foster shared responsibility. Many of these systems have endured, even if partially, as reminders that water can be conserved not just through engineering, but through cultural practices, spatial design, and relational care. Urban resilience today requires a shift—not just in tools, but in perspective. New models increasingly emphasize slowing water rather than draining it—through bioswales, rain gardens, recharge parks, and restored aquifers. This idea of *slow water*—allowing water to linger, percolate, and renew—offers a path toward restoring balance. Cities like Hyderabad, for example, have begun integrating such decentralized interventions alongside historical conservation programs.

Kumar et al. (2021) report that restored wetlands in Hyderabad have reduced nitrate levels by 35% and improved biodiversity by 25%, while Singh et al. (2020) note that aquifers historically buffered communities during droughts. These findings reflect a need to recon nect hydrological function with community values and place-based strategies. This chapter builds on these insights by exploring how traditional and modern approaches can converge in India's semiarid regions, particularly Telangana. While 87% of the world's wetlands have been lost, Telangana's remaining systems—if restored—can become models for future cities. Using two case studies, the Qutb Shahi Tombs Baoli and the Kapra Wetland, this study examines how ecological performance, cultural memory, and policy frameworks interact to shape urban water resilience. The novelty of this chapter lies in its layered approach—linking ecological restoration with cultural revitalization and participatory governance. It aims to show that urban water systems are not only technical assets, but shared narratives—embedded in rituals, landscapes, and everyday life. By reconnecting with these narratives, cities can learn to live within their ecological limits. The chapter argues that resilient urban futures depend not on domination of water, but on coexisting with its flows—making it visible again, and restoring the balance between design, ecology, and community.

6.2 Rationale of the Study

Research Hypothesis

Urban wetlands and aquifers, when effectively integrated into urban planning and conservation strategies, significantly enhance urban resilience by improving water quality, mitigating flood risks, and supporting ecological balance.

6.2.1 The Urgency for Urban Resilience

Rapid urbanization and climate change have intensified vulnerabilities in urban water systems, necessitating innovative, nature-based solutions to enhance resilience and water security. Urban wetlands and aquifers, as integral components of green infrastructure, provide a sustainable alternative to engineered water systems. Their ability to absorb stormwater, recharge groundwater, and regulate local climates positions them as critical assets in mitigating urban environmental challenges.

The 2030 Water Resources Group (2021) projects a 40% global gap between water demand and supply by 2030 unless transformative water management practices are adopted. The restoration of lakes in Bengaluru, for instance, has demonstrated a 30% reduction in peak stormwater runoff during heavy rainfall events (Nagendra 2018), showcasing the efficacy of wetland conservation in urban resilience. Furthermore, these systems alleviate the financial burden on urban infrastructure, reducing long-term costs associated with water treatment and flood management.

The strategic value of wetlands and aquifers lies in their dual role of addressing immediate urban needs and fostering long-term sustainability. By adopting these systems, cities can mitigate climate risks while improving urban livability and ecological health.

6.2.2 Addressing Knowledge Gaps

Despite the well-documented ecological benefits of wetlands and aquifers, their sociocultural and governance dimensions remain underexplored, particularly in regions like Telangana. Traditional water systems, such as stepwells and tanks, have historically played a dual role as water storage solutions and community hubs for cultural and religious activities. However, urbanization has disrupted these systems, leading to the erosion of their traditional knowledge and utility.

UNESCO (2016) highlights the cultural significance of Telangana's stepwells, emphasizing their role in water management and community engagement. Similarly, Rao and Singh (2019) note that the neglect of these heritage systems has diminished their functionality, underscoring the need for revitalization. By integrating traditional practices with modern technologies, cities can bridge ecological and cultural imperatives, fostering a holistic approach to sustainability.

Addressing these knowledge gaps requires interdisciplinary research and collaborative governance frameworks. Such efforts will ensure that wetlands and aquifers are recognized not only as ecological assets but also as cultural and socioeconomic resources in urban planning.

6.3 Materials and Methods

The methodology is structured into two stages: the first aims to understand the role of urban wetlands and aquifers in enhancing urban resilience; the second identifies their linkages with the goals and targets of the sustainable development goals (SDGs). This study employs a multidisciplinary approach, combining existing research on the urban wetlands and aquifers of Hyderabad (Sharma et al. 2021; Snyder 2019), a detailed literature review, field surveys, stakeholder interviews, and data analysis.

The first stage began with data collection and documentation of the built and natural components of the study area. Primary surveys were conducted at aquifers and stepwells, including the Qutb Shahi Tombs Baoli, along with interviews involving community members and conservationists to gather sociocultural and ecological insights. Secondary data, such as historical records, government reports, and satellite imagery, were analyzed to support the field data. Site measurements and GIS mapping were used to integrate ecological and hydrological findings. The mythological and cultural associations of stepwells, particularly in Hyderabad, were explored through religious texts and oral traditions. The geographical and hydrological features of wetlands and aquifers were examined using resources from the Central Ground Water Board (CGWB), the Ministry of Environment, Forest and Climate Change (MoEF&CC), and digital tools like GIS and Bhuvan. Secondary data were obtained from key resources such as the Telangana Ground Water Department (TGWD), which provides detailed reports on groundwater assessments and quality. The "Ground Water Year Book (2020–2021)" published by the Central Ground Water Board (CGWB) offers insights into groundwater levels and trends, forming a critical basis for hydrological analysis. The Hyderabad Metropolitan Development Authority (HMDA) was consulted for urban development dynamics, including master plans and development regulations, while the Hyderabad Metropolitan Water Supply & Sewerage Board (HMWSSB) contributed reports on water distribution and infrastructure projects. Any data gaps identified during secondary research were resolved through additional site visits and community interactions. The framework for analyzing these systems emphasized three main aspects: ecology, sociocultural practices, and hydrology.

In the second stage, an extensive literature review was conducted to explore the connections between urban wetlands, aquifers, and SDGs. Findings from the first stage were synthesized with relevant literature to map synergies between these traditional water systems and SDG goals.

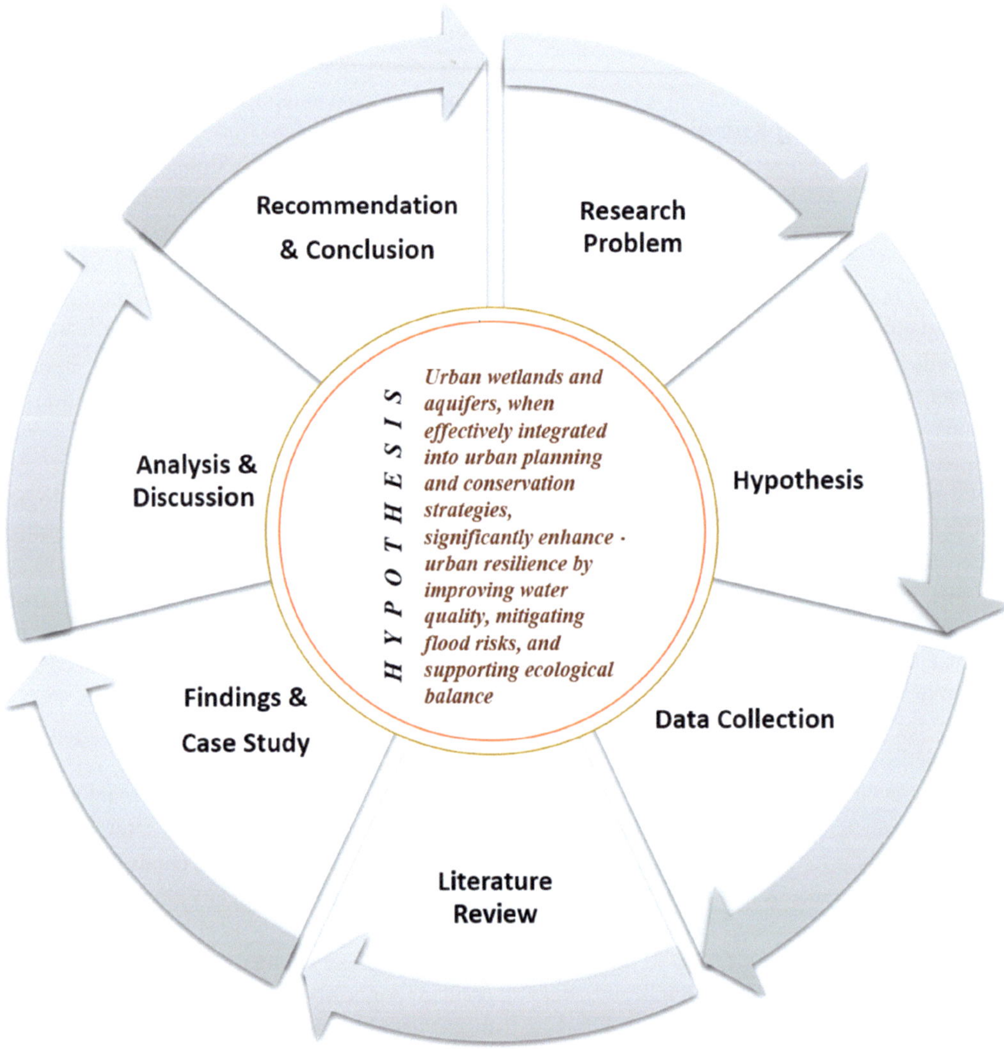

Fig. 6.1 Methodological framework for urban wetland and aquifer analysis. (Source: Author-generated schematic (Afreen Fatima 2025))

Key SDGs identified include SDG 6: Clean Water and Sanitation, SDG 11: Sustainable Cities and Communities, and SDG 15: Life on Land. By examining these linkages, the study demonstrates the role of urban wetlands and aquifers in achieving sustainable urban development. (See Fig. 6.1 for methodological framework and Fig. 6.2 for visual flowchart.)

This methodology provides a robust framework for assessing the socioeconomic, ecological, and cultural significance of urban wetlands and aquifers, emphasizing their potential to enhance urban resilience and sustainability.

6.4 Literature Review and Findings

A Qualitative Assessment Leading to Verification of Quantitative Inferences (Case Studies) Suggested as per the Research Hypothesis

Methodological insight: This chapter discusses the findings from a qualitative assessment of urban wetlands and their roles in urban resilience, aligning these findings with quantitative inferences derived from the research hypothesis. The qualitative review is anchored in a

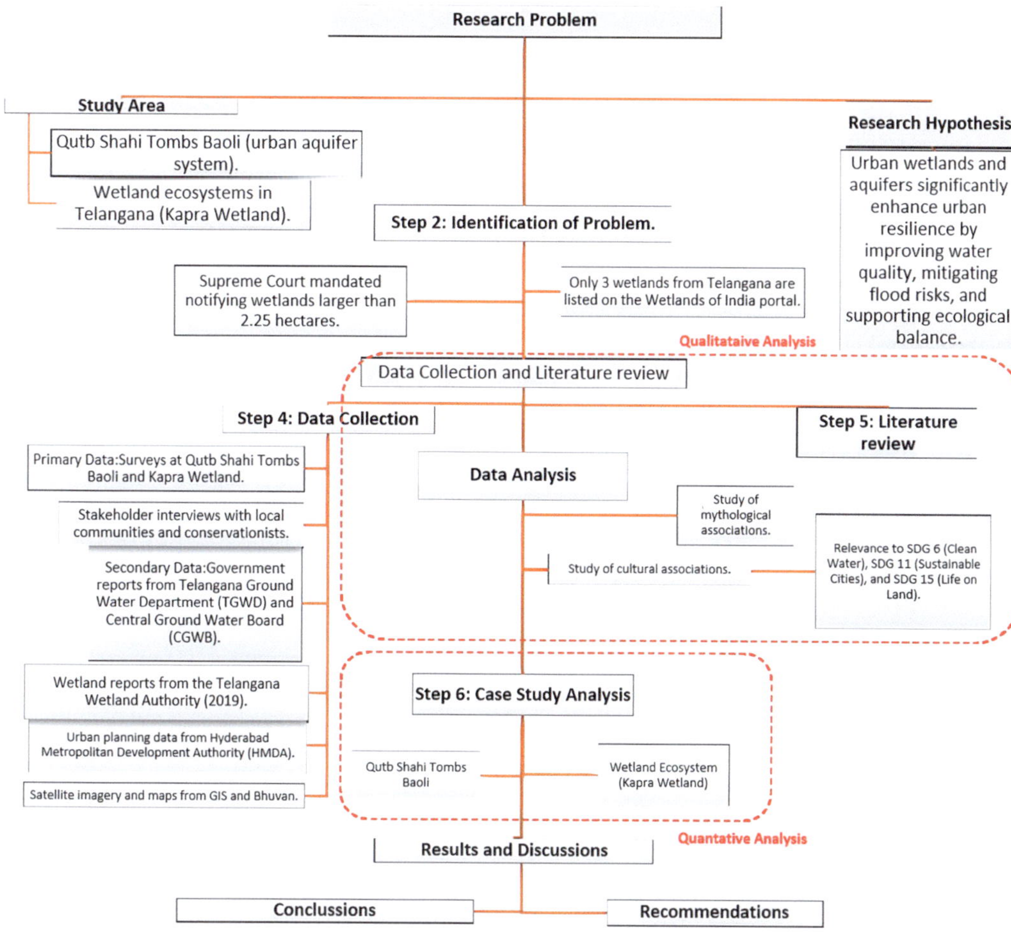

Fig. 6.2 Visual flowchart of research process. (Source: Author-generated graphic (Afreen Fatima 2025))

comprehensive analysis of theoretical frameworks, cultural narratives, and historical contexts. By examining these dimensions, the chapter establishes a foundation for validating quantitative data. Proposed case studies are identified and evaluated for their alignment with the hypothesis, emphasizing their empirical contributions to supporting the qualitative insights.

Findings The sustainability of wetlands and aquifers in Telangana is deeply rooted in their ecological, cultural, and socioeconomic functions, which collectively shape urban resilience. Their degradation due to urban sprawl, governance lapses, and insufficient community engagement underscores the urgent need for integrated conservation strategies.

Ecological Analysis

1. Wetland Biodiversity:
   Telangana's wetlands, such as Kapra Lake, support diverse flora like Neem (*Azadirachta indica*) and Banyan (*Ficus benghalensis*) and fauna like migratory birds. These species play crucial roles in maintaining ecological stability by enhancing soil quality and controlling pests (Kumar and Singh 2019). However, encroachments and pollution, including algal blooms and sedimentation, have significantly reduced wetland habitat quality (Telangana Wetland Authority 2024).
2. Aquifer Recharge and Flood Retention:
   Seasonal variations highlight the importance of wetlands like Kapra as natural aquifer recharge zones, especially in a semiarid

Table 6.1 Ecological insights on Telangana's wetlands and aquifers

Aspect	Key observations	Implications	References
Biodiversity	Loss of species diversity due to urban encroachments and pollution	Reduced ecological stability and degradation of wetland habitats	Telangana Wetland Authority (2024) and Kumar and Singh (2019)
Aquifer recharge	Seasonal recharge essential in Kapra Lake for flood mitigation	Critical for maintaining groundwater sustainability in semiarid zones	Rao and Singh (2019) and Telangana Wetland Authority (2024)
Pollution	Algal blooms and sedimentation due to solid waste inflow	Decline in water quality, affecting biodiversity and public health	Government of Telangana (2014)

Sources Compiled From: Government of Telangana (2014), Telangana Wetland Authority (2024), Kumar and Singh (2019) and Rao and Singh (2019)

Table 6.2 Cultural and historical dimensions of Telangana's water systems

Cultural element	Key insights	Implications	References
Stepwells	Qutb Shahi Tombs Baoli is a symbol of cultural and hydraulic ingenuity	Highlights the role of traditional systems in sustainable water use	UNESCO (2016) and Nagendra (2018)
Community practices	Rituals tied to water systems fostered collective responsibility	Can be revitalized to integrate cultural narratives into conservation	Ministry of Rural Development (2016)

Sources Compiled From: UNESCO (2016), Nagendra (2018) and Ministry of Rural Development (2016)

climate. These systems mitigate urban flooding by absorbing excess rainwater, as demonstrated in hydrological models by Rao and Singh (2019). However, urbanization has reduced their storage capacity and functionality (Table 6.1).

The ecological importance of Telangana's wetlands extends beyond biodiversity conservation. They act as natural sponges, regulating water cycles and mitigating urban heat island effects. Encroachments and declining recharge capacities highlight the necessity of integrating advanced hydrological mapping and natural restoration methods to reverse degradation trends.

Cultural and Historical Analysis

1. Stepwells as Cultural Heritage:
   Structures like the Qutb Shahi Tombs Baoli exemplify traditional hydraulic engineering, deeply intertwined with Telangana's cultural identity. These systems not only conserved water but also served as social gathering spaces and religious landmarks, emphasizing water's sanctity in local traditions (UNESCO 2016).
2. Community Practices:
   Historical records and interviews with heritage experts reveal that rituals associated with water systems were integral to fostering a collective responsibility for conservation (Nagendra 2018). Mission Kakatiya's success in restoring traditional tanks demonstrates the potential of blending cultural narratives with modern conservation techniques (Ministry of Rural Development 2016) (Table 6.2).

Stepwells and tanks symbolize a confluence of engineering, spirituality, and community-driven stewardship. However, modern urbanization has marginalized their cultural relevance. Reviving these water systems through community-led initiatives and educational programs could re-establish their role as symbols of sustainability and cultural pride.

Socioeconomic Analysis

1. Tourism and Economic Potential:
   Wetland and stepwell sites, such as the Qutb Shahi Tombs Baoli, have untapped potential for ecotourism, which could create employment and raise awareness about water

Table 6.3 Socioeconomic contributions of wetlands in Telangana

Aspect	Key findings	Implications	References
Tourism potential	Sites like the Qutb Shahi Tombs Baoli remain untapped for ecotourism	Revenue generation and increased public awareness on conservation	GHMC (2023)
Community engagement	Successful participatory efforts under Mission Kakatiya	Demonstrates the value of localized, community-driven interventions	Government of Telangana (2014)

Sources Compiled From: GHMC (2023) and Government of Telangana (2014)

Table 6.4 Governance and policy challenges for Telangana's wetlands

Governance aspect	Observations	Recommendations	References
Policy gaps	Weak enforcement of wetland protection laws	Develop stricter legal frameworks and ensure their implementation	Telangana Wetland Authority (2024)
Monitoring systems	Limited use of tools like GIS mapping	Integrate GIS and real-time tracking for better resource management	Rao and Singh (2019) and GHMC (2023)

Sources Compiled From: Telangana Wetland Authority (2024), Rao and Singh (2019) and GHMC (2023)

conservation. Current estimates suggest that a well-structured tourism model could boost local economies by 20–30% (GHMC 2023).

2. Community Engagement:
   Residents report diminished access to clean water due to pollution and declining wetland health. However, community-led projects like Mission Kakatiya show that participatory models can improve wetland functionality and socioeconomic benefits (Government of Telangana 2014) (Table 6.3).

The socioeconomic dimension of wetland management lies in its ability to create inclusive economic opportunities while addressing water scarcity. Ecotourism not only generates revenue but also fosters a sense of ownership among local communities. Strengthening partnerships with stakeholders could enhance the sustainability of these projects.

Governance and Policy Analysis

1. Policy Gaps:
   Reports from Telangana Wetland Authority reveal weak enforcement of wetland protection laws and the absence of monitoring systems for illegal encroachments (Telangana Wetland Authority 2024). The HYDRAA guidelines recommend integrating ecological and hydrological data into urban planning, but highlight limited implementation.
2. Monitoring Mechanisms:
   Advanced tools like GIS mapping have been proposed to track changes in wetland boundaries and aquifer recharge zones. However, their adoption remains inconsistent, hindering effective governance (Rao and Singh 2019) (Table 6.4).

Governance challenges highlight the need for a robust legal framework that enforces existing policies while promoting technological innovations. Incorporating community participation into governance models could enhance transparency and accountability, ensuring long-term conservation.

Eco-Cultural Linkages

The synergy between wetlands and stepwells exemplifies how ecological functions are deeply embedded in cultural practices. For instance, rituals at the Qutb Shahi Tombs Baoli reflect a historical understanding of water conservation. Reviving such practices can foster ecological awareness (Table 6.5).

Socioeconomic and Governance Integration

Economic opportunities like ecotourism hinge on strong governance and policy enforcement. Similarly, successful community engagement, as seen in Mission Kakatiya, requires institutional support to scale impact. Integrated governance can

Table 6.5 Integrated themes in wetland and aquifer management

Theme	Key observations	Implications	References
Eco cultural	Cultural and Ritual practices tied to water systems strengthen conservation	Offers a holistic approach to wetland sustainability	UNESCO (2016) and Kumar and Singh (2019)
Economic governance	Tourism potential depends on governance quality	Governance improvements can drive both ecological and economic gains	GHMC (2023) and Telangana Wetland Authority (2024)

Sources Compiled From: UNESCO (2016), Kumar and Singh (2019), GHMC (2023) and Telangana Wetland Authority (2024)

Table 6.6 Policy recommendations for wetland sustainability in Telangana

Recommendation	Proposed actions	Expected outcome	References
Ecological restoration	Implement nature-based solutions like floating treatment wetlands	Enhance biodiversity and improve water quality	Floating Treatment Wetlands Fact Sheet (2021)
Cultural integration	Develop educational programs on traditional water systems	Increase public awareness and cultural appreciation	UNESCO (2016)
Economic development	Launch ecotourism initiatives around heritage sites	Generate revenue while promoting conservation awareness	GHMC (2023)
Governance improvement	Enforce wetland protection laws and integrate GIS tools	Strengthen policy enforcement and monitoring systems	Rao and Singh (2019) and Telangana Wetland Authority (2024)
Community engagement	Expand participatory models like Mission Kakatiya	Build local capacity and ensure stakeholder involvement	Government of Telangana (2014)

Sources Compiled From: Floating Treatment Wetlands Fact Sheet (2021), UNESCO (2016), GHMC (2023), Rao and Singh (2019), Telangana Wetland Authority (2024) and Government of Telangana (2014)

ensure that socioeconomic benefits align with ecological goals.

Policy Recommendations

- Ecological Restoration:
  Implement large-scale wetland restoration projects using nature-based solutions, such as Floating Treatment Wetlands, to enhance biodiversity and water quality (Floating Treatment Wetlands Fact Sheet 2021).
- Cultural Revitalization:
  Develop programs that integrate traditional water systems into educational curricula, emphasizing their historical and cultural significance (UNESCO 2016).
- Economic Models:
  Launch ecotourism initiatives that leverage heritage sites like stepwells, generating revenue for local communities while promoting conservation (GHMC 2023).
- Governance Strengthening:
  Use GIS mapping and real-time monitoring systems to track encroachments and ensure compliance with wetland protection laws (Rao and Singh 2019).
- Community Engagement:
  Expand participatory models under Mission Kakatiya to include workshops, campaigns, and public consultations for broader impact (Government of Telangana 2014) (Table 6.6).

Wetlands and aquifers in Telangana represent dynamic systems that intertwine ecological, cultural, and socioeconomic dimensions. Addressing their degradation requires a holistic approach, integrating traditional wisdom with modern tools and participatory governance. By implementing these strategies, Telangana can position its water systems as global exemplars of sustainable urban resilience.

6.5 Results and Discussion

A Quantitative Assessment Proving Qualitative Inferences Based on the Research Hypothesis

This chapter analyzes quantitative data to provide measurable validation for the qualitative inferences established earlier. The analysis aims to bridge the gap between theoretical concepts and practical applications. Key findings are interpreted within the framework of the research hypothesis, demonstrating how empirical data corroborate and extend the qualitative insights. By integrating these approaches, the chapter emphasizes the synergy between qualitative and quantitative methodologies in achieving a holistic understanding of urban wetlands' ecological, cultural, and hydrological significance.

Case Study I: Telangana and Hyderabad City Mapped TWS

Understanding Hyderabad's water systems requires both a macro and a micro lens. At the macro level, the city-wide mapping of 74 stepwells reflects how hydrological strategies were historically decentralized, forming a dispersed yet resilient water infrastructure. At the micro level, sites like the Qutb Shahi Tombs complex offer rich, localized insights into how each baoli served spatially, culturally, and ecologically within its precinct. Together, they exemplify a scalable model of slow water infrastructure, blending urban planning, ritual, and recharge mechanisms. This section begins by examining the broader urban network before zooming in on the Qutb Shahi site as a focused case study.

Urban water resilience is shaped by regional climate, cultural memory, and topographic adaptation. Stepwells in Hyderabad are an example of local yet ingenious traditional water systems (TWS), sculpted to endure the semiarid ecology of the Deccan. Each geography lends a unique response to water, and in Hyderabad, the subterranean stepwell served as both storage and a cultural anchor. The mapping of 74 stepwells across the city was enabled through the 2023 Hyderabad Metropolitan Development Authority (HMDA) documentation and a funded heritage project led by Ar. Yeshwant Ramamurthy (Studio One, Hyderabad)—brought together diverse datasets to surface a forgotten yet crucial urban water legacy.

The mapping of 74 stepwells across Hyderabad provides critical evidence of a historical system that was designed to respect local geology, ecology, and culture—and to function in harmony with the natural behavior of water. (Refer to Fig. 6.3 for the spatial distribution of stepwells across Hyderabad.)

These stepwells, or baolis, were strategically located to collect seasonal rainfall and promote infiltration into the fractured rock aquifers of the semiarid Deccan landscape. By storing water at multiple levels and facilitating percolation, they enabled year-round groundwater recharge. Their clustering around topographic depressions, temples, and community nodes reflects an embedded hydro-ecological intelligence. These systems worked in tandem with urban wetlands to create a multilayered, decentralized network of "slow water."

Unlike large centralized infrastructure, these structures embody the principles of decentralization, distributed design, and place-based stewardship. In a region characterized by short but intense monsoons, high evaporation, and hydrogeologic variability, such site-specific strategies are more effective than uniform, engineered solutions. The legacy of these systems affirms that resilience is not only a technical outcome, but a cultural practice. As Erica Gies observes, "slowing water" is essential to restoring its power to recharge, renew, and connect—both hydrologically and socially (Gies 2022).

Mapping these structures today reveals a spatial archive of wisdom that prioritized:

- Respect for terrain and water flows
- Reconnection of surface and groundwater
- Scattered, scalable interventions
- Community participation in healing both land and relationships

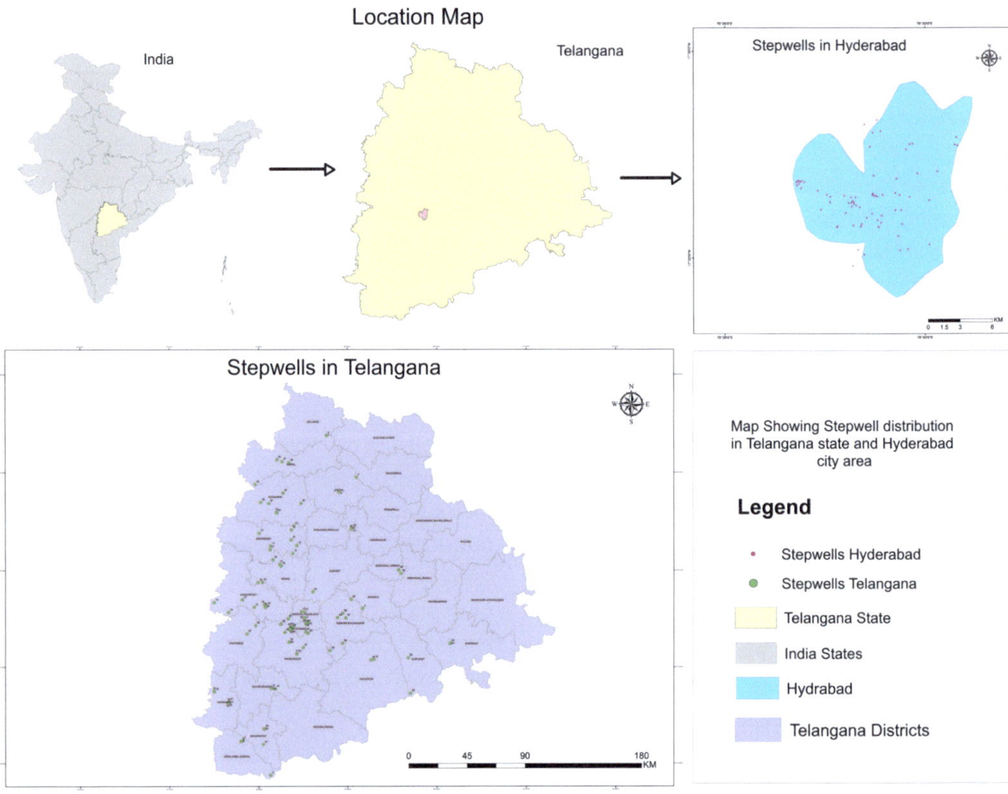

Fig. 6.3 Location map and distribution of stepwells in Telangana and Hyderabad. This map illustrates the spatial distribution of stepwells across Telangana and Hyderabad. Green dots represent stepwells in rural districts, while pink dots represent those within the Hyderabad city area. The sequence—from India to Telangana to Hyderabad—highlights the dense clustering of traditional water systems in the capital region. (Source: Author-generated map using GIS datasets, 2024. Author-generated GIS map using TGWD & HMDA datasets (2024))

Rapid real estate growth and unchecked urban sprawl across Hyderabad now threaten the ecological balance these stepwells once sustained (GHMC 2023; Telangana Wetland Authority 2024). Without active conservation, these subterranean systems risk erasure—yet they remain vital to climate adaptation strategies. Conservation demands collaborative frameworks that merge historical insight with contemporary water planning (Nanda 2023).

Reviving these systems through GIS documentation and integrated conservation policies is not only a heritage mandate but a practical response to urban water stress. These structures are not static remnants—they are active models for decentralized, equitable, and resilient water management in the twenty-first century (Fig. 6.4).

Case Study II: Kapra Wetland: A Struggling Urban Ecosystem

Kapra Wetland, located in Hyderabad, Telangana, spans approximately 10 hectares and has become a critical yet fragile urban ecosystem. Graded as "D" by Wetlands International, its water quality has severely declined due to pollutants and invasive species like water hyacinth, with oxygen levels at 40–60% of the required threshold, endangering aquatic life. Encroachments have claimed over 30% of its original area, compromising its capacity for aquifer recharge and flood mitigation during monsoons. Despite these challenges, the wetland remains a habitat for over 25 migratory bird species, showcasing its ecological significance. Additionally, unchecked wastewater inflow and

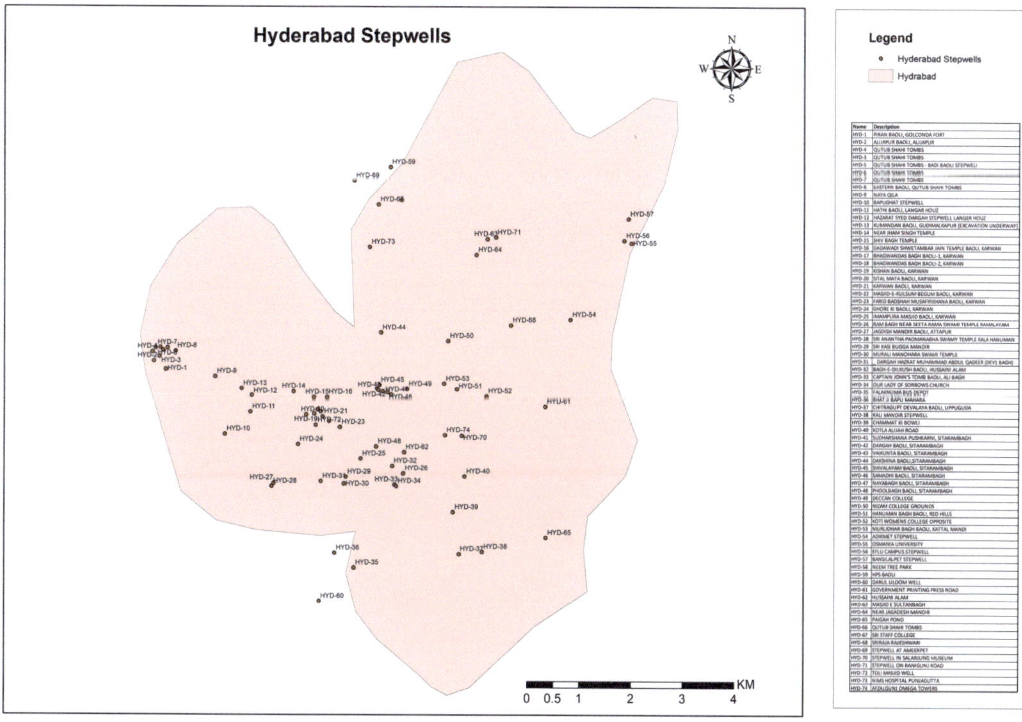

Fig. 6.4 Inventory of 74 Stepwells in Hyderabad. An urban-scale GIS visualization showing 74 historic stepwells, based on HMDA documentation. Each dot corresponds to a baoli indexed by location, function, and conservation status. (Source: Author-generated map using field data and HMDA documentation 2023)

weak enforcement of protective policies exacerbate its degradation. However, with strategic conservation efforts, including advanced monitoring tools like GIS mapping, ecotourism initiatives, and community-driven governance models, Kapra Wetland has the potential to evolve into a model for urban resilience and sustainability in Telangana (Table 6.7).

Kapra Wetland, a significant urban water body in Hyderabad, faces severe degradation due to pollution, encroachment, and a lack of systematic conservation measures. Despite these challenges, the wetland holds ecological importance as a habitat for biodiversity and a flood mitigation zone.

1. *Ecological Challenges:* The wetland, spanning 10 hectares, is burdened by pollution, invasive species, and encroachments, resulting in a drastic reduction in water quality and habitat health (Selvaraj et al. 2022).
2. *Biodiversity Support:* It remains an ecological hotspot, supporting over 25 migratory bird species and contributing to flood control during monsoon seasons (GHMC 2023).
3. *Encroachments and Urban Pressures:* The lake's original area of 113 acres has shrunk to 30 acres due to illegal constructions and blocked feeder channels (Kapra Municipality 2016).
4. *Community Efforts:* Social activists and locals have conducted weekly clean-up drives to revive the lake, underscoring the critical role of grassroots participation (The Hans India 2023).

The wetland's degradation highlights the pressing need for targeted conservation strategies. Comparisons with Ramsar-recognized wetlands demonstrate the potential of implementing advanced monitoring tools like GIS and community-driven governance frameworks to address challenges. Strategic interventions such

Table 6.7 Parameter Metrics for Kapra Wetland

Parameter	Dependent variable	Independent variable	Value classification (dependent)	Value classification (independent)
Hydrological efficiency	Biodiversity support	Wetland size	High	High
Groundwater sustainability	Seasonal recharge	Groundwater policies	High	High
Cultural integrity	Cultural site use	Heritage funding	High	Moderate
Economic impact	Local employment	Economic grants	Moderate	Moderate
Policy adherence	Encroachment reduction	Policy revisions	Moderate	Moderate
Community participation	Awareness campaigns	Community efforts	High	High
Ecological restoration	Wetland rehabilitation	Ecological funding	High	High
Technological advancement	Tech-driven monitoring	Tech integration	Moderate	High
Urban resilience	Flood prevention	Urban designs	High	High
Public health benefits	Disease control	Health programs	Moderate	Moderate
Historical significance	Cultural conservation	Cultural NGOs	High	Moderate
Community rituals	Local festivities	Festival management	High	High
Tourism potential	Recreation spaces	Public infrastructure	Moderate	High
Water retention capacity	Aquifer recharge	Rainfall dependence	High	Moderate
Flood mitigation	Stormwater management	Flood infrastructure	High	High

Source: Compiled from Telangana Wetland Authority (2024), GHMC (2023), Selvaraj et al. (2022) and SPI (2020)

as desilting, installation of sewage treatment plants (STPs), and wetland conservation policies can restore Kapra Wetland as a model for urban resilience. The wetland's alignment with Sustainable Development Goal (SDG) 15 (Life on Land) underscores its global significance in biodiversity conservation. Restoring Kapra Wetland involves not only ecological and hydrological improvements but also active community and governmental involvement. This integrated approach can position the wetland as a benchmark for sustainable urban water management in Telangana and beyond.

Hydrological and Ecological Efficiency

The hydrological and ecological roles of Kapra Wetland are vital for maintaining biodiversity and mitigating urban water challenges.

1. *Hydrological Efficiency:* The wetland serves as a natural aquifer recharge zone and flood mitigation area. Seasonal water retention plays a crucial role in maintaining groundwater levels in the semiarid region (Kumar and Singh 2019).
2. *Ecological Restoration:* The wetland's biodiversity includes native flora like Neem (*Azadirachta indica*) and fauna such as migratory birds. However, invasive species and pollution have drastically impacted its ecological health (Telangana Wetland Authority 2024).
3. *Impact of Urbanization:* Encroachments and sedimentation reduce the wetland's water retention and habitat capacity, necessitating restoration through desilting and ecological measures (Kapra Municipality 2016).

Unchecked urbanization threatens the hydrological and ecological functions of Kapra Wetland. Restoration measures such as constructing STPs, removing invasive species, and introducing floating treatment wetlands can enhance biodiversity and improve water quality. These efforts align with SDG 15, emphasizing the importance of preserving wetlands for regional and global ecological balance. Strengthening wetland restoration initiatives in Telangana can contribute to long-term biodiversity conservation and global conservation goals, reinforcing the ecological and cultural value of water systems.

Community Participation and Policy Adherence

Community involvement and effective policy enforcement are critical for the sustainable conservation of Kapra Wetland.

1. *Community Efforts:* Weekly clean-up drives involving 150 participants highlight grassroots initiatives to restore the lake. These activities reflect a growing local commitment to environmental stewardship (The Hans India 2023).
2. *Policy Gaps:* Weak enforcement of wetland protection laws and limited monitoring systems hinder effective governance. Only three out of 13,297 wetlands exceeding 2.25 hectares in Telangana are officially notified for protection (Telangana Wetland Authority 2024).
3. *Role of Authorities:* Agencies such as GHMC and the Irrigation Department have yet to implement proposed measures like desilting, phytoremediation, and bird sanctuary development, leaving restoration largely unaddressed (GHMC 2023).

The collaboration between communities and authorities can bridge gaps in policy implementation and resource management. Participatory governance frameworks that integrate community initiatives with technological tools like GIS mapping can strengthen wetland conservation. Additionally, stricter zoning laws and consistent enforcement are necessary to prevent encroachments and pollution. Empowering local communities while ensuring robust policy adherence can create a resilient framework for the sustainable management of urban wetlands like Kapra.

Recommendations

1. *Ecological Restoration:* Implement large-scale desilting and introduce nature-based solutions like floating treatment wetlands to improve water quality and biodiversity (Selvaraj et al. 2022).
2. *Policy Measures:* Enforce stricter zoning laws to declare the lake a "No Development Zone" and integrate GIS-based monitoring systems to track encroachments (Rao and Singh 2019).
3. *Community Engagement:* Conduct awareness campaigns and encourage local adoption of the wetland to prevent further degradation (The Hans India 2023).
4. *Technological Integration:* Use geospatial technologies for delineating the Full Tank Level (FTL) and hydrological studies to improve resource management (Kapra Municipality 2016).
5. *Collaborative Governance:* Strengthen coordination among GHMC, the Irrigation Department, and private stakeholders under CSR initiatives to finance and execute restoration projects (GHMC 2023).

Kapra Wetland exemplifies the challenges and opportunities in managing urban wetlands amidst rapid urbanization. Revitalization requires a multifaceted approach that combines ecological restoration, robust governance, and active community involvement. By integrating traditional wisdom with modern technologies, Kapra Wetland can serve as a model for sustainable urban water management, contributing to biodiversity conservation and urban resilience.

Case Study III: Qutb Shahi Tombs Complex*: Baoli Systems—Preserving Hyderabad's Historical Legacy*

Hyderabad, founded in 1591 by Mohammed Quli Qutb Shah along the banks of the Musi River, is a city deeply rooted in the synthesis of hydrological ingenuity and cultural heritage (Bilgrami and Wilmott 1908). Its water systems, including iconic stepwells and reservoirs, reflect an advanced understanding of sustainable water management and urban planning. Among the most notable early water sources was Hussainsagar Lake, built in 1562 by Hussain Shah Wali (Majumdar 1925). This reservoir initially spanned eight square miles and became a primary source of drinking water for Hyderabad through a slow-sand filtration system established in 1891 (Anon 1980).

The city expanded its hydrological framework with the construction of Himayatsagar (1927) and Osmansagar (1913) reservoirs to meet increasing water demands following the catastrophic floods of 1908 (Francis 1904). These tanks not only controlled flooding but also became pivotal sources for the twin cities of Hyderabad and Secunderabad. Later developments, such as the Singur and Manjira River projects, further fortified the city's water supply, albeit at ecological and social costs, including the submergence of 68 villages (Rao and Chandrakant 2010). Despite their engineering brilliance, these systems face contemporary challenges from overexploitation, pollution, and encroachments, underscoring the pressing need for rejuvenation and sustainable management (Table 6.8).

This image depicts the intricately designed baoli (stepwell) within the Qutb Shahi Tombs necropolis in Hyderabad, Telangana. Renowned for their architectural and hydrological ingenuity, these stepwells were awarded the "Award of

Table 6.8 Parameter Metrics for Qutb Shahi Tombs Baoli

Parameter	Dependent variable	Independent variable	Value classification (dependent)	Value classification (independent)
Hydrological efficiency	Aquifer recharge	Rainfall levels	High	High
Groundwater sustainability	Sustainable usage	Urban policies	High	High
Cultural integrity	Heritage restoration	Cultural revival	High	High
Economic impact	Tourism income	Funding allocation	High	High
Policy adherence	Regulatory enforcement	Policy drafting	Moderate	High
Community participation	Local awareness	NGO involvement	Moderate	Moderate
Ecological restoration	Revegetation efforts	Habitat restoration	High	High
Technological advancement	GIS utilization	Technology adaptation	Moderate	Moderate
Urban resilience	Resilience improvement	Climate initiatives	High	High
Public health benefits	Improved sanitation	Waste management	Moderate	Moderate
Historical significance	Architectural value	Governance support	High	High
Community rituals	Cultural ceremonies	Community leadership	High	Moderate
Tourism potential	Visitor engagement	Marketing campaigns	High	High
Water retention capacity	Seasonal water collection	Infrastructure updates	High	High
Flood mitigation	Reduced flood risks	Drainage systems	High	High

Source: Compiled from Selvaraj et al. (2022), Rao and Singh (2019), UNESCO (2016), and relevant case studies

Distinction" at the 2022 UNESCO Asia-Pacific Awards for Cultural Heritage Conservation. (Source: Aga Khan Trust for Culture. (2022). Bowlis in Qutb Shahi Tombs bag UNESCO award of distinction. *The Times of India*. Retrieved from https://timesofindia.indiatimes.com/)

The Qutb Shahi Tombs complex, situated across a sprawling 106-acre necropolis in Golconda, Hyderabad, stands as a testament to the architectural and hydrological ingenuity of the Qutb Shahi Sultanate (1518–1687) (Daulatabad: Fortified Water—Mate and Sonar 1987). (See Fig. 6.5 for the site plan of the seven baolis at Qutb Shahi Heritage Park.) Housing over 100 monuments, this UNESCO-recognized site merges Persian, Pathan, and Hindu architectural elements. Since 2013, the Aga Khan Trust for Culture has spearheaded conservation efforts, encompassing structural stabilization, stucco restoration, and the revival of traditional water systems, including baolis (stepwells) (Anon 1986).

The baoli systems at Qutb Shahi Tombs exemplify sustainable water management strategies. Their stepped designs maximize rainwater harvesting and aquifer recharge, which is crucial in Hyderabad's semiarid climate (Mahalingam 1975). (Illustrated in Fig. 6.6: Traditional water-lifting system used in Badi Baoli.) Such TWS line up with global best practices, such as Dutch polder systems, which integrate hydrological functionality with ecological conservation. These stepwells not only address groundwater depletion but also mitigate urban flooding, a growing concern in densely populated cities.

The baolis serve as cultural landmarks reflecting the socioreligious fabric of the Deccan Sultanate. These structures facilitated communal gatherings and rituals, embedding water systems

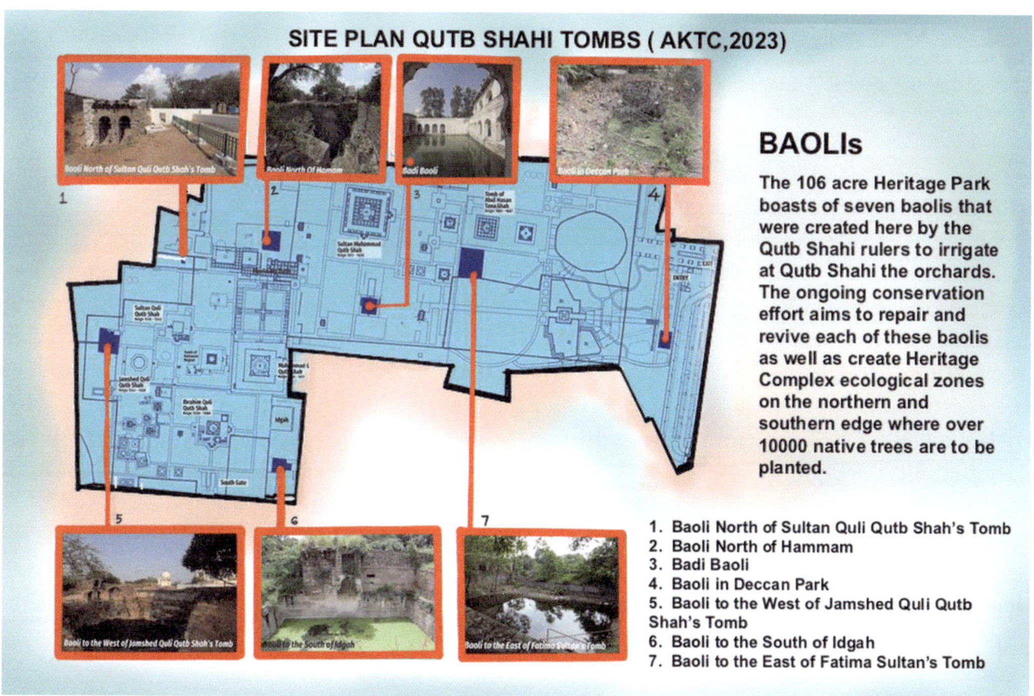

Fig. 6.5 Site Plan of Seven Stepwells at Qutb Shahi Tombs (106-acre Necropolis). This site plan presents the distribution and terrain-based alignment of seven baolis in the Qutb Shahi Heritage Park, reflecting hydrological foresight and ceremonial function. (Source: Aga Khan Trust for Culture 2023). Qutb Shahi Heritage Park Documentation

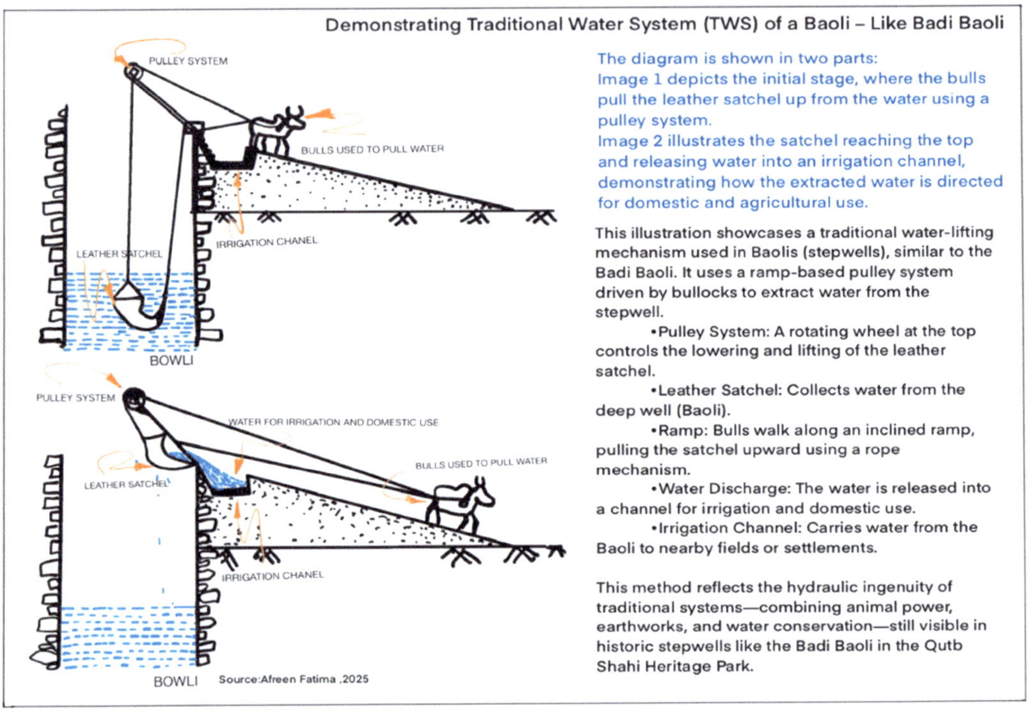

Fig. 6.6 Traditional water-lifting mechanism of Badi Baoli. A bullock-driven system using leather satchels and pulleys to draw water from deep wells, showcasing energy-efficient hydraulic engineering from historical Telangana. (Source: Author photograph and redrawing, 2025)

into the daily lives and identities of local communities (Narasimha Reddy and Subbalakshmi 1989). Economically, the site has leveraged its heritage-linked tourism potential, generating revenue that supports ongoing conservation efforts (Epigraphia Carnatica 1886).

Hydrological Efficiency and Groundwater Sustainability

The baoli systems exhibit exceptional hydrological efficiency and contribute significantly to sustainable groundwater management in Hyderabad's challenging environmental landscape. Matrix analysis categorizes aquifer recharge and rainfall levels as "High," emphasizing the efficacy of these stepwells in water harvesting. With the capacity to hold millions of liters, the baolis significantly bolster the site's water surplus and resilience (Nanda 2023; Bilgrami and Wilmott 1908; Reddy et al. 1989). Baolis epitomize a convergence of traditional water wisdom and urban sustainability. Their strategic geometry reduces surface runoff and facilitates deep infiltration, addressing critical groundwater concerns in semi-arid regions like Hyderabad. This aligns with decentralized water management approaches observed in the Govardhan Kunds, which similarly harness traditional methods to counter urban water scarcity (Bhattacharya 2015a, b; Dean et al. 2016a, b).

Revitalizing baolis serves as a cornerstone for sustainable groundwater practices, preserving hydrological heritage while addressing contemporary water challenges.

Cultural Integrity and Heritage Restoration

Baolis are not merely utilitarian structures but also cultural icons deeply rooted in the socioreligious traditions of the Qutb Shahi Sultanate and the broader Deccan region.

Heritage restoration and cultural revival efforts are rated "High," reflecting their central role in the matrix. Extensive restoration projects, including the seven-year initiative by the Aga Khan Trust, have rejuvenated these structures as cultural landmarks and operational water systems (Nanda 2023; UNESCO 2021). The baolis' integration into local traditions underscores their dual importance as functional water systems and symbols of collective memory. Like the Kunds, these stepwells blend cultural and hydrological significance, making them indispensable components of cultural heritage (Sharma et al. 2021; Rao 1930). Globally, the preservation of such water systems serves as a testament to the need for safeguarding cultural continuity alongside ecological functionality.

Preserving baolis ensures the perpetuation of historical narratives and ecological stewardship, fostering a shared sense of identity and responsibility.

Economic and Policy Impact
The economic potential of baolis, coupled with robust policy enforcement, is critical for their sustainable conservation and integration into urban systems.

The matrix ranks tourism income as "High" but regulatory enforcement as "Moderate," reflecting gaps in implementing conservation policies (Dean et al. 2016a, b; Rao and Chandrakanth 2010). Revenue from heritage tourism supports restoration but remains insufficient without stringent policy adherence (Nanda 2023).

While tourism revenue provides essential funding for conservation, the absence of comprehensive regulatory mechanisms risks overexploitation or neglect. Strengthening enforcement and aligning it with economic incentives can create a balanced framework for preserving baolis. This dual approach resonates with international examples where economic benefits are leveraged for sustainable heritage management (Hein 2019; Bhattacharya 2015a, b). Integrating economic incentives with rigorous policy frameworks ensures the longevity of baolis as both cultural and functional assets.

Ecological and Technological Integration
Baolis contribute to ecological restoration and offer opportunities for enhanced management through technological advancements. Revegetation and habitat restoration efforts are rated "High," while GIS utilization and technological adaptation are rated "Moderate." The restoration project implemented slope analyses and irrigation systems that utilize baoli water, supporting the plantation of over 11,000 native trees (Nanda 2023; Selvaraj et al. 2022). Baolis' ecological value extends to sustaining biodiversity and supporting afforestation. However, the integration of advanced technologies such as GIS and AI-driven predictive models could amplify their benefits. Case studies like the Dutch polders illustrate how data-driven ecological management can enhance resilience and sustainability (Hein 2019; Pandey et al. 2003).

Investing in technological innovations can elevate the functional and ecological contributions of baolis, positioning them as benchmarks for sustainable water management.

6.5.1 Traditional Systems in Telangana

Traditional water systems observed in Telangana and Hyderabad embody a remarkable amalgamation of indigenous engineering acumen and ecological sensitivity. These systems, including stepwells, tanks, and wetlands, illustrate a profound understanding of local hydrology, geography, and societal needs. Structures such as the stepwells at the Qutb Shahi Tombs and expansive tanks like Hussainsagar and Himayatsagar reflect a dual-purpose design that balances practical water management with cultural significance (Reddy et al. 1989; Rao 1930).

Developed during the Qutb Shahi and Asaf Jahi dynasties, these systems provided robust decentralized solutions for urban and peri-urban water needs, encompassing aquifer recharge and surface water storage. The intricate architectural and functional integration of these water systems underscores their role in shaping Hyderabad's urban resilience and cultural identity.

Moreover, these systems showcase resilience that transcends historical contexts. The strategic locations and adaptive designs of stepwells and tanks reflect a nuanced understanding of seasonal variations, groundwater dynamics, and community needs. These features are evident in the multifunctional roles of these structures—ranging from domestic water supply and irrigation to spiritual and social congregations. The duality of these systems, as both utilitarian infrastructures and cultural artifacts, underscores their holistic approach to water resource management.

6.5.2 Universal Water Values

- Hydrological Legacy: Traditional systems like baolis and wetlands embody sustainable water management principles that resonate globally.
- Cultural Continuity: These systems intertwine with rituals and histories, reflecting universal cultural values.
- Policy and Community Synergy: Combining stringent policies with grassroots engagement is critical for sustainable conservation.

6.6 Discussion

Urban wetlands and aquifers are pivotal in addressing critical urban challenges as integral components of green infrastructure. They provide multifunctional benefits, such as water filtration, flood mitigation, and groundwater recharge, which are essential for the resilience of urban ecosystems. Wetlands act as natural sponges, absorbing excess rainwater during monsoon seasons and mitigating flood risks in densely populated urban centers (Ramsar Convention 2021). Aquifers in semiarid regions like Telangana contribute to groundwater recharge, which is essential for hydrological cycles (Kumar et al. 2019). Despite their importance, urbanization and poor governance have led to the degradation of these systems. For example, in Telangana, only 3 out of 13,297 wetlands exceeding 2.25 hectares have been notified under protective frameworks (Telangana Wetland Authority 2024).

Case Studies

- Qutb Shahi Tombs Baoli: These stepwells serve as aquifer recharge zones and cultural landmarks. Restoration by the Aga Khan Trust for Culture demonstrates the potential of blending traditional engineering with modern conservation practices (Aga Khan Trust for Culture 2020).
- Kapra Wetland: This 10-hectare urban ecosystem in Hyderabad faces severe challenges, including encroachments, invasive species, and oxygen depletion, yet supports over 25 migratory bird species (GHMC 2024).

The degradation of these systems highlights governance lapses, including inadequate enforcement of protective policies and monitoring systems. However, the potential for revitalization is significant. For example, Kapra Wetland could be developed into an ecotourism hub, enhancing conservation efforts and generating revenue. Similarly, the heritage value of the Qutb Shahi Tombs Baoli could attract international attention and funding for sustainable restoration.

Opportunities also lie in technological integration, such as GIS mapping and community-driven governance models, which could improve monitoring and conservation practices. These efforts align with global frameworks like the Sustainable Development Goals (SDGs) 6, 11, and 15, emphasizing clean water, sustainable cities, and life on land. Addressing the challenges facing urban wetlands and aquifers requires integrated conservation strategies that combine traditional wisdom with modern tools. The Kapra Wetland and Qutb Shahi Tombs Baoli exemplify how these systems can be leveraged to enhance urban resilience and sustainability. These initiatives can serve as global exemplars of sustainable development, offering scalable solutions to the water stress and climate change challenges projected to affect half of the world's population by 2050 (United Nations Water 2023).

6.7 Conclusion

Traditional water systems, exemplified by Baolis and wetlands in Telangana and Hyderabad, represent a harmonious blend of cultural heritage and environmental stewardship. These structures, deeply rooted in historical practices, demonstrate how hydrological ingenuity can coexist with cultural expression. The stepwells and tanks constructed during the Qutb Shahi and Asaf Jahi dynasties illustrate an era of advanced understanding of water conservation and resource management. Here demonstrated, the spatial plan of seven baolis at Qutb Shahi Heritage Park, along with the traditional water-lifting system of Badi Baoli, visually demonstrates how past water practices were deeply ecological and community-centered. These systems, embedded in terrain and culture, serve as replicable blueprints for decentralized and resilient urban water management today. Their survival is a call not only for conservation but for reactivation in climate-adaptive planning.

In contemporary times, these systems continue to serve as beacons of sustainability, offering solutions to pressing urban challenges like water scarcity and ecological degradation. Their ability to integrate into the urban environment while preserving biodiversity and fostering community engagement highlights their enduring relevance. However, realizing their full potential requires a multifaceted approach that includes stringent policy reforms, technological advancements, and active community participation.

Moving forward, the revitalization of traditional water systems should be viewed not only as a conservation effort but also as an opportunity to reimagine urban water management strategies. By leveraging their historical significance and functional utility, these systems can pave the way for resilient and sustainable urban landscapes. Securing their preservation will ensure that future generations inherit a legacy that values the intricate balance between nature, culture, and innovation.

Acknowledgments The study acknowledges the support and contributions of several institutions and individuals. The Hyderabad Metropolitan Development Authority (HMDA) provided datasets on stepwells, enabling the analysis and mapping of traditional water systems. The School of Planning and Architecture (SPA), Bhopal, supported the research by providing infrastructure, library access, and academic resources. Thanks are extended to the Aga Khan Trust for Culture (AKTC) for sharing insights into the restoration of heritage-linked water systems, particularly the Qutb Shahi Tombs Baoli, which informed the integration of traditional knowledge with conservation practices. The author expresses gratitude to Prof. Ajay Khare, co-author and guide, for his mentorship and valuable contributions throughout the research. The study also acknowledges all participants for their informed consent, adhering to ethical research standards.

Conflicts of Interest The authors declare no conflicts of interest, affirming the impartiality of the study.

Informed Consent Consent was obtained from all individual participants involved, in adherence to ethical research practices.

Authors' Contributions All authors contributed equally to the conceptualization, execution, and outcomes of the study, reflecting a collective effort.

AI Disclosure This chapter was developed entirely by the authors and reflects original research and analysis. Grammar correction and language enhancement tools were utilized only to enhance language clarity, format compliance, and citation verification, while the intellectual content remains the sole creation of the authors.

Permission Disclosure All third-party materials incorporated in this chapter are licensed stock images, for which permissions are inherently available. Tables, figures, and other visual data presented within the manuscript have been generated exclusively by the authors and are entirely original. No additional permissions are required to use any materials included in this chapter.

Funding This research was conducted without external funding, ensuring its independence.

Data Availability The datasets analyzed during this study are available upon reasonable request from the corresponding author, ensuring transparency and reproducibility.

References

Aga Khan Trust for Culture. (2020). Qutb Shahi Heritage Park Annual Report 2019–2020. New Delhi: Aga Khan Trust for Culture (AKTC).
Anon (1980) Imperial Gazetteer of India, Mysore and Coorg. Superintendent of Government Press, Calcutta
Anon (1986) Gazetteer of the Bombay Presidency, Kolhapur. Government Central Press, Mumbai
Basant, Yadav Nitesh, Patidar Anupma, Sharma Niranjan, Panigrahi Rakesh K., Sharma V., Loganathan Gopal, Krishan Jaswant, Singh Suraj, Kumar Alison, Parker (2022) Assessment of traditional rainwater harvesting system in barren lands of a semi-arid region: A case study of Rajasthan (India) Journal of Hydrology: Regional Studies 42101149. https://doi.org/10.1016/j.ejrh.2022.101149
Bhattacharya R (2015a) Water wisdom in India: decentralized water management. Academic
Bhattacharya S (2015b) Traditional water harvesting structures and sustainable water management in India. Academic
Bilgrami SR, Wilmott C (1908) Historical and descriptive sketch of his highness the nizam's dominions. Clarendon Press
Dean AJ, Fielding KS, Newton FJ (2016a) Community knowledge about water: support for water-related policies. Water Int 41(7):934–947
Dean R et al (2016b) Cultural heritage and economic policy implications: global perspectives. Routledge
Epigraphia Carnatica (1886) Manual of the Kurnool District in the Presidency of Madras. Government Press
Francis W (1904) Madras District Gazetteer, Bellary. Superintendent Government Press
Fatima, A. A. Khare, A. (2025). Sustainable solutions through traditional synergies: The case of Indian stepwells. ISAR Journal of Multidisciplinary Research and Studies, 3(2), 45–65
Floating Treatment Wetlands Fact Sheet. (2021). Environmental Protection Agency (EPA) Region 5 Office. Washington, D.C.
GHMC (Greater Hyderabad Municipal Corporation) (2024) Urban Water Management and Heritage Conservation Annual Report. Hyderabad: GHMC
GHMC (2023) Urban water resilience through HYDRAA. GHMC
Gies E (2022) Water always wins: thriving in an age of drought and deluge. University of Chicago Press
Government of Telangana (2014) Mission Kakatiya program document. Government of Telangana, Hyderabad
Hein C (2019) Adaptive reuse of water systems: case studies from the Dutch polders. Cambridge University Press
Himayatsagar Project Report. (1927). Government of Hyderabad State Irrigation Department. Hyderabad
HMDA (2023) Hyderabad metropolitan development authority stepwell documentation report. HMDA, Hyderabad
Kapra Municipality (2016) Annual lake report. Kapra Municipality
Kumar D, Singh P (2019) Impact of traditional water systems on groundwater: a case study of Telangana. Water J 6(4):211–225
Kumar, A., Sharma, P., Meena, R. (2019). Urban water reuse and policy frameworks in Indian cities. Water Policy 21(6) 1203–1218
Kumar, A., Sharma, P., Meena, R. (2021). Nature-based solutions for sustainable urban drainage: Lessons from India. Environmental Science and Policy 122, 56–65
Mahalingam TV (1975) Administration and social life under Vijayanagar, 2nd edn. University of Madras
Majumdar RC (1925) Orissa in the making. University of Calcutta
Mate MS, Sonar JC (1987) Daulatabad: fortified water—a monument interpreted. Bull Deccan Coll Res Inst 46: 65–72
Ministry of Rural Development. (2016). MGNREGA: Convergence Guidelines on Water Conservation and Management. Government of India, New Delhi
Nagendra H (2018) The role of urban lakes in water management: lessons from Bengaluru. Environ Sci Pol 84:69–77
Nanda R (2023) Restoration practices at Qutb Shahi Tombs: hydrology and heritage conservation. Aga Khan Trust for Culture Publications
Narasimha Reddy D, Subbalakshmi V (1989) Technological, institutional, and ecological aspects of equity and efficiency in irrigation development. MIDS-ICSSR Interdisciplinary Seminar
Osmansagar Project Report. (1913) Government of Hyderabad State Irrigation Department. Hyderabad
Pandey S, Sinha R, Rao AS (2003) Ecological contributions of urban wetlands. Ecol Res J 18(2): 105–118
Ramsar Convention Secretariat. (2021) Ramsar Convention on Wetlands: Global Wetland Outlook 2021 Gland, Switzerland: Ramsar Convention Secretariat.
Rao VM (1930) Irrigation development in South India. University of Madras Press
Rao N, Chandrakanth MG (2010) Resources at the margin: case studies on tank irrigation. Springer
Rao S, Singh V (2019) Traditional tanks in South India: historical and hydraulic perspectives. Heritage Press
Reddy, V. R., Ratna Reddy, B., Uday Kumar, D. (1989) Water resource management in semi-arid Telangana: Historical perspectives. Indian Journal of Regional Studies, 15(3), 211–227
Rumana, Asad Josephine, Vaughan Iftekhar, Ahmed (2023) Integrated Traditional Water Knowledge in Urban Design and Planning Practices for Sustainable Development: Challenges and Opportunities Sustainability 15(16) 12434–https://doi.org/10.3390/su151612434
Selvaraj K, Raghunathan S, Joshi P (2022) Wetland conservation in India: challenges and innovations. Ramsar Publication

Sharma A, Kumar R, Singh S (2021) Govardhan Kunds and their multifunctional role in heritage conservation. Taylor & Francis

Singh, R., Kumar, A., Sinha, A. (2020). Integrated water management approaches for resilient urban ecosystems. Journal of Hydrology: Regional Studies, 31, 100–118

Snyder, M. (2019) Wetland restoration and floating island technology: A practical review. Ecological Engineering, 134, 1–12

SPI (Stockholm Partnership Initiative). (2020). Urban Water Resilience Framework. Stockholm: Stockholm Environment Institute

Telangana Wetland Authority (2024) Wetland health and policy challenges. Telangana Wetland Authority

The Hans India (2023) Desilt Kapra Lake immediately, urge social activists & locals. [cited 2025 Apr 17]. Available from: https://www.thehansindia.com/news/cities/hyderabad/hyderabad-desilt-kapra-lake-immediately-urge-social-activists-locals-809479

UNESCO (2016) Stepwells and water heritage in India: cultural and hydraulic significance. United Nations Educational, Scientific and Cultural Organization, Paris

UNESCO (2021) Conservation reports on world heritage sites: Qutb Shahi Tombs. UNESCO Publishing

United Nations Water. (2023) World Water Development Report 2023: Partnerships and Cooperation for Water. Paris: UNESCO

Water Resources Group (WRG). (2021). India 2030 Water Resources Group Report. Washington, D.C.: World Bank and WRG Secretariat.

Theme: Understanding Wetland Ecology and Importance Plant Diversity Assessment as a Robust Approach to Measuring Wetland Ecosystem Stability: A Case Study in and around Rudrasagar Lake, a Ramsar Site in Northeast India

7

Arpita Biswas, Pradip Debnath, Saptarshi Mitra, and Punarbasu Chaudhuri

Abstract

Wetlands, specifically located at the rural-urban interface, are increasingly threatened by human-induced pressures such as hydrological changes, habitat fragmentation, tourism expansion, and land-use patterns. These pressures can significantly alter wetland ecosystem services and cause serious challenges to ecological stability. The present study evaluated plant diversity as a strong indicator of ecosystem stability in and around Rudrasagar Lake, amidst increasing developmental pressures. Supported by geospatial mapping, extensive field surveys were conducted in five distinct areas of the wetland, representing a gradient of anthropogenic pressure. For qualitative and quantitative assessment, 30 quadrats were monitored from five distinct zones. Diversity metrics such as Shannon, Simpson, Margalef, and Pielou's indices were employed along with the Enhanced Vegetation Index (EVI) for assessing their relationship with ecosystem stability. One-way ANOVA (corresponding *p value* < 0.05) and linear regression analysis revealed strong positive correlations (r^2: Shannon $= 0.99$; Margalef $= 0.94$) between plant diversity and ecosystem stability. A total of 125 terrestrial and 22 aquatic plant species were observed. The open scrub zone represented the highest species diversity with a corresponding Shannon value of 3.1, while the human-induced zone showed the lowest diversity ($H' = 1.1$). Highly diverse communities in the Kemtali and Chandanmura areas achieve greater stability than the Rajghat and Neermahal areas with urban interface. This study highlights plant diversity as an important diagnostic tool for assessing wetland health and ecological resilience. The findings emphasise that areas rich in biodiversity exhibit greater ecological stability, emphasising the role of wetlands in supporting long-term urban and semi-urban sustainability.

Keywords

Biodiversity conservation · Habitat degradation · Ecological indicator · Vegetation stability · Wetland disturbance

Supplementary Information: The online version contains supplementary material available at https://doi.org/10.1007/978-3-032-06881-1_7.

A. Biswas · P. Chaudhuri (✉)
Department of Environmental Sciences, University of Calcutta, Kolkata, West Bengal, India

P. Debnath · S. Mitra
Department of Geography and Disaster Management, Tripura University, Agartala, Tripura, India

S. K. Chakraborty et al. (eds.), *Wetlands and Urban Resilience*, Sustainable Development Goals Series,
https://doi.org/10.1007/978-3-032-06881-1_7

7.1 Introduction

Plant diversity is a predominant biological indicator to measure and monitor ecosystem stability, particularly in fragile wetland ecosystem health. Wetlands serve as the most productive but threatened ecosystems globally (Xu et al. 2019). It provides robust contributions to ecosystem services consisting of diverse flora and fauna inhabitation, carbon sequestration, groundwater recharge, flood mitigation, and microclimate regulation (Gupta et al. 2020; Chakraborty et al. 2023). However, wetland ecosystems are facing unprecedented challenges from climatic uncertainties and global warming as a result of increasing anthropogenic activities (Pimparkar et al. 2023). This vulnerable situation draws immediate attention to the researchers in ecosystem monitoring and conservation efforts (Moomaw et al. 2018; Newton et al. 2020; Lazaro-Lobo and Ervin 2021). Recently, plant diversity assessment has become a powerful analytical approach to deciphering ecosystem dynamics and stability. Several diversity indices: Shannon-Wiener index, Simpson index, species richness and evenness were used for quantitative assessment of plant communities (Díaz et al. 2007; García-Palacios et al. 2018).

Rudrasagar Lake, a Ramsar site in northeast India, is an example of this vulnerability. Although the lake is largely surrounded by rural landscapes, increasing urban influences like uncontrolled tourism, expansion of rapid settlement, and the proliferation of brick kilns are continuously altering its ecological dynamics. These alterations have positioned Rudrasagar lake in the context of peri-urban wetlands, where pressures from urbanisation directly affect biodiversity patterns and ecological resilience. However, limited research has addressed how such transitional landscapes impact wetland health, particularly from a perspective of plant diversity. In the present context, assessing plant diversity emerges as a potential ecological tool. It not only reflects the entire scenario of the ecosystem but also serves as a predictive measure of resilience over the environmental stresses like habitat fragmentation, invasive species, and pollution (Rooney and McCann 2012; Geng et al. 2019). Moreover, higher plant diversity can enhance habitat heterogeneity by supporting complex trophic interactions, strengthening ecological stability (Lu et al. 2017). Furthermore, plants are also called "ecosystem engineers," which influence soil properties, hydrology, and microclimatic conditions that are common to both rural and urban environmental planning.

Conservation of wetlands is intrinsically associated with their plant diversity, as vegetation composition can structurally influence enormous ecological processes (Meli et al. 2014; Dubois et al. 2020a, b). Several scientific approaches have already been adopted to appraise wetland vegetation diversity, incorporating remote sensing techniques, extensive field surveys, and statistical analysis to quantify diversity and its temporal dynamics. The selection of appropriate diversity indices and sampling strategies is very important for obtaining reliable and comparable results across different wetland systems. How vegetation diversity contributes to ecosystem stability is very relevant in recent days with climate change and increasing environmental challenges (De Boeck et al. 2018). Current conservation practices increasingly require holistic, data-driven methods to determine how biodiversity enhances ecosystem services and resilience over the long term. In this context, measuring vegetation diversity in conjunction with spatial and statistical analysis provides a reproducible, affordable way to compare ecosystem health in various land-use categories, such as those for urbanising wetland interfaces.

The originality of this study is the use of plant diversity indices in estimating ecosystem stability, which provides quantitative measures in a peri-urban wetland that is facing immense anthropogenic pressure. While many studies investigate biodiversity or ecosystem stability in separate contexts, this study innovatively combines field vegetation data with geospatial analysis and ecosystem stability modelling. Furthermore, it fills an important research gap by linking plant diversity patterns to urban resilience, emphasising how biodiversity loss in rapidly transforming landscapes can impair ecosystem services with several statistical tools. The aim of this study is to examine plant diversity patterns across the anthropogenically influenced zones and how they influence the ecosystem stability of the study area. This research also aims to pinpoint the anthropogenic activities responsible for habitat fragmentation and vegetation loss. By examining the relations between diversity, degradation patterns, and stability, the current research encourages an increase in the understanding of wetland dynamics that can contribute to the adoption and implementation of conservation strategies.

7.2 Materials and Methods

7.2.1 Study Area

Rudrasagar Lake (150 km^2 with its sub-watershed), a Ramsar site since 2005, is located in the South Western part of Tripura. It is located in the Sonamura Sub-division along the Indo-Bangla transboundary (Fig. 7.1a) and lies within latitude 23°29′–23°31′ N and longitude 90°18′–90°19′ E. It particularly occupies a permanent surface area of 1.41 km^2 with a varying depth of 2–14 m. Its topographical elevation ranges between 19 and 30 m from the mean sea level (MSL) and yearly temperature varies from 9 to 38 °C. The lake is characterised by a humid and tropical monsoon climate with an annual average precipitation of 128 mm. The lake is seasonally inundated by perennial rivers: Noa Chera, Kematali Chera, and Durlavnarayan Chera, and drained through the artificial Kachhigang channel along the right bank of the Gomti river near Battali Bazar. It has enormous ecological importance for supporting diverse flora and fauna, and also harbours some rare and endangered species. This lake ecosystem is exceptionally sensitive to vulnerable conditions, as it can easily reciprocate anthropogenic activities. Several recreational activities have been identified for the presence of Lake Palace Neermahal, contributing to the areal shrinkage of the water body, eutrophication, plummeted depth, and biodiversity loss, all of which have long-term implications for wetland health (Debnath et al. 2023a, b).

Rudrasagar wetland of Tripura experiences a monsoonal climate with distinct rainy and dry seasons. As shown in Table 7.1, annual rainfall declines (from approximately 600 to 100 mm annual average) and the rainy season shrinks within the May to November period. Champion and Seth (1968) observed that the moist mixed deciduous forests occur in the south of the Brahmaputra River of northeast India with significant spatiotemporal heterogeneity. Deb (1981) classified the forests of Tripura based on climatic factors and divided them into (i) evergreen forest, (ii) moist deciduous mixed forest, (iii) swamp forests, (iv) bamboo and cane breaks, (v) Garjan forest, and (vi) Savannah and other Grassland. Climatic factors majorly determine and shape the vegetation cover in the region: the local spatio-temporal variability of rainfall affects vegetative growth.

7.2.2 Research Design

An extensive field survey has been conducted to document and monitor vegetation patterns along with their diversity. Anthropogenic pressure was considered during the selection of sample sites. The wetland area was divided into six distinct zones based on vegetation characteristics in terrestrial and aquatic landscape for qualitative and quantitative assessment. The zonation mechanism of the vegetation covers is demonstrated in Table 7.1. Based on anthropogenic disturbance, the aquatic vegetation landscape was characterised into two broad groups: Aquatic 1 is for normal aquatic vegetation and Aquatic 2 is for tourism-induced aquatic vegetation. Terrestrial landscape characterisation was followed by spinney, foraging zone, deforested zone, and tourism-induced area. The geographical attributes of the selected sample sites are represented in Table 7.1. Geospatial coordinates were demarcated using a portable GPS (Garmin eTrex 30), and field photographs were captured digitally with a Nikon D3000 camera (Fig. 7.2).

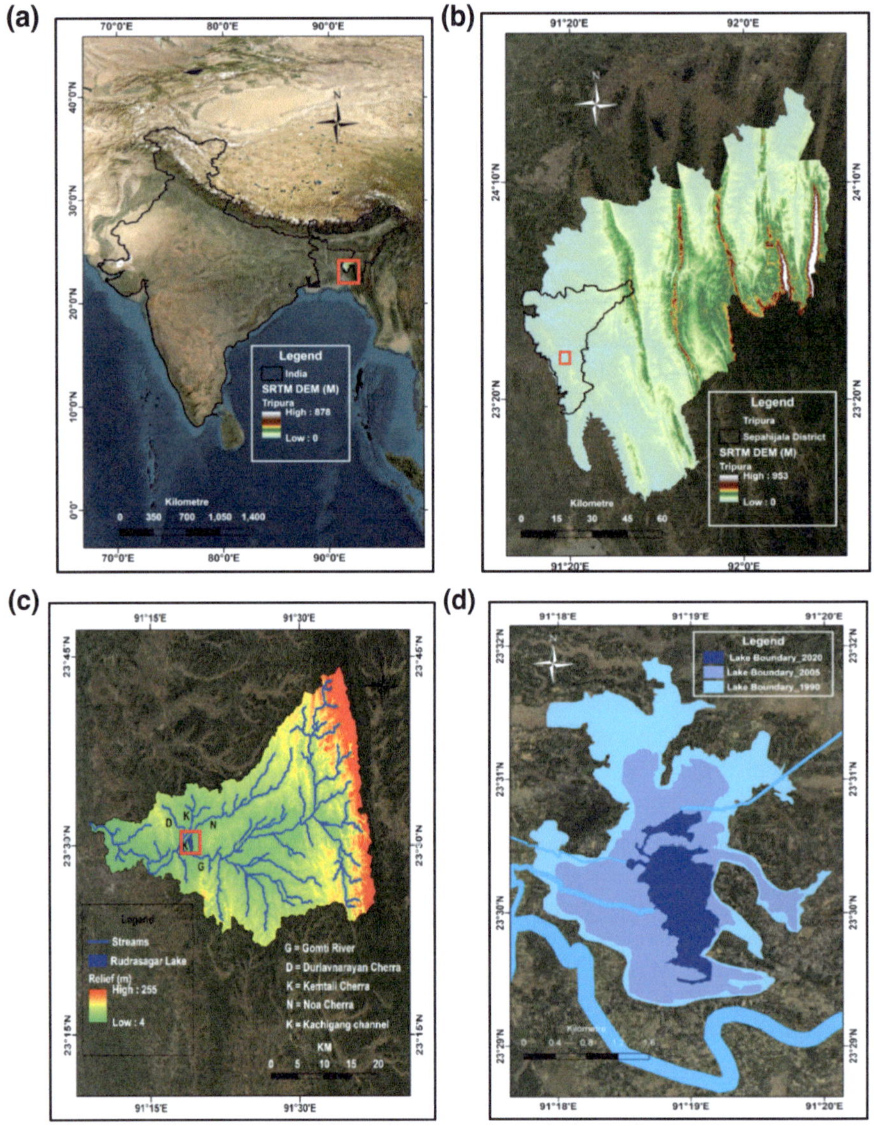

Fig. 7.1 Location map of the study area: (**a**) administrative divisions of India with an outline of Tripura, (**b**) high-accuracy photogrammetric Digital Elevation Model (DEM) of Tripura highlighting Sepahijala district illustrating the altitudinal variations ranging from 0 to 878 meters in the region, (**c**) delineating the watershed and catchment area of Rudrasagar Lake, including its drainage systems, and (**d**) spatial dynamics of the lake's surface area indicating areal shrinkage over time (1990–2020)

7.2.3 Data Curation

Primary data was obtained from the field survey to determine the species richness and diversity of each habitat. Vegetation documentation and monitoring surveys were administered in and adjacent to Rudrasagar Lake. The field survey was conducted between April 2022 and December 2023. A total of 30 (6*5) quadrats were randomly selected across six ecologically distinct zones for diversity assessment. Quadrat size typically ranges 1*1 m, ensuring comprehensive representation of terrestrial and aquatic vegetation. Quadrats used in the study were randomly placed within the study area, with the placement being determined by factors like accessibility and the

Table 7.1 Geo-coordinates (N latitude and E longitude) of the selected sample area

Zone	Quadrate 1	Quadrate 2	Quadrate 3	Quadrate 4	Quadrate 5	Features
Spinney (S)	23°30′ 43.18″ 91°18′ 34.48″	23°30′ 44.26″ 91°18′ 51.46″	23°31′ 10.96″ 91°18′ 30.63″	23°30′ 50.29″ 91°19′ 11.31″	23°30′ 38.42″ 91°19′ 24.85″	Small shrubs, bushes, big trees were observed in these quadrates
Aquatic (W) 1	23°30′ 23.66″ 91°18′ 40.83″	23°30′ 3.33″ 91°18′ 41.53″	23°29′ 50.83″ 91°18′ 52.58″	23°29′ 37.20″ 91°19′ 0.27″	23°30′ 16.97″ 91°19′ 10.02″	Undisturbed lake area
Foraging (F)	23°30′ 2.50″ 91°18′ 13.70″	23°29′ 47.43″ 91°18′ 30.59″	23°30′ 20.43″ 91°18′ 21.14″	23°29′ 29.01″ 91°18′ 42.17″	23°29′ 22.49″ 91°19′ 1.84″	Mainly used for grazing ground for livestock reared in the neighbouring localities
Deforested (D)	23°30′ 37.53″ 91°20′ 32.16″	23°30′ 18.56″ 91°20′ 23.40″	23°30′ 4.11″ 91°20′ 0.34″	23°30′ 32.18″ 91°19′ 46.34″	23°30′ 59.30″ 91°20′ 7.29″	Dense green area was deforested for settlements and urbanisation
Tourism (T)	23°29′ 41.40″ 91°19′ 17.84″	23°29′ 44.28″ 91°19′ 16.81″	23°29′ 43.63″ 91°19′ 14.16″	23°29′ 41.85″ 91°19′ 10.08″	23°29′ 39.14" 91°19′ 14.06″	Discharge of tourism waste, increasing area of tourist resort
Aquatic disturbed (A) 2	23°30′ 21.04″ 91°18′ 56.78″	23°30′ 14.01″ 91°18′ 59.71″	23°30′ 6.64″ 91°19′ 3.01″	23°29′ 58.61″ 91°19′ 5.59″	23°29′ 50.76″ 91°19′ 8.35″	Ferry route from Rajghat to Neermahal

Source: Computed by authors based on field data, 2022–2023

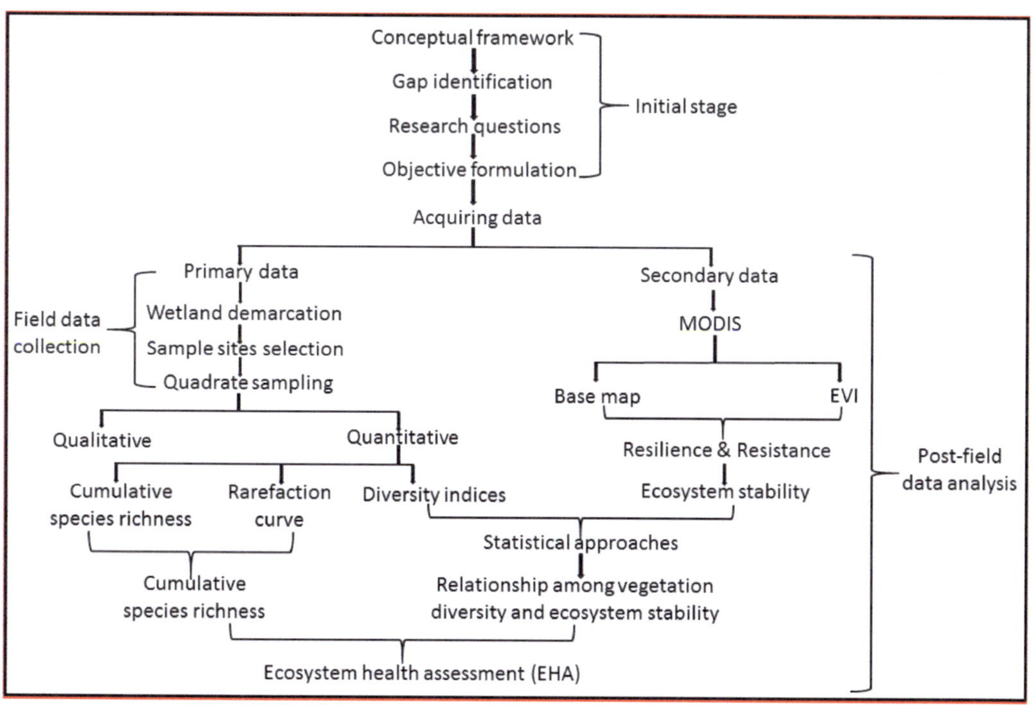

Fig. 7.2 Methodological framework adopted for this study

Fig. 7.3 Locations of the quadrates studied within the study area

presence of plant species. Figure 7.3 provides additional context about the spatial distribution of the quadrats designed for this study area. In each quadrat, all plant species were individually identified and documented following a standard survey format. This process and the observation continued until 80% co-occurrence of plant species was achieved. In this manner, the designated quadrats for the study showed that four species—*Barringtonia acutangula, Hevea brasiliensis, Lagerstroemia speciosa, and Syzygium* sp.—collectively accounted for more than 80% of the terrestrial plant species, while Ipomoea and Monochoria, along with water lettuce and water lilies, dominated the aquatic regions. Ecological zonal diversity of five distinct zones likely represents different ecological characteristics: open scrub zone, tourism-induced zone, terrestrial interfaces, aquatic interfaces, and transitional habitats. This sampling strategy enabled robust statistical analysis and captured the spatial heterogeneity of the wetland ecosystem. In this study, multispectral or hyperspectral satellite images were acquired from reliable sources. Landsat 8 with a 30 m resolution was acquired to analyse the spatio-temporal dynamics of green spaces from 2007 to 2021. A detailed description of the digital satellite images is provided in Table 7.2. Four distinct green space categories were identified based on their ecological characteristics: dense vegetative, grassland and shrubs, barren land, and aquatic plant species. Table 7.3 indicates the EVI-based classification of these green space categories with their corresponding EVI value ranges. Based on the vegetation cover assessment, several key features can be observed. EVI feature ranges indicate that dense vegetative areas have the highest greenness and biomass, with EVI values between 0.3 and 0.93. In contrast, barren land shows marginal EVI values ranging from 0.84 to 1.8, signifying sparse or non-existent vegetation cover in the study area.

Table 7.2 Comprehensive description regarding Landsat satellite images utilised for this study

Satellite	Sensor	Acquisition date	Projection	Spatial resolution	No. of bands	WRS path/ Raw
Landsat 8	Operational Land Imager and (OLI) and Thermal Infrared Sensor (TIRS)	10.15.2017	WGS84 UTM Zone 46 N	30 m	11	136/44
		10.12.2022				

Source: United States Geological Survey (USGS), 2022

Table 7.3 Green space categorisation

Vegetation category	EVI features	Characteristics
Dense vegetative	0.3–0.93	Areas with a high density of healthy vegetation, likely dominated by dense canopy cover, tall, and blooming plants
Grassland and shrubs	0.18–0.3	Presence of grasslands, low-growing vegetation cover, and shrubby areas with moderate greenness and biomass
Barren land	0.84–1.8	Sparsely vegetated or unvegetated areas, such as exposed soil, rock, or sand, with very low to negligible greenness
Water body	– 0.22–0.084	Surface water bodies like lakes, rivers, or wetlands, having an aquatic plant species including water lilies, water lettuce, azolla, and Eichhornia, commonly called water hyacinths

Source: Computed by the authors, 2022

7.2.4 Methods

Ecosystem stability was computed by assessing its two components, resistance and resilience (Geng et al. 2019). Resistance measures the degree to which an ecosystem resists change when faced with environmental perturbations, while resilience quantifies the ability of an ecosystem to recover from a disturbance (De Keersmaecker et al. 2015; Hodgson et al. 2015). Ecosystem stability was calculated by Geng et al.'s (2019) formula:

$$E_S = r_1 + r_2 \tag{7.1}$$

Here, E_s denotes ecosystem stability, whereas r_1 and r_2 stand for ecosystem resistance and resilience.

Ecosystem stability varied between 0 and 1, where a higher value indicates towards more stable ecosystem. Resistance and resilience quantification to short-term climate anomalies, which was based upon the autoregressive (AR1) model using the equation (Geng et al. 2019):

$$Vt = \alpha * Tt + \beta * Pt + \gamma * Rt + \delta * V(t-1) + \varepsilon \tag{7.2}$$

where *Vt, Tt, Rt,* and *Pt* were denoted for addressing standardised EVI anomaly, standardised air temperature anomaly, standardised solar radiation anomaly, and standardised precipitation anomaly, respectively, at time *t*. $Vt-1$ was referred to as the standardised one-month-lagged EVI, α, β, γ, and δ were the model coefficient discrepancies, while ε was the residual errors. The coefficient α is used to measure the immediate plant response sensitivity to instantaneous temperature fluctuations, and the value approaching 1 indicates higher sensitivity to temperature changes. As resistance is inversely proportional to sensitivity, temperature resistance, water availability, and solar radiation anomalies can be represented by $1-\alpha$, $1-\beta$, and $1-\gamma$, respectively.

7.2.4.1 Vegetation Diversity

Various diversity indices such as Shannon-Weiner index, Simpson index, Pielou's evenness index, and Margalef index were utilised to quantify plant diversity. Relative abundance (R_a) was obtained from the percentage of each species contributing to the total individual by following the formula Woldemariam et al. (2018):

$$Ra = \frac{N_i}{N} * 100 \quad (7.3)$$

where N_i is the number of individuals of each species and N is the total number of species. Shannon-Weiner index (H') was used to analyse species diversity in various habitats based on their abundance using the following formula:

$$H' = \sum_{i=1}^{S} Pi * \ln(Pi) \quad (7.4)$$

where s is the number of observed species in the community and Pi is the proportion of individuals in the ith species. Simpson's dominance Index (Ds) was used for measuring the degree of concentration by the following formula:

$$Ds = 1 - \frac{\sum ni(ni-1)}{N(N-1)} \quad (7.5)$$

where n is the number of individuals in a single species and N is the number in total population. Furthermore, Pielou's Index (Ev) was used to assess the evenness of species diversity in the observed community and was derived by:

$$Ev = \frac{Sw}{\ln(H)} \quad (e)$$

here, Sw is denoted for the Shannon index, and H is the species richness. Margalef index as a richness index was calculated following the equation:

$$M = \frac{S-1}{\ln N} \quad (7.7)$$

here, S and N indicated the richness and delimited study area, respectively. Cumulative species richness curve and rarefaction curves were employed to analyse the adequacy of sampling effort with the species richness and depict the relationship among the number of individuals sampled and with corresponding observed species, respectively.

7.2.4.2 Vegetation Index

Widely accepted vegetation indices such as the Normalized Difference Vegetation Index (NDVI) and Enhanced Vegetation Index (EVI) were employed to quantify vegetation abundance and health status. For this study, we used the Enhanced Vegetation Index (EVI) as a proxy for above-ground biomass, and the greenness of vegetation was considered a good indicator for evaluating large-scale spatial patterns of vegetation stability; Seddon et al. 2016).

$$\text{EVI} = G * \frac{\rho_{\text{NIR}} - \rho_{\text{red}}}{\rho_{\text{red}} - C_1 * \rho_{\text{red}} - C_2 * \rho_{\text{blue}} + l} \quad (7.8)$$

This notation typically indicates the formula for the Enhanced Vegetation Index (EVI), which is a modified version of NDVI. The key components of the EVI are: G is a gain factor, typically set to 2.5, which is used to scale the overall index value; ρNIR, ρRed, and ρBlue represent the near-infrared, red, and blue band reflectance; $C1$ and $C2$ are coefficients used to correct for the influence of atmospheric aerosols. L is a soil adjustment factor, typically set to 1, which helps reduce the impact of soil background on the vegetation signal.

For calculating the degree of vegetation cover over the different time intervals, the annual rate and the percentage of change were measured by using Eqs. (7.2) and (7.3), illustrated by the following mathematical notation:

$$pc = \frac{p^2 - p^1}{p^1} * 100 \quad (7.9)$$

$$rc(\text{ha/yr}) = \frac{p^2 - p^1}{t^i} \quad (7.10)$$

here, pc represents the per cent of change; rc determines the rate of change; $p1$ stands for the initial observed area or vegetation cover at the earlier time interval; $p2$ represents the final observed area of vegetation cover at the later time interval; t^i represents the time interval between $p1$ and $p2$ in years (yr).

7.2.5 Statistical Analysis

The current study analyses the ecosystem stability across several geographical locations in the study area. First, Kolmogorov-Smirnov test for normality was analysed for all the variables to find the distribution pattern, indicating whether the variables were following normal distribution, which helped to ensure the validity of the further statistical analysis. One-way analysis of variance (ANOVA) was performed to know the difference in ecosystem stability among the selected zones, following Eq. 7.11:

$$F = \frac{\text{MSb}}{\text{MSw}} \tag{7.11}$$

where, F indicated the coefficient of ANOVA, while MSb and MSw indicated the mean sum of squares between the groups and within the groups, respectively, at a significant level $p < 0.05$. A linear regression model was performed with Eq. 7.12 to analyse the dependency of vegetation diversity on ecosystem stability:

$$Y = \alpha X + \beta \tag{7.12}$$

where Y and X were dependent (ecosystem stability) and independent (diversity indices) variables, respectively. α indicated the slope of the regression line, while β indicated the Y-intercept. Combining the application of ANOVA, normality testing, and linear regression curves utilising the ecological data, this study can develop an understanding of ecological systems with its variables. All the statistical analyses have been done using GraphPad Prism (v.10.1.0).

7.3 Results

7.3.1 Vegetation Diversity Pattern

A total of 125 terrestrial and 22 aquatic plant species from 53 and 16 families belonging to 24 and 14 orders, respectively, were observed in the study area. The cumulative species richness curve corresponding to sampling zones is presented in Fig. 7.4a, which indicated that 80% of the observed species were covered after the completion of the fourth sampling zone quantification, while 90% of all species were covered after completing the fifth sampling zone, and a similar result was consistent in the previous study. Four vegetation diversity indices including the Margalef index, Simpson index, Shannon index, and Pielou's index were evaluated in selected zones. The Shannon-weiner index and Margalef index had an analogous trend, indicating higher species richness in the spinney zone (S) following the aquatic zone (W). Figure 7.4b represents the lowest species richness observed in tourism tourism-driven zone (T) followed by the aquatic disturbed zone (A). The lowest species evenness and highest species dominance were observed in the aquatic disturbed zone, as the zone was dominated by *Euryale ferox* and common water hyacinth. The quadrats studied in the Kemtali, Chandanmura, and Rangamura areas, which fell under the spinney zone, showed elevated species diversity.

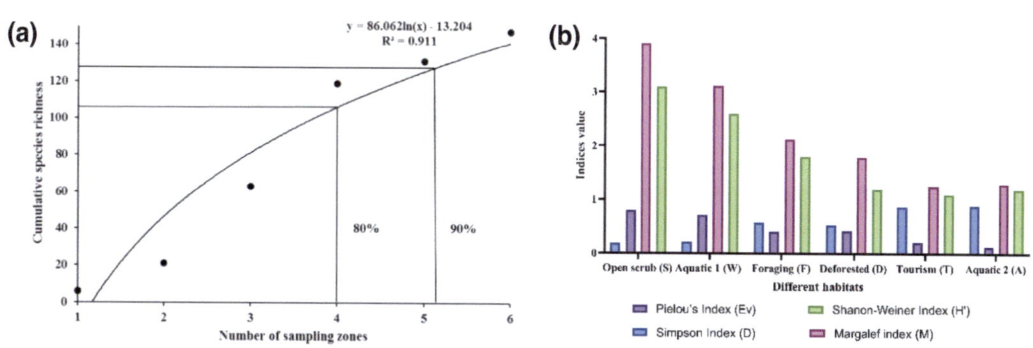

Fig. 7.4 Species diversity pattern: (**a**) cumulative species richness curve across the zones and (**b**) spatial variation pattern of vegetation distribution indices across different vegetation habitats

Indranagar area demonstrated comparatively lower species diversity than the spinney zones, as this area fell under the foraging zone (F). Several herbs and shrubs were observed in those areas. Rarefaction curves are generally used to assess species richness with the number of increased sampling units. The intrinsic analysis of these rarefaction curves demonstrated distinct spatial heterogeneity among species composition and confirmed that the effort of sampling across the entire study area was sufficient to accumulate the majority of species diversity within this wetland ecosystem, particularly. The varying plateaus and steepness of each curve in all zones indicated variation in species accumulation and efficiency of sampling completeness. The point at which a curve reaches a clear plateau indicates that the maximum number of species are present, while curves without distinct plateaus can be considered undiscovered species. Each curve with different colours represented the species accumulation as many individuals were sampled in different quadrates of a particular zone. In the aquatic 1 (W) zone, the rarefaction curves showed a steep upward slope, which indicated that new species were being observed at a rapid pace (Fig. 7.5a). Q1 was observed to have the steepest curve, indicating higher species richness with increasing sampling effort. Conversely, in Q5, lower species diversity was observed with the flattened curve. Similarly, in Fig. 7.5b, the rarefaction curve for aquatic disturbed zone (A) species richness across five distinct quadrates was observed. In this zone, all the curves were sigmoidal, but the differentiation in steepness indicated the varying level of species accumulation. Q3 was observed to be the steepest curve, while Q1 was flattened, indicating lower diversity. In comparison with aquatic zone 1, these curves are flatter with lower asymptotes, indicating lower diversity, which explains the higher disturbance in these quadrants. Figure 7.5c indicates almost even distribution in all quadrants of the foraging ground. Q4 and Q5 showed the steepest curve, followed by Q3, suggesting a greater species diversity in those quadrates. In Fig. 7.5d, quadrate 5 showed the highest total population, and a dominant species, Areca palm, was observed. In the spinney zone (Fig. 7.5e), the highest species diversity was observed, and all the curves for Q1–5 were steep. Figure 7.5f depicts the rarefaction curve of the tourism-induced region with low species diversity and less steep curves than the other five regions, which may suggest an anthropogenic disturbance in those regions.

7.3.2 Pattern of Ecosystem Stability

This Ecosystem stability model combined both resistance and resilience, two important characteristics of ecosystems in response to vegetation patterns. Short-term climate anomalies have been considered based on the Enhanced Vegetation Index (EVI). In the Rudrasagar lake area, the spatial variation of resilience and resistance was observed significantly. Ecosystem stability varied spatially with these two components varying across the study area (Li et al. 2018; Willis et al. 2018), and distribution patterns differed significantly with other influencing factors. Resistance was higher in almost all zones in the study area, while tourism-induced areas showed low resilience. The spatial distribution of ecosystem stability was very much similar to the pattern observed in resilience. Resilience was observed to be higher in the Kemtali, Chandanmura, and Rangamura areas, as was the species diversity. Spinney areas showed the highest ecosystem stability (Fig. 7.6). The result of one-way ANOVA indicated a significant difference among the five zones ($p < 0.0001$). Ecosystem stability was highly dependent on both the components, resilience and resistance.

7.3.3 Effect of Vegetation Diversity on Ecosystem Stability

Analysis of the linear regression model revealed significant insights by demonstrating the intensity of the linear relationship among each diversity index with ecosystem stability (Table 7.4). These linear regression models can provide unique attributes of the ecosystem's condition.

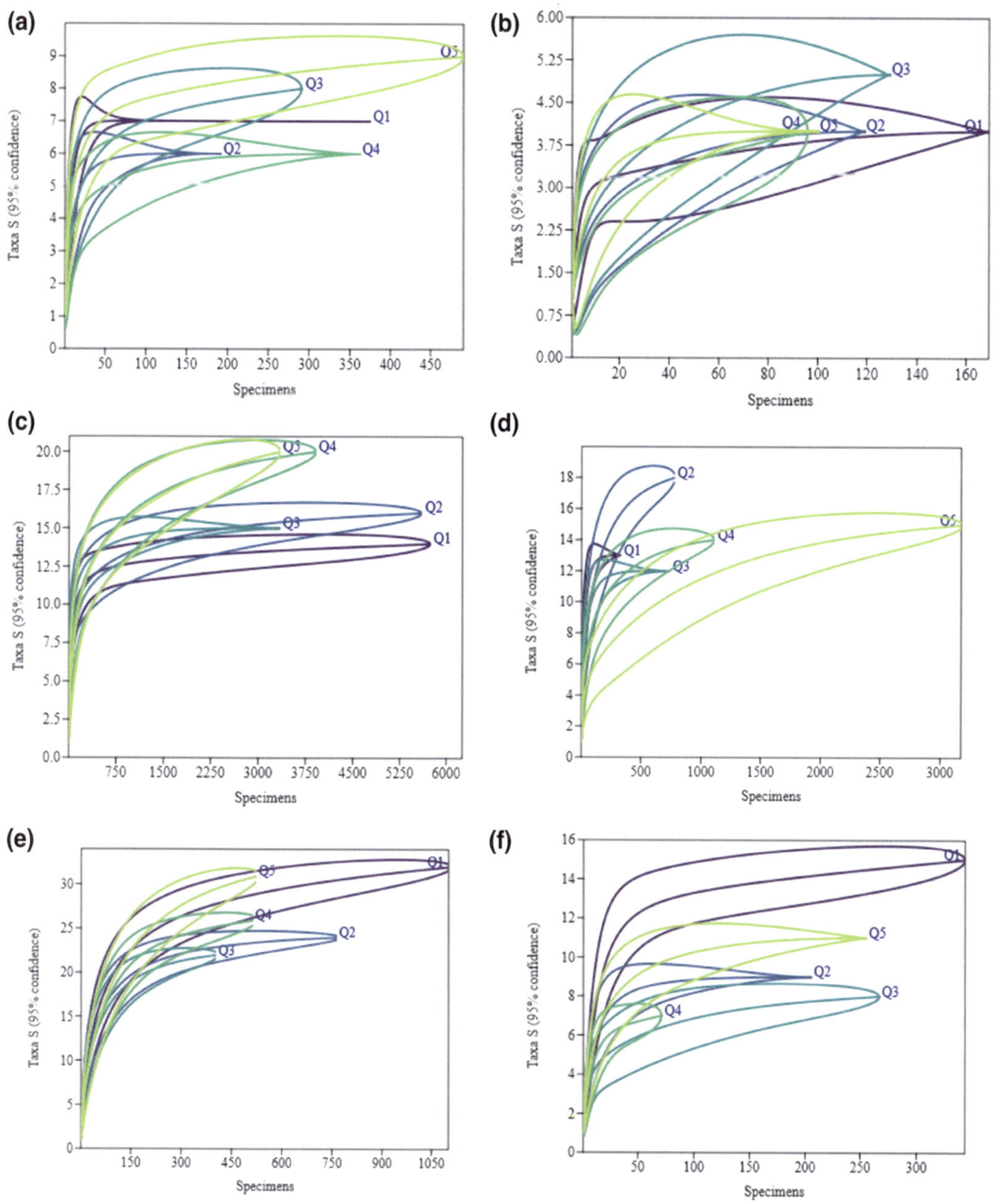

Fig. 7.5 Rarefaction curve for (**a**) aquatic vegetation, (**b**) aquatic disturbed zone in terrestrial area, (**c**) foraging area, (**d**) deforested land, (**e**) spinney area, and (**f**) tourism-induced area

Shannon Index showed the strongest correlation ($r^2 = 0.99$) with a significant *p value* (<0.0001). This may suggest that about 98.76% of the variation in ecosystem stability can be explained by the Shannon's diversity index strongly supports the dependence of ecosystem stability on plant diversity. Similar strong results were observed with the Margalef index ($R^2 = 0.94$ and $P = 0.001$). This regression model can explain 93.90% of the variation in ecosystem stability with the Margalef index. Margalef index mainly focused on species richness, and the result

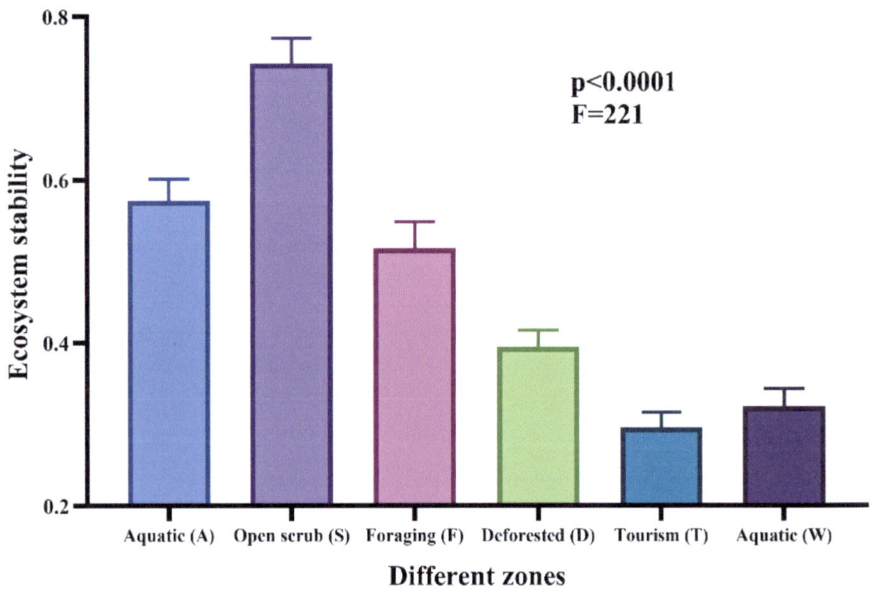

Fig. 7.6 Spatial variability of ecosystem stability among different vegetation zones

Table 7.4 Different diversity index models for assessing ecosystem stability

Indices name	Equation	R^2 value	P value
Shannon index	$Y = 0.2667*X + 0.009387$	0.9876	<0.0001
Simpson index	$Y = -0.6622*X + 0.8581$	0.7990	0.0163
Pielou's index	$Y = 0.7867*X + 0.1482$	0.8404	0.011
Margalef index	$Y = 0.2055*X + 0.03637$	0.9390	0.0014

Source: Computed by author, 2024

demonstrated that areas with higher species richness can exhibit higher ecosystem stability. The regression model species dominance and evenness can emphasise how they contribute to the ecosystem stability. The lowest correlation was observed with the Simpson dominance index ($R^2 = 0.799$ and $P = 0.016$). A relatively weak relationship supported the statement that species dominance indicates lower ecosystem stability as the overall community is diverse. Similar patterns of both the Simpson's and Margalef indices suggested that species richness and dominance patterns are important factors in ecosystem stability. The negative coefficient of Simpson's index may confirm that species dominance is negatively correlated with ecosystem stability. The lower R^2 value and higher P value suggested that evenness and richness were not the best predictor model of ecosystem stability, where other influential factors may also affect ecosystem stability. Disturbed areas like tourism-induced zones and aquatic disturbed areas were characterised by lower diversity values and demonstrated lower stability. These results eventually indicated how the zones with greater diversity and richness tend to increase ecosystem stability, while disturbed areas were observed with reduced diversity and interlinking compromised stability. This may support ecological balance and suggest that more diverse systems are generally more stable and resilient to stress (Fig. 7.7).

7.4 Discussion

The current study examined the spatial variation of plant diversity in different areas in the periwetland areas and its connection to ecosystem

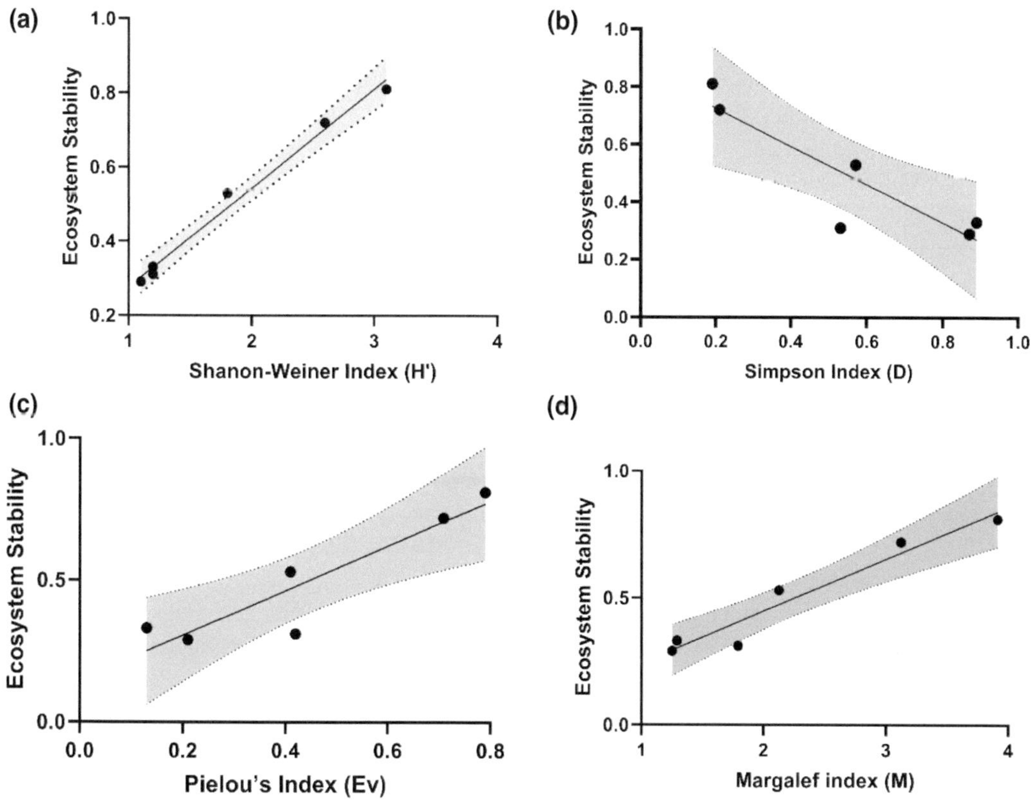

Fig. 7.7 Linear correlations among several diversity indices with ecosystem stability: (**a**) Shannon index, (**b**) Simpson's index, (**c**) Pielou's index, and (d) Margalef index

stability. The four vegetation diversity indices showed similarities but varied spatially with the alteration of the zones. The linear correlation among plant diversity and ecosystem stability revealed through these multiple diversity indices analysis provides significant information on wetland ecosystem dynamics, besides raising important insights about the complexity of these dynamic relationships. The findings of this chapter supported and challenged a few existing ecological research, offering new perspectives on how different attributes of plant diversity contribute to ecosystem stability.

Among the regions, the spiny or open scrub area (S) showed the highest values for the Shannon-Weiner (H′), Margalef (M), and Simpson (D) indices, indicating that this region harbours a highly diverse and well-structured plant community (Fig. 7.4b). Almost even species distribution and lower dominance in the said zone indicated a balanced and resilient ecosystem (Fig. 7.6). Conversely, tourism (T)-induced and aquatic (W)-disturbed areas revealed the lowest diversity, indicating potential ecological disturbances or degradation for anthropogenic interventions (Figs. 7.4b and 7.6). Open scrub area showed the highest stability (0.75 ± 0.02), while disturbed zones exhibited (0.27 ± 0.03), reinforcing the concept of higher plant diversity resilience and stability in urban ecosystems. Similar results were observed in the study of Rooney and McCann (2012), where higher vegetation diversity in temperate marshes positively enhances the stability. Similarly, research on urban wetlands of China and North America (Geng et al. 2019; Lazaro-Lobo and Ervin 2021) has exhibited that ecological restoration frameworks promoting native plant diversity significantly improve resilience against urban stress. The diverse linear correlation strength of different diversity indices highlighted the importance of using multiple metrics for comprehensive and reliable assessment while indicating the complexity of ecosystem stability (Donohue et al. 2016). This finding of a negative slope in the regression model of Simpson's index indicated the inverse relation with ecosystem stability, which contradicted the findings of De Mazancourt et al. (2013). A strong positive correlation of ecosystem stability with Shanon-weiner ($R^2 = 0.89$) emphasised the importance of diversity in maintaining stable ecosystems of urban wetland, while a negative relation with Simpson's index possibly indicates reducing overall ecological resilience.

A few species, such as *Bambusa* sp., *Barringtonia acutangula, Trewia nudiflora, Trema orientalis, Syzygium* sp., and *Syzygium cumini,* were observed to be very common around the vicinity of the Rudrasagar Lake (Figs. 7.8 and 7.9). In recent days, the wetland area has been shrinking, and also the water quality of the lake is deteriorating. The pollution level of the lake is also increasing (Biswas et al. 2024) for several anthropogenic activities. Plant diversity and abundance are also decreasing with increasing urbanisation and settlements around the lake vicinity. Hence, ecosystem stability might decline more rapidly than linear regression models would predict. The observations and findings of this study can have important implications for diversity maintenance and also wetland conservation and management.

7.4.1 Vegetation Coverage Dynamics

Spatial and temporal transformation of vegetation coverage from 2017 to 2022 is illustrated in Fig. 7.10 and Table 7.5, where deep green colour indicates dense vegetative cover, light green colour indicates grassland and shrubs, and grey colour indicates barren land. Compared to 2017, the dense vegetation cover area has reduced in 2022 due to anthropogenic disturbances. NDVI-based assessment demonstrated that the study area had

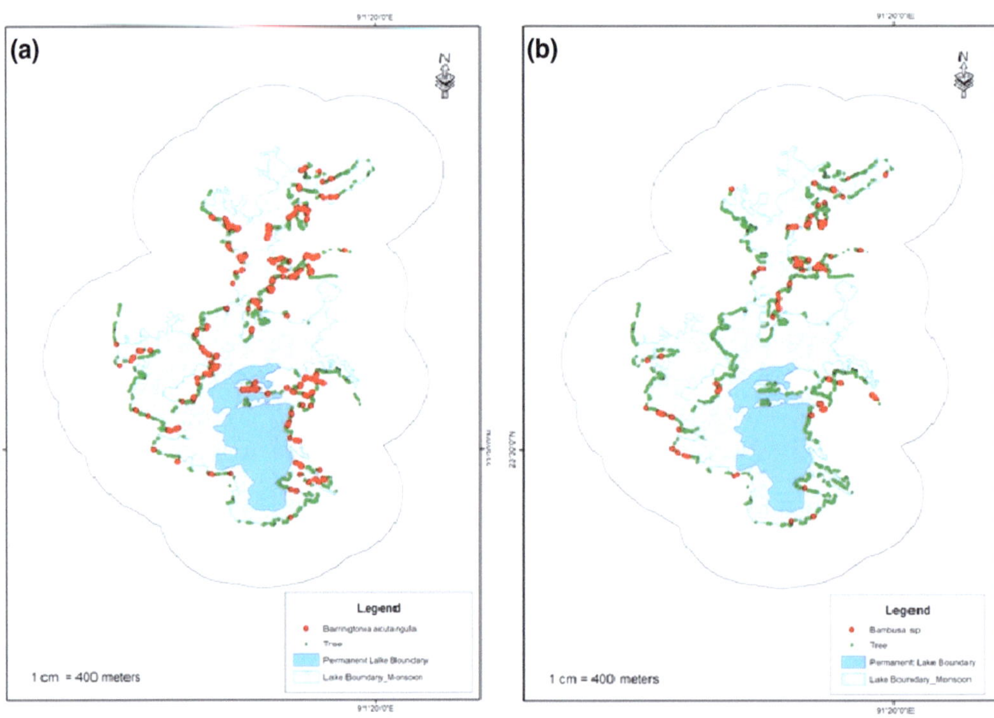

Fig. 7.8 Distribution of (**a**) *Barringtonia acutangula* and (**b**) *Bambusa* sp. around the Rudrasagar Lake

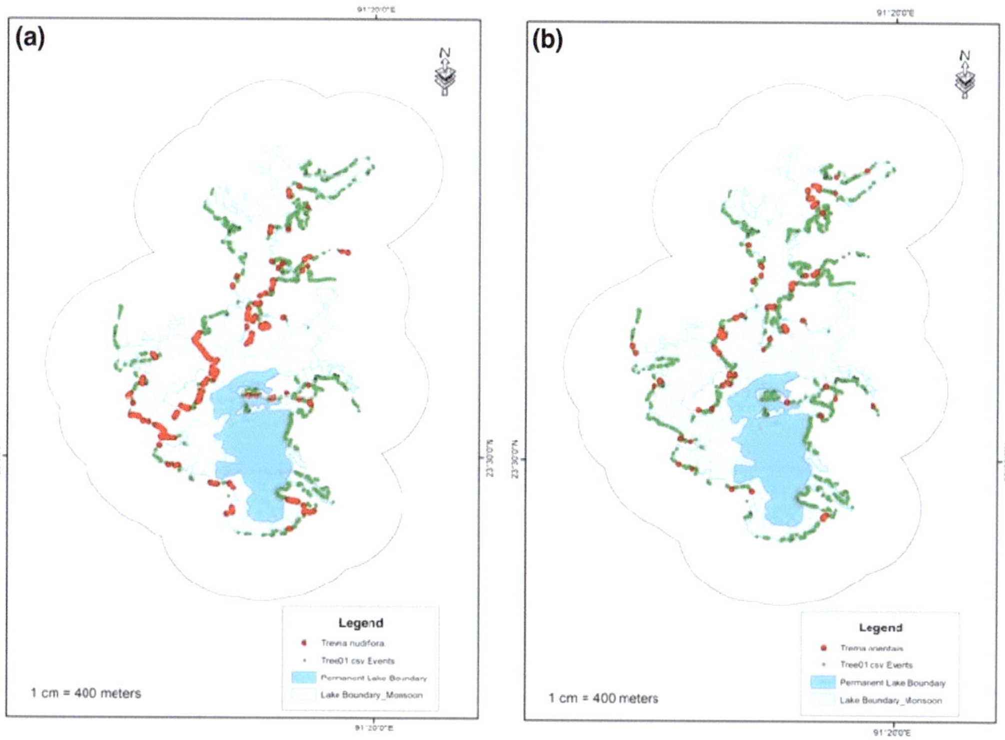

Fig. 7.9 Distribution of (**a**) *Trewia nudiflora* and (**b**) *Trema orientalis* around the vicinity of the Rudrasagar Lake

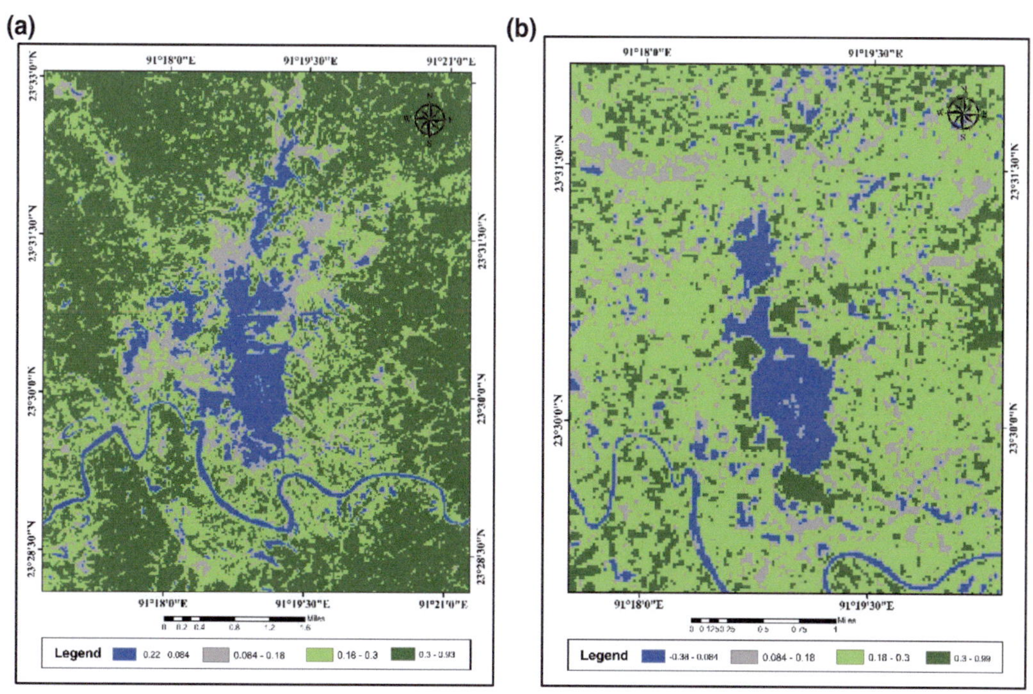

Fig. 7.10 Vegetation covers dynamics adjacent to the lake area (**a**) 2017 and (**b**) 2022

Table 7.5 NDVI-based assessment for green space categorisation

Vegetation category	Timeline 2017 Area (km^2)	Timeline 2022 Area (km^2)	Percentage (%)	Rate of change (%)
Dense vegetative	1072.062	731.385	−31.778	−68.135
Grassland/shrubs	338.571	668.227	97.367	65.931
Barren land	99.675	144.716	45.188	9.008
Water body	76.203	42.183	−44.644	−6.804

Source: Computed by the authors, 2022

undergone significant vegetation cover changes from 2017 to 2022. The data shows that the area classified as Dense vegetative decreased by 31.78% over the 5-year period, with an annual rate of change of—68.14%. This substantial decline in dense vegetation suggests a potential loss of ecologically valuable habitats and ecosystem services within the study region. In contrast, the Grassland/shrubs category experienced a notable increase of 97.37%, with an annual rate of change of 65.93%. This expansion of grassland and shrubland areas may indicate a shift in the landscape, potentially driven by factors such as land-use changes, climate variability, or management practices. The barren land category also saw a significant increase of 45.19%, with an annual rate of change of 9.01%. This suggests a potential conversion of vegetated areas to more degraded, unproductive land cover. Lastly, the Water body category decreased by 44.64%, with an annual rate of change of −6.80%. Similar findings were observed by Aronson et al. (2017) and Li et al. (2020) that emphasised ecosystem fragmentation around wetland and landscape homogenisation as a result of rapid urbanisation. Diversity loss due to urban expansion and encroachment was also found in the study by Seto et al. (2012). This decline in water bodies could be linked to factors such as drought, water management policies, or infrastructure development and may have implications for water resource availability and aquatic ecosystem health. These findings highlight the dynamic nature of the vegetation cover and land-use patterns within the study area over the 5-year period. Understanding these changes is crucial for informed decision-making and the development of effective land management strategies to promote sustainable development and ecosystem conservation (Lambin and Meyfroidt 2010; Piao et al. 2014). Likewise, Biswas et al. (2024) signify in their study that anthropogenic pressure transforms the cultural and aesthetic value of this wetland. Rampant settlements adjacent to the lake are the primary driving force behind the increased rate of decline in dense vegetation cover. The present study also shows how the plant diversity conservation and urban resilience are interdependent, as the open scrub and foraging zones offer better stability against environmental stresses. These outcomes are not only locally relevant, but they are also important for global wetland research. Similar studies on urban wetland in Brazil and Thailand were underscored where plant diversity was established as a factor that enhances flood resilience and improves ecosystem services (Takeda et al. 2013; Dubois et al. 2020a, b).

The mapping of plant species diversity is the primary task of biodiversity assessment and can provide scientific knowledge for conserving biodiversity. However, it is a big challenge to map the spatial distribution of tree species along the periphery of Rudrasagar Lake (Fig. 7.11). Bharati (2019) has identified that almost 46% of the catchment area was occupied by forests, broadly categorised into three categories: tropical evergreen, semi-evergreen, and moist deciduous.

With the help of the map above, we have tried to show the spatial distribution of some important trees. Supplementary Table 7.1 and Fig 7.1 represent the diverse dominant vegetation types identified and documented during the field study. These highly productive terrestrial tree species are *Heveabrasi liensis (523),*

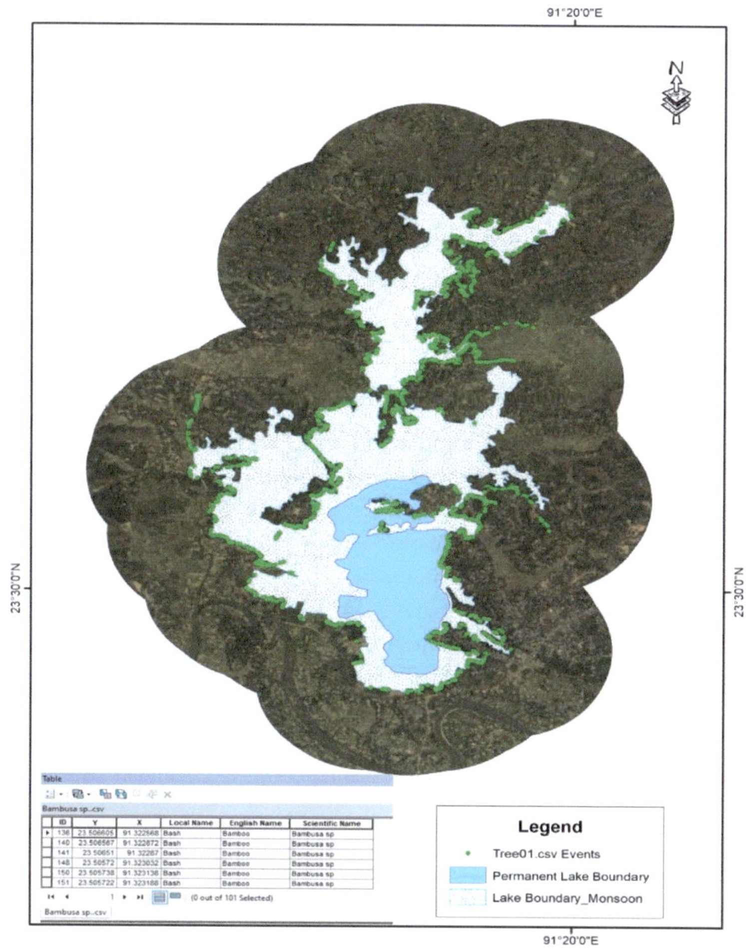

Fig. 7.11 Distribution of terrestrial tree species on the Rudrasagar Lake vicinity

Barringtonia acutangula (372), Syzygium sp. *(351), Trewia nodiflora (274), Lagerstroemia speciosa (235), Syzygium cumini (163), Microcos paniculata (157), Mangifera indica (123), Trema orientalis (110), Bambusa* sp. *(101), Artocarpus heterophyllus (79), Dipterocarpus turbinatus (68), Gmelina arborea (66), Phoenix sylvestris (60), Ziziphus mauritiana (60), Polyalthia longifolia (57), and Borassus flabellifer (52).* The other main and highly productive plant species found on the lake basin are Tectona grandis, Shorea robusta, Gmelina arborea, Artocarpus chaplasha, Vitex peduncularis, Ziziphus jujuba, and Azadirachta indica, along with six bamboo species.

7.5 Limitations of the Study

Although the current study provides valuable insights into the role of plant diversity in the stability of wetland ecosystems, several limitations must be acknowledged. Quantitative assessments in the field of ecological landscape have been a current research trend. However, due to the lack of standard quantitative methodological references, the ecological approach in the current chapter still has some limitations. Adopting a unified framework for measuring plant diversity and ecosystem stability will be helpful for future research that would facilitate reproducibility and cross-site comparisons. Some ecological design methods also do not receive full consideration for their impact. For example, the relationship between the water quality and the plant diversity pattern was not specified in this chapter. Incorporating analysis of physico-chemical properties of wetland water alongside biodiversity assessments will enhance the establishment of interdependence. The land use land cover changes (LULC) in the study area were not administered in this chapter to support the decreasing dense plant patches for settlements. Integrating high-resolution remote sensing data and GIS-based LULC change detection will provide strong evidence of urbanisation, habitat change, and its impact on vegetation types. Integration of high-resolution remote sensing data and GIS-based LULC change detection will give substantial evidence of urbanisation, habitat change, and its impact on vegetation types. Although some limitations still exist, this study may still provide a research idea on the importance of sustainable management of wetlands and their surrounding ecosystems. This chapter could indicate the reasons for implementing ecological design methods for protecting the vegetation diversity as well as biodiversity to improve the ecosystem service functions of any wetland in both rural and urban landscapes. Future research could offer practical field strategies for increasing plant resilience.

7.6 Recommendations

To improve plant diversity and species richness within a wetland ecosystem, a multifaceted scientific approach is very much required. Maintaining proper hydrology with its natural flow regime is very important as it supports a wide variety of both aquatic and terrestrial plant species. To avoid the direct influx of contaminated water and sewage mixing into the lake, implementation of a secondary artificial water body is necessary. Identification of non-native, invasive species is crucial to control their growth and mitigate their unwanted competition. Introducing native plant species to the wetland ecosystem is also very crucial for maintaining heterogeneity. Anthropogenic activities within the lake vicinity should be inhibited to ensure healthy growth and regeneration of plant species. Moreover, regular monitoring and implementation of several management strategies can restore wetland diversity, which leads to enhanced ecosystem stability.

7.7 Conclusions

This chapter integrates a combination of approaches, including biodiversity matrices and spatial analysis, along statistical modelling. Implementation of several diverse approaches in combination with a mathematical model makes this chapter very comprehensive and reliable for measuring vegetation responses to climate variability in Rudrasagar Lake. This chapter may act as a blueprint for future research, which may establish a standardised and quantitative method for analysing ecological design. Quantitative assessment of vegetation diversity patterns indicated the Shannon diversity index and Margalef species richness index as reliable indicators of ecosystem stability. The impact of anthropogenic activities on plant diversity has a long-term effect on ecosystem stability. In this way, the impact on the biodiversity of wetlands can be more comprehensively explored. The importance of a targeted scientific approach for wetland landscape protection and planning of sustainable management practices design needs to be addressed accordingly. Higher tree abundance and species heterogeneity, coupled with increased regeneration in largely undisturbed areas compared to mild and highly disturbed sites, suggested the requirement for adequate protection in the study site. It is, therefore, crucial to establish and maintain policies that prevent further degradation of the northeast Indian Ramsar site.

Acknowledgements The authors highly acknowledge Mr. Gouranga Chakraborty, Manager of Rudrasagar Udhastha Fisherman Samabay Samiti Limited (RUFSS), for providing valuable information and support during the field study. Dr. Stabak Roy and Mr. Sankar Prasad Gayen are also highly acknowledged for their Technical support (geospatial mapping). The authors sincerely acknowledge Mr. Ranjan Das for his assistance in the identification and categorisation of plant species by their botanical names.

Statements and Declaration: Funding This study was supported by the Department of Bio-Technology (DBT), NER-BPMC, Govt. of India (Sanction No.: BT/PR25459/NER/95/1341/2017 dated 14/10/2019).

Competing Interest The authors declared that there were no conflicts of interest in the manuscript and did not harm the environment during the study period.

Authors' Contribution All the authors actively contributed to this work. A. Biswas and P. Debnath conducted a field survey. A. Biswas conceptualised and drafted the manuscript, analysed the data, and prepared Figs. (7.2, 7.4, 7.5, 7.6 and 7.7). P. Debnath prepared the maps (Figs. 7.1, 7.3, 7.8, 7.9, 7.10 and 7.11) and helped in drafting manuscripts. S. Mitra and P. Chaudhuri supervised the entire study and critically reviewed the chapter for improvement. All the authors have carefully gone through the results and discussion section and agreed to communicate the manuscript in this current format.

AI Disclosure Analyses and conceptions in this chapter were presented with the assistance of different analytical software. They were often used to perform calculations, data representations, and help with the interpretation of results and discussion. However, the research design, data curation, and overall analysis were conducted by the authors themselves. Analytical software was used as a mechanical tool to enhance the research process, but all conclusions and insights presented in this chapter are the product of reasoning and decision-making. The authors declare that no AI assistance was used in the research and preparation of this chapter.

Data Availability All the primary data were generated in a field study and a laboratory. The analysed data of this study are included in this published chapter. The administrative boundary of the study area was collected from https://www.diva-gis.org/datadown

References

Aronson MFJ, Lepczyk CA, Evans KL, Goddard MA, Lerman SB, MacIvor JS (2017) Biodiversity in the city: key challenges for urban green space management. Front Ecol Environ 15(4):189–196. https://doi.org/10.1002/fee.1480

Bharati H (2019) Habitat ecology, productivity and fish diversity of rudrasagar lake, tripura (Doctoral dissertation, Central institute of fisheries education)

Biswas A, Debnath P, Roy S, Bhattacharyya S, Mitra S, Chaudhuri P (2024) Spatio-temporal variation in water quality due to the anthropogenic impact in Rudrasagar Lake, a Ramsar site in India. Environ Monit Assess 196(7):1–23. https://doi.org/10.1007/s10661-024-12736-6

Chakraborty SK, Sanyal P, Ray R (2023) Ecosystem services and values of wetlands with special reference with East Kolkata wetlands. In: Wetlands ecology: eco-biological uniqueness of a Ramsar site (East Kolkata Wetlands, India). Springer, Cham, pp 227–255

Champion HG, Seth SK (1968) A revised survey of the forest type of India. Government of India Publications, New Delhi

De Boeck HJ, Bloor JM, Kreyling J, Ransijn JC, Nijs I, Jentsch A, Zeiter M (2018) Patterns and drivers of biodiversity–stability relationships under climate extremes. J Ecol 106(3):890–902. https://doi.org/10.1111/1365-2745.12897

De Keersmaecker W, Lhermitte S, Tits L et al (2015) A model quantifying global vegetation resistance and resilience to short-term climate anomalies and their relationship with vegetation cover. Glob Ecol Biogeogr 24(5):539–548. https://doi.org/10.1111/geb.12279

De Mazancourt C, Isbell F, Larocque A, Berendse F, De Luca E, Grace JB et al (2013) Predicting ecosystem stability from community composition and biodiversity. Ecol Lett 16(5):617–625

Deb DB (1981 & 1983) The flora of Tripura State, vol. I & II. Today and Tomorrow's Printers and Publishers, New Delhi

Debnath P, Roy S, Bharadwaj S, Hore S, Nath H, Mitra S, Ciobotaru AM (2023a) Application of multivariable statistical and geospatial techniques for evaluation of water quality of Rudrasagar wetland, the Ramsar site of India. Water 15(23):4109. https://doi.org/10.3390/w15234109

Debnath P, Roy S, Mitra S (2023b) Utilisation of geospatial approach and WAWQI for sustainable wetland management: a case study of Rudrasagar wetland, the Ramsar site of India. In: Simhachalam A, Ghosh SK (eds) Geoinformatic applications in land, water and infrastructure development, pp 6–36

Diaz S, Lavorel S, De Bello F, Quétier F, Grigulis K, Robson TM (2007) Incorporating plant functional diversity effects in ecosystem service assessments. Proc Natl Acad Sci 104(52):20684–20689. https://doi.org/10.1073/pnas.0704716104

Donohue I, Hillebrand H, Montoya JM, Petchey OL, Pimm SL, Fowler MS et al (2016) Navigating the complexity of ecological stability. Ecol Lett 19(9):1172–1185

Dubois R, Proulx R, Pellerin S (2020a) Ecological uniqueness of plant communities as a conservation criterion in lake-edge wetlands. Biol Conserv 243:108491. https://doi.org/10.1016/j.biocon.2020.108491

Dubois C, da Silva EV, Brondízio ES (2020b) Urban wetlands and flood mitigation in São Paulo: integrating ecological infrastructure into urban planning. Urban Ecosyst 23(6):1315–1331. https://doi.org/10.1007/s11252-020-00982-3

García-Palacios P, Gross N, Gaitán J, Maestre FT (2018) Climate mediates the biodiversity–ecosystem stability relationship globally. Proc Natl Acad Sci 115(33): 8400–8405. https://doi.org/10.1073/pnas.1800425l1

Geng S, Shi P, Song M, Zong N, Zu J, Zhu W (2019) Diversity of vegetation composition enhances ecosystem stability along elevational gradients in the Taihang Mountains, China. Ecol Indic 104:594–603. https://doi.org/10.1016/j.ecolind.2019.05.038

Gupta G, Khan J, Upadhyay AK, Singh NK (2020) Wetland as a sustainable reservoir of ecosystem services: prospects of threat and conservation. In: Restoration of wetland ecosystem: a trajectory towards a sustainable environment. Springer, pp 31–43

Hodgson D, McDonald JL, Hosken DJ (2015) What do you mean, 'resilient'? Trends Ecol Evol 30(9): 503–506

Pimparkar, A. M., Patil, S. N., Patil, B. D., & Kadam, A. K. (2023) Comparative assessment of wetland water quality from rural and urban area of Aurangabad District, Maharashtra, India using water quality index. HydroResearch, 6, 269–278

Lambin EF, Meyfroidt P (2010) Land use transitions: socio-ecological feedback versus socio-economic change. Land Use Policy 27(2):108–118

Lázaro-Lobo A, Ervin GN (2021) Wetland invasion: a multi-faceted challenge during a time of rapid global change. Wetlands 41(5):64

Li D, Wu S, Liu L et al (2018) Vulnerability of the global terrestrial ecosystems to climate change. Glob Chang Biol 24(9):4095–4106. https://doi.org/10.1111/gcb.14327

Li X, Zhou W, Ouyang Z (2020) Forty years of urban expansion in Beijing: what is the relative importance of physical, socioeconomic, and neighborhood factors? Appl Geogr 115:102136. https://doi.org/10.1016/j.apgeog.2019.102136

Lu X, Kelsey KC, Yan Y, Sun J, Wang X, Cheng G, Neff JC (2017) Effects of grazing on ecosystem structure and function of alpine grasslands in Qinghai–Tibetan Plateau: a synthesis. Ecosphere 8(1):e01656

Meli P, Rey Benayas JM, Balvanera P, Martínez Ramos M (2014) Restoration enhances wetland biodiversity and ecosystem service supply, but results are context-dependent: a meta-analysis. PLoS One 9(4):e93507. https://doi.org/10.1371/journal.pone.0093507

Moomaw WR, Chmura GL, Davies GT, Finlayson CM, Middleton BA, Natali SM et al (2018) Wetlands in a changing climate: science, policy and management. Wetlands 38(2):183–205

Newton A, Icely J, Cristina S, Perillo GM, Turner RE, Ashan D et al (2020) Anthropogenic, direct pressures on coastal wetlands. Front Ecol Evol 8:144

Piao S, Nan H, Huntingford C, Ciais P, Friedlingstein P, Sitch S et al (2014) Evidence for a weakening relationship between interannual temperature variability and northern vegetation activity. Nat Commun 5(1):1–7

Rooney N, McCann KS (2012) Integrating food web diversity, structure and stability. Trends Ecol Evol 27(1):40–46

Seddon AWR, Macias-Fauria M, Long PR et al (2016) Sensitivity of global terrestrial ecosystems to climate variability. Nature 531(7593):16986. https://doi.org/10.1038/nature16986

Seto KC, Güneralp B, Hutyra LR (2012) Global forecasts of urban expansion to 2030 and direct impacts on biodiversity and carbon pools. Proc Natl Acad Sci 109(40):16083–16088. https://doi.org/10.1073/pnas.1211658109

Takeda S, Tanaka N, Kuraji K (2013) Effectiveness of a natural wetland for urban stormwater management in Bangkok. Landsc Ecol Eng 9(1):155–163. https://doi.org/10.1007/s11355-012-0195-4

Woldemariam, W., Mekonnen, T., Morrison, K., & Aticho, A. (2018) Assessment of wetland flora and avifauna species diversity in Kafa Zone, Southwestern Ethiopia. Journal of Asia-Pacific Biodiversity, 11(4), 494–502

Willis KJ, Jeffers ES, Tovar C (2018) What makes a terrestrial ecosystem resilient? A complex set of biotic and abiotic factors determines the resilience of an ecosystem. Science 359(6379):988–989. https://doi.org/10.1126/science.aar5439

Xu T, Weng B, Yan D, Wang K, Li X, Bi W et al (2019) Wetlands of international importance: status, threats, and future protection. Int J Environ Res Public Health 16(10):1818. https://doi.org/10.3390/ijerph16101818

Policy Regulation and Innovative Strategies for Overcoming Contemporary Challenges to Combat Climate Crisis by Sustainable Approach Toward Urban Wetland Ecosystem

8

Sumanta Bhattacharya, Bhavneet Kaur Sachdev, and Arkadyuti Seth

Abstract

The expanding rate of urbanization and the increasing effect of global climate change exert significant stress on urban wetland habitats that are critically important as water suppliers, habitat providers, and climate modulators. Such ecosystems continue to experience pollution, loss of habitat, and changes in land use, which make the formulation of sound policy regulations and advanced techniques in the protection and sustainable management of these ecosystems all the more essential. Indeed, this chapter aims to discuss the challenges facing today's urban wetlands and consider how to overcome such challenges via optimal policies, interdisciplinary cooperation, and climate-resilient policies. This one focuses on effective incorporation of wetlands into the built environment through concepts such as green infrastructure and the use of multifunctional space and green space. In addition, the study explores the application of nature-based solutions, for instance, the restoration of natural water regimes and the use of plant species, to increase the climate resilience of wetlands. In addition, it establishes an understanding of a community related to an urban wetland, an engagement of stakeholders, and the financial instruments related to the sustainable management of an urban wetland. Using the models such as the payments for ecosystem services and public-private partnership, this chapter demonstrates how urban wetlands can be financed for financial viability while enhancing climate change mitigation and adaptation. Finally, the chapter concludes that national, regional, and local amalgamation is an effective way of tackling the problems of climate change and the sustainable management of wetland ecosystems in urban areas.

Keywords

Urbanization · Climate-resilience policies · Wetlands · Sustainable management · Green space · Ecosystem

8.1 Introduction

Living facilities like flood control, water treatment, carbon storage and consumption, and the habitats of organisms also depend on the urban wetland. Conversely, due to the increased urgency of urbanization and climate change, these important systems are under pressure. Current urban wetlands are under pressure due to conversion for other uses, pollution, and

S. Bhattacharya (✉)
Asian International University, Imphal, Manipur, India

B. K. Sachdev
Suresh Gyan Vihar University, Jaipur, India

A. Seth
Sai Nath University, Ranchi, India

S. K. Chakraborty et al. (eds.), *Wetlands and Urban Resilience*, Sustainable Development Goals Series,
https://doi.org/10.1007/978-3-032-06881-1_8

deterioration, thus indicating that they are unable to perform their duties. In this respect, the governance of urban wetlands is not only an environmental issue but a major strategic factor in the attainment of climate change mitigation and promotion of sustainable urbanism. Urban wetlands as a philosophy is not limited to the market conceiving of spaces within the city simply as "greenspace" but rather an approach of including wetlands within the fabric of urbanism as well as its management systems. Hence, wetlands, especially in urban areas, play multiple important roles that address climate change mitigation and adaptation. Some of these are to use water through absorption and filtration of stormwater, thus minimizing flood incidents during rainfall and CO_2 isolation to address the issue of the heat island effect in cities. Further, urban wetlands fulfill the function of carbon sources, hence assisting the city in undertaking climate measures. The other issue that is evident, especially for an expanding city, is the one concerning wetland areas. Most wetlands have been developed for human settlement and infrastructure development through processes like draining and filling, hence making cities. Not only do urban wetlands have an increased importance as climate change exhibits itself through sea level rise, increased rain events, and temperature fluctuations, but the concerns of environmental quality and public health are also protected. Nevertheless, wetlands remain vulnerable to conversion for other uses, and if they are not protected and restored, they can cease to exist and, in doing so, diminish their functionality in supplying these services. It is clear that management of the challenges that surround wetlands in urban areas needs a revolution in thinking in the management of urban ecosystems. Classically, city design has considered wetlands as wastelands or even negative assets. However, there is a newly updated appreciation for the environmental, social, and even the economic roles of wetlands, hence their incorporation into sustainable development within cities. It becomes clear from the case of urban wetlands that these areas play a critical role in the development of climate-resistant cities. They protect against natural disaster and health of people in cities. With respect to the above observations, the following critical challenges constrain the optimal application of urban wetland systems as the strategic option for the management of the climate crisis: Poor policy regulation of urban wetlands; Contemporary issues ailing urban wetlands; and the inability to devise and implement workable strategies that can help foster the optimal management of urban wetland systems. In this case, proper polices and regulations in the management natural environment will be fundamental in avoiding further degradation of these important ecosystems. Thus, while designing modern protocols, urban planners, scientists, and policymakers have to work on creating new strategies in the area of wetlands' preservation and restoration that would use the ideas of modern technology and people's participation. These strategies should also consider the dynamic and interconnected nature of urban wetland ecosystems and their responsibilities not only for climate change and adaptation in the region but also for environmental management at large (Alikhani et al. 2021).

Some of the policy regulations should cover the ways through which the wetland ecosystems can be placed as a priority in the urban sprawl, and how the wetlands can be protected against conversion and pollution activities. Wetlands in urban environments should be incorporated into green infrastructure, with objectives such as regulating stormwater, preventing the release of pollutants, and providing recreational areas for urban residents. Moreover, the actions that have been taken for the restoration of wetlands should aim to improve the ecosystem services provided by wetlands to support their climate change resilience. These restoration projects cannot focus only on the return to the previous state of the wetlands but need to adopt the principles of ecological restoration, which posits for the restoration of the natural hydrological regime of these affected water bodies, raising the biological diversity on them, and at the same time, their potential to sequester carbon dioxide from the atmosphere (Bassi et al. 2014).

Besides policy and regulation, other aspects have interesting development potential for wetland restoration and management in urban areas. Modern tools such as remote sensing, geographic information systems (GIS), and the use of drone technology can effectively capture the implications of wetland health and, in effect, help in the planning of management strategies. They can also help to practice adaptive management—response to climate change effects for wetlands by using real-time results. Moreover, finding such finances for the long-term wetland conservation efforts can also be found through new approaches like partnership with the private sectors, carbon trade, tourism trade, etc. Of vital significance is the role of community participation in steering the management of wetland resources. The inhabitants of the local context, especially those within close proximity to wetland zones of the country, have rich information about the geography. These corrective and monitoring exercises also promote the ownership of the natural resource management-related initiatives by these communities, while at the same time guaranteeing that the applied corrective measures are culturally appropriate. The information calls for public awareness campaigns regarding the utilization of wetlands and the importance of preserving them (Guo et al. 2017).

The purpose of this research is to review the state of the art of the use of urban wetlands to mitigate the impacts of climate change, mainly focusing on the management and restoration processes. The objectives include getting acquainted with the current issues that affect wetlands in urban areas such as this one; preventing urban wetland conversion, pollution, and climate change impacts; and studying the ensuing policy measures and new methods of combating the challenges. Thus, the study aims at reviewing the best approach to incorporating UWs into the city planning process for climate change resiliency and improved ES provisioning. Furthermore, it aspires to give guidelines on the policy and technological practices that should be adopted with regard to urban wetland conservation and management.

8.2 Methodology

The current research utilizes face and content validation quantitative research approach in addition to a qualitative approach, as depicted in Fig. 8.1. Here, it reviews the literature in urban wetland ecosystems, climate change, and policies. Literature of wetland restoration projects in urban areas will be reviewed to determine the best practices and the difficulties encountered. Although urban planners, environmental scientists, and policymakers may already have some minimal standards regarding the management of wetlands, we will endeavor to find out how they carry out the practice in detail. Meanwhile, the use of geographic information systems and remote sensing will be employed in the evaluation of the wetland health and dynamics. Therefore, the data analysis will focus on assessing the efficiency of policies and new measures concerning sustainable wetland management.

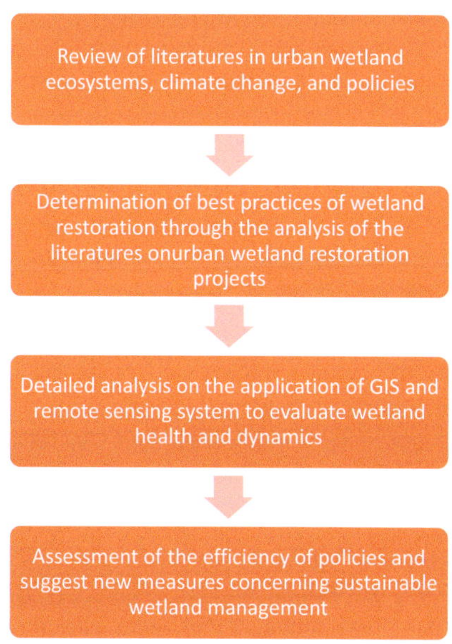

Fig. 8.1 Methodological flow chart

8.3 Incorporation of Wetlands into Urban Planning

The inclusion of the wetlands in the management and design of cities is essential in attaining sustainable urban centers, and also in managing the impact of change in climate. Such spaces are ecosystem services; as cities become larger and develop further, it becomes crucial to maintain them. Wetlands, which can control water systems, contribute to species variety and act as buffers against climate change in cities, form part of this green infrastructure. But the incorporation of wetlands into the overall development of urban environment practices is a great challenge, which needs informed strategies as well as a different kind of approach toward the overall management of urban space. Urban zoning laws, approaches toward integrating environmental protection and management into the uncontrollable growth of cities and urbanization, as well as wetland-friendly urban design, are some of the major steps that help in wetland protection and bringing into city planning (Pandey and Ghosh 2023). Figure 8.2 depicts the processes through which wetlands can be incorporated into urban planning.

Urban zoning laws are basic in maintaining existing wetlands and incorporating wetlands that have been restored into the urban structure. A protective function can also be identified with the help of zoning: it is impossible to fill or build on wetlands for the sake of expanding city territories. This way, cities can also limit the encroachment of wetlands, which are an important resource and can be easily transformed to commercial, residential, or industrial use because of zoning laws. Local zoning ordinances are able to identify certain wetlands to be conservation zones, whereby development activity is either prohibited or very much regulated. As such, this approach can avoid the loss and further deterioration of wetlands, and let these very important ecosystems continue to perform their important roles that include acting as water purification systems, flood management, and carbon sinks. Besides the protection of existing wetlands, zoning ordinances regulating the development of

Fig. 8.2 Incorporation of wetland in urban planning

built-up urban areas can be employed in the integration of restored wetlands into redevelopment projects. Physical redevelopment of urban wetlands, which have been degraded, is one of the critical approaches to increasing urban resilience to climate change. As wetlands are restored, there are specific ways that they can be brought back into the urban context to provide value both to the environment and the inhabitants of urban centers. This process is eased by zoning laws with the inclusion of restored wetlands in the development plan for a city. This may comprise designing wetland restoration activities to fit into the overall urban development plans for the production of structures such as roads, buildings, or halls. By developing areas that promote wetland rehabilitation and their application to urban areas, cities can enhance the general environmental status as a result of the increasing population density (Mitra 2024).

One of the main issues that urban officials face when implementing wetland policies is the accommodation of the spread of built-up areas while conserving wetlands. That is why, as the cities expand, urbanization often happens at the cost of valuable ecosystems—wetlands in this case. That said, there are ways of resolving this dilemma through management of planning strategies to favor the development of the city without the elimination of green areas. The first is to guarantee that wetlands incorporate green urban US infrastructure. Green infrastructure means a system of natural and semi-natural elements that sustain ecosystems' goods and services and enhance urban sustainability. As a part of this infrastructure, wetlands are important for stormwater management, as they help minimize the temperature difference between the city and the surrounding environment, and act as habitat for the local wildlife. Thus, SES collaboration can enhance the incorporation of wetlands in green infrastructure networks of cities to avoid the loss of valuable habitats and water conservation in the course of urban development. The inclusion of wetlands within GSI necessitates understanding and working with different disciplines of urbanism, ecological systems, and approaches to climate change. Another stakeholder is the planner, who should ascertain specific participating wetland sites and ensure that they link with other sites occupied by other land use types, including parkland, forests, and rivers. This connectivity is indispensable for preserving species and creating conditions for their displacement within the territories of cities. This is the case that wetlands can be used naturally as the formal road networks that offer habitats for the wild creatures and at the same time social amenities for the people. Hence, urban wetlands are a strategic asset for the management of a city that seeks to develop strategies for combating the effects of climate change (Li et al. 2022).

The fourth important component associated with the integration of wetlands in city design is wetland-appropriate design for cities, which considers both the environmental benefits of wetland systems and the human requirements for utilizing space. This way, they should consider how wetlands can be made to be useful for both ecological and social aspects in urban design. For example, Wetlands can be designed to function as multipurpose areas and serve as recreational places for people, and also serve other functions such as conserving bio-diversity as well as enhancing climate resilience. Some aspects of wetland are pedestrian paths, informative kiosks, or areas for meetings and other purposes public can both actively interact with the urban wetlands and at the same time receive valuable services offered by wetlands. Such facilities can become meaningful sources of irritation in connection with the environment and improve people's quality of life in large cities (Ahn and Schmidt 2019).

Wetland-friendly urban design can also relate to the development of spaces that encourage the support of related biota and climate-change mitigation. Wetlands support diverse species, most of which are considered to be of profound importance in sustaining balance in ecosystems. Urban design must ask how wetland can be rehabilitated or created in order to support the Plantae and Animalia kingdoms. Furthermore, wetland has a significant function in climate change adaptation since it reduces the consequences of climate change. They help as natural barriers against floods, act as water absorption facilities, and also regulate temperature in the region. These functions, if incorporated in wetland design in urban areas, will help the cities to adapt to climate change while at the same time offering the public various services. Occasionally, it may be necessary to solve the problem of creating or restoring wetlands within densely urbanized areas. Under this, it can incorporate unused plots of land, converting industrial areas with low activity, or integrating wetlands with roofs and other urban structures. Such strategies enable the cities to optimize the use of the limited space and simultaneously harvest the benefits in terms of the environmental standards of the built-up zones. Also, while implementing wetlands in urban environments can increase the aesthetic and recreational value and at the same time, promote increased mental health, as well as environmental education (Kim et al. 2011).

Cumbersome challenges are inherent even in the integration of wetlands into this strategic planning process. Organizing it involves a team, which includes urban planners, decision-makers, environmental groups, and the residents of any affected region. The particular aspects of communication management in an organization have a significant impact on the preservation process and the improvement of the situation in the wetlands while taking into consideration the necessity of urbanization. Therefore, it is important that the Endowment engage the public as the lead planner for its wetland projects to ensure that wetlands that are designed and managed meet the public's needs while also sustaining ecosystem function (Devuyst 2000).

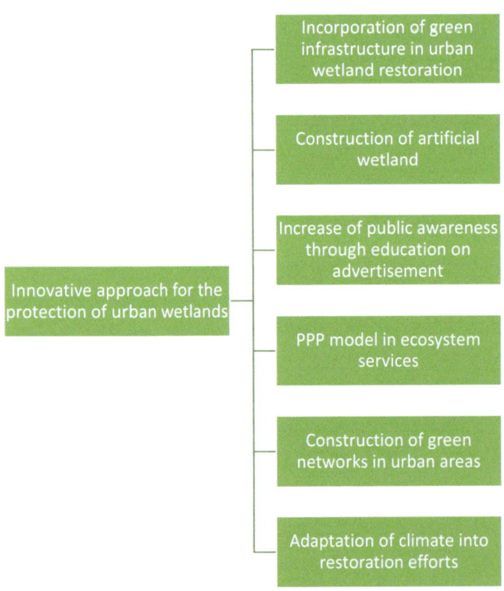

Fig. 8.3 Innovative approach for the protection of urban wetlands

8.4 Result

This chapter identifies technological solutions as critical to enhancing stewardship of wetlands in smart cities, given the myriad challenges of urbanization and climate change interventions. Figure 8.3 depicts the innovative approaches that can be taken for the protection of urban wetlands. Due to the increasing pollution, invasion, and climatic disasters affecting wetlands in urban areas, the use of integrated technologies as a way of maintaining and improving the functionality of wetlands will act as a useful endorsement. These technologies provide improved ability not only for monitoring but also for the restoration processes, improved management, and increased awareness toward wetland conservation needs. For smart wetland management, the most remarkable technological advancement is that of remote sensing technologies. New methods, such as the use of satellite imagery and drones, have enhanced the assessment of wetland health and the level of decay in such habitats. Spatial remote sensing has a high spatial extent at a relatively coarse grain that can be employed for wetland evaluation, to map conversion, and for water quality assessment. This way, environmental managers can establish signs indicating degradation including reduced maturity of the plants, or fluctuations in water levels, and can take corrective action. On the other hand, drones afford a more detailed and local view of the wetland and can be used to survey small patches or parts of the wetland that may not be easily reached on foot. Such technologies allow for continuous collection of data, and as such, the conditions of wetlands can be assessed for management or restoration purposes (Lawler et al. 2013).

The other technology applicable to smart wetland management is geographic information systems (GIS). GIS facilitates the management of wetlands by identifying essential areas for conservation/ restoration by mapping the wetland systems. As applied in wetland planning, GIS makes it possible to superimpose data layers for different aspects, for instance, land use, water quality, and species presence, to come up with a suitable map for wetland management. They are very instrumental when it comes to pointing out the spatial pattern of wetland areas that are likely to harbor a connection of different habitats to allow species replenishment. With GIS, one is also able to evaluate the effects of the expansion of urban areas on wetlands, as well as evaluate where to develop the next urban areas without destroying the wetlands. Functions of spatial analysis provided by GIS actually make it the most important tool used in decision-making for resource allocation in wetlands (Premalatha et al. 2010).

Another technological innovation in smart management of wetlands relates to the use of sensors for measurement of water and ecosystem health. Self-calibrating probes for measuring the pH, dissolved oxygen, turbidity, and nutrient levels of water can be placed in the wetlands to obtain current data on water status. Such information is invaluable for determining the state of wetlands, especially where pollution or high nutrient loads predominate. This allows managers to identify unfit algal formations, pollution hazards, or excessive nutrient surges that could consequently degrade the wetlands. The inclusion of sensor networks in wetland management systems means that the changing of the environment will be documented much better, and new threats can be countered much faster (Chen et al. 2023).

AI and ML are gradually being applied to strengthen wetland practices. Remote sensing, GIS, and sensors are techniques used in the collection of large amounts of data on the environment, and AI is capable of extrapolating aspects of this data that might not be easily recognized. For instance, it can estimate the future changes in wetland conditions using the past records and factors such as weather and landscape changes. These predictive capabilities provide for the development and implementation of solution strategies prior to the environmental issue reaching disastrous proportions. Also, AI can be applied to the creation of automated solutions for the management of existing wetlands, for example, to regulate invasive species or fine-tune reseeding methods for local plant species. One of the major strengths of AI is its capacity to analyze large volumes of information within the shortest time and with great accuracy—a strength that can prove invaluable to wetland management processes (Kim et al. 2024).

Even restoration methods are not left in the cold when it comes to technology. For example, enabling bioengineering techniques along with geospatial information systems and aerial imagery has brought about the enhancement of restoration interventions' effectiveness. Many wetland restoration projects involve hydrologic, plant, and wildlife restoration, and thus, the use of data to manage the restoration designs and procedures can provide precision tools. It also noted the ways in which drones could be used to plant vegetation or spread seed over huge areas of wetland, and the way in which GIS could assist in choosing the most advantageous zones for rehabilitation work. These technological tools make it possible to restore wetlands more effectively and even in hard-to-reach or difficult-to-access areas by generalizing outcomes to prevailing conditions of wetland site (Semeraro et al. 2021).

Also, the increased usage of citizen science applications provides support for the involvement of local communities in wetland conservation. Applications on mobile devices and innovative databases enable residents and visitors, scientists, and other conservation practitioners to feed into data on the wetland state. These platforms give great data from numerous contributors and enrich the range and quality of the monitoring. Citizen science is especially amenable to large cities, where perhaps it would be too expensive to send professional teams for the regular monitoring of all wetland sites. These technologies help increase the public's awareness of wetland conservation problems and would ensure that local communities will be more inclined to participate in wetland management. Moreover, the encouragers of public engagement include the innovations of the human interface through technologies like VR and AR. Thus, these technologies may be helpful to provide the audience with awareness of wetland value and its contribution to climate change issues. The use of technology such as VR and AR can also feature virtual tours of wetlands, hence people may rarely visit these regions. It also publicly mobilizes support for conservation and policy, and other measures for the protection of wetlands (Buytaert et al. 2014).

It explores the management of wetlands in relation to the enhancement of technological improvements that enable the development of desirable and sustainable urban systems. The values and benefits of wetlands are widely known today to act as mitigation measures in climate change, including the prevention of floods, water purification, and carbon storage. Through the latest technologies like remote sensing, GIS, sensors, AI, and platforms for community engagements, these cities can effectively manage the wetlands, enhance the restoration processes, and make sure that the wetlands' future is secure. In view of climate change affecting urban areas in the future, technology will be even more relevant to the management of wetlands to support cities by providing tools to protect these ecosystems and improve urban sustainability. These unique and valuable ecosystems can persist to deliver enhanced ecological functions in addition to supporting wildlife populations and mitigating climate change by improving urban wetland management through technology (Stefanakis 2019).

8.5 Nature-Based Solutions and Climate Adaptation

The application of nature-based solutions in this scenario is considered more effective through mitigating the implications of climate change and attaining sustainable resource use. These solutions restore the natural landscape in order to solve environmental issues, whilst the outcomes tend to include such advantages as increased bio-diversity, augmented ecosystem services, and higher climate change mitigation. Nature-based solutions are especially important for climate change adaptation in the context of urban wetlands, as they are iconic ecosystems with multiple benefits for the ecosystems and urban societies, as depicted in Fig. 8.4. For example, wetlands, which are sometimes called "nature's kidneys," perform a number of important functions including the regulation of water levels, water filtration, and the storage of carbon, as well as the support of species diversity. The second is that the nature-based solutions such as the wetland restoration and conservation provide urban centers with effective mechanisms to adapt to the impacts of climate variation at relatively lower costs (Knegt et al. 2024).

It is, however, important to note that more sensitive to climate change effects, including the greenhouse effect, floods, heat waves, and storms, among others due to increased urbanization. These problems become worse with city development because natural habitats and open spaces are replaced by structures, and impervious surfaces decrease the ability of urban areas to cope with the stormwater, reduce surface temperature, or improve habitat quality. These challenges can be met through nature-based solutions where what has been destroyed through climate change is replaced through the process of natural conservation such as the wetlands. Wetlands act as natural barriers to rain and floods, helping to regulate floodwaters and minimize flooding in urban areas. They also play an important role in temperature regulation by reducing the impacts of the urban heat island effect through the action of transpiring organisms (Shivanna 2022).

Wetland also acts as carbon pools and sequester carbon dioxide from the atmosphere. Gradually, climate change goals have made them critical in the CAR-sink, mainly due to their contribution to carbon storage. Cities can achieve society's carbon reduction ambitions while at the same time reintroducing lost wetland habitat and conserving existing aquatic ecosystems. The additional ability of wetlands to store carbon also positions them as a critical resource in the Natural Infrastructure for Climate Resilience framework, as such solutions are seen to have two benefits: removing carbon from the atmosphere and increasing urban resilience (Moomaw et al. 2018).

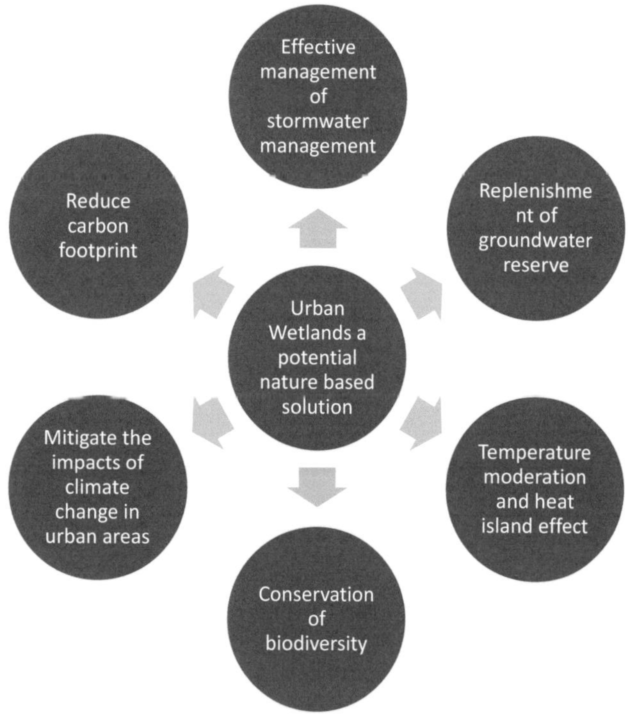

Fig. 8.4 Urban wetlands: a potential nature-based solution

While the welfare of Los Angeles is already highlighted culturally, socially, and economically, there are also social and economic benefits that can be derived from the Wetlands. Through the multiple benefits of sustainable WS in urban areas such as aesthetic, recreational, and educational values, wetland restoration can improve the living standards of citizens. Such places can benefit psychological assistance and quality of living, as these elements bring the feel of nature to the urban environment that is growing rapidly. Additionally, the recreational function expressed through the aesthetic appeal of urban wetlands can create economic impacts that help support sustainable urban growth strategies. Wetland restoration, therefore, becomes a way through which cities can design landscapes that are useful on several fronts and for different people (Díaz-Pinzón et al. 2022).

Another advantage of NBS in urban wetlands is that it provides support for biodiversity. Wetlands are one of the most diverse in terms of the number of species that inhabit them, for example, birds, amphibians, fish, as well as some plant species, and migratory birds. However, urban wetlands can help develop the habitats of these species, thereby increasing the overall volume of bio-diversity. This biological diversity is not only necessary for sustaining the health of the ecosystems but is also part of providing climate-resistant urban systems. In heterogeneous ecosystems, there is more capacity to change and respond to variations in their environment and other complicated interactions for ecosystem services, which include pollination, pest control, and nutrient cycling, among others (Alikhani et al. 2021).

Nature-based solutions also support attainment of broader environmental and sustainability objectives, as well as objectives described in the United Nations Sustainable Development Goals (SDGs). Wetlands initiatives for restoration and conservation make them fit well within the framework of the selected SDGs such as Climate change, Life on land, Life below water, and Sustainable cities. It is now possible to incorporate wetlands into urban systems to facilitate change toward more sustainable, resilient, and climate-urbanism. Moreover, the use of nature-based solutions in the wetland management also contributes to the development of more equal cities since those benefits also include such factors as access to green areas and equality and social respect for more vulnerable population groups, as well as environmental justice and social cohesion. A number of advantages are associated with the application of nature-based solutions for climate adaptation through wetland management: First, the approach of wetland management contributes to climate change adaptation through enhancing ecosystem functions, stabilizing habitats, improving soil conditions, and increasing water runoff capacity WSC-The successful implementation of the use of nature-based solutions to manage wetlands for climate change adaptation depends on effective planning, stakeholder participation, and long-term strategies. Coproducts with continuing education: To initiate a management strategy that will foster the creation of design that can sustain the balance between urban development and restoration of natural ecosystems, urban planners, policymakers, and environmentalists have to seriously establish the most appropriate strategies to use in addressing future goals. Other threats include competition for the use of land, pollution, and alteration of wetlands through habitation and development. Another reason for engaging stakeholders is that their support is key to making and sustaining wetland restoration in the future. Communications and outreach related to wetlands and their functions in climate change mitigation are critical components of European wetland conservation and management. This is primarily because for nature-based solutions to be integrated in the planning of cities, the city must have policies on the preservation of wetlands. These policies should consist of encouraging land-use regulations that protect wetlands from being degraded, support wetland restoration initiatives, and ensure reliability in green urban infrastructure networks of wetlands. Furthermore, cities should spend on the identification and observation to evaluate how efficient nature-based solutions are, as well as the adjustment of measures where necessary. To this end, emerging technologies like remote sensing, geographic information systems (GIS and sensors) can assist the city in detecting the state of wetlands, as well as changes in ecosystem status across time (Anderson and Gough 2022).

8.6 Community Participation and Stakeholder Engagement

Stakeholder involvement and public outreach are core requirements of wetland conservation and resultant climate change mitigation. Triad for decision-making does not only increase the effectiveness of wetland restoration but furthermore assures that the project outcome will be sustainably achieved and in the best interest and benefit of different stakeholders involved and the society at large. Engaging the communities in the management of wetlands makes them take full responsibility for the systems since they are well aware of the need of maintaining such valuable resources. Wetlands are commonly found in urban and peri-urban areas and are highly vulnerable to land conversion, pollution, and other developmental threats. People who inhabit the areas adjacent to the wetlands are themselves knowledgeable in terms of history, functions, and problems as to the natural habitat in question. Getting people involved in wetland management also opens up a way for the application of folklore knowledge, which supports scientific work and aids in the recognition of community wetland conditions. Such an approach enhances the implementation of management strategies since they are a product of both scientific research and floor knowledge (Hernandez et al. 2024). Figure 8.5 depicts the steps required for effective community participation and stakeholder engagement.

Civil society, nongovernmental organizations, policymakers, and the public are also important partners in the overall wetland management, and this is because they form part of the stakeholders who are engaged in all the wetland management processes. Through stakeholder participation in the wetlands conservation and restoration planning and management, it becomes easy to

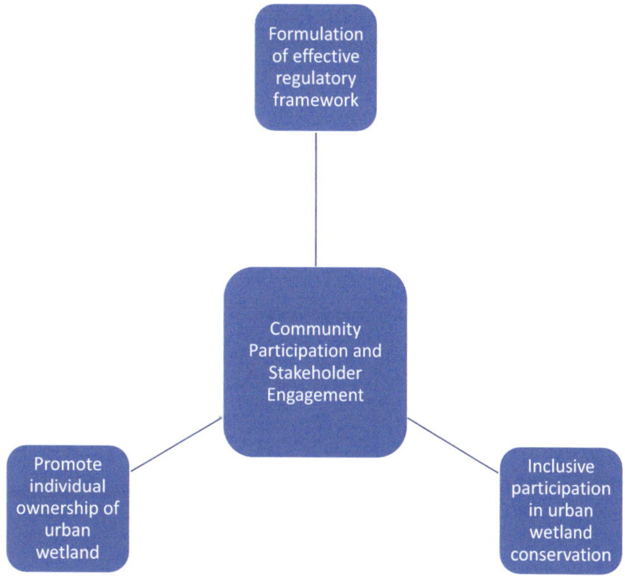

Fig. 8.5 Community participation and stakeholder engagement for the implementation of urban wetlands

find out where there are opportunities for cooperation, conflict, as well as where support is likely to come from. Government agencies provide the political/legal framework, and other organizations with environmental backgrounds give the required information and guidance on restoration methodology for wetland conservation and management. It also requires partnerships with local stakeholders like the businesses in the tourism sector, agriculture, or real estate companies whose self-interests would be extremely important for their contribution to local environmental initiatives (Kadoma et al. 2023).

Involvement of the community and stakeholders in wetland conservation also promotes ownership of individuals, which in effect supports a conservation culture. Those who form positive attitudes toward Wetlands and participate in the conservation process will hence exhibit favorable behavior in avoiding polluting wetlands among other social ills, avoid encroachment on wetlands, and champion environmental policies that protect wetlands. Publicity and sensitization are other means of enhancing public understanding of the role of wetlands in climate mitigation and thus should be promoted (de Oliveira et al. 2024).

8.7 Discussion

Economic valuation and financing mechanisms are necessary components to effective management and conservation of urban wetlands. The environmental importance of wetlands is clear to most people, but the economic benefits are rarely highlighted. However, when a wetland is given an economic value, it is easier to justify the cost of protection and possibly restoration. Economic valuation makes it possible to assign a concrete value to the multiple functions that wetlands perform for society in terms of flood control, water treatment, climate change mitigation, and support for species. These are required services for the health of urban dwellers and improvement of cities' coping capability in the face of climate change. That is why understanding of this value enables policymakers and stakeholders to better decide where to spend money for wetland protection and restoration. Hence, there are different techniques for economic valuation of wetlands, which are market valuation, contingent valuation, and cost-benefit analysis. Market-based assessment of wetland ecosystem services can be defined with reference to the cost of substituting them through artificially provided or engineered services. For instance, whilst the expense of putting up floodgate structures or water treatment facilities can be used as the cost that reflects the value of wetlands in preventing floods or enhancing water quality, respectively. The last of the primary methods, contingent valuation, entails surveys to establish the variables that people would pay for wetland conservation in order to establish the social benefits of these systems. Cost-benefit analysis is a technique that aims at comparing the cost of wetland restoration or preservation with the benefit likely to accrue from such processes, thus providing the rationale for the likely cost of financing such projects. Therefore, beyond the use of economic valuation, there is a need to encourage the application of sustainable financing frameworks to support wetland management and conservation. Government-impairing and/or grants remain inadequate funding mechanisms to address the increasing needs of wetland rehabilitation and conservation. Thus, new ways of financing are sought: PPCs, eco-tourism, and PES. Public-private partnerships are a useful cooperative venture where those in government organizations, the private sector, and nongovernmental organizations can collaborate in order to combine finances, employees, and knowledge to ensure protection of wetlands. Tourism begins to provide revenue derived from the sale of environmental services; eco-tourism begins to pull revenue through the sale of environmental services such as wetlands. PES programs, as with other compensation-based programs, offer monetary rewards and other incentives for land owners and communities to preserve wetland through recognition of their value in delivering ecosystem services (Yang et al. 2023).

The concept of adaptive management is beginning to be used more widely in the context of wetland ecosystems to build climate change resilience. It is also an unstructured, cyclic approach to management of strategies where adjustments can be made due to the emergence of new information, different conditions, and new issues. Bearing this in mind, adaptive management is most effective for wetland conservation and restoration as it gives much attention to possible climate change affecting its strategies. Wetlands are functions of climate, for instance, water availability, temperature, and the rate of change of climate variables such as water availability, temperature, and the frequency of flooding and droughts. Consequently, adaptive management offers a mechanism through which society learns and responds to these kinds of changes, which allows wetlands to continue generating beneficial ecosystem services (Williams and Brown 2014).

Monitoring and evaluation is one of the most important of all tenets of adaptive management. When it comes to wetlands, it refers to the gathering of information on numerous environmental factors, including the amount of water, state of vegetation, kinds and number of species, as well as water quality. These indicators approximately measure the trajectories of the ecosystem and help managers to detect new threats or stressors that may appear. For instance, if a wetland has started to show signs of degrading water quality, then adaptive management means that the problem will be identified and the solution implemented. Monitoring can also be used in the evaluation of the success in restoration endeavors and adaptability for change, as monitored data informs the interventions about the relevance of the overall goal of wetland restoration for improved climate change resilience (Oakes et al. 2022).

Climate change adds complexity and variability to an organization, and as such, the management processes must be adaptive. Adaptive management suggests decision-making under conditions of uncertainty, so that management strategies can afford to be all "no regrets" were climate projections to shift. For instance, wetlands could be increased or restored to carry out their inherent functions of offering flood management, carbon storage, and improving water quality, in addition to several other uses. Such actions are useful to undertake regardless of the exact effects that climate change is going to have. Also, the concept of adaptive management allows for the use of what should be named a sort of prognosis, which makes it possible to contemplate a number of possible climate futures and create strategies potent in all these scenarios.

Adaptive management also ensures the involvement of all the parties to the management process such as the government, people, nongovernmental organizations, and researchers. Involving these stakeholders in the decision-making and management processes means that people who are most knowledgeable and informed about certain issue or problem will contribute their views to the management's decision-making process, which is particularly important in the contemporary business world as the management has to deal with issues such as climate change that are not easily resolvable by traditional, one-size-fits-all solutions. Partnership also fosters the exchange of information and assets that can enable stakeholders to deal with new development and take coordinated action. For instance, local people possess inadequate and outdated knowledge of wetland systems that research would benefit from when formulating adaptive management plans. On a similar note, an examination of cross-cutting issues that link wetlands with other sectors like urban planning and agriculture provides strategies to incorporate wetland management in resilience for climate change (Roelich and Giesekam 2018).

The use of adaptive management in wetlands can also pose some challenges in that short-term interventions may compete with long-term visions. Essential wetland functions may also take several years to develop after degradation, and the long-term effectiveness of restoration in increasing species variety and climate stability might be unnoticed. Such is the reason why it becomes hard to negotiate for grants and political championing of wetland management. However, what adaptive management may be able to do is to track the progress of the restoration in increments such that these stakeholders are not discouraged. For stakeholders to remain committed to wetland restoration, adaptive management needs to set clear and achievable goals, and report progress periodically. The adaptation is needed for the management of wetlands as well as for the sake of the population that has to adapt to climate change. Cities that rely on wetlands to regulate floods, clean water, and otherwise, to perform myriad functions can apply adaptive management to strengthen the impact of climate change. For instance, place-based adaptive management can assist in the appropriate use and targeting of wetlands in the management of storm water, flood risks, or heat island effect, where wetlands afford cooler climates. The efficiency of wetland connection to wider climate adaptation policies shows that cities are less exposed to climate impacts and the well-being of their people is improved (Maltby 2022).

8.8 Sustainable Urban Wetland Restoration Models

Appropriate sustainable development models, as depicted in Fig. 8.6, in urban wetland restoration are of particular importance to determine if the wetlands in urban areas can continue to offer essential services to support climate adaptation, biological diversity conservation, and human welfare. These models are centered on the re-establishment of ecosystem services of wetlands with a special emphasis on wetlands and urbanization processes. In order to use the potential ability of wetlands to stimulate the process of natural pollution degradation in subsequent intercalation and restore disrupted natural systems to their natural state effectively and sustainably, planning should be made

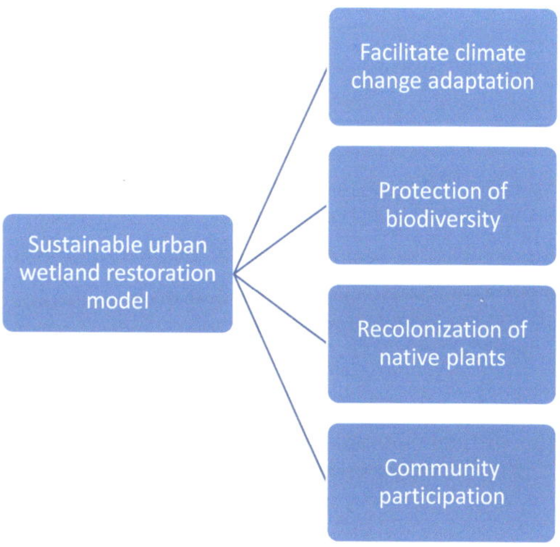

Fig. 8.6 Sustainable urban wetland restoration model

comprehensively and regard to the participation of multi-disciplinary experts; and ecological restoration science should incorporate existing indigenous knowledge for solving modern environmental problems, as well as taking advantage of cultural methods, tools, and practices in the human and natural worlds as resource. Another important characteristic of sustainable models for urban wetland restoration relates to the connectivity of wetlands with the green infrastructure of cities. This also involves raising awareness that wetlands are an integral part of the city's Natural Asset Base, which contributes to adaptive life-supporting ecosystem infrastructure, including nutrient retention and water and flood regulation. These services can be integrated into restoration models by turning the wetlands into multiple functional areas through restoration to support the planning of urban facilities. To illustrate, restored wetlands may be made into a park system, with provision of trails for hiking, blinds for bird watching, and other facilities for ecological learning and public enlightenment (Alikhani et al. 2021).

Another aspect of sustainable restoration within urban wetlands is reliance on native plant species and the replication of hydrological characteristics of natural systems. The proposed models include the recolonizing of native plants and the restoration of the natural water regime; these models seek to increase the level of ecosystem of wetlands by minimizing maintenance expenses and intentional interventions. This approach promotes conservation of the bio-diversity, enables the growth of native stock, and offers long-term ecological gains. Also, working with native creatures means that the artificial and restored wetlands are well-suited to the climate of the region and are less vulnerable to shifts in climate (Zivec et al. 2023).

As part of conducting the study, there is a need to implement the element of community participation in the sustainability of the wetland restoration project. Getting community participation in restoration enhances ownership, thus increasing chances of sustainability of restoration programs. Many of the restoration programs involve the local culture and history about wetlands, which goes a long way in helping to understand how these wetlands could be restored in a way that would support the local culture and society. Local people are also a potential conservation partners when it comes to wetland health and have been observed to participate in activities that include volunteering, educating the public, and backing conservation activities. Another factor that may be sensitive to the success of wetland restoration and management is financial sustainability. Government alone or donor-based funding restoration models may not be enough to guarantee a sustainable future. Government or donor-based restoration funding models alone may not be sufficient to guarantee a sustainable future. Therefore, new sources of funding—such as payments for ecosystem services (PES), eco-tourism, and public–private partnerships—are being explored for the restoration and management of wetlands (Feng et al. 2024).

8.9 Policy Frameworks for Wetland Protection and Restoration

Wetland conservation and management policies play a crucial role in the conservation of wetland ecosystems in urban, as well as rural areas. Wetlands are regulated by numerous services that include water purification, flood regulation, carbon storage, and habitat for species. But rapid urbanization, pollution, and climate change have led to enormous pressure on these ecosystems that require effective and well-coordinated policies for the protection and restoration of these ecosystems. As this chapter has illustrated, the focal considerations in good framework policies for wetland conservation are linked to the incorporation of wetlands into environmental and urban land-use policies. Proximity land-use changes, conversion of wetlands to other uses, and increases in infrastructure all pose a major threat to many urban wetlands. There is therefore a need for policies to ensure that wetlands are taken into consideration in planning, zoning, and urban development projects. Specific planning and zoning tools can be incorporated, which prevent the draining of wetland areas and require that any construction of buildings undertaken in a wetland is achieved in a manner that will not devastate this natural resource. Third, policies should include the requirement to remediate degraded wetlands when undertaking urban regeneration or redevelopment, to guarantee that green urban infrastructure is included in the design of the urban environment to support human and ecological uses (Alikhani et al. 2021).

National environmental policies also have their equivalent role in wetland conservation. Specifically, wetlands are subject to protection under national legislation of the respective country as embodied in environmental protection acts, water management and legislation, as well as policies on biological diversity. It means these laws can supply the legal basis of wetland conservation and management, including legal procedures of wetland restoration, prohibition of unlawful activities including reclamation or pollution, as well as legal delimitation of the wetland protection area. Governments should also meet the provisions of international treaties including the Ramsar Convention, which seeks to conserve and sustainably use of wetlands of international significance. Government policies that support wetlands' protection within the frame of the convention encourage worldwide understanding and preservation of wetlands and involve local and regional cooperation in the management process. Wetlands also must remain a part of climate-change adaptation programs and strategies. Coastal wetland, as one of the major types of environment, has a natural protective function against climate change hazards like floods, storm surges, and a rise in the sea level. According to national and regional climate adaptation strategies, wetlands need to be reconstructed and safeguarded to maximize their climate resilience. The government should also encourage the incorporation of wetlands in urban climate change control measures and bring wetlands into the battle against urbanization. Green bonds, climate Finance, and carbon credit programs are some of the important financial instruments under climate change adaptation that can support the funding requirement for the protection and conservation of wetlands.

As has been mentioned earlier, the local governments have the specific duty of protecting and conserving wetlands. National policies are coordinated by local authorities, and these are capable of fashioning studied regulations that fit local realities concerning complex issues such as the wetlands in urban areas. Local policy frameworks should further aim at the identification, mapping, and management of wetlands in relation to the urban scenario. Further, they should promote the participation of those communities, stakeholders, as well as the private players in the management of these wetlands. Participation is effective in encouraging local people to be involved in conservation and the necessity of their provision of various needs will be met. There is therefore the need for governments, businesses, and civil society to do all they can to ensure that adequate resources are committed to wetland restoration activities. They include payments for ecosystem services (PES), public-private partnerships, eco-tourism, etc., which may be used for mobilizing the required funds. Moreover, the formulation of economic rewards for sustainable utilization of the wetlands includes encouraging people to practice eco-tourism, or otherwise, the sale of products that are derived from the wetlands taking care not to harm the wetland; for example, sales of fruits in the forest without harming the trees.

8.10 Conclusion

Urban wetland conservation and rehabilitation are essential for improving climate change adaptability, as well as biological diversity, and promoting sustainable urban development. Having reviewed the details of wetlands, it is clear that with increasing complexities and problems related to climate change, storm surges, contaminated water supply, wetlands provide critical services to the ever-expanding urban cities including flood control, water treatment, and carbon storage. Nevertheless, the wanted ecosystems are under pressure from urbanization, pollution, and other human impacts, reinforcing the demand for proper policy solutions and unusual approaches. It means that the conservation of wetlands is possible through the inclusion of wetlands in urban planning, through cross-sectoral cooperation, as well as involving as many people as possible from all over the world in these processes. In addition, knowledge and technologies in adaptive management, technological implementation, and climate-smart solutions are crucial to sustain the wetland ecosystem in view of dynamic environmental challenges. Proper institution, financial solutions, and appropriate legal processes are the ways toward the protection and restoration of urban wetlands toward a sustainable urban future. In conclusion, only comprehensive and integrated with the involvement of state, industry, and civil society wetland management is capable of not only preserving these natural assets but also minimizing climate change effects, improving the quality of life of cities' inhabitants, and actively implementing rational use of natural resources under today's conditions for future generations.

References

Ahn C, Schmidt S (2019) Designing wetlands as an essential infrastructural element for urban development in the era of climate change. Sustainability 11(7):1920. https://doi.org/10.3390/su11071920

Alikhani S, Nummi P, Ojala A (2021) Urban wetlands: a review on ecological and cultural values. Water 13(22):3301. https://doi.org/10.3390/w13223301

Anderson V, Gough WA (2022) A typology of nature-based solutions for sustainable development: an analysis of form, function, nomenclature, and associated applications. Land 11(7):1072. https://doi.org/10.3390/land11071072

Bassi N, Kumar MD, Sharma A, Pardha-Saradhi P (2014) Status of wetlands in India: a review of extent, ecosystem benefits, threats and management strategies. J Hydrol Reg Stud 2:1–19. https://doi.org/10.1016/j.ejrh.2014.07.001

Buytaert W, Zulkafli Z, Grainger S, Acosta L, Alemie TC, Bastiaensen J, De Bièvre B, Bhusal J, Clark J, Dewulf A, Foggin M, Hannah DM, Hergarten C, Isaeva A, Karpouzoglou T, Pandeya B, Paudel D, Sharma K, Steenhuis T, Tilahun S, Van Hecken G, Zhumanova M (2014) Citizen science in hydrology and water resources: opportunities for knowledge generation, ecosystem service management, and sustainable development. Front Earth Sci 2:26. https://doi.org/10.3389/feart.2014.00026

Chen S-L, Chou H-S, Huang C-H, Chen C-Y, Li L-Y, Huang C-H, Chen Y-Y, Tang J-H, Chang W-H, Huang J-S (2023) An intelligent water monitoring IoT system for ecological environment and smart cities. Sensors 23(20):8540. https://doi.org/10.3390/s23208540

De Knegt B, Breman BC, Le Clec'h S, Hinsberg AV, Lof ME, Pouwels R, Roelofsen HD, Alkemade R (2024) Exploring the contribution of nature-based solutions for environmental challenges in the Netherlands. Sci Total Environ 929:172186. https://doi.org/10.1016/j.scitotenv.2024.172186

de Oliveira M, Morrison T, O'Brien KR, Lovelock CE (2024) Governance of coastal wetlands: beyond the community conservation paradigm. Ocean Coast Manag 255:107253. https://doi.org/10.1016/j.ocecoaman.2024.107253

Devuyst D (2000) Book reviews: "better environmental decisions: strategies for governments, businesses, and communities", Ken Sexton, Alfred A. Marcus, K. William Easter & Timothy D. Burkhardt (Eds). J Environ Assess Policy Manag 2(1):167–168. https://doi.org/10.1142/S1464333200000096

Díaz-Pinzón L, Sierra L, Trillas F (2022) The economic value of wetlands in urban areas: the benefits in a developing country. Sustainability 14(14):8302. https://doi.org/10.3390/su14148302

Feng J, Wen Y, Zhang H, Duan W, Hao H (2024) Wetland restoration, household income, and livelihood structure of farmers. Front Sustain Food Syst 8:1256115. https://doi.org/10.3389/fsufs.2024.1256115

Guo M, Li J, Sheng C, Xu J, Wu L (2017) A review of wetland remote sensing. Sensors 17(4):777. https://doi.org/10.3390/s17040777

Hernandez CL, Sharpe LM, Jackson CA, Harwell MC, DeWitt TH (2024) Connecting stakeholder priorities and desired environmental attributes for wetland restoration using ecosystem services and a heat map analysis for communications. Front Ecol Evol 12:1290090. https://doi.org/10.3389/fevo.2024.1290090

Kadoma A, Perry M, Renaud FG (2023) Stakeholders' perceptions of wetland conservation and restoration in Wakiso District, Uganda. Environ Dev Sustain. https://doi.org/10.1007/s10668-023-04008-z

Kim KG, Lee H, Lee DH (2011) Wetland restoration to enhance biodiversity in urban areas: a comparative analysis. Landsc Ecol Eng 7:27–32. https://doi.org/10.1007/s11355-010-0144-x

Kim YI, Park WH, Shin Y, Park J-W, Engel B, Yun Y-J, Jang WS (2024) Applications of machine learning and remote sensing in soil and water conservation. Hydrology 11(11):183. https://doi.org/10.3390/hydrology11110183

Lawler JJ, Spencer B, Olden JD, Kim SH, Lowe C, Bolton S, Beamon BM, Thompson L, Voss JG (2013) Mitigation and adaptation strategies to reduce climate vulnerabilities and maintain ecosystem services. Clim Vulnerability:315–335. https://doi.org/10.1016/B978-0-12-384703-4.00436-6

Li Z, Ma L, Gou D, Hong Q, Fai L, Xiong B (2022) The impact of urban development on wetland conservation. Sustainability 14(21):13747. https://doi.org/10.3390/su142113747

Maltby E (2022) The wetlands paradigm shift in response to changing societal priorities: a reflective review. Land 11(9):1526. https://doi.org/10.3390/land11091526

Mitra S (2024, February 1) Governing urban wetlands in India: a pathway to sustainable urbanisation. Observer Research Foundation. https://www.orfonline.org/research/governing-urban-wetlands-in-india-a-pathway-to-sustainable-urbanisation

Moomaw WR, Chmura GL, Davies GT, Finlayson CM, Middleton BA, Natali SM, Perry JE, Roulet N, Sutton-Grier AE (2018) Wetlands in a changing climate: science, policy and management. Wetlands 38:183–205. https://doi.org/10.1007/s13157-018-1023-8

Oakes LE, Peterson St-Laurent G, Cross MS, Washington T, Tully E, Hagerman S (2022) Strengthening monitoring and evaluation of multiple benefits in conservation initiatives that aim to foster climate change adaptation. Conserv Sci Pract 4(6): e12688. https://doi.org/10.1111/csp2.12688

Pandey B, Ghosh A (2023) Urban ecosystem services and climate change: a dynamic interplay. Front Sustain Cities 5:1281430. https://doi.org/10.3389/frsc.2023.1281430

Premalatha M, Abbasi T, Abbasi SA (2010) Applications of GIS in wetland management: an overview. Res J Chem Environ 14:87–103

Roelich K, Giesekam J (2018) Decision making under uncertainty in climate change mitigation: introducing multiple actor motivations, agency and influence. Clim Pol 19(2):175–188. https://doi.org/10.1080/14693062.2018.1479238

Semeraro T, Turco A, Arzeni S, La Gioia G, D'Armento R, Taurino R, Medagli P (2021) Habitat restoration: an applicative approach to "biodiversity heritage relicts" in social-ecological systems. Land 10(9):898. https://doi.org/10.3390/land10090898

Shivanna KR (2022) Climate change and its impact on biodiversity and human welfare. Proc Indian Natl Sci Part A Phys Sci 88(2):160–171. https://doi.org/10.1007/s43538-022-00073-6

Stefanakis AI (2019) The role of constructed wetlands as green infrastructure for sustainable urban water management. Sustainability 11(24):6981. https://doi.org/10.3390/su11246981

Williams BK, Brown ED (2014) Adaptive management: from more talk to real action. Environ Manag 53(2): 465–479. https://doi.org/10.1007/s00267-013-0205-7

Yang L, Zhang Z, Zhang W, Zhang T, Meng H, Yan H, Shen Y, Li Z, Ma X (2023) Wetland park planning and management based on the valuation of ecosystem services: a case study of the Tieling Lotus Lake National Wetland Park (LLNWP), China. Int J Environ Res Public Health 20(4):2939. https://doi.org/10.3390/ijerph20042939

Zivec P, Sheldon F, Capon SJ (2023) Natural regeneration of wetlands under climate change. Front Environ Sci 11:989214. https://doi.org/10.3389/fenvs.2023.989214

9 Unpacking the Role of Urban Wetlands in Urban Resilience and Sustainable Transitions: An Expose from the City of Harare in Zimbabwe

Nyasha Chaminuka

Abstract

Climate change represents the greatest threat to humanity, with its effects being especially pronounced in urban areas. This is due to the elevated greenhouse gas emissions produced by various human activities (Tirivangasi HM, Dzvimbo MA, Chaminuka N, Mawonde A, Sci Afr 20:e01710, 2023). Wetlands present opportunities for carbon sequestration, which can unintentionally assist in mitigating and adapting to climate change. Aquatic ecosystems absorb carbon dioxide from the atmosphere, which aids in lessening climate change impacts. Natural wetlands are vital in curbing the accumulation of greenhouse gases such as carbon dioxide in the air (Junk WJ, An S, Finlayson CM, Gopal B, Kvet J, Mitchell SA, Mitsch WJ, Robarts RD Aquat Sci 75:151–167, 2013). Safeguarding wetlands enhances ecosystem resilience and promotes biodiversity (Wetlands International 2020). These environments provide a wide range of advantages, especially in addressing climate change and thereby improving urban resilience (Ramsar Convention, Updated information on Zimbabwe's Ramsar Sites Available at: https://www.ramsar.org/news/updated-information-on-zimbabwes-ramsar-sites, 2016). Wetlands can alleviate the adverse effects of climate change by enhancing water quality, managing floods, and storing water (Bilge Ozturk et al. 2022; Cicek et al. 2022). The research employed a qualitative framework, utilizing a case study approach. Comprehensive interviews were conducted with urban inhabitants, city planners, climate advocates, and officials from the Ministry of Environment and Tourism. Despite the considerable advantages offered by wetlands, they have experienced significant degradation due to human activities. These ecologically sensitive areas in Harare have been greatly affected by illegal settlements and waste disposal. Furthermore, uncontrolled urban agriculture has resulted in crop planting in wetlands, further disrupting these ecosystems. The study emphasizes the urgent need for a change in strategy to prioritize wetland conservation effectively for sustainable and resilient cities.

Keywords

Climate change · Carbon sequestration · Biodiversity · Greenhouse gases · Sustainability

N. Chaminuka (✉)
Faculty of Applied Social Sciences, Department of Development Studies, Zimbabwe Open University (ZOU), Gweru, Zimbabwe

S. K. Chakraborty et al. (eds.), *Wetlands and Urban Resilience*, Sustainable Development Goals Series,
https://doi.org/10.1007/978-3-032-06881-1_9

9.1 Introduction

Climate change continues to be a significant obstacle to achieving Sustainable Development Goals and urban resilience more than ever. This is due to the understanding that this phenomenon adversely affects human health, food security, and intensifies natural disasters (Musasa 2022). Beyond being a worldwide environmental concern, climate change poses a significant risk to efforts aimed at reducing poverty and fostering economic growth, potentially reversing numerous development achievements attained in recent years (Marondedze 2022). Climate change exacerbates pre-existing vulnerabilities by diminishing livelihood resources, thereby undermining the resilience of cities against climate-related impacts (Pereira et al. 2024). The ongoing challenges posed by climate change in Zimbabwe have been highlighted in various studies by researchers such as Chikodzi et al. (2013), Gutsa (2014), Murombo et al. (2019), and Brazier (2015). Zimbabwe as a whole, and Harare specifically as the capital, faces the impact of climate-related disasters such as drought and extreme heat (Masundire and Ncube 2024). Researchers express concerns about the inadequacies of responses to climate-induced natural disasters (Brazier 2015, 2017). Urban areas are significant contributors to climate change due to industrial activities, vehicle emissions, and high levels of pollutants (Tirivangasi et al. 2023). The connection between urbanization and climate change is closely associated with greenhouse gas emissions (Conway et al. 2019; Chirisa 2024). This adversely affects Sustainable Development Goal 13, which calls for immediate action to combat climate change and Goal 11 which calls for inclusive and resilient cities (Cohen et al. 2021). Zimbabwe is experiencing the troubling consequences of climate change (Dzvimbo and Zhanda 2020). The consequences of these factors severely affect livelihoods, individual health, and disrupt socio-economic stability (Chikodzi et al. 2013; Brazier 2015). Climate change is deeply intertwined with various issues, including food and water security, economic development, and local governance (Mudombi and Chigara 2024). Ongoing droughts serve as a clear sign of climate change, and as a developing country, Zimbabwe is anticipated to face the most severe effects of climate change (Dzvimbo and Zhanda 2020).

To improve urban resilience and create sustainable cities, it is essential to protect wetlands due to their vital role (Chirisa 2024). In numerous regions of Africa, wetlands serve as a crucial source of livelihoods through agriculture and livestock farming (McCartney et al. 2015; Bayram and Arabulan 2025). Despite their significant value, over 70% of wetlands globally have been lost since the 1970s (Convention on Wetlands 2021). Considering the various challenges faced in urban areas, this study intends to examine how wetlands aid in enhancing urban resilience in Harare. Furthermore, the research aims to identify the main threats to wetlands in Harare and assess the effectiveness of conservation efforts directed at these vital ecosystems. The study is novel as it aims to investigate the nexus between urban wetlands and sustainable transitions, an area that remains underexplored in development discourse.

9.2 Materials and Methods

The research utilized a qualitative methodology. This methodology is naturalistic, enabling the examination of events within their genuine contexts (Tenny et al. 2022). It offers a way to explore and comprehend the significance that individuals or groups attribute to social or human issues (Moser 2018). The qualitative approach focuses on depicting and elucidating social phenomena (Korrstjens and Moser 2018). Opinions, feelings, and experiences, when viewed comprehensively, are central to the qualitative approach. Furthermore, this method is more appropriate for studying novel phenomena (Kohler et al. 2023).

The research utilized a case study research design. In a case study, the investigator offers an in-depth portrayal of issues, and the method is comprehensive as it captures events within their real-life setting (ibid). A case study method was employed to gain more profound insights from the experiences of residents and officials in Harare. This approach aids in examining how a complex array of factors converges to create a specific outcome. The case study method is particularly adaptable as it incorporates nearly all data collection techniques (Ibid). This approach is focused on particular instances and contexts; therefore, in this research, it assisted in exploring the views of residents, officials from the Ministry of Environment and Tourism, the Environmental Management Agency, and extension officers. The exploratory case study offered valuable insights and lessons learnt from the city of Harare.

The researcher interacted with Harare residents by conducting interviews that featured open-ended questions. Key informant interviews were conducted with representatives from the Ministry of Environment and Tourism, the Environmental Management Agency, officials from Harare City Town Planning, and extension officers. A total of 60 interviews were carried out in Budiriro, Glenview, and Epworth. Three focus group discussions were held with community members from Budiriro, Glenview, and Epworth. Furthermore, field observations and mapping activities were undertaken. A desktop review was also carried out to validate both primary and secondary data.

9.3 Results and Discussion

The research was conducted in Harare, the capital city of Zimbabwe. Harare has an estimated population of 1,849,600, with approximately 2,487,209 residents in its metropolitan area (Zimstat 2022). This metropolitan area comprises the municipalities of Chitungwiza and Epworth (Fig. 9.1).

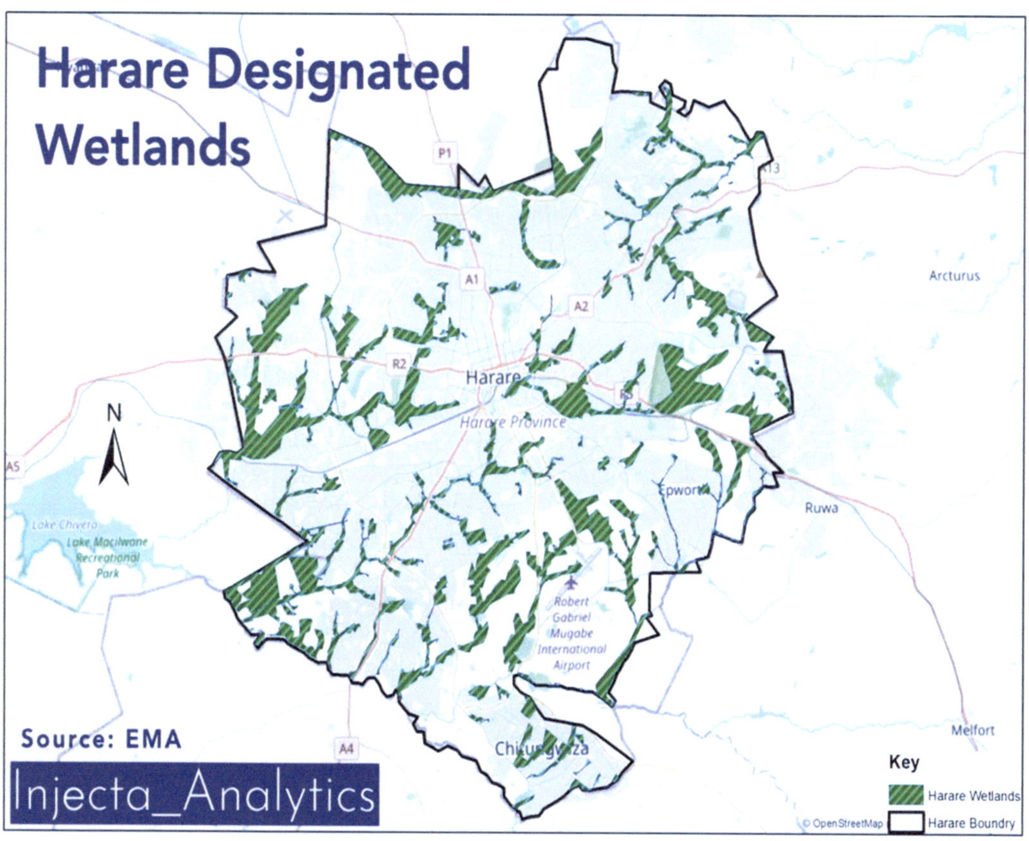

Fig. 9.1 Map of Harare showing wetlands. (Source: EMA)

The city has undergone significant urbanization, leading to increased housing shortages and the establishment of human settlements in wetland areas. In 2020, the urbanization rate was recorded at 32.24% (Statista 2022). Unscrupulous land barons exploit this situation by illegally allocating land in wetlands to desperate home seekers (Dzvimbo et al. 2023). Settlements in wetlands, particularly along the Manyame and Mukuvisi rivers, have intensified environmental degradation due to pollution and the drying up of these wetlands (Marondedze 2022). Establishing human settlements in such ecologically sensitive regions jeopardizes the pursuit of resilient and sustainable urban environments.

In order to learn from Harare residents' experiences regarding the importance of wetlands in mitigating climate change and strengthening urban resilience, the discussion took a qualitative approach. A sample of 60 participants was reached through data saturation. The participants were purposively sampled, and interviews were conducted from September 2024 to November 2024. Since the participants were chosen through resident associations, they were chosen based on the criterion that they were information-rich. They were drawn from the Environmental Management Agency (EMA), Ministry of Environment and urban residents settled in wetlands. Table 9.1 shows profiles of respondents interviewed in the city of Harare. The study aimed to analyze the factors behind the encroachment of wetlands, the role of wetlands in strengthening urban resilience, and the obstacles encountered in their conservation.

The abbreviation R stood for respondents. The following codes were applied to the respondents:

Table 9.1 Categories of respondents and reasons for encroaching into Harare wetlands

Interview code	Description
Category 1	Categorization according to the reason for invading wetlands
R01-R15	Housing and settlement
R16-R30	Urban farming
R31-R45	Industrial activity
R46-R60	Extraction of water

60 respondents (R) coded as R1-R60. Respondents 01-15 indicated that they moved into wetlands to settle due to housing shortages. Respondents R16-R30 indicated that they were exploiting wetlands for urban farming to enhance food security and supplement their household income. Respondents R31-R45 indicated that they exploited wetlands for setting up industries, and respondents R46-R60 used wetlands for the extraction of water. These results align with a study conducted by Muguti (2020), which demonstrated that wetlands are essential for replenishing urban water reservoirs (Fig. 9.2).

9.4 The Importance of Urban Wetlands in Addressing Climate Change Mitigation and Building Resilient Cities

Since wetlands reduce flooding during tropical storms, wetlands are essential for mitigating and adapting to climate change (Rangoei et al. 2013; Dinsa and Gemeda 2019). Investing in wetland conservation is essential for creating climate-resilient urban areas (Liang 2021). Additionally, wetlands contribute to carbon sequestration through their vegetation and soils. The effectiveness of wetlands in absorbing greenhouse gases (GHGs) varies based on the management practices employed, which can differ from one city to another (Kolka et al. 2016). The soils of wetlands have a heightened capacity to capture and store carbon, aiding in climate change mitigation (Lal 2008). A significant quantity of carbon in wetlands is stored as peat over long periods. The evaluation of peat soil involves factors such as the content of Soil Organic Carbon (SOC) and its thickness (United States Department of Agriculture 2014). Investing in wetland conservation improves the livelihoods of urban households through agricultural activities (Gandidzanwa and Togo 2024).

To strengthen the city's resilience, metropolitan councils can implement a planning-oriented strategy that creates various supportive options to boost urban resilience against current urban challenges such as climate change, rapid

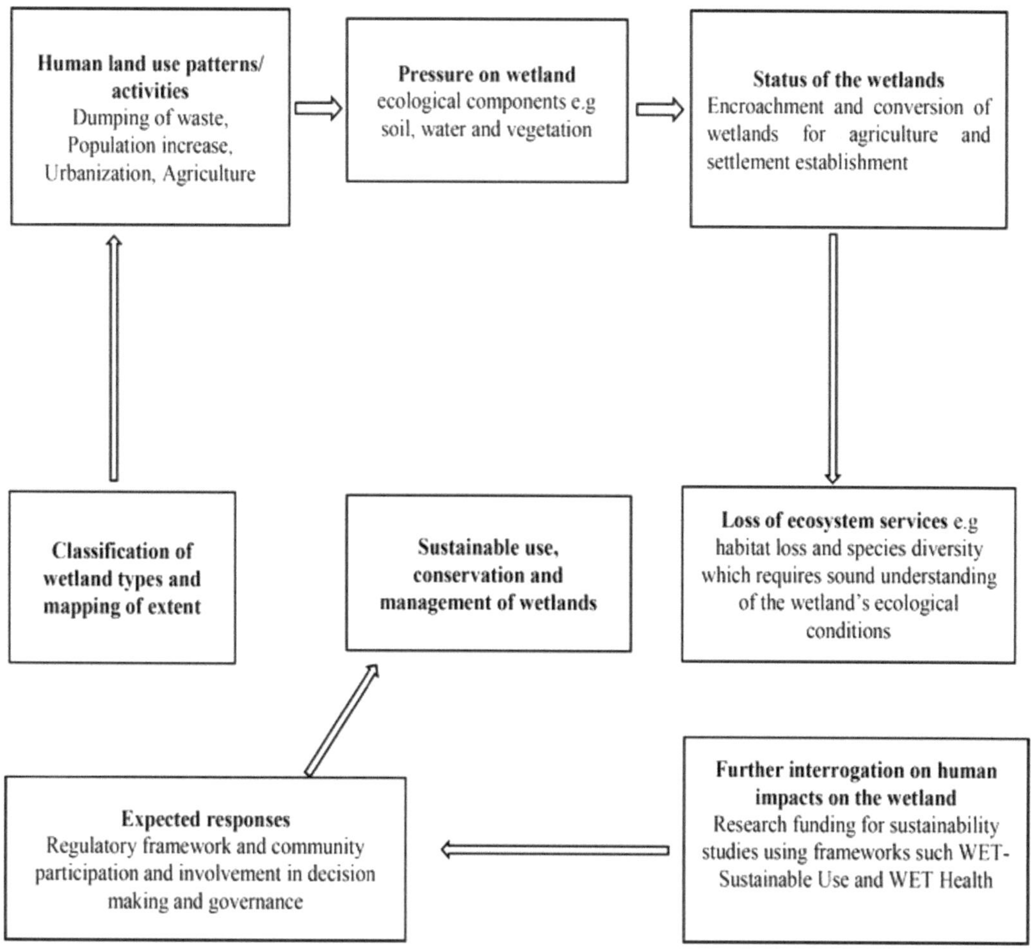

Fig 9.2 Proposed framework for sustainable wetland management. (Source: Musasa et al. 2022)

urbanization, and human settlements in wetlands (Magdi 2022). Key spatial and temporal patterns, along with important factors related to urban resilience, will improve the ability to adapt to urban risks like flooding, disease outbreaks, and the possibilities for sustainable urban growth (Chirisa 2024).

9.5 Prospects of Local Participation in Wetland Conservation for Climate Change Mitigation and Adaptation

Reaching out to the local community is crucial as it cultivates a sense of ownership and enhances community involvement. When local residents engage in environmental conservation initiatives, it promotes sustainable development (Silipiwe and Takarani 2020). Urban councils and other local authorities are essential in safeguarding wetlands. The city of Harare is currently focused on managing its wetlands, which have increasingly come under threat from human activities (Mwonzora 2021). Although urban councils are meant to act as custodians of land in cities, they are not fully implementing policies and laws for environmental protection (Nemutamvuni et al. 2020). The Harare City Council is primarily tasked with regulating land use, particularly in ecologically sensitive areas like wetlands (Chirisa 2024). Zimbabwe's constitution provides local authorities and communities with the right to protect the natural environment, including wetlands. Through the Urban Sustainability Framework, cities have a crucial role in enhancing wetlands in urban settings (Uprise Zimbabwe 2024). The ongoing degradation of urban wetlands reflects ineffective governance structures in cities, leading to the growth of environmentally harmful practices (Matamanda and Chinozvina 2020). Urban councils have encountered difficulties in fully complying with Master Plans, by-laws, and Local Environmental Plans (Community Water Alliance 2021). The City Capability Framework highlights the need to incorporate sustainability indicators into urban planning (Gibberd 2017).

The Zimbabwean government has embraced the definition provided by the Ramsar Convention, describing wetlands as any areas of marsh, fen, peatland, or water, whether these are natural or man-made, permanent or temporary, containing static or flowing water, whether it is fresh, brackish, or saltwater, and also includes the riparian land surrounding the wetland (Government of Zimbabwe 2003: 11). Zimbabwe has committed to the adoption and implementation of international agreements, including the Ramsar Convention, the Convention on Biodiversity, and the Sustainable Development Goals.

The pursuit of sustainable urban development is in line with several Sustainable Development Goals (SDGs), specifically Goals 6, 11, 13, 14, and 15. Goal 13, which emphasizes Climate Action, advocates for collaborative efforts to mitigate the effects of climate change, including the protection of wetlands (Strokal 2021). This creates opportunities for cities to incorporate nature conservation into urban planning or adopt biophilic designs (Mwonzora 2021). Valuable lessons can be learned from Brazil and the United States, both of which have established comprehensive laws aimed at wetland conservation (Mwonzora 2021). However, these laws have faced challenges as they have often been employed as a reactionary measure to stop environmental degradation rather than taking a proactive approach (Matamanda 2018).

There exists extant research on the degradation of wetlands in Zimbabwe, yet it has primarily concentrated on the various human activities that contribute to wetland loss (Cunlife 2020). Nevertheless, the literature tends to overlook the crucial issue of the involvement of local authorities and communities in safeguarding the environment (Muziri et al. 2019; Sharai et al. 2020). This oversight is evident in the limited number of studies examining the role of urban councils in creating and enforcing environmental protection laws (Kalanzi 2015). Consequently, there is a noticeable gap in scholarly evidence that explores the role of local authorities in wetland conservation.

The global importance of wetlands in creating sustainable and resilient urban environments has been acknowledged, as highlighted by the Millennium Ecosystem Assessment (2005), Ramsar Convention on Wetlands (STRP 2009), and the United Nations Framework Convention on Climate Change (UNFCCC 2017; Dayathilake et al. 2021). In Harare, community members have united to contribute to the critical mission of preserving wetlands. The Budiriro Water Foundation, a community organization, has been actively educating residents about the dangers linked to farming on wetlands, which are vital for the water supply. Furthermore, the Conservation Society of Monavale (COSMO) was founded to promote the safeguarding of the Monavale vlei. This location is part of the Manyame basin, which serves as a crucial habitat for numerous reptiles, mammals, and bird species (Fig. 9.3).

The community has been proactive in tackling biodiversity decline, engaging in restoration initiatives and outreach efforts. Other significant Community-Based Organisations (CBOs) include the Harare Wetlands Trust (HWT) and the Dzivarasekwa Residents Trust. The central government has developed policies and regulations like the Environmental Management Act to set standards for environmental protection. While the government is recognized for formulating laws and policies, it has been criticized for its lack of effective implementation of these regulations such as the National Wetlands Policy of 2022. Zimbabwe hosted the COP 15 on wetlands in Victoria Falls from 23 to 31 July under the theme ‘Protecting Wetlands For Our Common Future’. Countries

Fig. 9.3 Photograph of Monavale Vlei being conserved by the local community. (Image by Philippa Marett)

were urged to synchronize their efforts in the restoration, protection, and efficient management of wetlands.

9.6 Opportunities Presented by Wetlands in the City of Harare

Wetlands offer significant potential for improving the quality of life and sustainable livelihoods of people living in urban areas. They can be utilized for sustainable development, allowing residents to engage in activities like controlled urban farming. This approach supports the achievement of sustainable development via the Sustainable Development Goals (Chaminuka and Dube 2017; Chaminuka et al. 2021). Furthermore, urban agriculture enhances food security since individuals can grow fresh produce close to their homes. Besides enhancing food security, urban farming in the city of Harare contributes to sustainable livelihoods and employment creation (Chirisa 2024).

Wetlands are essential components of the ecosystem. They help reduce the effects of pollution and serve as habitats for wildlife. Additionally, wetlands absorb atmospheric carbon dioxide, which contributes to mitigating global warming. Wetlands provide a variety of advantages, such as supplying water in urban settings, filtering water, and assisting in the capture of carbon dioxide (Musasa et al. 2023). Their significance goes beyond just the regulatory aspects of carbon capture. In Harare City, wetlands are essential for replenishing the water that flows into Lake Chivero, which subsequently supplies water to the city's inhabitants (Cunlife 2020).

Furthermore, wetlands add aesthetic value to cities by serving as urban green spaces, providing chances for recreation and tourism (Manivel 2025). This is illustrated by Monavale Vlei in Harare, which attracts tourists eager to experience its stunning scenery. Tourists partake in birdwatching in specially designated areas for this activity (BirdLife International 2015). Furthermore, the urban wetlands in Harare contribute positively to the city's overall atmosphere. In a world increasingly influenced by climate change, marked by higher temperatures, heatwaves, sea-level rise, and reduced rainfall combined with persistent droughts, wetlands serve as essential water sources (Schuijt 2002). Many cities are facing water shortages, underscoring the critical importance of conserving wetlands, especially in urban regions that depend on them for their water supply (White et al. 2020). This conservation is essential for fostering urban sustainability by ensuring an uninterrupted water supply (Finlayson and Gardner 2020). Wetlands usually enhance resilience to climate change by acting as buffers against both droughts and floods, sequestering carbon, and, when left undisturbed, helping to cool the climate in the long run (Lee et al. 2022).

In Harare, the advantages of wetland conservation are clear, particularly in light of the adverse effects that wetland loss has had on the water table, leading to increased dryness and water scarcity in the city (Gweshe 2020; Cunlife 2020). Despite the numerous values, functions, and advantages that wetlands offer, they face significant threats, not only in Zimbabwe but around the globe (Ramsar Convention on Wetlands 2018). Hence, it is accurate to assert that wetlands represent some of the most endangered ecosystems on the planet.

Recently, the Harare City Council has made the alarming choice to allocate residential plots to urban residents in delicate wetland areas, including Glen Norah, Budiriro, Mabvuku, Warren Park, Waterfalls, Belvedere, Msasa Park, and Ruwa. This strategy has led to what many describe as 'disorderly growth' (*Newsday*, 26 September 2019). The construction of homes and buildings in these crucial wetlands has significantly heightened the risk of flash floods, as evidenced by the devastating events that occurred in certain parts of Budiriro and Chitungwiza in 2021 (*The* Herald, 3 February 2021). Considering the essential role wetlands play in flood management, it is crucial that urban planners and professionals reassess these policies to safeguard our communities from the increasingly severe impacts of such developments. The ongoing intrusion into these ecosystems not only puts lives at risk but also jeopardizes the future resilience of our urban areas.

9.7 Threats to Wetlands in the City of Harare

Wetlands are increasingly becoming under immense pressure due to anthropogenic activities. Approximately 50% of the global population has settled in coastal areas and wetlands over the past hundred years (Matamanda and Chinozvina 2020; Davies et al. 2020). Notable cities like Washington DC and Lagos were founded in wetland regions (Byram and Alaburan 2025). The city of Harare has faced encroachment on its wetlands in areas such as Mufakose, Warren Park, Southlea Park, Epworth, and Seke due to the rise of informal settlements. Additionally, the establishment of the Long Chen Plaza, a prominent shopping mall, by the Chinese is located on a wetland. The invasion of wetlands in urban regions is a widespread issue globally, and Harare is no exception to this trend (Tirivangasi et al. 2023). The expansion of human settlements into wetland areas has heightened the risk of flooding; for instance, in March 2019 when Cyclone Idai struck Zimbabwe, many residents were rendered homeless as rainwater inundated their homes, which was described as nature's response to human-induced degradation. The development of human settlements and housing in wetland regions leads to an increase in flash flood occurrences, as these locations typically have elevated levels of moisture (Fig. 9.4).

In addition, the Mukuvisi wetland that provides water to the Mukuvisi nature reserve and nearby regions has decreased in size over the last 30 years (Mhlanga 2022). The situation is worsening rapidly as over 115,000 hectares of land have been lost in Harare and its surrounding areas due to informal settlements in wetlands (Kudumba 2022). The growth of the Masasa Industrial area and the unchecked urban farming along the banks of the Mukuvisi river have intensified the drying of the wetland (ZELA 2021). Mufakose and Marimba residential neighborhoods were divided by a wetland that was subdivided for housing stands and has also been converted into a landfill. The area's aesthetic appeal has diminished significantly, and it has also led to its desiccation, as indicated by local residents who were interviewed. Many community members expressed their concerns over the ongoing disappearance of wetlands, which previously acted as the city's ecological lungs. The increase in human settlements encroaching upon wetlands has emerged as a worldwide issue and a major environmental challenge (Matamanda et al.

Fig. 9.4 Photograph showing a house constructed in a wetland and resultant flooding. (Source: Blessing Chigwenhembe of VOA)

2018). In addition to population growth and elevated poverty rates, inadequate enforcement of environmental conservation policies worsens the degradation of the environment (Marambanyika 2015). Communities also demonstrate a lack of awareness regarding environmental conservation, which poses serious risks to the sustainable management of ecosystems. Human activities persist in jeopardizing ecosystems and intensifying the loss of biodiversity. The city of Harare has lost more than 30 wetlands due to activities such as the construction of roads and illegal settlements (UNDP 2020). A case in point is the area between Budiriro 1 and Budiriro 5, which used to be separated by a wetland but was invaded by Budiriro 3 extension. The unchecked growth of human development continues to present a significant danger to wetlands and intensifies environmental degradation (ZELA 2021). Many cities in Africa are still facing challenges related to urban growth and infrastructure development, resulting in significant pressure on the environment (Matamanda and Chinozvina 2020). The speed at which wetlands are disappearing in the city of Harare is concerning, as shown in Table 9.2 (Fig. 9.5).

9.8 Uncontrolled Urban Farming

Even though urban farming offers possibilities for sustainable livelihoods in cities, it poses a significant risk to wetlands. This is due to the fact that this practice leads to soil erosion and the resulting silt buildup in nearby streams (Marondedze 2022). Urban farming, when left unchecked, can lead to significant adverse effects on both the environment and public health. Excessive use of chemicals like fertilizers, insecticides, and herbicides interferes with ecological systems (Musasa et al. 2023). The application of insecticides and herbicides presents significant dangers, contributing to water pollution and eutrophication that jeopardize our ecosystem (Thomas et al. 2022). The implications of dependence on these chemicals are extensive, causing elevated runoff that speeds up eutrophication, putting at risk the health of individuals consuming contaminated water and food. In addition, this reliance on chemicals increases plant toxicity and soil acidity, frequently surpassing tolerable levels for plants (Musasa et al. 2023). Eutrophication happens when an excess of nutrients throws the ecosystem balance off in both land and water environments (Somay-Altas and Gokcen-Yilmaz 2023). Regrettably, unregulated urban farming, particularly close to stream banks, utilizes fertilizers and pesticides, which exacerbates the contamination of essential water resources. Unsustainable urban agriculture can cause significant damage to our environment (Chaminuka and Dube 2021). Critics contend that farming should primarily be confined to rural locations, as its expansion into city areas diminishes their aesthetic appeal and threatens the quality of life in urban environments (Thomas et al. 2022). The proximity of urban and peri-urban agriculture to dense populations heightens the risk of both infectious and chronic diseases. Urban farming can contribute to environmental contamination

Table 9.2 Percentage loss of wetland area in Harare's residential suburbs

Wetland area	% loss of wetland area
Prospect	15,7
Kuwadzana	30
Mbare	30,8
Warren Park	30,8
Dzivarasekwa	32,5
National Sports Stadium	36,4
Tafara	46,2
Budiriro	49
Houghton park	71
Epworth	78,8
Total	49,6

Source: Moyo and Cunliffe (2020)

Fig. 9.5 Photograph showing the wetland which separated Budiriro 1 and 5 before being turned into Budiriro 3 extension. (Source: Dialogue Earth 2024)

through the use of agrochemicals, like pesticides and herbicides, as well as runoff from livestock waste, fostering conditions that support the proliferation of disease-carrying organisms (Nzvere et al. 2024). Anthropogenic activities have led to the deterioration of urban wetlands due to illegal settlements, sand extraction, and unregulated urban farming, which have caused siltation (Hettiarachich et al. 2021).

Additionally, crops cultivated near municipal waste disposal sites are considerably at a higher risk of contamination, which endangers public health (Marondedze 2022). The noncommunicable health hazards linked to urban agriculture are alarming, including neurological damage from pesticide exposure, lead contamination in crops, and respiratory issues in infants caused by nitrates present in irrigation water sourced from wastewater (Thomas et al. 2022). Heavy metals and pathogens not only endanger consumers who eat tainted produce but also present dangers to individuals who come into contact with contaminated soil and water during farming practices (Ninsheka et al. 2024). Infectious diseases such as malaria and schistosomiasis are usually transmitted by vectors like mosquitoes, houseflies, and snails, raising significant public health concerns regarding urban agriculture (Byram and Arabulan 2025). Moreover, agrochemicals pose risks to farmers' health due to direct contact—whether through spraying, inhaling toxic residues, or consuming harmful substances present on crops, resulting in allergies and respiratory illnesses. It is essential that we acknowledge these dangers and promote sustainable agricultural practices to protect human health and the environment (Ibid). Also linked to this practice is streambank cultivation in urban areas. The issue has become increasingly evident, particularly in the Chivero lake catchment region, which encompasses areas like Warren Park (Magombeyi 2017). Additionally, urban agriculture is linked to the excessive application of agrochemicals like fertilizers, herbicides, and pesticides, leading to the pollution of wetland ecosystems and jeopardizing the quality of the natural environment (Byram and Alaburan 2025). The creation of canals and ditches for irrigation disrupts wetland hydrological systems. These actions alter the soil's ability to retain moisture, which hastens the desiccation of wetlands (Fig. 9.6).

Numerous wetland regions in Harare are at a significant risk of being converted into housing projects. Although some wetlands have already been changed into commercial establishments, a substantial number still confront major threats (Ruzvidzo 2020). As of 2020, an estimated 43–49% of wetlands have been impacted by human activities (Toriro 2020). This troubling scenario prompts inquiries about why this pattern persists, especially considering that the Harare City Council (HCC) and the Environmental Management Agency (EMA) appear to be permitting it to happen without action. The perpetuation of this trend can be attributed to several factors, such as uncontrolled urban growth, infrastructure development initiatives, poor waste management practices, and agricultural encroachment into wetland territories (Muziri et al. 2019). Researchers warn that urban wetlands in Zimbabwe, including those in Harare, are approaching extinction due to the mounting pressures of urban land use (Murungweni 2013).

Without a substantial shift in the strategies employed by local authorities, the ongoing deterioration of these essential ecosystems appears unavoidable. The persistent decline of wetlands in areas such as Kuwadzana, Warren Park, Belvedere, Malbereign, Epworth, Mabvuku, and Waterfalls highlights this pressing need (Newsday, 18 January 2021). It is crucial that authorities implement the Built Environment Sustainability Tool (BEST) to manage the interplay between the built environment and natural ecosystems effectively (Gibberd 2015). If action is not taken, there is a risk of causing irreparable harm to these vital wetlands and the advantages they offer.

Politically influential individuals have used their power and influence to grab and parcel out land in wetlands. An example of this was the seizure of the Hillside wetlands in Harare by former cabinet Minister Petronella Kagonye (Gweshe 2020). Corruption and political mudslinging have clouded the functionality of Urban

Fig. 9.6 Photograph showing an offplot of beans along the Mukuvisi river. (Source: Author's field observation)

Councils, especially in light of illegal land acquisitions by land barons and politicians (Chiweshe 2020). This phenomenon is not recent; it has been an established practice in the distribution of both residential and commercial lots by council officials and elected representatives (Muchadenyika and Williams 2017; Uchena 2019). A news article from Al

Jazeera dated January 11, 2021, reported that "Harare City Council is experiencing an increase in illegal settlements, which have arisen as influential businessmen, referred to as land barons, allocate land without appropriate planning and authorization" (Masundire and Ncube 2024).

As Harare continues to grow, the demand for housing is increasing. This has led to a concerning trend where land barons are taking over wetlands, often in cooperation with local urban councillors, to provide for those seeking housing (Muguti 2020; *Newsday*, 18 January 2021). Even more troubling is that these and other irregularities in settlement patterns contravene city regulations related to housing and infrastructure development, especially the City Master Plan (Masundire and Ncube 2024). Analysts have identified shortcomings within the city bylaws, particularly concerning the City Master Plan. They bemoaned that Legislation related to urban land use is outdated and largely neglects wetlands and their vital contributions to Harare, resulting in their degradation. The absence of an updated Master Plan reveals that Harare's land-use policies fail to prioritize the protection of wetlands, which continues to jeopardize water supplies. It's particularly significant that the Master Plan for Harare has not seen an update since 1992 (Gweshe 2020). The ongoing deficiencies in the bylaws continually adversely affect both the environment and the infrastructure of the city.

The Constitution of Zimbabwe (Government of Zimbabwe 2013) and related laws, including the Regional Town and Planning Act (Government of Zimbabwe 1976) and the Urban Councils Act (Government of Zimbabwe 1996), explicitly assign local governments the responsibility of efficiently managing urban land. The ongoing interference from various Ministers of Local Government has undermined this important function. Such meddling not only erodes the autonomy and independence of local authorities but also obstructs effective urban planning and development (Marondedze 2022). It is crucial to reinstate the intended independence of local governments to guarantee responsible and responsive management of urban land, ultimately benefiting communities nationwide (Magaisa 2020). The present situation has led to successive Ministers of Local Government favoring land barons and cooperatives involved in the development and allocation of residential plots in fragile wetland areas (Newsday, September 26, 2019). Many examples highlight the alarming trend of illegal and irregular settlements rising in urban areas such as Kuwadzana, Glen Norah B, Budiriro, Chitungwiza, Mufakose, Westlea, Crowborough, and Epworth, all occurring under the watchful eye of the city council (Ruwende 2018; Kutaura 2018). As highlighted by Mr. Precious Shumba, Director of the Harare Residents Trust, "these stands are distributed through biased and overtly corrupt channels, which have continuously benefitted corrupt councillors" (Kutaura 2018).

Although the argument presented is valid, it is crucial to highlight that a complicated network of stakeholders significantly influences the settlement of home seekers in areas classified as wetlands across Harare. These stakeholders include local land barons, housing cooperatives, political party leaders, city officials, land developers, and powerful politicians (The Herald, 27 November 2020. The situation is further complicated by the tendency of local government officials in Zimbabwe to comply with the Minister's instructions regarding this unlawful seizure of urban land. These powers effectively render the Minister an authoritarian figure over local authorities (Magaisa 2020). The Ministry of Local Government Public Works and National Housing is currently distributing residential stands to Legislators in Harare, a move that some view as a strategy to ensure loyalty in light of increasing political tension. This action is perceived as an attempt to quell an impeachment motion directed at the current president of Zimbabwe, Emmerson Mnangagwa (ZimEye 10 April 2025). Urban local governance has been affected by political conflicts between opposition-led urban councils and the Ministers of Local Government, who typically come from ZANU PF (Magaisa 2020).

9.9 Effects of Destruction of Urban Wetlands on Urban Sustainability and Resilience

The destruction of urban wetlands has debilitating effects on urban resilience and the capacity to recover from natural and man-made shocks. Resilience is the ability of a community or society to resist, absorb, and recover from the effects of a disaster or external shocks in a timely manner (van Breda 2018). Resilience is a function of the society's capacity to organize itself prior to, during, and post times of disaster based on resources and social capital (Torres et al. 2019; Walker 2019). Resilience is also the ability to limit, recover through survival, adapt, and change during difficult situations (Ibid). It pertains to the extent of change or disruption that a system can withstand without transitioning into a different state that possesses distinct structural and functional characteristics and provides various ecosystem services that are beneficial to humans (Salvia and Quaranta 2020). The disappearance of urban wetlands has profound implications for human life in city landscapes and urban resilience. Climate change and its variations negatively impact agricultural production (Lemi and Hailu 2019). In Zimbabwe, the variability in food supply caused by climate change has led to unpredictable food prices, which keep increasing and are becoming unaffordable for many individuals. Zimbabwe's food systems are being affected by climate change, posing a significant issue given that the nation's economy relies heavily on primary sectors like agriculture and mining for its GDP growth. The impacts of climate change are extensive and affect all areas without discrimination (Marondedze 2022). Vulnerable urban populations, including individuals with disabilities and the elderly, are particularly affected as they struggle to care for themselves (Kosanic et al. 2022). Many of these individuals live in informal settlements near flood-prone areas or on steep slopes vulnerable to mudslides. These living situations lead to low living standards and restricted availability of food and nutrition, giving rise to what has been referred to as "silent hunger" (Dubbeling et al. 2019).

Water shortages have become a persistent issue, with certain areas experiencing prolonged periods without access to running water, sometimes lasting for months or even years. This situation endangers residents, exposing them to waterborne illnesses as they are compelled to utilize unsafe water sources (Gandidzanwa and Togo 2022). When the water supply is inadequate, people often rely on contaminated water, leading to water-related health issues like cholera, dysentery, and diarrhea (Mudombi and Chigara 2024). Many are forced to fetch water from unprotected sources and resort to makeshift sanitation solutions, as toilets remain unavailable due to the water scarcity (Dungumaro 2007; Dzvimbo et al. 2017).

9.10 Challenges to Sustainable Wetland Management in the City of Harare

The efforts of the Zimbabwean government to safeguard wetlands in Harare and its neighboring areas face considerable challenges that demand urgent attention. The political rivalry currently exists between ZANU-PF and the CCC, which has emerged as the largest opposition party in Zimbabwe since 2023. Land barons have exploited corruption and political ties to appropriate land in Harare, with wetlands also being affected (Gweshe 2020). The focus has transitioned from being merely administrative to fundamentally political, necessitating a delicate balance among various interrelated interests from political factions, businesses, and entrenched power dynamics, which ultimately hinder effective environmental planning. Corruption and misuse of authority have resulted in inappropriate land distributions, where land barons exploit corruption and political connections to appropriate and subdivide land in the wetlands of Harare (Gweshe 2020).

The previously mentioned challenges have been worsened by the absence of a unified policy on urban wetland management in Zimbabwe (Parliament of Zimbabwe 2018). A significant hurdle in implementing effective strategies for the sustainable preservation of wetlands in both rural and urban areas of Zimbabwe was the lack of a National Wetlands Policy until 2022 (Nzvere 2024). However, this void should not be seen as an excuse for the troubling trend of wetland degradation and the continuous encroachment on wetland areas in the City of Harare (Mudombi and Chigara 2024). Establishing a Wetlands Policy would provide clear directives for both local and national governments in their efforts to conserve and restore these essential ecosystems. This practice is commonplace in various other regions, like South Africa (Pasquini and Cowling 2015). Nonetheless, it is still uncertain whether the development of a wetlands policy, along with current mapping initiatives aimed at generating a detailed inventory of wetlands (*The Herald*, 24 June 2021), will result in effective protection at all levels of government in Zimbabwe. This uncertainty is further intensified by the clear lack of dedication at local government levels to adhere to existing city regulations, including the City Master Plan and the Regional Town and Country Planning Act (Community Water Alliance 2021). Tackling this issue is vital for protecting our wetlands and ensuring a sustainable future for both nature and our communities.

9.11 Challenges and Limitations

The study primarily employed qualitative methods, resulting in a lack of quantifiable data that inadvertently affected its reliability. The research area is particularly delicate, as it pertains to political elites and land barons engaged in the invasion of wetlands. Several respondents were individuals who had experienced disasters related to living in wetlands. Conducting interviews with them evoked memories of the hardships they had faced.

9.12 Conclusion

Wetlands play a crucial role in both mitigating and adapting to climate change due to their multiple functions. Preserving these wetlands can help reduce the impacts of climate change in various ways, including improving carbon storage, providing water resources, and serving as tourist attractions. Wetlands play a crucial role in enhancing urban resilience against the increasing challenges in urban areas. By incorporating nature-based approaches like wetlands, urban areas can successfully manage problems such as flooding, extreme heat, and loss of biodiversity. In addition to environmental advantages, green infrastructure promotes healthier communities, improves air and water quality, and boosts the overall quality of life in urban settings. In metropolitan locations, wetlands encourage the growth of vegetation, which filters pollutants like carbon dioxide, nitrogen dioxide, and dust particles from the air and improves air quality. This is crucial for building sustainable and resilient cities. In particular, trees provide oxygen that enhances the air quality in cities and absorb carbon dioxide, which helps slow down climate change. The relationship between wetlands and urban resilience in the face of climate change is rooted in their common objective of reducing climate-related risks while improving the adaptability and sustainability of urban areas. Urban wetlands utilize nature-based strategies to tackle climate issues, developing systems that enhance and support conventional urban infrastructure. This collaboration enhances cities' resilience against climate-related pressures, including extreme weather occurrences, rising temperatures, and floods. However, wetlands are increasingly threatened by human activities such as urban agriculture, the growth of informal settlements, and the subdivision of land in these sensitive ecological regions. Protecting urban wetlands is essential for promoting sustainable development and building sustainable cities, as emphasized in the Sustainable Development Goals. Numerous challenges have overshadowed the sustainable management of urban wetlands. These challenges include the political infighting between ZANU-PF and the opposition CCC, which has turned wetland management into a contentious issue. In addition, the absence of effective wetland management policies and adherence to urban master plans has resulted in the emergence of human settlements within urban wetlands. The sustainability and resilience of cities are on shaky ground due to rampant human activities. The United Nations Sustainable Development Goal 11 emphasizes the responsibility to ensure that cities are safe, resilient, and sustainable, while Sustainable Development Goal 6 focuses on providing adequate water supply and sanitation as a fundamental human right by 2030. However, achieving this goal may be challenging due to the rapid growth of urban areas that are often unplanned and uncontrolled as a result of political and socioeconomic factors. The hope of achieving sustainable, resilient cities will remain elusive if wetland ecosystems are not managed sustainably.

9.13 Recommendations

The study recommends an urgent call to domesticate the International Conventions on Wetlands Conservation. Integrating wetlands into city Master Plans and creating greenbelts around wetlands in the city of Harare is critical for building resilient and sustainable cities. There is also an urgent need to expedite the implementation of the National Wetlands Policy.

Acknowledgments The author is grateful to officials from the Ministry of Environment and Tourism and the Environmental Management Agency for providing data during this study. Sincere gratitude is also extended to Harare residents who participated in the study and representatives of residents' associations.

Potential Conflict of Interest The author declares that there is no conflict of interest.

Informed Consent Informed consent was obtained from all respondents in this study.

AI Discloser The author declares that no AI was used in drafting this research.

Funding The research was not supported by any funding.

References

Bayram BA, Arabulan S (2025) Evaluation of the feasibility of urban agriculture in Edirne City Center. Pamukkale Univ J Soc Sci Instit 67:291–311

Birdlife International (2015) Position on Climate Change

Brazier A (2015) Climate change in Zimbabwe, facts for planners and decision makers. Konrad-Adanauer-Stiftun

Brazier A (2017) Climate change in Zimbabwe: a guide for planners and decision makers. Konrad-Adanauer-Stiftung, Harare

Chaminuka N, Dube E (2017) Urban Agriculture as a food security strategy for urban dwellers: a case study of Mkoba residents in the city of Gweru, Zimbabwe People. Int J Soc Sci 3(2):26–45. ISSN 2454-5899. Date of Publication: 15th July, 2017. https://doi.org/10.20319/pijss.2017.32.2645

Chaminuka N, Dube E, Kabonga I, Mhembwe S (2021) Enhancing urban farming for sustainable development through sustainable development goals. Springer Change 9(7):503–511

Chaminuka N Dube E, Kabonga I & Mhembwe S (2021) Enhancing Urban Farming for Sustainable Development Through Sustainable Development Goals.Sustainable Development Goals for Society Vol 2 Springer

Chatiza K (2020) The State of By-Laws and their contribution to urban resilience in Zimbabwe

Chikodzi D, Murwendo T, Simba FM (2013) Climate Change and Variability in Southeast Zimbabwe: Scenarios and societal oppoortunities. American Journal of Climate Change 2:36–46

Chirisa I (2024) Wetlands preservation in urban areas. The quest for resilience building in Harare, Zimbabwe

Chiweshe MK (2020) Money, power and the complexities of urban land corruption in Zimbabwe. Afr J Land Policy Geospatial Sci 3(5):131–141. Climate Change to 2050

Cohen L, Manion L, Morrison K (2017) "Research Methods in Education"

Community Water Alliance (2021) Water, wetlands and life: the inseparable triplets, 2 February 2021. Available at: https://kubatana.net/2021/02/02/water-wetlands-and-life-the-inseparable-triplets/

Convention on Wetlands (2021) Global wetland outlook: special edition 2021. Secretariat of the Convention on Wetlands, Gland. Available at: www.global-wetland-outlook.ramsar.org/

Conway G, Badiane O, Glatzel K (2019) Food for All in Africa: Sustainable Intensification for African Farmers

Cunlife R (2020) Current status of Harare wetlands. Harare Wetlands trust disabled populations in the context of climate and environmental change. Curr Opin. Available at https://kubatana.net/wp-content/uploads/2020/07/Current-Status-of-Harare-Wetland-Report-July-2020.pdf

Davies GT, Finlayson CM, Pritchard DE, Davidson NC, Gardner RC, Moomaw WR, Okuno E, Whitacre JC (2020) Towards a universal declaration of the rights of wetlands. Mar Freshw Res 72(5):593–600. https://doi.org/10.1071/MF20219

Dayathilake DDTL (2021) "Estimation of Soil CarbonStocks" in Urban Wetlands

Dinsa TT, Gemeda DO (2019) The Role of Wetlands for Climate Change Mitigation and Biodivesity Conservation. Journal of Applied Sciences and Environmental Management

Dubbeling M, Veenhuizen RV, Hallida J (2019) Urban agriculture as a climate change and disaster risk reduction strategy. Field actions science reports. The journal of field actions Special Issue 20. Urban agriculture: another way to feed cities

Dungumaro EW (2007) Socio economic differentials and availability of domestic water in South Africa. Phys Chem Earth 32:1141–1114

Dzvimbo MA Dzvimbo MM, Mashizha TM (2017) The link between rural institutions and rural development; Reflectionson smallholder farmers and donors in Zimbabwe

Dzvimbo M, Zhanda K (2020) Policy making prospects and the challenges of climate change and the urban energy sector in Zimbabwe. J Urban Syst Innov Resil Zimbabwe 2(1):203

Dzvimbo MA, Mawonde A, Chaminuka N, Zhanda K (2023) Rethinking social housing provisions and public health in emerging low-income peri-urban human settlements, an expose from Harare, Zimbabwe. In: The Palgrave handbook of global social problems. https://doi.org/10.1007/978-3-030-68127-2_149-1

Finlayson C, Gardner RC (2020) Towards a Universal Declaration of the rights of Wetlands

Gandidzanwa CP, Togo M (2022) Adaptive responses to water, energy, and food challenges and implications on the environment: an exploratory study of Harare. Sustainability 14(16):10260

Gandidzanwa CP, Togo M (2024) Impacts of water, energy, and food nexus challenges on household vulnerability: the case of Harare City, Zimbabwe. Environ Res Lett 19(2024):114038

Gibberd J (2015) Measuring Capability for Sustainability: The Built Environment Sustainability tool (BEST). Building Research and Information

Gibberd J (2017) Strengthening sustainability planning: the city capability framework. Procedia Eng 198:200–211

Government of Zimbabwe 1976

Government of Zimbabwe 1996

Government of Zimbabwe 2013

Gutsa I (2014) Climate Change and Policy Making in Zimbabwe. In Search of evidence based policy making? https://doi.org/10.15174/au.2014.594

Gweshe E (2020) Time to Wake Up to the reality of wetlands degradation. Newsday 12 March

Hettiarachchi M, Community Water Alliance (2021) Water, wetlands and life: the inseparable triplets. Environ Sustain 55:101159

Junk WJ, An S, Finlayson CM, Gopal B, Kvet J, Mitchell SA, Mitsch WJ, Robarts RD (2013) Current state of knowledge regarding the world's wetlands and their future under global climate change: a synthesis. Aquat Sci 75(1):151–167. https://doi.org/10.1007/s00027-012-0278-z

Kalanzi W (2015) Local governments and wetland conservation in Uganda: contributions and challenges. Aust J Public Adm 50(1):157–171

Kohler T, Rumyantseva M, Welch C (2023) Qualitative restudies: research designs for retheorizing. Organ Res Methods 28(1). https://doi.org/10.1177/10944281231216323

Kolka R, Scott D, Bridgham, Chien-Lu (2016) Soils of peatlands:histosols and gelisols

Korstjens I, Moser A (2018) Series: practical guidance to qualitative research. Part 4: trustworthiness and publishing. Eur J Gen Pract 24(1):120–124

Korstjens I, Moser A (2018) Series: Practical Guidance to Qualitative Research. Part4. Trustworthiness and Publishing.European Journal of General Practice

Kosanic A, Petzold J, Martín-López B, Razanajatovo M (2022) An inclusive future: disabled populations in the context of climate and environmental change Curr Opin Environ Sustain 55:101159

Kudumba BT (2022) Transforming urban policy to combat wetland degradation in Harare. Open J Soc Sci 10(11):479–505

Lal R (2008) Carbon sequestration. Philos Trans R Soc B 363:815–830. https://doi.org/10.1098/rstb.2007.2185

Lee TS, Randall LA, Kahal NL, Kinas HL, Carney VA, Rudd H, Baker TM, Sanderson K, Creed IF, Moehrenschlager A, Duke D (2022) A framework to identify priority wetland habitats and movement corridors for urban amphibian conservation. Ecol Solut Evid 3(2):e12139. https://doi.org/10.1002/2688-8319.12139

Lemi T, Hailu F (2019) Effects of climate change variability on agricultural. J Environ Sci Nat Res 17(1):IJESNR.MS.ID.555953

Liang Z (2021) Assessment of the construction of a climate resilient city: an empirical study based on the difference in differences model. Int J Environ Res Public Health 18:2082. https://doi.org/10.3390/ijerph18042082

Magaisa AT (2020) Big Saturday read: urban councils—a case of responsibility without power, 21 November 2020. Available at: https://www.bigsr.co.uk/single-post/big-saturday-read-urban-councilsa-case-of-responsibility-without-power

Magdi S (2022) Towards a resilient city: an adaptive planning model for a healthy and happy city. MSA Eng J 1(4):126-144

Manivel R (2025) Green infrastructure: enhancing environment resilience and sustainability frameworks and case studies on green infrastructure and urban resilience to climate change. PSG College of Science and Arts

Marambanyika T (2015) An analysis of the impacts of human activities and management strategies on wetland processes in Southern Zimbabwe. PhD thesis, University of Kwazulu Natal, Pietermaritzburg

Marondedze AK (2022) The geography of contemporary urbanization and its effects on landscape sustainability in Harare Metropolitan Province. Freie Universität Berlin

Masundire M, Ncube RK (2024) Socio-ecological conundrums towards the wise use of wetlands in the Harare Metropolitan Province, Zimbabwe. Ngenani 2(1 & 2):206–226

Matamanda A, Chirisa I, Mukamuri B Kaduwo P (2018) Harare's Deteriorating Wetlands: Why Sound Policies and Legislations Are Not Enough

Matamanda AR, Chinozvina QL (2020) Driving forces of citizen participation in urban development practice in Harare, Zimbabwe. Land Use Policy 99:105090

McCartney M, Rebelo LM, Sellamuttu S (2015) Wetlands, livelihoods and human health. In: Finlayson C, Horwitz P, Weinstein P (eds) Wetlands and human health. Wetlands: ecology, conservation and management, vol 5. Springer, Dordrecht. https://doi.org/10.1007/978-94-017-9609-5_7

Mhlanga D and Mhlanga D (2022) Impact of Human activities on Wetlands Provisioning and Cultural Services

Millennium Economic Assessment 2003

Moyo L, Cunlife R (2020) Loss of Wetlands in Harare 2008–2019

Muchadenyika D, Williams LJ (2017) Politics and the Practice of Planning;The case of Zimbabwean Cities

Mudombi TP, Chigara B (2024) The housing provision and environmental protection dilemma a cause-impact analysis of urban housing development on Harare's wetlands. Oikos The Zimbabwe Ezekiel Guti University bulletin of Ecology Science Technology Agriculture Food Systems Review and Advancement

Muguti T (2020) Harare Provincial Coordinators Presentationon WetlandsManagement . Workshop Organised by Harare Wetlands Trust

Murombo T (2019) Climate Change in Zimbabwe. Towards a Low Carbon Energy Industry Forthcoming in Climate Change Law in Zimbabwe: Concepts and Insights (Konrad Adenauer Foundation)

Murungweni FM (2013) Effect of land use change on quality of urban wetlands: a case of Monavale Wetland in Harare. Geoinfor Geostat 5:2

Musasa T, MarambanyikaT (2022) Assessing the Sustainability of Land uses in Driefontein and Intunjambili Wetlands, Zimbabwe. Scientific African Volume 16 July 2022

Musasa TCR, Muringaniza K, Dakarai Mhlanga E (2023) Perspective chapter: the impact of human activities on wetlands' provisioning and cultural Services in Epworth, Zimbabwe. IntechOpen. https://doi.org/10.5772/intechopen.1001064

Muziri C, Banhire T, Matamanda AR (2019) Wetlands under the impact of urban development in Zimbabwe. In: Chirisa I (ed) The sustainability ethic in the management of the physical, infrastructural and natural resources of Zimbabwe. Langaa RPCIG, pp 1–12

Mwonzora G (2021) Local governance and wetlands management: a tale of Harare City in Zimbabwe. Article in urban Forum 2021

Nemutamvuni K, McKay TJM, Tantoh HB (2020) Active citizenry, community-based organisations and the protection of urban wetlands: the case of Colbyn, Tshwane, South Africa. Glob Ecol Conserv 24:e01244

Newsday, 26 September 2019, (2019)

Newsday, 18 January 2021, (2021)

Ninsheka L, Ssemakula E, Tiyo C, Kalibwani R, Kityo R, Mugizi W, Byamukama W (2024) Effects of urban agriculture on the socio-economic status of farmers in cities of Sub-Sahara Africa. A case of Zambia, South Africa, and Nigeria: a review. East Afr J Agric Biotechnol 7(1):3344. https://doi.org/10.37284/eajab.7.1.1709

Nzvere C, Chirubu L, Masunda T (2024) Environmental literacy, the key to sustainable conservation of wetlands in urban areas. Smart Cities Reg Dev J 8(1): 59–74

Nzvere C, Chirubvu L, Masunda T (2024) Environmental literacy the key to sustainable conservation of wetlands in urban areas

Ozturk B, Kavlak O, Cabuk N & Cetin M (2022) The nature values of constructed wetlands in climate change

Parliament of Zimbabwe (2018)

Pasquini L and Cowling RM (2015) What enables Local Governments to mainstream climate change adaptation? Lessons learned from South Africa

Pereira P, Inacio M, Pinto L, Kalinauskas M, Bogdzevic K, Zhao W (2024) Mapping ecosystem services in urban and peri-urban areas. A systematic review. Geogr Sustain 5(3):491–509. https://doi.org/10.1016/j.geosus.2024.06.002m.6198

Ramsar (2016) Updated information on Zimbabwe's Ramsar Sites. Available at: https://www.ramsar.org/news/updated-information-on-zimbabwes-ramsar-sites

Ramsar (2018) The Convention on Wetlands

Ruwende I (2018) Council Demolishes illegal structure. The Herald 29 October 2018

Ruzvidzo S (2020) Urban development and wetland conservation. United Nations Development Programme. Available at: https://www.zw.undp.org/content/zimbabwe/en/home/blog/urban-developmentand-wetland-conservation.html

Salvia R, Quaranta G (2020) Food chains and ecosystem services through a resilience lens. Cooperative approaches for a changing world. In: Mattas K (ed) Sustainable food chains and ecosystem. Springer, Cham, pp 17–34. https://doi.org/10.1007/978-3-030-39609-1_2

Schuijt K (2002) Land and water use of wetlands in Africa: economic values of African Wetlands, Interim Report IR-02-063. International Institute for Applied Systems Analysis, Austria

Science and Technical Review Panel (STRP) 2009 Resolution x.1 The Ramsar Strategic Plan 2009-2015

Sharai S, Tawanda T, Gladman C (2020) Threats and conservation strategies on urban wetlands: a case of Monavale and surrounding areas in Harare, Zimbabwe. Am J Environ Prot 9(2):36–43. https://doi.org/10.11648/j.ajep.20200902.12

Silipiwe S, Tarakini T, Chibememe G (2020)Threats and Conservation Strategies on Urban Wetlands; A case of Monavale and Surrounding Areas in Harare

Somay-Altas M, Gokcen-Yilmaz S (2023) Evaluation of the geogenic-effected groundwater antimony pollution, in the settlement area, utilizing the water quality index and health risk assessments: a case study from Kiraz Basin, Izmir Turkey. ACS Earth Space Chem 7(5): 1120–1130. https://doi.org/10.1021/acsearthspacechem3c00031

Strokal V (2021) Transboundary rivers of Ukraine: perspectives for sustainable development and clean water. J Integr Environ Sci 18(1):8

Tenny S, Brannan M, Brannan G (2022) Qualitative study. StatPearls

Thomas PS, Kombe WJ, Lupala A (2022) Effects of stakeholders' perception of urban agriculture on the governance of urban agriculture in the wards of Daraja Mbili and Lemala in Arusha City, Tanzania. Int J Soc Sci Res Rev 5(8):349–364

Tirivangasi HM, Dzvimbo MA, Chaminuka N, Mawonde A (2023) Assessing climate change and urban poverty in the context of the COVID 19 lockdowns: rethinking personality and societal challenges in Zimbabwe. Sci Afr 20:e01710. https://doi.org/10.1016/j.sciaf.2023.e01710. Epub 2023 May 11. PMID: 37223655; PMCID: PMC10172155

Toriro P (2020) Urban planning impact on wetlands management in Harare, 9 November 2020. Presentation made in Harare and livestreamed

Torres P, Wood G Designing for Digital Playing out. Proceedings of CHI Conference on Human Factors in Computing Systems

The Source (2020) Annual Review of Wetlands International

UNDP (2020) Project Implementation Review (PIR) Vietnam's Wetland Protected Areas

Uprise Zimbabwe (2024) Urban and peri-urban resilience through investment for sustainable ecosystems in Zimbabwe

UNFCCC(2017) COP 23

Walker T (2019) The ArcGIS Online Wetlands 2019 Extent Report

White MP, Lewis R, Gascon M, Roberts B Fleming LE (2020) Blue Space, health and Well being. A narrative Overview and synthesis of potential benefits

Zimbabwe Environmental Law Association(ZELA) 2021

Zimbabwe National Statistics Agency (ZIMSTAT) (2022) Zimbabwe Population and Housing census. Fertility Thematic Report

Assessment of Water Quality and Pollution Dynamics in Sagardighi: An Urban Wetland of West Bengal

10

Bappa Sarkar

Abstract

Wetlands are one of the most important elements of any urban area, acting like the kidneys of a city. Due to urban expansion and land use changes, wetlands are shrinking, and the quality of water is deteriorating at a faster rate. In this study, various water quality parameters—physical, chemical, and biological—were analyzed and compared with (BIS ISDWS, New Delhi 6:2–3, 2012) standards. Different statistical techniques and the Water Quality Index (WQI) were employed to assess the status and spatiotemporal patterns in the water quality of the Sagardighi Wetland, West Bengal. The data were collected from the Water Pollution Control Board, West Bengal, from 2012 to 2024. A total of 13 physicochemical parameters (pH, EC, Ca, Mg, Na, K, N, Cl, PO43, SO_4, TDS, $CaCO_3$, turbidity) and 4 microbial parameters (COD, BOD, DO and total coliform) were chosen for the study. The WQI was calculated using the weighted arithmetic water quality index method followed by Horton (J Water Pollut Control Fed 37:300–306, 1965) and developed by Brown et al. (Water Sew Works 117(10):339–343, 1970) and CODI, BODI, DO index were applied. In premonsoon time, WQI ranged from 63.19 (2010) to 17.77 (2024), whereas in the monsoon period, the WQI was 52.96 (2010) to 92.42 (2024), indicating poor water quality. The total coliform count was 7000 in 2024, indicating significant contamination, whereas COD was 28 mg/L and BOD was 7 mg/L, which are near the permissible limits. This study will enhance our knowledge about the spatiotemporal changes in urban water quality and provide future roadmaps that will be helpful for policymakers and concerned authorities in developing sustainable water quality management strategies.

Keywords

Urban wetland · WQI · Water quality · Sagardighi

10.1 Introduction

Wetlands play a vital role in any urban area. Wetlands serve numerous functions, including hydrological roles such as regulating water flow within watersheds and purifying and recycling surface and groundwater. They also perform biogeochemical functions, such as storing, and cycling nutrients, organic compounds, and metals, alongside being biologically significant as some of the most productive and diverse ecosystems on the planet (Manral et al. 2012). Wetland ecosystems are often called "nature's

B. Sarkar (✉)
Department of Geography, Dinhata College, Dinhata, West Bengal, India

S. K. Chakraborty et al. (eds.), *Wetlands and Urban Resilience*, Sustainable Development Goals Series,
https://doi.org/10.1007/978-3-032-06881-1_10

kidneys" due to their resemblance to human kidneys. They naturally filter pollutants and purify water, which is crucial in maintaining environmental health. Rapid growth and diversified activities influenced urban wetlands (Sharma et al. 2024). The rapid increases in urbanization have led to the decay of wetlands causing serious water pollution and ecological imbalance (Joshi N, 2020 et al.) (He et al. 2014; Jantz et al. 2005; Zhang et al. 2007; Koch 2023; Wang 2024). Wetlands play a great role during the rainy seasons, they act as a flood controller of the urban area and they reduce the urban waterlogging. Several living animals, plants and other organisms are staying and helping to balance the ecology and biodiversity. Different types of migratory birds are found during the winter seasons. But with the dynamic changes in urban landscape urban wetlands are decaying at a faster rate (Mitsch et al. 2023). Wetlands are suffering, and water quality is degraded. The water quality of Delhi's urban wetlands is poor and unsuitable for drinking. Concerning irrigation water quality, water from Bhalswa Lake is unsuitable, whereas water from other wetlands is marginal to suitable (Joshi et al. 2021). Some urban wetlands have declined biochemical oxygen demand and chemical oxygen demand (Chandra et al. 2010; Wang 2024). The physicochemical parameters also vary from one season to another. Lower values of the majority of ions in winter as compared to summer indicate the dilution effect (Dixit et al. 2021). The wetland has reached in nutrients but some external anthropogenic intrusion may increase the water pollutants. Some wetlands are influenced by anthropogenic activities which cause the increase of sulfur, chloride, iron, and lead (Ranjan et al. 2017). Most of the urban wetlands are unfit for human consumption (Teshome et al. 2020).

Cooch Behar Municipality, established in 1884, currently hosts 35 urban wetlands, with Sagardighi being the most prominent due to its size, location, and role in flood control and recreation. Despite their ecological and social significance, these wetlands are facing increasing degradation. It acts as an urban flood controller and a recreation site of this city. Although water pollution is a chief matter of apprehension in the Cooch Behar municipality, very little work was conducted in the urban wetlands of Cooch Behar. However, one scientific investigation on bio-physical-chemical analyses of Sagar Dighi wetland (Miraj and Bhattacharya 2017; Das et al. 2016) has been reported. Bio-physical-chemical parameters were discussed elaborately by Miraj and Bhattacharya (2017). They found that the Sagardighi of Cooch Behar is comparatively in moderate condition, but almost all the parameters are prone to cross the critical pollution level if necessary action is not taken in time. Sagardighi, the largest and most centrally located urban wetland in Cooch Behar, has experienced significant ecological degradation over the years. Once a vital water body for the town's ecological balance, it now faces multiple environmental threats due to growing urban pressures. This wetland's natural capacity to act as a flood buffer and groundwater recharger has weakened, primarily due to altered hydrology and poor maintenance. Without immediate restoration and sustainable management efforts, the wetland risks turning into a heavily polluted, biologically inactive water body, thereby losing its socio-environmental and ecological value.

Sagardighi is a vital urban wetland in Cooch Behar that supports biodiversity, regulates floods, and serves as a recreational and cultural space. However, it faces increasing threats from pollution, urban encroachment, and ecological degradation. A focused study is needed to assess water quality, identify root causes of degradation, and guide effective conservation and management strategies. The present study aims to analyze the trend of water quality for aquatic life and human consumption since 2010, and to investigate the root causes of urban wetland degradation.

10.2 Study Area

Sagardighi Wetland is located in the heart of Cooch Behar Municipality, West Bengal. The area of this wetland is 14.60 hectares, which was excavated in 1807 AD by Maharaja Harendra Narayan. It holds historical and ecological importance, attracting numerous migratory bird species during the winter season. The latitudinal and longitudinal extensions are 26°19′12″N to 26°19′21″N and 89°26′21″E to 89°26′28″E, respectively (Fig. 10.1). Most of the administrative buildings, constructed between 1880 and 1920, now serve as key administrative offices. This wetland is one of the busiest and attractive sites of this Heritage city.

10.3 Methodology

In this study, 13 physicochemical parameters such as pH, EC, Ca, Mg, Na, K, N, Cl, PO43, SO_4, TDS, $CaCO_3$, and Turbidity and 4 biochemical parameters such as COD, BOD, DO, and total coliform were selected for the study (Fig. 10.2). The data were collected from the Water Pollution Control Board, West Bengal, from 2012 to 2024 (Fig. 10.2).

Water Quality Index (WQI) method is used based on Horton (1965) and Brown et al. (1970) in this study. The permissible limit was identified following the Bureau of Indian Standards (BIS 2012) for drinking water, which has been utilized while calculating WQI,

$$WQI = \sum wiqi\ n\ \mathrm{i} = 1 / \sum wi\ n\ \mathrm{i} = 1 \quad (10.1)$$

where WQI = Water Quality Index.

wi = assigned weight of parameter i (Table 10.1).
qi = water quality value of parameter i.
n = number of parameters.

The following formula is used to calculate the value of qi (Brown et al. 1970):

$$\mathrm{qi} = 100\ [(Vi\text{–}\mathrm{Vid})/(Si\text{–}\mathrm{Vid})] \quad (10.2)$$

where Vi = observed value of the parameter i.

Si = standard permissible value of the parameter i.
Vid = ideal value of the parameter i in pure water. The ideal values of all parameters are 0 for drinking water, except dissolved oxygen and pH (Tripathy and Sahu 2005).

Further, to derive the pH quality rating, the following formula was applied:

$$q\mathrm{pH} = 100\ [(V\mathrm{pH}\text{–}7.0)/(8.5\text{–}7.0)] \quad (10.3)$$

where VpH = observed value of pH.

The quality rating of dissolved oxygen was calculated using the following formula:

$$qDO = 100\ [(VDO\text{–}14.6)/(5.0\text{–}14.6)] \quad (10.4)$$

where VDO = observed value of dissolved oxygen.

The WQI values were classified into five categories, such as excellent quality (WQI = <25), good quality (WQI = 25–50), poor quality (WQI = 51–75), very poor quality (WQI = 76–100), and unsuitable for drinking (WQI = >100) (Table 10.2).

Biological Oxygen Demand (BOD) Index was calculated based on the following formula:

Photo plates 1 and 2: Sagardighi Wetland, Cooch Behar, West Bengal

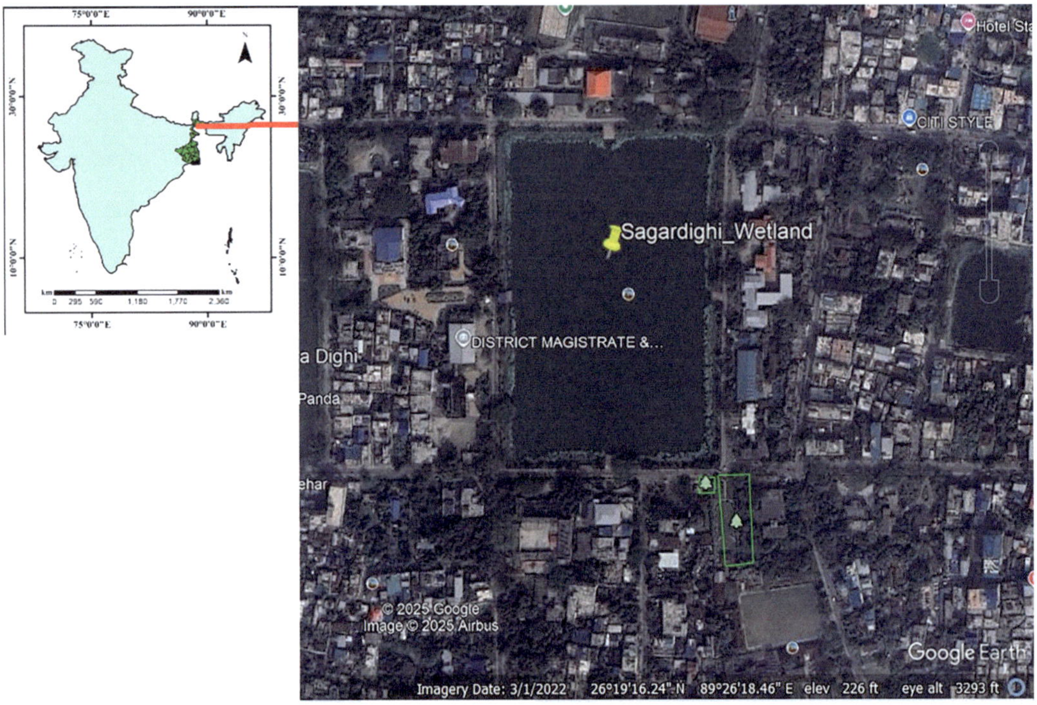

Fig. 10.1 Study area, Sagardighi Wetland, Cooch Behar

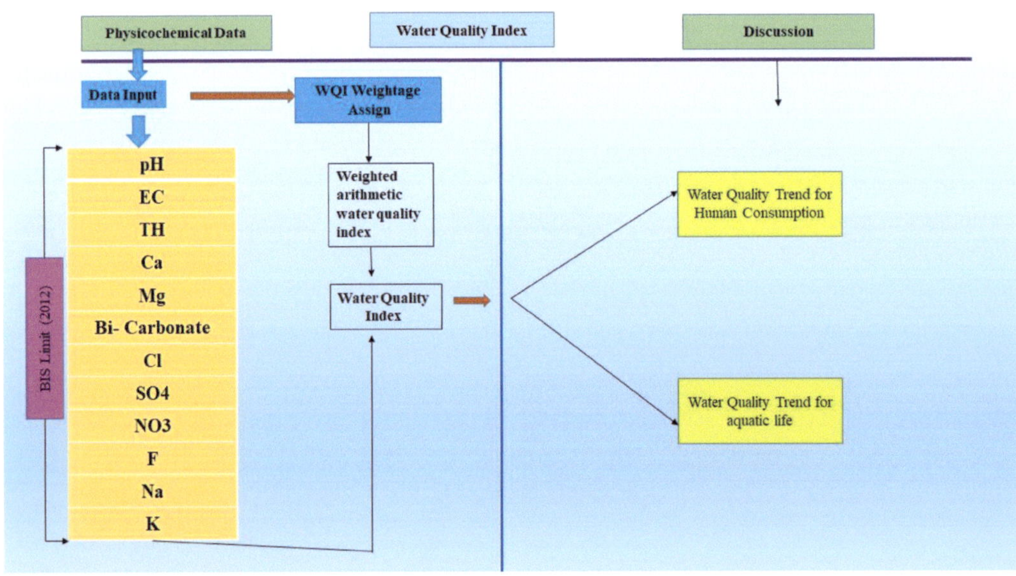

Fig. 10.2 Methodological flowchart

Table 10.1 Weight of the parameters

Parameter	Permissible limit (BIS 2012)	Weight
pH	8.5	0.082325299
EC	300	0.00233255
Ca	75	0.009330201
Mg	30	0.023325501
Na	200	0.003498825
K	55	0.012723001
N	45	0.015550334
Cl	250	0.00279906
PO43	1	0.69976504
SO_4	200	0.003498825
TDS	500	0.00139953
$CaCO_3$	200	0.003498825
Turbidity	5	0.139953008

Table 10.2 Water quality range

WQI range	Status
0–25	Excellent
25–50	Good
50–75	Poor
75–100	Very poor
More than 100	Unsuitable

Misra and Patel (2001)

$$\text{BODI} = \frac{\text{Observed BOD}}{\text{Permissible BOD}} \times 100 \qquad (10.5)$$

Dissolved Oxygen Index (DOI) was calculated based on the following formula:

$$\text{DOI} = \frac{\text{Observed DO}}{\text{Permissible DO}} \times 100 \qquad (10.6)$$

Chemical Oxygen Demand (COD) Index was calculated based on the following formula:

$$\text{CODI} = \frac{\text{Observed COD}}{\text{Permissible COD}} \times 100 \qquad (10.7)$$

10.4 Results and Discussion

10.4.1 Physicochemical and Microbial Trend

10.4.1.1 Biochemical Oxygen Demand (BOD)

The total oxygen needed by aerobic microorganisms to fully break down organic waste in a water body is referred to as biochemical oxygen demand (BOD). Therefore, BOD serves as a measure of organic pollution, where higher values signify greater levels of organic contamination (Patel et al. 1983). During the monsoon period, the Biochemical Oxygen Demand (BOD) declined from 3.35 in 2010 to 2.4 in 2015 and further dropped to 1.4 in 2020, whereas in 2024 the BOD value reached 2.4, which indicated a declining trend during the monsoon seasons. The fluctuations of BOD may occur due to high runoff and urban drainage water involvement (Minch et al. 1962; Effendi et al. 2018). BOD index values decreased from 111.67 in 2010 to 80.00 in 2024, indicating a slight improvement but still beyond the permissible limit (Table 10.1). BOD levels decreased from 111.67 in 2010 to 80.00 in 2024, showing some improvement but still significantly higher than the permissible limit, indicating organic matter pollution. In contrast, the premonsoon period water quality is much better than the monsoon period. In contrast, the premonsoon period shows a steady improvement in water quality. The BOD value decreased markedly from 4.5 in 2010 to 2.6 in 2015 and then further to 1.2 in 2020 (Table 3), where it stabilized through 2024. This consistent decline in BOD during the premonsoon period indicates good management of wetlands by the municipality; some restrictions and management policies were taken by the Cooch Behar municipality.

10.4.1.2 Dissolved Oxygen (DO)

The total amount of oxygen dissolved in a water body is known as dissolved oxygen (DO), and its concentration is influenced by the physical, chemical, and biological processes occurring within the water body. A dissolved oxygen (DO) level of 4–6 mg/L is considered the optimal range for maintaining good water quality and supporting aquatic life (Bora and Goswami 2017). A dissolved oxygen (DO) relatively stable trend was observed during the pre- and post-monsoon period from 2010 to 2024. In 2010 DO value was 7.8 and 7.1 in 2015; however, it rose to 7.2 in 2024, indicating marginal improvement in water quality. In premonsoon time, the DO value was recorded 9 in 2010, indicating excellent quality, but it decreased to 7.2 in 2015 and rose to 8.2 in 2024. DO index was declined from 180 in 2010 to 164 in 2024. Though it decreased, it still remains above the permissible limit of DO (Table 10.3). It reflects reduced organic pollution. These trends highlight the varying dynamics of oxygen availability across seasons and years, with signs of recovery in recent years.

10.4.1.3 Chemical Oxygen Demand (COD)

Organic pollution in water can be assessed by measuring a key parameter known as chemical oxygen demand (COD). Higher levels of organic pollution were indicated by COD measurements, which rose from 30 mg/L in 2010 to 46.8 mg/L in 2015 during the monsoon season. As a result of improved pollution management measures, COD dropped to 29 mg/L by 2020 and 30 mg/L by 2024. Comparably, during the premonsoon period, COD was 48 mg/L in 2010, but it drastically decreased to 23 mg/L in 2020 and stayed steady at 30 mg/L in 2024, indicating steady progress in lowering the amount of organic matter in water. The Chemical Oxygen Demand Index (CODI) was stable from 2010 to 2024 (12.0) during monsoon seasons and decreased from 19 in 2010 to 12 in 2024, indicating improved water quality. For the last 5 years, the municipality corporation has been trying to implement some restrictions and manage Sagardighi for better water quality.

10.4.1.4 Total Coliform

The Total coliform is a significant indicators of water quality analysis and contamination, as its presence suggests the possibility of harmful pathogens or fecal pollution in the water. Total coliforms should be <5000 in water for healthy aquatic life (BIS 2012). Total coliform reduced from 5400 in 2010 to 140 in 2024 during premonsoon and remained the same in monsoon periods, indicating water quality improvement for living animals at Sagrdighi with a permissible limit by 2024 (Table 10.3).

10.4.2 Geochemical Parameters Trends

The electric conductivity (EC) increased from 50.6 to 86.6 during 2010–2015 in monsoon periods; however, it declined in 2024, i.e., 40, indicating a significant improvement in water quality. In the premonsoon season, EC

Table 10.3 Physicochemical and microbial trend of Sagardighi Wetland, Cooch Behar

Index parameters	Permissible limit	Monsoon				Premonsson			
		2010	2015	2020	2024	2010	2015	2020	2024
BOD	<3 mg/l	111.67	80.00	46.67	80.00	150.00	86.67	40.00	40.00
DO	>5 mg/l	156.00	142.00	142.00	144.00	180.00	144.00	144.00	164.00
COD	<250 mg/l	12.00	18.72	11.60	12.00	19.20	9.60	9.20	12.00
Total coliforms	<5000 (MPN/100 ml)	1400.00	220.00	160.00	140.00	5400.00	180.00	140.00	140.00

increased from 52.9 in 2010 to 65 in 2015, indicating higher salinity levels, but it declined to 46.21 in 2010 and 42.87 in 2024. These patterns imply that although EC fluctuated significantly during the monsoon season, it gradually and steadily improved during the premonsoon period, most likely as a result of improved management of water resources and initiatives to lower saline levels. Higher conductivity concentration in water suggests the presence of dissolved particles from a variety of anthropogenic and natural sources (Karmakar et al. 2024; Pimparkar et al. 2023). EC does not directly impact health, but it influences the parameters of water (Sarkar 2024).

Nitrogen concentration was increased from 0.15 in 2010 to 0.8 in 2020; however, by 2024, it sharply declined to 0.04 during the monsoon season. In premonsoon, the nitrogen level was recorded at 0.146 in 2010 and increased to 0.435 in 2015 and 0.77 in 2020, but in 2024 it sharply declined to 0.08. It indicates significant changes to the way nitrogen pollution is managed or better methods for managing water quality during the monsoon season. Nitrate toxicity causes aquatic species to swim differently, grow more slowly, and eventually die. It results in histological changes to the brain, esophagus, and gills of fish (Banerjee et al. 2023; Baker et al. 2017).

In 2010, the pH was 7.25 to reach 8.79 in 2015 during monsoon periods, indicating more alkaline conditions; however, in 2024 it reduced to 7.38, indicating a stabilization toward mild alkaline. In the premonsoon period, the pH started at 6.85 in 2010, reflecting slightly acidic water conditions. By 2015, it increased to 7.56, crossing into mildly alkaline territory, and remained stable at 7.5 in 2020. By 2024, the pH had risen to 7.86, indicating a further shift toward alkaline conditions, possibly due to reduced acidifying pollutants or changes in water management practices. The data on water temperature highlights notable seasonal and temporal variations during the monsoon and premonsoon periods from 2010 to 2024 (Table) In 2010, Ca was recorded at 4.12 mg/l and it suddenly increased to 9.62 mg/l in 2015, 12.02 in 2020 indicating a rise in dissolved ca due increases of runoff but in 2024 it suddenly decline at 8 mg/l, reflecting a possible reduction in minerals due to improved water management. In the premonsoon period, calcium levels began at 5.71 mg/L in 2010, rising significantly to 11.2 mg/L in 2015. This trend reversed slightly in 2020, with levels dropping to 9.62 mg/L, and further declined to 5.6 mg/L in 2024, nearing the initial levels observed in 2010. The gradual reduction in calcium concentration after 2015 may indicate improved control over mineral inputs, reduced agricultural runoff, or natural stabilization of water chemistry.

During the monsoon period, chloride concentrations increased from 2.99 mg/L in 2010 to a peak of 14.67 mg/L in 2020, reflecting a rise in saline intrusion or anthropogenic inputs such as wastewater discharge. However, by 2024, chloride levels declined to 5.87 mg/L, indicating potential improvements in water management. In the premonsoon period, chloride levels showed similar trends, starting at 5.91 mg/L in 2010, peaking at 19.57 mg/L in 2020, and reducing to 5.91 mg/L in 2024, suggesting improved water quality management over time. Magnesium concentrations in the monsoon period increased steadily from 0.98 mg/L in 2010 to a peak of 5.18 mg/L in 2020, likely due to higher runoff carrying magnesium-rich minerals. By 2024, levels had reduced to 2.91 mg/L, indicating stabilization. During the premonsoon period, magnesium levels began at 1.94 mg/L in 2010, decreased to 0.58 mg/L in 2015, rose to 3.46 mg/L in 2020, and slightly declined to 2.92 mg/L in 2024, reflecting fluctuating mineral dynamics.

In the monsoon period, phosphate concentrations increased gradually from 0.037 mg/L in 2010 to 0.149 mg/L in 2020, peaking at 0.5 mg/L in 2024, indicating nutrient enrichment potentially from agricultural runoff. During the premonsoon period, phosphate levels showed a decrease from 0.054 mg/L in

2010 to 0.016 mg/L in 2020, followed by a slight increase to 0.05 mg/L in 2024, reflecting reduced nutrient loading over time. Potassium levels remained stable at 2 mg/L during the monsoon period in 2010 and 2015 but dropped to 0.3 mg/L in 2020 before rising again to 2 mg/L in 2024. In the premonsoon period, potassium levels showed a gradual decline from 2 mg/L in 2010 to 1.1 mg/L in 2020 and stabilized at 1 mg/L in 2024, indicating reduced agricultural or organic inputs.

Sodium concentrations during the monsoon period fluctuated slightly, starting at 2.5 mg/L in 2010, increasing to 3 mg/L in 2015, and then declining to 2.3 mg/L in 2024. In the premonsoon period, sodium levels were 1.19 mg/L in 2010, peaked at 3 mg/L in 2015, dropped significantly to 0.6 mg/L in 2020, and rose to 1.8 mg/L in 2024, reflecting variations in ionic inputs over time. The majority of freshwater systems typically contain modest levels of sulfate, however regions with sulfate-containing ores or human activity may have higher levels (Davies 2007). Sulfate concentrations during the monsoon period increased significantly from 0.055 mg/L in 2010 to 8.62 mg/L in 2020, before declining to 3.45 mg/L in 2024, suggesting reduced industrial or agricultural runoff. In the premonsoon period, sulfate levels rose from 0.088 mg/L in 2010 to a peak of 15.67 mg/L in 2020 and declined to 2.4 mg/L in 2024, indicating improved water management practices.

In the monsoon period, TDS levels decreased steadily from 154 mg/L in 2010 to 20 mg/L in 2024, reflecting a significant improvement in water quality. During the premonsoon period, TDS started at 38 mg/L in 2010, rose slightly to 68 mg/L in 2020, and dropped to 28 mg/L in 2024, suggesting similar positive trends. Calcium carbonate concentrations increased in the monsoon period from 14.14 mg/L in 2010 to 48 mg/L in 2015 and 2020, followed by a reduction to 32 mg/L in 2024. In the premonsoon period, levels rose from 22 mg/L in 2010 to 38 mg/L in 2020 but decreased to 26 mg/L in 2024, indicating improved mineral stabilization. Turbidity during the monsoon period rose from 17.4 NTU in 2010 to 24.3 NTU in 2015 but decreased to 19.6 NTU in 2024, reflecting reduced sediment loads. In the premonsoon period, turbidity dropped significantly from 20.8 NTU in 2010 to 4.75 NTU in 2020, followed by an increase to 13.1 NTU in 2024, indicating some variability in sediment inputs (Tables 10.4 and 10.5).

10.4.3 Water Quality Index

The WQI values were classified into five categories, such as excellent quality (WQI = <25), good quality (WQI = 25–50), poor quality (WQI = 51–75), very poor quality (WQI = 76–100), and unsuitable for drinking (WQI = >100). In 2010, the WQI value was 63.19, indicating poor water quality, but it reduced to 25.87 in 2015 and 17.77 in 2020, indicating a declining trend, which reflects excellent quality of water for aquatic life, and it rose to 45.32 good quality, indicating fit for aquatic life and human consumption also during the premonsoon period. WQI was 52.96 in 2010 and reached 88.22 in 2015, slightly declined to 65.99, and suddenly increased to 92.42 in 2024. Indicating the different types of pollutants are involved during the monsoon period due to urban waterlogging, flooding, and some mismanagement of the surrounding area of Sagardighi. During the premonsoon period, the correlation matrix reveals significant relationships among water quality parameters. pH strongly correlates with nitrogen (1.00) and turbidity (0.96) while showing an inverse relationship with TDS (−0.77). Electrical conductivity (EC) is highly dependent on calcium, magnesium, phosphate, and $CaCO_3$ (all >0.95), but inversely related to TDS (−1.00). Calcium and magnesium exhibit perfect correlation (1.00), indicating a shared source, while both positively relate to chloride, phosphate, and sulfate (>0.90). Sodium correlates positively with turbidity (0.90) but negatively with calcium (−0.16) and magnesium (−0.17). Potassium shows an antagonistic

Table 10.4 Water quality parameters during monsoon and premonsoon in Sagardighi wetland

	Monsoon				Premonsoon			
Parameter	2010	2015	2020	2024	2010	2015	2020	2024
BOD	3.35	2.4	1.4	2.4	4.5	2.6	1.2	1.2
EC	50.6	86.6	86	40	52.9	65	46.21	42.87
DO	7.8	7.1	7.1	7.2	9	7.2	7.2	8.2
N	0.15		0.8	0.04	0.146	0.435	0.77	0.08
PH	7.25	8.79	7.8	7.38	6.85	7.56	7.5	7.86
Temp	32	32	29	29	21	24	19	30
Ca	4.12	9.62	12.02	8	5.71	11.2	9.62	5.6
Chloride	2.99	6.84	14.67	5.87	5.91	4.99	19.57	5.91
COD	30	46.8	29	30	48	24	23	30
Mg	0.98	3.84	5.18	2.91	1.94	0.58	3.46	2.92
Phosphate	0.037	0.139	0.149	0.5	0.054	0.038	0.016	0.05
K	2	2	0.3	2	2	1	1.1	1
Na	2.5	3	2.2	2.3	1.19	3	0.6	1.8
Sulfate	0.055	2.96	8.62	3.45	0.088	0.97	15.67	2.4
TDS	154	70	70	20	38	37	68	28
$CaCO_3$	14.14	48	48	32	22	30	38	26
Turbidity	17.4	24.3		19.6	20.8	7.07	4.75	13.1
Total coliforms	70,000	11,000	8000	7000	270,000	9000	7000	7000

Source: Water Pollution Control Board, West Bengal

Table 10.5 WQI of Sagardighi, Cooch Behar

Year	Premonsoon (WQI)	During monsoon (WQI)
2010	63.19	52.96
2015	25.87	88.22
2020	17.77	65.99
2024	45.31	92.42

relationship with calcium (−0.73) and chloride (−0.95). TDS, inversely related to most parameters, requires monitoring, as its reduction correlates with higher ionic concentrations (Table 10.6).

During monsoon periods, the correlation matrix reveals significant relationships among the parameters. pH strongly inversely correlates with potassium (−0.95), while EC is positively influenced by calcium (0.63) and sodium (0.73) but negatively by magnesium (−0.95). Calcium and magnesium positively impact chloride and sulfate, both of which are highly correlated (0.99) and contribute significantly to TDS. Nitrogen correlates positively with TDS (0.91) and chloride (0.84) but inversely with phosphate (−0.98). Turbidity negatively correlates with calcium (−0.83), nitrogen (−0.82), and $CaCO_3$ (−0.92). These findings indicate the need to manage sodium, potassium, and nutrient levels for maintaining water quality and ensuring effective irrigation (Table 10.7).

The degradation of the Sagardighi Wetland is primarily driven by seasonal pollution, especially during the monsoon, when urban waterlogging and flooding carry a high load of pollutants into the wetland. Poor urban management, including inadequate drainage, encroachment, and lack of wetland protection, further exacerbates the issue (Kumar et al. 2021). Chemical and nutrient pollution from domestic waste, runoff, and possible agricultural sources has led to elevated levels of nitrogen, phosphate, chloride, and sulfate, increasing the risk of eutrophication (Zarkami et al. 2022; Zamani et al. 2025). High turbidity and sedimentation also reduce water clarity,

Table 10.6 Pearson's linear correlation matrix between different parameters in Sagardighi Wetland during the premonsoon period

	PH	EC	Ca	Mg	Na	K	N	Chloride	Phosphate	Sulfate	TDS	CaCo3	Turbidity
PH	1.00												
EC	0.78	1.00											
Ca	0.55	0.95	1.00										
Mg	0.54	0.95	1.00	1.00									
Na	0.74	0.16	−0.16	−0.17	1.00								
K	0.16	−0.49	−0.73	−0.75	0.79	1.00							
N	1.00	1.00	1.00	1.00	−1.00	−1.00	1.00						
Chloride	0.17	0.74	0.91	0.92	−0.54	−0.95	1.00	1.00					
Phosphate	0.72	1.00	0.98	0.97	0.06	−0.57	1.00	0.80	1.00				
Sulfate	0.18	0.75	0.92	0.93	−0.53	−0.94	1.00	1.00	0.81	1.00			
TDS	−0.77	−1.00	−0.96	−0.95	−0.14	0.50	−1.00	−0.75	−1.00	−0.76	1.00		
$CaCO_3$	0.77	1.00	0.96	0.95	0.14	−0.50	1.00	0.75	1.00	0.76	−1.00	1.00	
Turbidity	0.96	0.58	0.30	0.28	0.90	0.43	1.00	−0.12	0.50	−0.10	−0.57	0.57	1.00

Significance levels = ***0.01; **0.05; *0.1

Table 10.7 Pearson's linear correlation matrix between different parameters in Sagardighi Wetland during the monsoon period

	PH	EC	Ca	Mg	Na	K	N	Chloride	Phosphate	Sulfate	TDS	$CaCO_3$	Turbidity
PH	1.00												
EC	−0.25	1.00											
Ca	0.23	0.63	1.00										
Mg	0.21	−0.95	−0.40	1.00									
Na	0.33	0.73	0.38	−0.84	1.00								
K	−0.95	0.04	−0.53	−0.09	−0.38	1.00							
N	0.11	0.10	0.78	0.21	−0.28	−0.37	1.00						
Chloride	0.08	−0.43	0.32	0.69	−0.72	−0.21	0.84	1.00					
Phosphate	−0.26	0.07	−0.70	−0.36	0.33	0.49	−0.98	−0.89	1.00				
Sulfate	0.21	−0.44	0.35	0.70	−0.66	−0.34	0.85	0.99	−0.92	1.00			
TDS	−0.12	−0.18	0.47	0.48	−0.64	−0.09	0.91	0.96	−0.91	0.93	1.00		
$CaCO_3$	0.41	−0.09	0.71	0.36	−0.25	−0.61	0.95	0.85	−0.99	0.89	0.84	1.00	
Turbidity	−0.63	−0.12	−0.83	−0.10	−0.13	0.83	−0.82	−0.58	0.86	−0.66	−0.57	−0.92	1

Significance levels = ***0.01; **0.05; *0.1

affecting aquatic ecosystems. Additionally, the fluctuating water quality index (WQI) values over the years highlight inconsistent management and the urgent need for continuous monitoring and effective conservation efforts.

10.5 Conclusion

The Sagardighi Wetland plays a crucial role in the ecological and hydrological balance within the Cooch Behar Municipality. It serves as an important natural reservoir, supporting biodiversity, regulating the microclimate, and providing water for various human and aquatic needs. Most of the water quality parameters indicate unhealthy conditions for aquatic life as well as human consumption from 2010 to 2020. Different kinds of construction activities and urban water intrusion increase the pollution agents. From 2020 onward, major steps were taken to protect Municipal corporations from different restrictions implemented from 2020 to 2024. As a result, most of the water quality parameters were found within permissible limit. In premonsoon time, WQI ranged from 63.19 (2010) to 17.77 (2024), whereas in the monsoon period, the WQI was 52.96 (2010) to 92.42 (2024), indicating poor water quality. Still, the monsoon period water quality decays in nature. Proper management and planning are required during the monsoon period. The restoration efforts for the Sagardighi Wetland have shown significant progress in improving water quality over the past 4 years. However, the persistent challenges during the monsoon period highlight the need for a more robust, integrated water management strategy to ensure the long-term sustainability of this vital ecological asset. Regular water quality assessments should be conducted to identify and address emerging issues. Engage local stakeholders in wetland conservation through education and incentives. Implement sustainable urban drainage systems (SUDS) to minimize runoff pollution.

References

Baker JA, Gilron G, Chalmers BA, Elphick JR (2017) Evaluation of the effect of water type on the toxicity of nitrate to aquatic organisms. Chemosphere 168. https://doi.org/10.1016/j.chemosphere.2016.10.059

Banerjee P, Garai P, Saha NC et al (2023) A critical review on the effect of nitrate pollution in aquatic invertebrates and fish. Water Air Soil Pollut 234:333. https://doi.org/10.1007/s11270-023-06260-5

Bis ISDWS (2012) Bureau of Indian Standards. New Delhi 6:2–3

Bora M, Goswami DC (2017) Water quality assessment in terms of water quality index (WQI): case study of the Kolong River, Assam, India. Appl Water Sci 7:3125–3135

Brown RM, McClelland NI, Deininger RA, Tozer RG (1970) A water quality index—do we dare. Water Sew Works 117(10):339–343

Chandra R, Nishadh KA, Azeez PA (2010) Monitoring water quality of Coimbatore wetlands, Tamil Nadu, India. Environ Monit Assess 169:671–676. https://doi.org/10.1007/s10661-009-1206-0

Davies TD (2007) Sulphate toxicity to the aquatic moss, *Fontinalis antipyretica*. Chemosphere 66(3):444–451

Dixit A, Siddaiah NS, Joshi P (2021) Hydrogeochemical assessment of wetlands of Gurugram, Haryana, India: implications for natural processes and anthropogenic changes. Arab J Geosci 14:199. https://doi.org/10.1007/s12517-020-06423-2

Effendi H, Sabila MF, Setiawan Y (2018) Correlation between water quality and land use change in Ciliwung watershed. Nat Environ Pollut Technol 17(1):139–144

Horton RK (1965) An index number system for rating water quality. J Water Pollut Control Fed 37(3):300–306

Joshi P, Siva Siddaiah N, Dixit A (2021) Urban wetlands of Delhi, India: water quality and pollution status. Chem Ecol 37(2):104–131

Karmakar D, Sarkar B, Islam N (2024) Drinking water quality in Cooch Behar Municipality, West Bengal, India: assessment using WQI and public perception. Environ Qual Manag 34(1):e22226

Koch S, Bianchi V, Grutka SA, Aygadoux Martins CT, de Almeida Alves AA, Gonzalez da Silva JA et al (2023) Diagnosis of water quality in an urban wetland. Environ Soc Manag J/Rev Gest Soc Ambient 17(2):1–13

Kumar S, Agarwal A, Villuri VGK, Pasupuleti S, Kumar D, Kaushal DR et al (2021) Constructed wetland management in urban catchments for mitigating floods. Stoch Env Res Risk A 35:2105–2124

Manral U, Raha A, Solanki R et al (2012) Hydrological characteristics and flood plain vegetation of human impacted wetlands: a case study from Okhla Bird Sanctuary, national capital region, India. Asian J Conserv Biol 1:110–119

Minch VA, Egan JT, Sandlin M (1962) Design and operation of plastic filter media. J Water Pollut Control Fed 34:459–469

Mishra PC, Patel RK (2001) Quality of drinking water in Rourkela, Outside. Int J Environ Pollut 8:165–169

Mitsch WJ, Zhang L, Griffiths LN, Bays J (2023) Contrasting two urban wetland parks created for improving habitat and downstream water quality. Ecol Eng 192:106976

Patel SG, Singh DD, Harshey DK (1983) Pamitae (Jabalpur) sewage polluted water body, as evidenced by chemical and biological indicators of pollution. J Environ Biol 4:437–449

Pimparkar AM, Patil SN, Patil BD, Kadam AK (2023) Comparative assessment of wetland water quality from rural and urban area of Aurangabad District, Maharashtra, India using water quality index. HydroResearch 6:269–278

Prerna Joshi N, Siddaiah S, Dixit A (2020) Urban wetlands of Delhi, India: water quality and pollution status. Chem Ecol. https://doi.org/10.1080/02757540.2020.1836164

Sarkar B (2024) Assessment of drinking water quality using GIS technique and water quality index in West Bengal, India. Dev Environ Sci 16:559–573

Sharma P, Sarkar R, Deka JP et al (2024) Assessing water quality of Deepor Beel, Assam, NE India, using water quality index: a case of Ramsar wetland. Arab J Geosci 17:20. https://doi.org/10.1007/s12517-023-11818-y

Tripathy JK, Sahu KC (2005) Seasonal hydrochemistry of groundwater in the Barrier Spit system of the Chilika Lagoon, India. J Environ Hydrol 8(2):165–169

Wang S, Shentu H, Yu H, Wang L, Wang J, Ma J et al (2024) Impact of urbanization and land use on wetland water quality: a case study in Mengxi town. Urban Clim 55:101855

Zamani-Ahmadmahmoodi R, Bayati S, Abdollahi K, Mafi-Gholami D, Joorabian Shooshtari S, Gharahi N et al (2025) Water quality and eutrophication status of the Zarivar Wetland (Iran). Appl Water Sci 15(2):24

Zarkami R, Abedini A, Sadeghi Pasvisheh R (2022) Analysis of the eutrophication in a wetland using a data-driven model. Environ Monit Assess 194(12):882

11 Transformed Wetlands and Urban Resilience: A Case Study from Jiadhal Wetlands, Dhemaji, Assam, India

Rajib Pegu, Hrykpb Borah, Parashmoni Deka, Abhinab Borah, and Dhanjit Nath

Abstract

Wetlands are one of the sources of an ecosystem. It can provide resilience to natural hazards like floods, drought, fire, storms, landslides, and erosion. India now has 80 Ramsar sites covering an area of 1.33 million hectares. In Assam, the only recognized Ramsar site is Deepor Beel located on the south-western edge of Guwahati city in the Kamrup district. There are a lot of big and small wetlands in the Dhemaji district of Assam, India. Most of the Jiadhal Wetlands have disappeared or are in the process of disappearing. This loss harms the key function performed by the wetlands. The methodology used for this chapter is Google Earth Engine, Erdas Imagine for Image Processing, ArcGIS, and QGIS for data visualization, and Python Programming for Data Handling and Analysis. This chapter will highlight the changing structure of the Jiadhal Wetland through the years. This chapter will emphasize the conservation challenges and the threat to the wetland due to the conversion of agricultural land or use for buildings, settlement, public infrastructures, proper drainage, flood, etc. Large sections of the wetlands are turned into agricultural land and aquaculture to increase fish productivity, and the area is urban resilience now.

Keywords

Changing structure · Issues · Challenges · Threats · Conservation measures · Urban resilience

11.1 Introduction

Wetlands are essential habitats that support a variety of plants and animals, including endangered species, helping to conserve biodiversity (Zedler and Kercher 2005). One of the most productive ecosystems in the world is wetlands (Ghermandi et al. 2008). People all around the world greatly value wetlands, considering the various ecological benefits they provide, including their economic advantages, biodiversity, existence values, and means of subsistence (Groot et al. 2018). The life of countless species of birds, fish, amphibians, reptiles, mammals (Chakraborty et al. 2023), and invertebrates depends on water and wetland vegetation (Mitsch and Gosselink 2000). They are indispensable in

R. Pegu (✉)
Assistant Professor, Department of Geography, Sankardeva Mahavidyalaya, Pathalipahar, Lakhimpur, Assam, India

H. Borah
TERI School of Advanced Studies, New Delhi, India

P. Deka · D. Nath
Gauhati University, Guwahati, Assam, India

A. Borah
Mahapurusha Srimanta Sankaradeva Vishwavidyalaya, Nagaon, Assam, India

S. K. Chakraborty et al. (eds.), *Wetlands and Urban Resilience*, Sustainable Development Goals Series,
https://doi.org/10.1007/978-3-032-06881-1_11

the water cycle, flood control, water filtering, and biodiversity. By storing carbon on Earth, they help slow climate change (Mitsch et al. 2013).

Wetlands are ecosystems in which there is a permanent or seasonal water saturation of the land and act as a transitional area between terrestrial and aquatic habitats. These are environments in which dry ground meets or gets saturated with water, and they can be identified by the existence of water-dependent plant and animal species (Shine and Klemm 1999). The United States Geological Survey defines wetlands as transitional areas linking habitats on land and underwater where the water table is normally at or close to the surface, or where the ground is occasionally covered by shallow water (U.S. Geological Survey 2024. Their companies are also significant to human civilization and economic activity (Shine and Klemm 1999).

As of November 2024, wetlands totalling more than 257,317,367 hectares were formally recognized as Ramsar Sites under the Ramsar Convention. In environmental studies, such spots have been appreciated for their significance worldwide. Approximately 1,358,067 hectares of wetlands were officially recognized as Ramsar Sites in India by the Ramsar Convention. In Assam, there are an estimated 1392 wetlands containing 22,896 fisheries of various types; only 394 (30.38%) among those wetlands are officially recognized as spanning an area of over 70,000 hectares, according to the Assam government, Directorate of Economics and Statistics, 2006.

Wetlands, among the most biodiverse and productive ecosystems globally (Ghermandi et al. 2008), are vanishing at an unprecedented rate, with a 35% global decline since 1970 (Ramsar Convention 2024). Even though 80 Ramsar sites totalling 1.33 million hectares have been designated in India, uncontrolled human activities such as urbanization, agricultural encroachment, and hydrological alterations pose existential risks to wetlands (Choudhury et al. 2020; Bassi et al. 2014). This situation is best illustrated by the Jiadhal Wetlands in Assam's Dhemaji District, which lost 66 km^2 of water bodies and 122 km^2 of vegetation between 2015 and 2024 as a result of frequent floods, sedimentation from the Brahmaputra River system, and fast changes in land use. In addition to endangering local livelihoods that depend on wetland resources, these pressures disrupt important ecological services, such as flood mitigation, groundwater recharge, and biodiversity support (Zedler and Kercher 2005; Groot et al. 2018).

Advanced geospatial technologies that integrate granular spatiotemporal analysis are frequently absent from existing studies on Indian wetlands (Xu et al. 2011). For example, although Shine and Klemm (1999) highlight the policy-science gaps in wetland management, few frameworks take into account the synergistic effects of sedimentation, dynamic hydrology, and socioeconomic forces that are specific to floodplain ecosystems like Jiadhal. According to Reis et al. (2017), the identification of degradation thresholds and adaptive management potential is further limited by the lack of long-term, high-resolution remote sensing evaluations.

This study addresses these gaps by employing a multidisciplinary approach:

1. LULC categorization using Google Earth Engine's Random Forest algorithms (87–93% accuracy) to measure urbanization and agricultural growth (2143 km^2 in 2015 to 2270 km^2 in 2024).
2. Water-body fragmentation (2000–2024) was mapped using the Modified Normalized Difference Water Index (MNDWI), which showed that anthropogenic water abstraction caused an intermittent recovery in 2020 and then sharp reductions in 2022–2024.
3. Water/non-water pixels are dynamically separated by OTSU thresholding, emphasizing the expansion of sandbars driven by sedimentation (254 km^2 in 2015 to 358 km^2 in 2024).

This study aims to quantify structural changes over time, identify conservation issues, and assess the impact of anthropogenic activities to examine the spatiotemporal dynamics of the Jiadhal Wetland in Assam, India. The goal of the research is to offer practical advice for sustainable wetland management and policy changes that will reduce ecological deterioration and increase resilience to

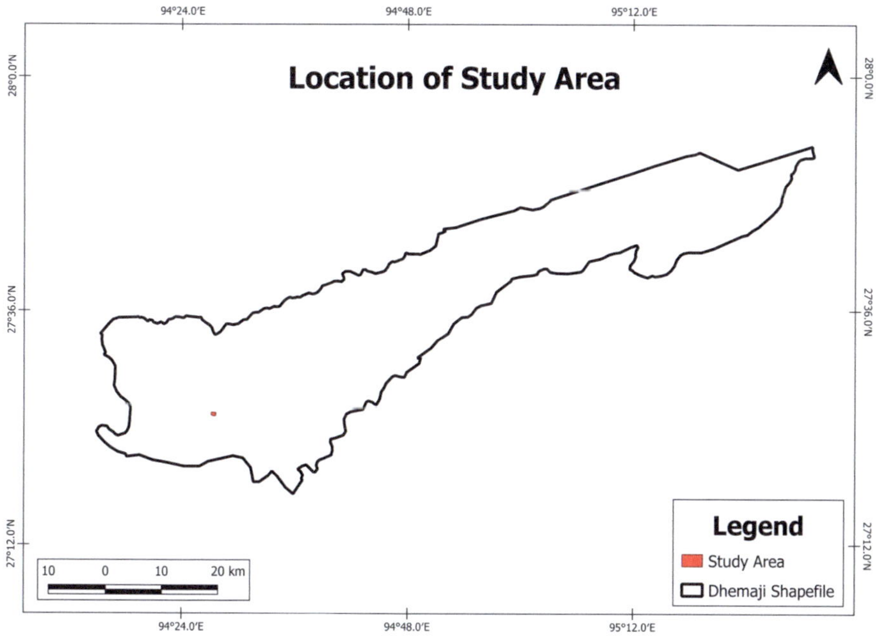

Fig. 11.1 Dhemaji district

stressors brought on by climate and human activity.

The objective of the Study is:

(i) To investigate how the Jiadhal Wetland's structure has changed over time through spatial and temporal analysis
(ii) To identify the key conservation challenges currently threatening the ecological integrity of the wetland
(iii) To examine the influence of human activities on the wetland, with a particular focus on the impact of the growing human population on the wetland's health and sustainability

Study Area

The Dhemaji district came into being in 1989. This study focuses on the Jiadhal Wetland Region in the Dhemaji District of Assam, India. The Dhemaji district comprises of erstwhile Dhemaji and Jonai sub-division and parts of Machkhowa mouza and Bordoloni. The district is bordered by arch-shaped Arunachal hills on the North and the East and stretches to the Brahmaputra River with Subansiri on one side and the river Siang on the other. Geographically situated between the 94° 12′18″ E and 95°41′32″ E longitudes and 27° 15′27″ N and 27°57′16″ N latitudes, the district covers an area of 3237 Sq. Km is a basically plain area lying at an altitude of 104 m above the Mean Sea Level (Fig. 11.1).

Although the Dhemaji district contains a lot of plain land that can be used for farming, a lot of it is waterlogged from rainwater and excessive floodwater because of poor land use planning and haphazard development. While some of these places stay as ditches, cesspools, or wet "beel" areas, others dry up in the winter. They have many "beels" and marshes that are between 10 and 20 feet high and covered in a dense population of weeds, water hyacinth, water lily, etc. Furthermore, flash floods brought on by embankment breaches have sandcast certain places (Fig. 11.2).

The Jiadhal Wetland covers an area of 53 ha (0.53 sq. km) and is situated in the Dhemaji subdistrict's southwest section. The latitudinal extent of the wetland is between 27°23′34″ N and 27°25′12″ N and the longitudinal extent is between 94°27′7″ E to 94°42′28″ E. The ongoing and recent developmental activities have

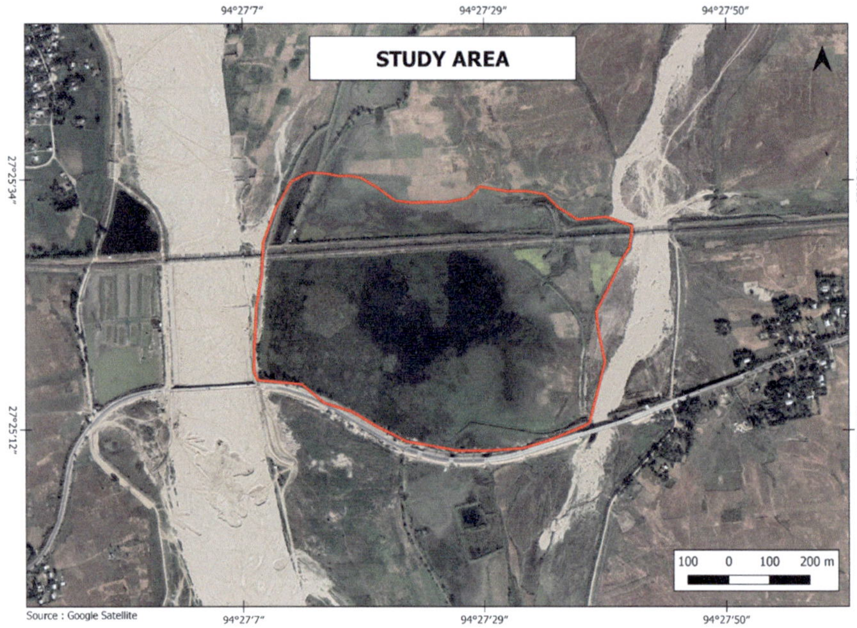

Fig. 11.2 Study area

significantly impacted the wetlands and the biodiversity within and surrounding the area. The impact of developmental activities on wetlands and biodiversity is a pressing concern, in order to keep the environment in balance, these ecosystems are essential.

11.2 Materials and Methods

Figure 11.3 illustrates the data used from various sources to identify the changing structure of the Jiadhal Wetland and to analyse the threats to Jiadhal Wetland. Land use and land cover mapping (LULC) of the Dhemaji district and the wetland region was done to understand the changes in the LULC cover over time. For LULC classification, USGS Landsat 8 data was used. To calculate the Normalized Difference Water Index (NDWI) to understand the water content in the wetland, USGS Landsat 7 and Landsat 8 and Sentinel 2 MSI data were used. Landsat 7 and Landsat 8 data have a 30 m spatial resolution, and Sentinel 2 has 10 m spatial resolution. For the extraction of Water Pixels, the OTSU threshold method was used to separate the water and non-water pixels. The Threshold was applied to the MNDWI values of Sentinel-2 images of the study area. All of these datasets have a very good resolution and therefore it was used to analyse the changes in the wetland over time. Computations of LULC, NDWI, and Water Pixel extraction using Otsu threshold were done in Google Earth Engine, and QGIS and ArcGIS were used to create the final map layouts.

Figure 11.3 illustrates the step-by-step process followed to identify structural changes in the wetland over time.

11.2.1 Land Use Land Cover Classification

Supervised land use and land cover classification of the study area and the Dhemaji district was done using the Random Forest Classifier algorithm in the Google Earth Engine code editor. The land use land cover classification was done for 3 years (2015, 2020, 2024) with a 5-year interval.

The Random Forest (RF) classifier is a widely used machine learning algorithm available in

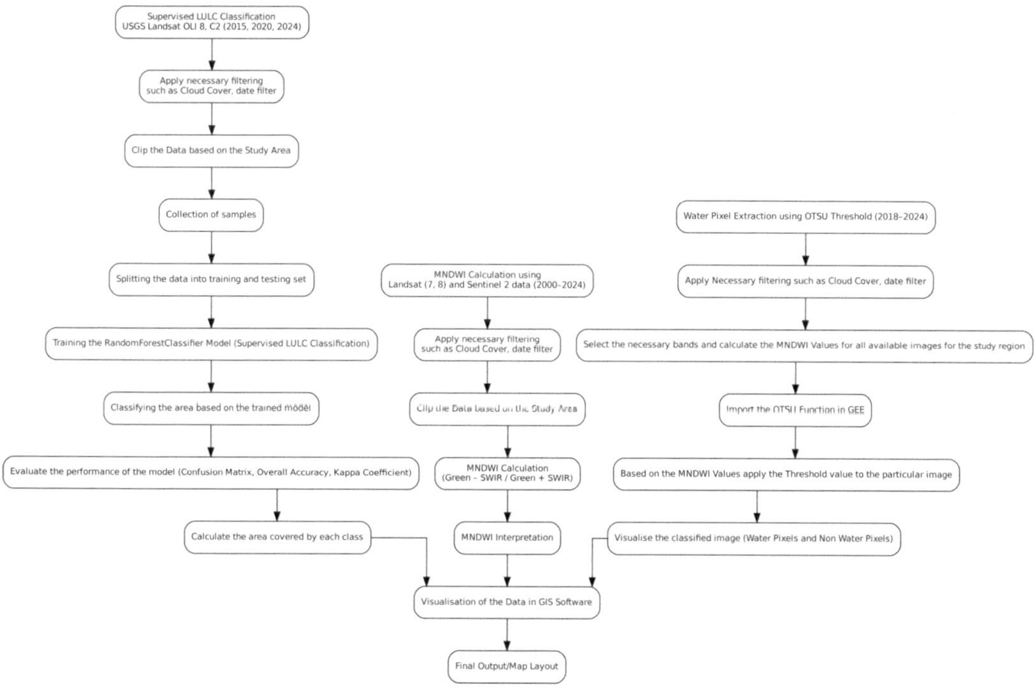

Fig. 11.3 Methodology flowchart

Google Earth Engine (GEE) for performing supervised classification of geospatial data. It is particularly effective for handling high-dimensional datasets, such as satellite imagery, due to its robustness, accuracy, and ability to manage nonlinear relationships. Random Forest operates by constructing a multitude of decision trees during training and aggregating their predictions to produce a more accurate and stable output. Each decision tree in the ensemble is trained on a random subset of the training data, and a random subset of features is considered for splitting at each node. This randomness reduces the risk of overfitting and ensures diversity among the trees, resulting in better generalization. In GEE, the RF classifier is implemented through the ee.Classifier.smileRandomForest function. First, labelled training data are prepared, which includes features (e.g. spectral bands, vegetation indices) and corresponding class labels. The RF algorithm then learns patterns and relationships from the training data to classify each pixel in the image into a specific class. One of the key strengths of the RF classifier in GEE is its ability to handle missing data and automatically assess feature importance, aiding in the identification of the most relevant predictors for classification. It also allows for parameter tuning, such as the number of trees (numberOfTrees) and the number of variables to split at each node (variablesPerSplit), to optimize classification performance. Random Forest is widely applied in GEE for tasks like land use land cover mapping, forest monitoring, and crop classification. Its compatibility with GEE's scalable cloud processing makes it a preferred choice for large-scale, high-resolution geospatial analysis.

The samples for five land cover classes (Water, Agricultural Land, Vegetation, Built-Up Area, Sandbar) were taken, and 80% of the samples were used for training the model, while 20% of the samples were used for validation. 50 trees were taken in the Random Forest classification in order to prevent the risk of overfitting.

The accuracy of the classification was calculated based on the confusion matrix. A confusion matrix is a table used to evaluate the performance of a classification model by comparing predicted and actual class labels. It summarizes predictions into four categories: True Positives (TP), True Negatives (TN), False Positives (FP), and False Negatives (FN), providing insights into the model's accuracy, precision, and recall.

The Kappa coefficient, a statistical measure used to evaluate the agreement between predicted and actual classifications while accounting for agreement occurring by chance, was also calculated. It ranges from −1 to 1, where:

- 1 indicates perfect agreement
- 0 represents agreement equivalent to random chance
- Negative values indicate worse-than-random agreement

The formula for Kappa is:

$$\kappa = \{P_o - P_e\}/\{1 - P_e\}$$

where

- Po: Observed agreement (proportion of correctly classified instances)
- Pe: Expected agreement (proportion of agreement by chance)

11.2.2 Modified Normalized Difference Water Index (MNDWI)

Modified Normalized Difference Water Index (MNDWI) was calculated for the wetland area for 6 years (2000, 2005, 2010, 2015, 2020, 2024) with a 5-year interval. The MNDWI is a satellite-derived index used to detect and monitor water bodies. It is designed to enhance water features while suppressing signals from vegetation, soil, built-up areas, and other land cover types. It is particularly useful in urban environments and for detecting water features in areas with mixed land cover. The MNDWI was calculated in the Google Earth Engine code editor using Landsat 7 data (2000–2010), Landsat 8 data (2012), and Sentinel 2 MSI data (2020–2024).

Formula:

$$\text{MNDWI} = \frac{(\text{Green} - \text{SWIR})}{(\text{Green} + \text{SWIR})}$$

where

- Green: Reflectance in the green band (e.g. Band 3 in Landsat 7 or Band 3 in Landsat 8)
- SWIR: Reflectance in the shortwave-infrared band (e.g. Band 5 in Landsat 7 or Band 6 in Landsat 8)

MNDWI values range from −1 to 1.

- Positive values generally indicate water bodies.
- Negative or near-zero values represent non-water features like vegetation, urban areas, or dry soil.

Although MNDWI is designed to enhance water features while suppressing signals from other land cover types, it has certain limitations, such as:

- Performance can vary with highly turbid water or mixed pixels containing water and non-water features.
- Atmospheric conditions and sensor-specific differences may require careful preprocessing.
- Confusion can occur in areas with high soil moisture or sparse vegetation.

MNDWI is widely used in remote sensing applications for accurate water body delineation and to assess the health and extent of the wetlands.

11.2.3 OTSU Threshold for Water Pixel Extraction

Otsu thresholding is a widely used image segmentation technique that automatically determines the optimal threshold to separate pixels into two classes, such as foreground and background or water and non-water, based on their intensity values. It was introduced by Nobuyuki Otsu in 1979. The method assumes that the image histogram (a graph showing the frequency of pixel intensity values) is bimodal, meaning it has two distinct peaks corresponding to two classes. The goal is to find the threshold value that minimizes the intra-class variance (variation within each class) or equivalently maximizes the inter-class variance (separation between the two classes).

Steps for OTSU Threshold

Step 1: Divide pixels into two classes

The pixel intensities in the image are separated into two classes based on a threshold T:

- *Class 1* ($C1$): Pixels with intensity values below T
- *Class 2* ($C2$): Pixels with intensity values above T
- *Step 2: Calculate probabilities of each class*

The probabilities of a pixel belonging to $C1$ or $C2$ are calculated from the histogram:

- $\omega1(T)$: Proportion of pixels in $C1$
- $\omega2(T)$: Proportion of pixels in $C2$

$$w_{1(T)} = \sum_{\{i=0\}}^{\{T\}} p(i), w_{2(T)} = \sum_{\{i=T+1\}}^{\{L\}} p(i).$$

where

- $p(i)$ is the normalized frequency of intensity i in the histogram.
- L is the maximum intensity value.
- *Step 3: Compute Mean Intensities of Each Class*

The mean intensity of pixels in each class is computed as:

- For class 1 ($\mu1(T)$):

$$\mu_{1(T)} = \left\{\sum_{\{i=0\}}^{\{T\}} i \cdot p(i)\right\} / \left\{w_{1(T)}\right\}$$

- For class 2 ($\mu2(T)$):

$$\mu_{2(T)} = \left\{\sum_{\{i=T+1\}}^{\{L\}} i \cdot p(i)\right\} / \left\{w_{2(T)}\right\}$$

Step 4: Calculate Inter-class Variance

The inter-class variance (σ_b^2) quantifies the separation between the two classes and is defined as:

$$\sigma_b^2 = w_{1(T)} \cdot w_{2(T)} \cdot \left[u_{1(T)} - \mu_{2(T)} \right]^2$$

Here:

- $\omega1(T)$ and $\omega2(T)$ are the class probabilities.
- $\mu1(T)$ and $\mu2(T)$ are the mean intensities of each class.
- $[\mu1(T) - \mu2(T)]$ measures the difference between the class means.
- *Step 5: Optimize threshold*

The optimal threshold T is the value that *maximizes* σ_b^2, ensuring the greatest distinction between the two classes. The algorithm iterates through all possible thresholds T to find the one that gives the highest inter-class variance.

Step 6: Apply threshold

Once the optimal threshold T is identified:

- Pixels with intensity $\leq T$ are assigned to class 1 (e.g. water).
- Pixels with intensity $>T$ are assigned to class 2 (e.g. non-water).

This stepwise procedure enables Otsu thresholding to automatically and adaptively segment images without the need for manual intervention. It is particularly effective when the pixel intensity histogram is bimodal, as in the case of water and non-water regions in remote sensing data.

To extract water pixels from the wetland, the Otsu thresholding method was applied to the Modified Normalized Difference Water Index (MNDWI) values derived from Sentinel-2-multispectral data. This method effectively separates water and non-water pixels by dynamically determining the optimal threshold based on the MNDWI value distribution. The threshold was computed using the OEE Library in Google Earth Engine, and water pixels were extracted by applying this threshold. The extracted water pixel data were analysed to observe temporal changes in the wetland structure from 2018 to 2024.

11.3 Results and Discussion

The LULC will show how the wetland has gone change through different phases and urbanized.

11.3.1 Land Use Land Cover Classification

11.3.1.1 LULC 2015

The analysis of the land use land cover (LULC) classification of the district for the year 2015 (Fig. 11.4) indicates that agricultural land was the dominant land cover type, covering 66.20% of the district's total area. This highlights the district's dependency on agriculture as a primary land use. Vegetation accounted for 11.61%, followed by Built-Up areas at 7.75%, and Sandbars, which covered 7.78%. Water bodies made up a smaller portion, contributing 6.58% to the total land cover (Table 11.1).

Such a distribution reflects the region's socio-economic activities, primarily driven by agriculture, and underscores the relatively limited extent of urbanization as indicated by the modest proportion of built-up areas. The presence of sandbars and water bodies suggests dynamic hydrological and fluvial processes, likely influencing the district's wetland systems.

The wetland-specific land use land cover classification (Fig. 11.5) revealed significant insights into the changing structure and dynamics of these ecologically critical areas. Monitoring such changes is essential for sustainable management and conservation, particularly as wetlands are vital for biodiversity, water regulation, and ecosystem services.

The classification process achieved an overall accuracy of 87%, with a Kappa Coefficient (*k*) of 0.826, indicating a strong agreement between the classified and reference data. This level of accuracy validates the reliability of the classification and provides confidence in its application for

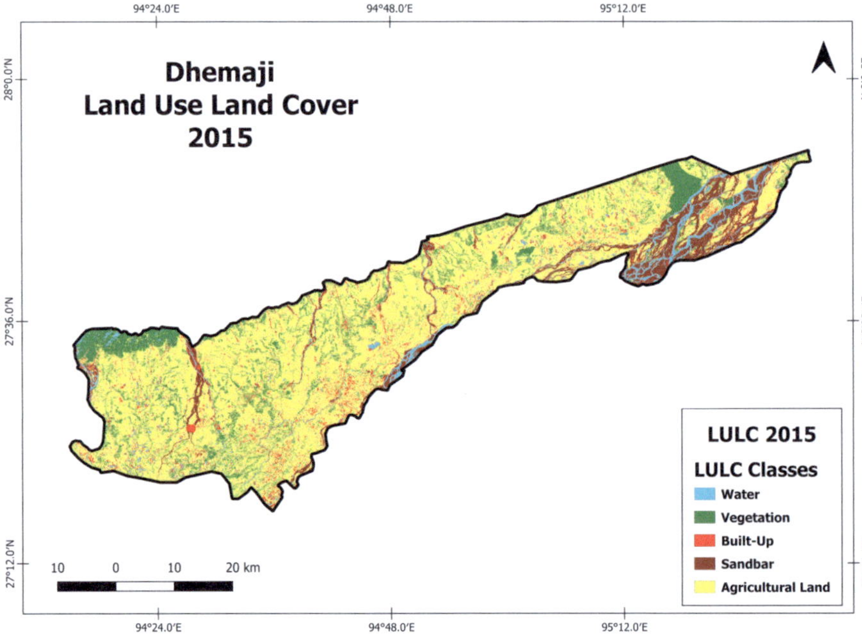

Fig. 11.4 LULC 2015 (Dhemaji District)

monitoring land use changes over time (Tables 11.2 and 11.3).

11.3.1.2 LULC 2020

The analysis of the land use land cover (LULC) classification of the district for the year 2020 (Fig. 11.6) indicates that agricultural land was the dominant land cover type, covering 68.08% of the district's total area, while vegetation accounted for 10.41%, followed by Sandbar at 10%, and built-up areas, which covered 6.27%. Water bodies made up a smaller portion, contributing 5.22% to the total land cover.

A significant decrease in water bodies was observed compared to the previous year, suggesting potential environmental stress or the expansion of agricultural activities. This decline may also be linked to the increased sandbar area, indicating ongoing natural sedimentation processes. Vegetation cover has diminished, likely due to escalating human activities and deforestation. In contrast, built-up areas in the district have reduced by approximately 50 km^2, while agricultural land has expanded by nearly 100 km^2, reflecting a shift in land use priorities or it could also be due to natural events such as flooding which may have reclaimed built-up spaces.

Table 11.1 Land use land cover area, 2015

Land cover type	Area (km^2)	Area (%)
Water	213	6.58
Vegetation	376	11.61
Built-up	251	7.75
Sandbar	254	7.78
Agricultural land	2143	66.20
Total	3237	100

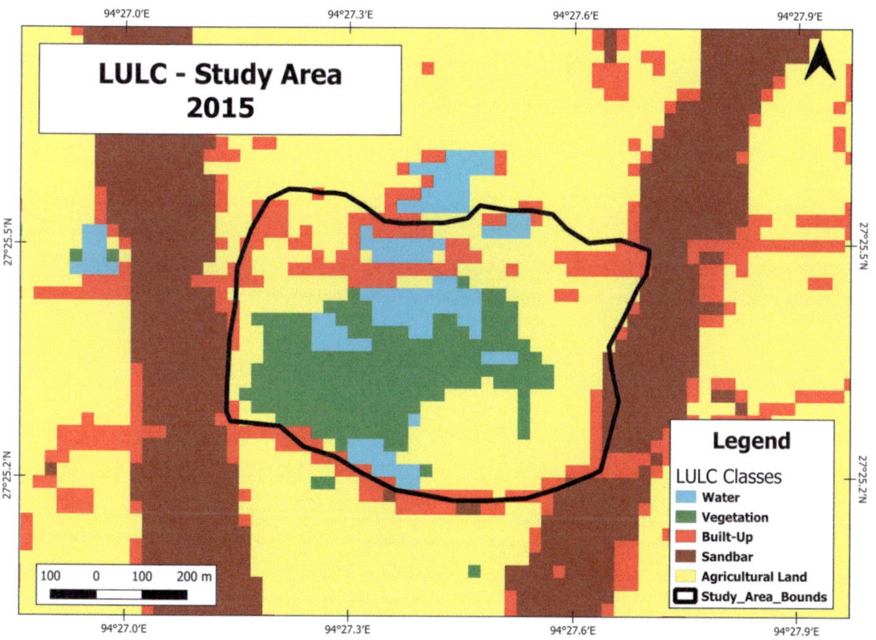

Fig. 11.5 LULC 2015—study area

Table 11.2 Classification summary table (2015)

Metric	Water bodies	Vegetation	Built-up areas	Sandbars	Agricultural land	Overall/Kappa
Producer's accuracy	1.00	1.00	0.64	0.95	0.86	–
User's accuracy	1.00	1.00	0.58	0.95	0.89	–
Overall accuracy	–	–	–	–	–	87.32%
Kappa coefficient	–	–	–	–	–	0.826

Table 11.3 Confusion matrix (2015)

Class	Water bodies	Vegetation	Built-up areas	Sandbars	Agricultural land	Total (row sum)
Water bodies	5	0	0	0	0	5
Vegetation	0	7	0	0	0	7
Built-up areas	0	0	7	1	3	11
Sandbars	0	0	1	19	0	20
Agricultural land	0	0	4	0	24	28

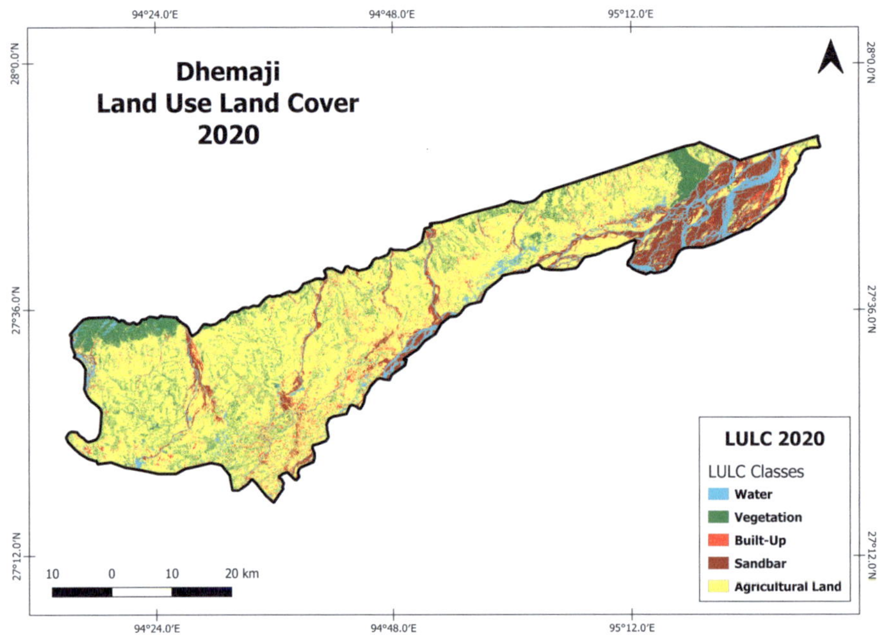

Fig. 11.6 LULC 2020

Table 11.4 Land use land cover area, 2020

Land cover type	Area (km^2)	Area (%)
Water	169	5.22
Vegetation	337	10.41
Built-up	203	6.27
Sandbar	324	10.00
Agricultural land	2204	68.08
Total	3237	100

Further investigation is needed to understand the underlying causes of this reduction (Table 11.4).

The wetland has experienced an increase in water coverage, replacing areas previously dominated by vegetation and other land use types. This rise in water levels is likely attributed to favourable meteorological conditions during the observed period, which may have replenished surface water bodies. Further analysis of water content in the wetland areas has been conducted using the Modified Normalized Difference Water Index (MNDWI), providing deeper insights into these changes. Understanding these structural transformations is critical for biodiversity conservation, as wetlands serve as habitats for diverse species and contribute to ecological balance (Fig. 11.7).

The classification process attained an overall accuracy of 87.27% and a Kappa Coefficient of 0.833, reflecting a strong agreement between the classified data and reference observations (Tables 11.5 and 11.6).

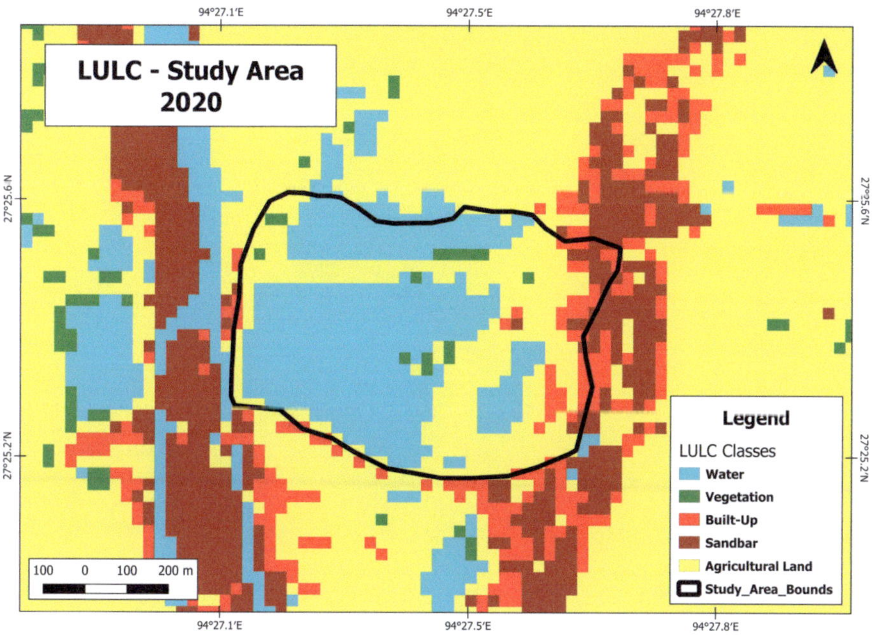

Fig. 11.7 LULC 2020 study area

Table 11.5 Classification summary table (2020)

Metric	Water bodies	Vegetation	Built-up areas	Sandbars	Agricultural land	Overall/Kappa
Producer's accuracy	1.00	1.00	0.67	0.80	0.86	–
User's accuracy	1.00	0.67	0.50	0.92	0.95	–
Overall accuracy	–	–	–	–	–	87.27%
Kappa coefficient	–	–	–	–	–	0.830

Table 11.6 Confusion matrix (2020)

Class	water bodies	Vegetation	Built-up areas	Sandbars	Agricultural land	Total (row sum)
Water bodies	10	0	0	0	0	10
Vegetation	0	6	0	0	0	6
Built-up areas	0	0	2	1	0	3
Sandbars	0	0	2	12	1	15
Agricultural land	0	3	0	0	18	21

11.3.1.3 LULC 2024

The analysis of the land use land cover (LULC) classification of the district for the year 2024 indicates that agricultural land was the dominant land cover type, covering 70.12% of the district's total area, while Sandbar accounted for 11.05%, followed by Vegetation at 7.84%, and built-up areas, which covered 6.42%. Water bodies made up a smaller portion, contributing 4.54% to the total land cover (Fig. 11.8 and Table 11.7).

The LULC data for the year 2024 show significant changes in land cover types over the period. There has been a consistent decline in the area covered by water bodies, with a reduction of approximately 66 km^2 from 2015 to 2024. This decrease may be due to reduced rainfall, urban expansion, or agricultural activities affecting water bodies. Vegetation cover has steadily decreased over the years. In total, about 122 km^2 of vegetation was lost from

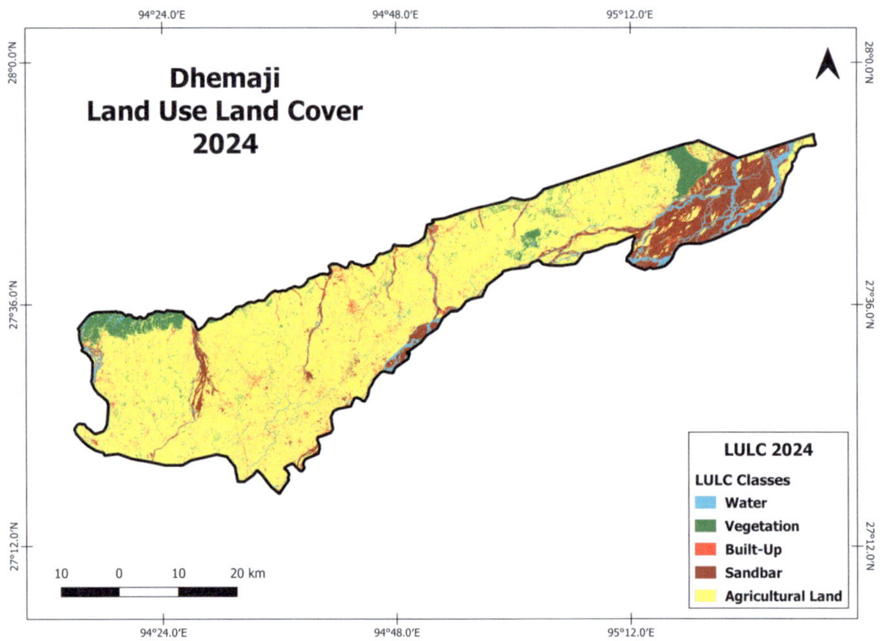

Fig. 11.8 LULC Dhemaji 2024

Table 11.7 Land use land cover area, 2024

Land cover type	Area (km^2)	Area (%)
Water	147	4.54
Vegetation	254	7.84
Built-up	208	6.42
Sandbar	358	11.05
Agricultural land	2270	70.12
Total	3237	100

2015 to 2024, which may be attributed to deforestation and agricultural expansion. Built-up areas showed a slight increase from 2020 to 2024 (5 km^2). This indicates moderate urban development but not a significant increase in built-up areas. Sandbar areas have increased significantly, growing by 104 km^2 from 2015 to 2024. This may reflect natural sedimentation processes, changes in water flow, or decreased vegetation that supports sandbar stability. The increasing sandbar area could be due to sedimentation processes or changes in the hydrological system, which might have been influenced by changes in water levels or land management practices. Agricultural land has consistently increased over the years, growing by 127 km^2 from 2015 to 2024. This increase likely reflects the expansion of agriculture into other land use types, including vegetation and water bodies (Fig. 11.9).

The wetland has remained relatively stable in terms of the water content, while some portions have been covered by sandbars. Further analysis of water content in the wetland areas has been conducted using the Modified Normalized Difference Water Index (MNDWI), providing deeper insights into these changes. These findings highlight the need for continued monitoring to ensure the sustainable management of these vital ecosystems.

The classification process achieved an overall accuracy of 93.40% and a Kappa Coefficient of 0.907, indicating a high level of agreement between the classified results and the reference data (Tables 11.8 and 11.9).

The Land Use Land Cover (LULC) analysis from 2015 to 2024 highlights notable trends in the district's landscape. These findings underscore the importance of sustainable land management to balance agricultural demands with ecological preservation. The observed changes in wetlands

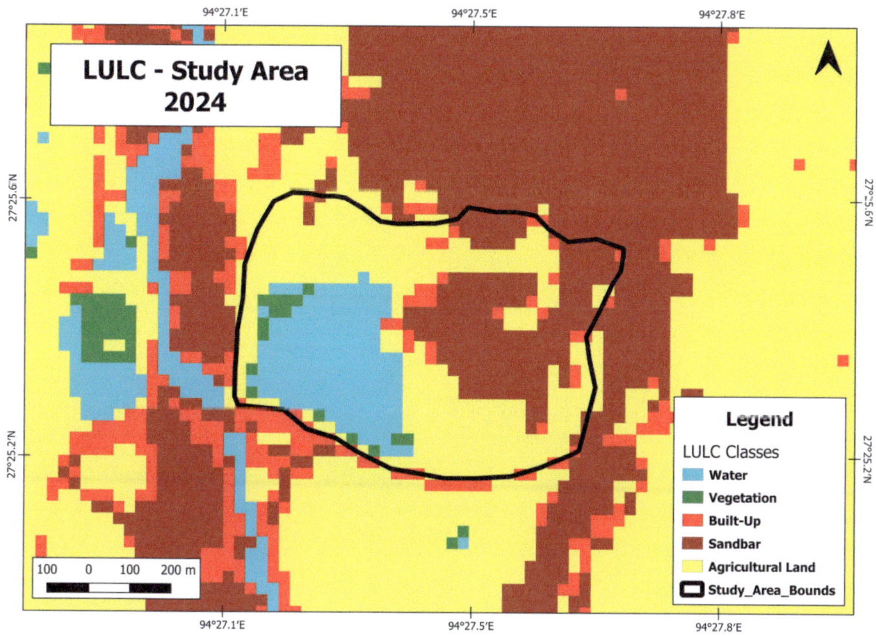

Fig. 11.9 LULC 2024 study area

Table 11.8 Classification summary (2024)

Metric	Water bodies	Vegetation	Built-up areas	Sandbars	Agricultural land	Overall/Kappa
Producer's accuracy	1.00	0.88	0.67	0.90	1.00	–
User's accuracy	1.00	1.00	0.57	0.96	0.95	–
Overall accuracy	–	–	–	–	–	93.41%
Kappa coefficient	–	–	–	–	–	0.907

Table 11.9 Confusion matrix (2024)

Class	Water bodies	Vegetation	Built-up areas	Sandbars	Agricultural land	Total (row sum)
Water bodies	12	0	0	0	0	12
Vegetation	0	7	0	0	1	8
Built-up areas	0	0	4	1	1	6
Sandbars	0	0	3	26	0	29
Agricultural land	0	0	0	0	36	36

and vegetation signal potential biodiversity loss and water resource challenges, emphasizing the need for targeted conservation efforts. Accurate LULC classifications provide critical insights for policymakers to address socio-environmental challenges and ensure long-term ecological sustainability.

11.3.2 Modified Normalized Difference Water Index (MNDWI)

The Modified Normalized Difference Water Index (MNDWI) is a remote sensing index used to enhance the detection of surface water bodies in satellite imagery. Positive MNDWI values,

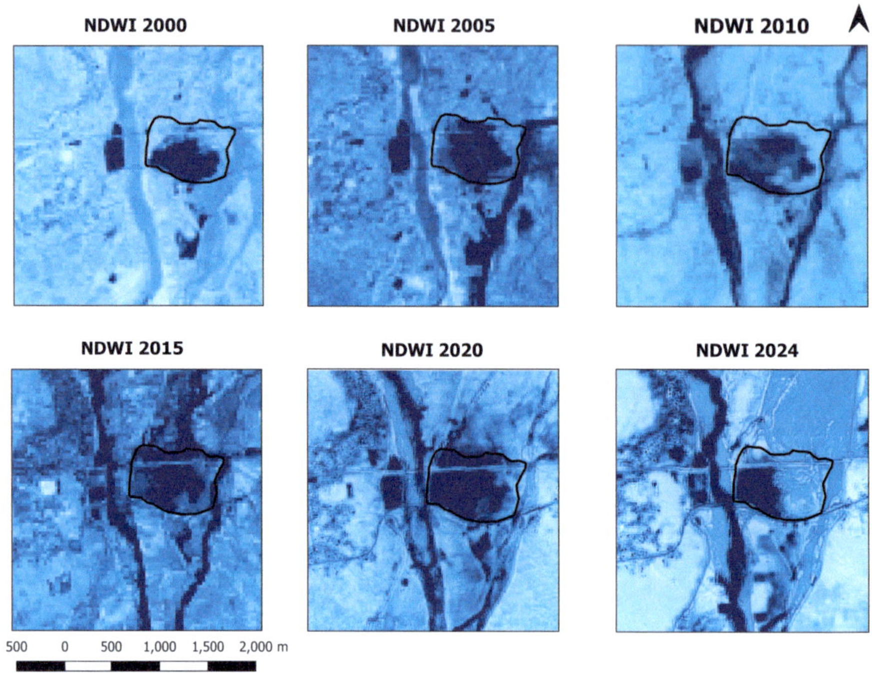

Fig. 11.10 MNDWI—2000 to 2024

typically ranging from 0.1 to 1, indicate water presence. Higher positive values generally correspond to clear and deeper water. Negative values (e.g. −0.5 to 0) or values near zero are often associated with built-up areas, vegetation, or other non-water features like soil or sand (Fig. 11.10).

The MNDWI values range from −0.4 to 0.4 in these images, highlighting areas dominated by water (positive values) and non-water features like vegetation and built-up areas (negative values). The darker areas represent a higher MNDWI value, indicating the presence of water, whereas the lighter areas represent a lower MNDWI value, indicating non-water features in the area.

Interpretation of Wetland Changes

The visual patterns in the images suggest significant alterations in the wetland structure over the 24-year period:

2000–2005: The 2000 and 2005 images reveal the presence of distinct water bodies with relatively high MNDWI values, indicating stable wetland areas. However, there may be early signs of fragmentation, as smaller water patches appear disconnected, hinting at the onset of anthropogenic pressures or natural shifts.

2005–2010: Between 2005 and 2010, a noticeable shrinkage in the extent of water-dominated areas is observed. This could be due to increased sedimentation, reduction in water availability during dry seasons, or human activities such as encroachment for agriculture or settlement expansion. Linear features with lower MNDWI values may represent the narrowing of water channels or vegetation encroachment along the edges.

2010–2015: The 2015 image suggests further degradation, with prominent reductions in waterbody sizes and connectivity. This indicates a heightened threat to the wetland's ecological functions, possibly driven by increased human settlement or changes in upstream hydrology due to deforestation or infrastructural development. The dominance of non-water features in parts of the image signifies a decline in wetland health.

2015–2020: By 2020, some recovery of water areas is evident, likely due to restoration efforts or climatic variations resulting in higher rainfall or water inflows. However, this recovery seems uneven, with water still confined to specific pockets, and the overall fragmentation persists.

2020–2024: The 2024 image indicates mixed trends. While some areas show improved water retention, others continue to decline. This variability could be attributed to differential impacts of flood events, human interventions like dam construction or sand mining, and climatic changes affecting water availability.**Key Threats to Wetland Conservation**

Wetland degradation in the Dhemaji region is driven by several interlinked factors. One of the primary threats is *encroachment and land use change*, particularly the expansion of agricultural activities, which significantly reduces the wetland area. Additionally, *sedimentation and siltation* pose major challenges, as rivers in Assam—including those that feed into the Dhemaji wetlands—carry heavy sediment loads that lead to the gradual filling and shrinkage of water bodies. *Hydrological modifications* such as the construction of embankments and dams upstream further disrupt the natural flow of water, negatively impacting the wetland's hydrology and its ability to retain water. *Climate change* also plays a crucial role, with changing precipitation patterns and the increasing frequency of extreme weather events like floods and droughts accelerating wetland degradation. Moreover, the spread of *invasive vegetation*, often identifiable by low-MNDWI patches along the water margins, reduces open water areas and contributes to the decline in biodiversity.

The MNDWI values reflect a dynamic and fragile wetland system under threat from both anthropogenic and climatic factors. The visible fragmentation and reduction in water-dominated areas underscore the urgent need for integrated wetland management strategies, including sediment control, hydrological restoration, and strict regulation of land use changes.

11.3.3 Water Pixel Extraction USING OTSU Threshold

These maps represent the proportion of pixels representing water in the wetland area. A period of 7 years, starting from 2018 to 2024, has been taken into account. The threshold values within 0.05–0.1 were used to distinguish between the water and non-water pixels. The red areas show the water pixels, whereas the Black areas show areas covering non-water. These maps provide insights into the changing structure of the wetland and the threats associated with it (Fig. 11.11).

In 2018, there was very little water availability in the study area. The water coverage appears to be sparse and fragmented. The major portion of the wetland is classified as non-water, indicating less availability of water in the area. This could be because of minimal water retention, low precipitation conditions, Higher evaporation, and also due to anthropogenic activities such as land cover changes, water abstraction from surface water bodies. Poor hydrological functioning of the wetland indicates its vulnerability to climate and human activities.

By 2019, although there had been no visible significant changes in the quantity of water in the wetland, the water seemed to have contracted or become confined. This could be due to the changes in the Land cover system, or utilization of water from the bordering areas. Continued stress on the wetland system, with reduced spatial extent of water bodies.

In 2020, a significant increase in the water pixels was observed, indicating high availability of water in the wetland. This could be due to high precipitation, flooding, and better water retention due to various natural and manmade activities. A temporary recovery of wetland hydrology improves its ecological and hydrological functions.

In 2021, there have been no significant changes as compared to 2020, but compared to the previous years, there has been an increase in the water in the wetland. This indicates that the water holding capacity or the water retention capacity of the wetland has improved, and there

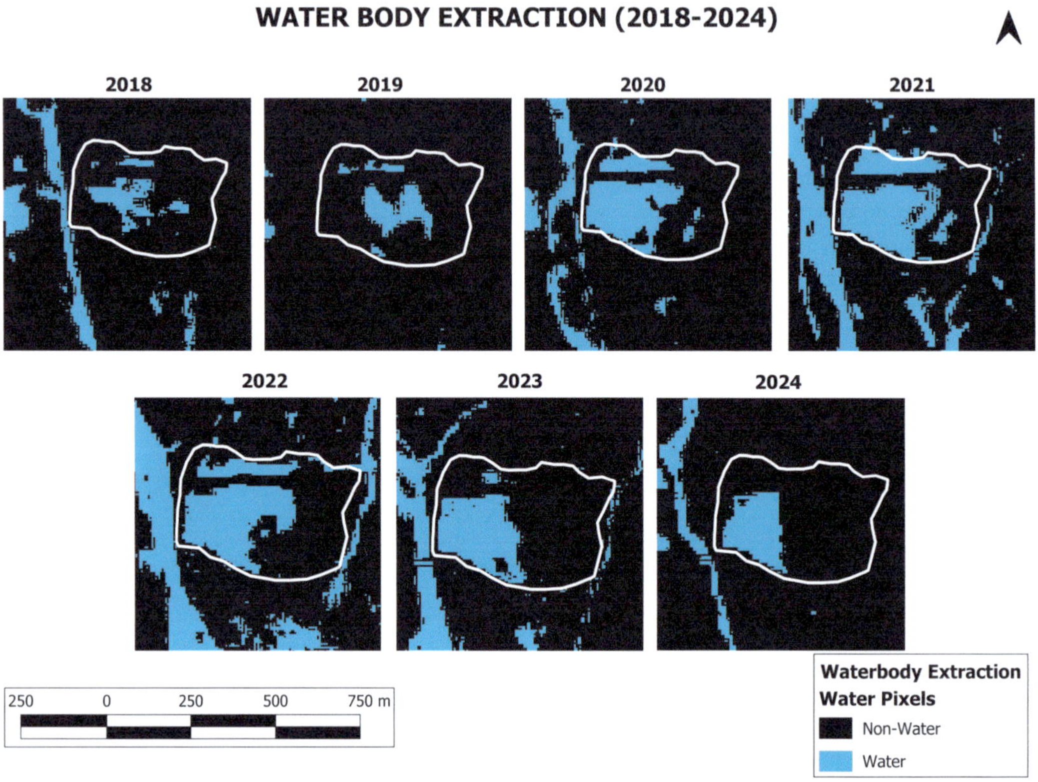

Fig. 11.11 Water pixel extraction using OTSU threshold

is a higher availability of water in the wetland. The wetland displayed resilience, with an enhanced ability to store water.

From 2022 to 2024, there has been a significant reduction in the water availability in the wetland. This could mainly be due to the increased anthropogenic activities in the area, the conversion of land for agricultural purposes, and increased water abstraction from the wetland system. As this region is mostly dominated by agricultural land, and as the proportion of agricultural land has shown an increase, one of the major reasons could be the conversion of these lands for agricultural purposes and the use of water for irrigation purposes, which is reducing the water availability in the wetland region. The wetland is under serious threat, with declining water availability impacting its ecological balance.

The wetland has experienced dynamic changes over the years, with periods of recovery and decline. However, the recent trend of reduced water availability (2022–2024) signals an urgent need for action to prevent further degradation. A combination of restoration, sustainable management, and community involvement is critical to securing the wetland's future and preserving its ecological and socio-economic benefits.

Impact of Human Activities on Jiadhal Wetland

The area of the Jiadhal Wetland has been reduced from in 2015 to area in 2024 due to natural and human activities. Drastic changes in the wetland occurred due to the encroachment of human activities. The Jiadhal river may not have changed its course till now in the study area. But the sediments that have been brought down by the rivers have led to the change in the structure of the wetland.

The human encroachment in the wetland is mainly of agricultural activities, community fishing grounds, settlements, etc. The wetland is filled up, tilled, and agriculture is cultivated in those areas.

Recommendations

Wetland restoration efforts should focus on implementing conservation programs aimed at rehabilitating degraded wetland areas. Introducing native vegetation can significantly improve water retention and help reduce soil erosion. To ensure long-term sustainability, it is essential to establish regular monitoring programs that assess wetland health through remote sensing and field surveys. A comprehensive wetland management plan must be developed, incorporating guidelines for sustainable water use. Additionally, strict regulations on water abstraction, particularly for agricultural and industrial purposes, should be enforced. Promoting the use of alternative water sources can help alleviate pressure on the wetland ecosystem.

11.4 Conclusion

Wetland is a product of nature. Proper protection and conservation should be undertaken by the local community. NGO, SHG, and the government should come forward and create awareness among the people about the benefits and threats of the wetland. Moreover, Gaon Panchayat and the administration of the district to come forward with a proper plan and programme to preserve the wetlands. The analysis of Dhemaji's wetland and land use dynamics (2000–2024) highlights significant transformations. Agricultural land expanded substantially, while water bodies and vegetation declined due to urbanization, sedimentation, and encroachment. MNDWI analysis reveals fragmentation and shrinkage of water bodies, with temporary recoveries followed by consistent declines in 2022–2024, signalling poor hydrological functioning. The wetlands are urbanized and resilient to the floods in the coming days. The wetland areas are now covered with markets, settlements, fisheries, etc. For many years, the Jiadhal Wetland was able to sustain with nature.

Acknowledgements We are grateful to the local communities of Jiadhal, Dhemaji, for providing us with relevant data and information regarding wetlands in the study area.

Potential Conflicts of Interest On behalf of all the authors, the corresponding author states that there is no conflict of interest.

Informed Consent Informed consent was obtained from all individual participants included in the study.

Authors' Contribution All the authors have contributed equally to this work.

Data Availability Statement The datasets during and/or analysed during the current study are available from the corresponding author on reasonable request.

Funding The research was not supported by any funding.

References

Bassi N, Kumar MD, Sharma A, Pardha-Saradhi P (2014) Status of wetlands in India: a review of extent, ecosystem benefits, threats and management strategies. J Hydrol Reg Stud 2:1–19. https://www.sciencedirect.com/science/article/pii/S221458181400010X

Chakraborty SK, Sanyal P, Ray R (2023) Diversity and classification of wetlands in international and national perspectives. In: Wetlands ecology: eco-biological uniqueness of a Ramsar site (East Kolkata Wetlands, India). Springer, Cham, pp 167–226. https://link.springer.com/chapter/10.1007/978-3-031-09253-4_3

Choudhury M, Sharma A, Chakravorty A, Dutta J (2020) Wetlands are in peril—a case study of Son Beel Wetland of Assam, India. In: Natural resources and their ecosystem services. HSRA Publication, p 102. https://www.researchgate.net/profile/Abhijit-Mitra-4/publication/352056332_Services_of_Natural_Resources_-_BY_Dr_ABHIJIT_MITRA-/links/60b73c6da6fdcc476bdf0682/Services-of-Natural-Resources-BY-Dr-ABHIJIT-MITRA.pdf#page=119

De Groot D, Brander L, Finlayson CM (2018) Wetland ecosystem services. In: The wetland book: I: structure and function, management, and methods. Springer Netherlands, pp 323–333. https://research.vu.nl/en/publications/wetland-ecosystem-services

Ghermandi A, Van den Bergh JC, Brander LM, De Groot HL, Nunes PA (2008) The economic value of wetland conservation and creation: a meta-analysis.

Fondazione Eni Enrico Mattei (FEEM), Milano. https://papers.ssrn.com/sol3/papers.cfm?abstract_id=1273002

Reis V, Hermoso V, Hamilton SK, Ward D, Fluet-Chouinard E, Lehner B, Linke S (2017) A global assessment of inland wetland conservation status. Bioscience 67(6):523–533. https://academic.oup.com/bioscience/article-abstract/67/6/523/3859715

Shine C, De Klemm C (1999) Wetlands, water, and the law: using law to advance wetland conservation and wise use (No. 38). IUCN. https://books.google.com/books?hl=en&lr=&id=vbdHSAuybRkC&oi=fnd&pg=PP15&dq=%E2%80%A2%09Shine,+C.,+%26+De+Klemm,+C.+(1999).+Wetlands,+water,+and+the+law:+using+law+to+advance+wetland+conservation+and+wise+use+(No.+38).+IUCN.&ots=ggmbAa9HAd&sig=4ekGiBgQCNOGpucM0i5jRW0-S1Y

U.S. Geological Survey (2024). https://www.usgs.gov/faqs/what-are-wetlands#:~:text=Wetlands%20are%20transitional%20areas%2C%20sandwiched,is%20covered%20by%20shallow%20water

Xu C, Sheng S, Zhou W, Cui L, Liu M (2011) Characterizing wetland change at landscape scale in Jiangsu Province, China. Environ Monit Assess 179: 279–292. https://link.springer.com/article/10.1007/s10661-010-1735-6

Zedler JB, Kercher S (2005) Wetland resources: status, trends, ecosystem services, and restorability. Annu Rev Environ Resour 30(1):39–74. https://www.annualreviews.org/content/journals/10.1146/annurev.energy.30.050504.144248

Mitsch, W. J., & Gosselink, J. G. (2000). Wetlands (3rd ed.). John Wiley & Sons. https://www.wiley.com/en-us/Wetlands%2C+3rd+Edition-p-9780471366165

Mitsch, W. J., Bernal, B., & Hernandez, M. E. (2013). Wetlands, carbon, and climate change. Landscape Ecology, 28(4), 583–597. https://doi.org/10.1007/s10980-012-9758-8

The Convention on Wetlands. (2024). Review of Phase 2: The Ramsar Strategic Plan 2016–2024. Ramsar Secretariat. https://www.ramsar.org/sites/default/files/2024-03/phase_2_document_review_research_report.pdf

Part IV

The Hidden Treasures: Unveiling the Socioeconomic Benefits of Urban Wetlands

Assessment of Eutrophication and Its Impact on Human Health: A Case Study of Barasat City

12

Sanjana De, Trishita Ghosal, Snehasish Mandal, Debasmita Routh, Kasturi Mukherjee, Manoj Kumar Singh, Pratik Dash, and Abhidipta Paul

Abstract

Eutrophication is a phenomenon characterized by the excessive accumulation of minerals and nutrients in a water body, leading to the unbalanced proliferation of aquatic plants and algae. This proliferation, known as an algal bloom, obstructs the penetration of light and oxygen into the water column, resulting in reduced oxygen levels, a condition referred to as hypoxia or anoxia. These altered environmental conditions significantly disrupt the aquatic ecosystem. Human activities, particularly those associated with agriculture, urbanization, and wastewater discharge, accelerate the process of eutrophication. Various parameters such as nitrate and phosphate concentrations, electrical conductivity, dissolved oxygen levels, total dissolved solids, pH levels, and temperature are routinely measured to assess the extent of eutrophication in water bodies. In a study conducted in the Barasat city, ten water samples were collected and analyzed from different land use categories to investigate the relationship between anthropogenic activities and the level of eutrophication. The results revealed that water bodies primarily utilized for domestic purposes or sewage disposal exhibited signs of eutrophication or even hyper-eutrophication. Furthermore, the study included a household survey to evaluate the main uses of these water bodies and assess the potential impact of direct human contact with eutrophic waters on human health. This comprehensive approach aimed to deepen our understanding of the intricate interplay between human activities, water quality, and the ecological consequences of eutrophication in aquatic ecosystems.

Keywords

Eutrophication · Water pollution · Waterborne diseases · Cyanobacteria

12.1 Introduction

Freshwater, a vital yet increasingly scarce resource, is under growing pressure due to human activities. Eutrophication refers to the excessive growth of plants, spurred by nutrient enrichment from human sources, which disrupts the availability of freshwater and also the aquatic ecosystems. This phenomenon has escalated alongside the surge in global water consumption,

S. De · T. Ghosal · D. Routh · K. Mukherjee (✉)
Department of Geography, Adamas University, Kolkata, West Bengal, India

S. Mandal · M. K. Singh (✉) · A. Paul
Department of Biotechnology, Adamas University, Kolkata, West Bengal, India
e-mail: manoj.k.singh@adamasuniversity.ac.in

P. Dash
Khejuri College, Vidyasagar University, Midnapore, West Bengal, India

S. K. Chakraborty et al. (eds.), *Wetlands and Urban Resilience*, Sustainable Development Goals Series,
https://doi.org/10.1007/978-3-032-06881-1_12

which has increased sevenfold since the early twentieth century. Consequently, ensuring water storage and maintaining water quality have become paramount concerns (Ramesh et al. 2013).

The topic of eutrophication and water quality is intricately linked to several Sustainable Development Goals (SDGs) set forth by the United Nations. Efforts to mitigate eutrophication to improve water quality directly contribute to Goal 6 (Clean Water and Sanitation) by addressing water pollution and ensuring access to safe drinking water. Sustainable agricultural practices, aimed at reducing nutrient runoff and pollution, align with Goal 12 (Responsible Consumption and Production). These efforts also support Goal 14 (Life Below Water) by conserving marine and freshwater ecosystems affected by eutrophication. Additionally, addressing soil erosion and promoting sustainable land management practices relates to Goal 15 (Life on Land). Improving water quality not only enhances environmental sustainability but also contributes to Goal 3 (Good Health and Well-being) by reducing waterborne diseases associated with poor water quality, ultimately fostering healthier communities and ecosystems (THE 17 GOALS | Sustainable Development). The World Health Organization (WHO) highlights that water-related diseases account for 80% of global health issues, with developing countries bearing a significant burden. Waterborne diseases like diarrhea, malaria, and dengue fever cause substantial morbidity and mortality, emphasizing the criticality of addressing water quality challenges.

Eutrophication, characterized by excessive plant growth due to human-induced nutrient enrichment, stands as a predominant challenge for many surface waters today. This phenomenon vividly showcases human alterations to the biosphere, impacting aquatic ecosystems globally from the Arctic to the Antarctic. Anthropogenic inputs of nutrients, particularly nitrogen (N) and phosphorus (P), contribute significantly to water body enrichment. This process drives oligotrophic water bodies toward mesotrophic, eutrophic, and ultimately hypertrophic states. These eutrophication states (from oligotrophic to hypertrophic) have been classified based on nitrate and phosphorous content, as shown in Table 12.1.

Table 12.1 Trophic classification of water bodies

Trophic category	Total nitrate (mg/m^3)	Total phosphorous (mg/m^3)
Ultraoligotrophic	<200	<5
Oligo-mesotrophic	200–400	5–10
Meso-eutrophic	400–650	10–30
Eutrophic	650–1500	30–100
Hyper-eutrophic	>1500	>100

Source: Wetzel (1975)

Each stage presents distinct characteristics, from clearwater with minimal organic matter to highly productive water bodies with abundant nutrients, fostering excessive plant growth and ecosystem disturbances:

1. *Ultraoligotrophic/Oligotrophic:* Clearwater with minimal organic matter, sediment, and low biological activity.
2. *Oligo-mesotrophic/Meso-eutrophic:* Water containing more nutrients, leading to turbidity, and thriving phytoplankton populations.
3. *Eutrophic:* Water rich in nutrients, exhibiting high biological productivity. Some species may face competition due to nutrient abundance.
4. *Hyper-eutrophic:* Highly productive water bodies, almost reaching wetland status. Many clearwater species struggle to survive in such nutrient-rich environments.

During the initial stages of the formation of the Earth, at that time water bodies are in the state of oligotrophy, which means having nutrient deficiency. Enriching the water with mineral nutrients, such as nitrogen (N) and phosphorus (P), causes the transformation of water bodies from oligotrophic to mesotrophic, eutrophic, and finally hypertrophic stages.

Mesotrophic and eutrophic phases show medium and rich levels of nutrients and cause water quality problems. However, in the case of the hypertrophic phase, it indicates excessive enrichment (minerals) of aquatic ecosystems which makes the water body fully unusable for fisheries, recreation, industry, and drinking because of the increased growth of undesirable algae and aquatic weeds and the amount of dissolved oxygen decreases due to the death and decomposition of these algae and aquatic weeds. Associated production of cyanobacteria (blue-green algae) occurs in drinking water supplies which may pose a serious health hazard to animals and humans.

Eutrophication in water bodies is primarily caused by human activities. While some water bodies may naturally have nutrient-rich soils, leading to eutrophication.

The impacts of fertilizers on eutrophication are profound. Anthropogenic activities, particularly the excessive use of phosphorus and nitrogen in agriculture, significantly contribute to eutrophication of nitrogen through runoff from agricultural lands, categorized as "nonpoint" source of pollution. Khan and Ansari 2005; Vitousek et al. 1997).

Phosphorus binds to soil particles and primarily moves with soil erosion, accumulating as sediments in water bodies (Jones 2020). Nitrates, on the other hand being water-soluble, easily move with surface runoff into rivers or percolate into groundwater. Studies show that a significant portion of nitrogen applied as fertilizer remains in soils, contributing to soil residues, erosion, leaching, or atmospheric release (Iverson 2009). The global phosphorus budget indicates an imbalance between phosphorus input (fertilizer) and output (removal), leading to phosphorus accumulation in soils and water ecosystems, notably in agricultural areas (Jarvie et al. 2018; Schindler 1977; Sharpley et al. 1994; Schroeder et al. 2004; Bashkin and Howarth 2002; Djodjic et al. 2005).

Modern agricultural practices have intensified fertilizer use and land management, significantly fueling pollution, with nitrogen and phosphorus fertilizer use rising dramatically from 1960 to 1990 (Kuang et al. 2005).

Eutrophication is also caused by the municipal and industrial wastewater discharges, nitrogen leaching from septic tanks, and stormwater runoff, categorized as "point sources" of nutrient pollution (Ramesh et al. 2013). Additionally, fossil fuel combustion, notably from coal-fired power plants and vehicles, releases nitrogen oxides, contributing significantly to nutrient pollution (EIA 2008).

Population expansion drives increased agricultural demand and fossil fuel consumption, indirectly contributing to nutrient pollution. Nitrogen fertilizers contribute to atmospheric deposition, ammonia emissions, and nitrogen oxide production, leading to long-distance nutrient transport and increased phytoplankton productivity (Constant et al. 1992).

The proliferation of harmful algal blooms, particularly cyanobacteria, can lead to the production of toxins such as microcystins and anatoxins, posing serious health risks to individuals exposed through recreational activities, drinking water sources, or consumption of contaminated seafood (Paerl et al. 2018). These toxins can cause a range of health problems, including liver damage, gastrointestinal issues, skin irritation, and neurological disorders, presenting a direct threat to human well-being (Carmichael 2008). Addressing eutrophication not only protects aquatic ecosystems but also plays a crucial role in safeguarding human health and ensuring access to safe and sustainable water supplies.

Managing eutrophication, a pervasive environmental issue stemming from nutrient over-enrichment in aquatic ecosystems, necessitates a comprehensive and integrated approach. Physical methods such as dredging, which involves the removal of nutrient-rich sediments from water bodies, can mitigate internal nutrient cycling and reduce the accumulation of phosphorus and nitrogen that fuel algal blooms (Wagner 2004). Aeration techniques are employed to increase dissolved oxygen levels, limiting the formation of anoxic environments that release phosphorus and nitrogen from sediments (Wagner 2004). Chemical interventions like algicides, particularly copper-based compounds, are used to control algal growth by disrupting their photosynthetic processes (Wagner 2004). Phosphorus-binding agents, including aluminum-based flocculants, are applied to bind phosphorus and prevent its release into water, thereby reducing internal nutrient loads (Wagner 2004; Lurling and van

Oosterhout 2013). Biological control methods such as phytoremediation utilize aquatic macrophytes like Eicchornia crassipes and Salvinia auriculata to absorb excess nitrogen and phosphorus, contributing to water purification efforts (Petrucio and Esteves 2000). Duckweeds (Lemnoideae), known for their rapid growth and nutrient absorption capabilities, offer another bio-control mechanism against eutrophication (Ozimek 1990). Public awareness campaigns and legislative measures are crucial in fostering community engagement and implementing sustainable practices to reduce nutrient inputs from agricultural runoff, industrial discharges, and urban sources, thereby safeguarding water quality and ecosystem health (Kadlec and Bevis 2009). Ongoing research and adaptive management strategies are essential for addressing the dynamic challenges posed by eutrophication and ensuring long-term environmental sustainability.

In this chapter, the author aims to examine the phenomenon of eutrophication and its impact on water quality, particularly in freshwater ecosystems. This includes analyzing the drivers and processes leading to eutrophication, evaluating the consequences of nutrient enrichment on aquatic ecosystems, and human health.

The novelty of this study lies in integrating satellite-derived chlorophyll data with ground-level eutrophication and health impact assessments in a rapidly urbanizing city. The specific objectives include (i) evaluating the nutrient enrichment and eutrophication status across selected urban ponds, (ii) assessing the link between LULC patterns and eutrophication, and (iii) examining health risks associated with human use of eutrophic waters.

12.2 Rationale of the Study

Despite numerous studies on eutrophication, a significant knowledge gap remains in understanding its direct implications on human health, particularly within the context of rapidly urbanizing peri-urban environments in India. Most existing research tends to focus either on the ecological impacts of eutrophication or on water quality parameters in isolation. However, there is limited literature that holistically integrates water quality assessments with health surveys and land use analysis to provide a comprehensive understanding of how eutrophication affects communities that directly rely on these water sources for daily use. Barasat city, with its fast-paced urban growth, mixed land use patterns, and dense population, presents a unique case where traditional pond usage coexists with mounting anthropogenic pressures. Yet, localized studies combining spatial analysis (e.g., satellite imagery and GIS) with on-ground water quality testing and household health assessments remain scarce. Addressing this gap is critical for designing targeted interventions and sustainable urban water management strategies.

12.3 Study Area

Barasat is one of the oldest cities of West Bengal which is situated within the Kolkata Agglomeration belt (Fig. 12.1). Before the partition of India in 1947, the town was located on the eastern side of the Sealdah-Bongaon railway track. However, after partition, the western part of the city along the Sealdah-Bongaon railway track was gradually converted into a residential area. This occurred owing to the huge influx of Hindu refugees from erstwhile Pakistan. The city continued to grow more and more following the creation of Bangladesh in 1971. With a 125% population growth rate between 1991 and 2011, the fast pace of population growth in Barasat has become more noticeable in recent decades. The nonurban landscape of this area is being altered continuously and being converted into urban land. The rapid, hasty urbanization associated with population growth has become a significant challenge in this area (Mukherjee and Das 2018; Mukherjee et al. 2023, 2024). Thus, for analyzing the level of eutrophication in different ponds of the city and the impact of their usage, Barasat can be a good case study region.

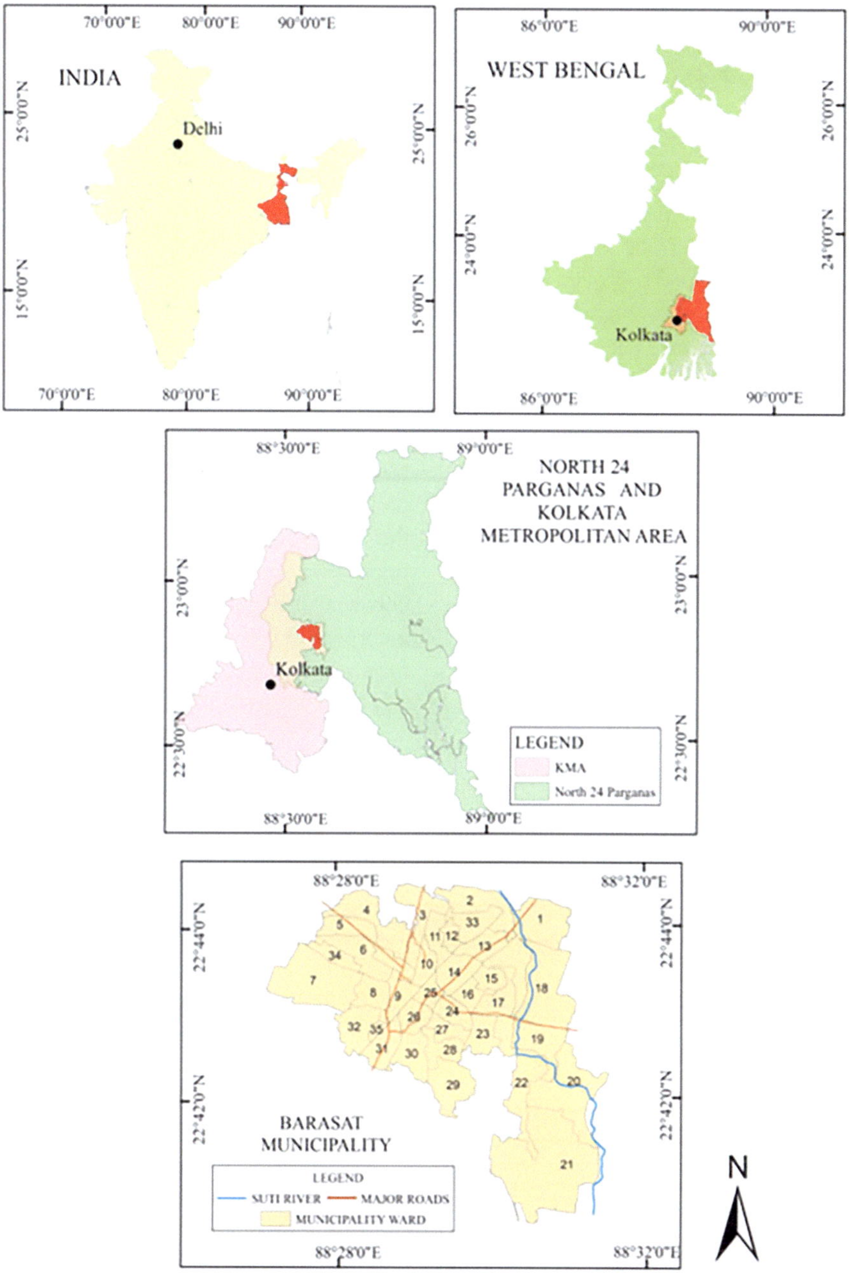

Fig. 12.1 Location map of the study area

12.4 Data and Method

12.4.1 Primary Data Collection

The water sample and the demographic data were gathered through water collection and a questionnaire survey, respectively (Fig. 12.2). Ten ponds were selected randomly from all over the city (Fig. 12.3). The collection took place during the pre-monsoon period to capture baseline data. Primary demographic data have been collected from the households of Barasat surrounding the ponds, targeting the residents who were using the pond water regularly for different purposes.

12.4.2 Remote Sensing Data Collection

From the United States Geological Survey (USGS) website, Landsat 8 OLI satellite images (138 Path and 44 Row) of 30 meters of spatial resolution for 2020 have been downloaded. Image only without the cloud has been selected. The image was assigned the Universal Transverse Mercator (UTM) (Zone 45) and World Geodetic System, depending on the topographical map (scale 1:50000) (WGS84). After rectification on the GIS platform, the map of Barasat was utilized to subset the satellite images. The masked images were then used to classify into six land use land cover classes (vegetation, homestead with plantations, built-up, open space, water bodies, and agricultural/cropland) (Mukherjee et al. 2022) and to determine the chlorophyll content of the pond for 2020 (April) for this study. The NIR and Green bands are used by the following Formula 12.1 to calculate the chlorophyll content index (Hunt et al. 2011).

$$\text{Green Chlorophyll Index} = \left(\frac{\text{NIR}}{\text{Green}}\right) - 1 \quad (12.1)$$

12.4.3 Accuracy Assessment

Accuracy assessment in classification involves using the confusion matrix to detect and categorize errors systematically. The key metric, overall accuracy, measures the proportion of correctly classified pixels to the total validation points, offering a comprehensive evaluation of classification performance across all classes. Additionally, the kappa coefficient, ranging from 0 to 1, provides a quantitative measure of agreement between the classified and reference images, incorporating commission and omission errors for a more nuanced assessment.

12.4.4 GIS-Based Buffer Analysis on LULC

In conducting the analysis of land use and land cover (LULC) surrounding the ponds, a GIS-based buffer analysis was utilized to create a 1 km buffer zone around each pond. This buffer zone facilitated the assessment of LULC characteristics within the immediate vicinity of the ponds. The Zonal Histogram technique was then employed to quantify the area covered by each LULC category within these buffer zones.

Fig. 12.2 Water sample collection

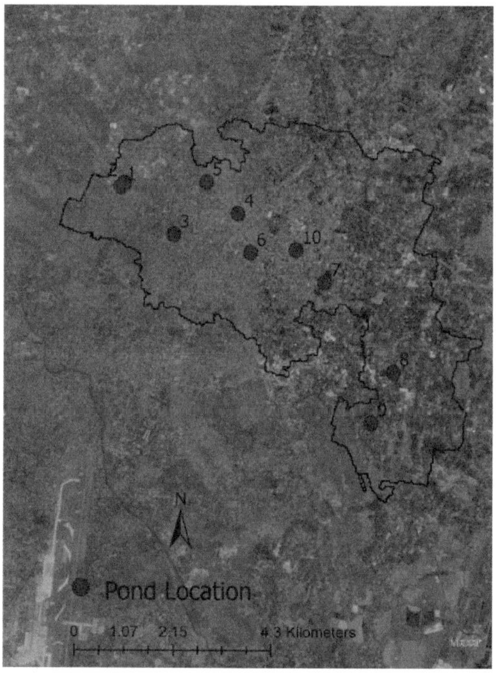

Fig. 12.3 Location of ponds

12.4.5 Determination of Nitrate Concentration

The method utilized the oxidation of diphenylamine by nitrates in the sample, resulting in a blue coloration (ΛMax = 590 nm) in an acidic mixture. Nitrate levels were determined using a calibration curve with standard solutions (0.6, 1.25, 2.5, 5, 10, and 100 mg KNO_3 mL^{-1}) (Aoun et al. 2013).

12.4.6 Determination of Phosphate Concentration

The method involved the formation of molybdophosphoric acid, reduced to the intensely colored complex, molybdenum blue. This method is highly sensitive and detectable even below 0.1 mg phosphorus per liter, with analysis at a wavelength of 650 nanometers. Calibration curves were constructed using standard solutions (0.25, 0.5, 1, 2, and 3 mg KH_2PO_4 mL^{-1}).

12.4.7 Measurement of DO, TDS, EC, and pH

Aquasol AM-AL-01 was utilized for TDS, EC, and pH measurements, while a Lutron DO-5510 Oxygen Meter was employed for DO measurements during sample collection.

12.4.8 Microscopic View of Algae in the Water Sample

After collecting the water samples, it was observed under the microscope at different at 1×, 10×, and 40× magnification (Fig. 12.4).

12.4.9 Quantifying the Effect of LULC on Water Parameters

Partial Least Squares (PLS) Regression is a recent technique that generalizes and combines features from principal component analysis and multiple regression which can handle multicollinearity and nonlinear relationships between variables more effectively. It is particularly useful when we need to predict a set of dependent variables (in this case, different water parameters) from a large set of independent variables, i.e., predictors (in this case, different types of land use). It originated in the social sciences, specifically the economy (Wold 1985). It is recommended in cases of regression where the number of explanatory variables is high, and where it is likely that there is multicollinearity among the variables, i.e., that the explanatory variables are correlated. In this context of analyzing the impact of land use on crime, PLS Regression is used to identify the specific land use variables that have the greatest impact on crime levels. In Partial Least Squares Regression (PLSR), the VIPs (Variable Importance for the Projection) score quickly identifies which explanatory variables contribute the most to the models and the impact of the independent variable (different types of land uses) on the dependent variable (water parameters). It will

Fig. 12.4 Water sample preservation for microscopic view of algae

help in the analysis of the impact of individual land use on parameters. VIP score is calculated for each predictor variable and is based on its contribution to the prediction of the dependent variable. The score is estimated as the sum of the squared loadings of a predictor variable across all latent variables in the model. The squared loadings represent the contribution of each predictor variable to each latent variable, and the sum across latent variables reflects the overall contribution of the predictor variable to the model. XLSTAT provides a complete PLS regression method to model and predict the data in Excel. The entire methodology has been shown below (Fig. 12.5):

12.5 Result

12.5.1 Accuracy Assessment

Accuracy assessment was performed on the GIS platform with the help of validation points collected with the help of Google Earth Pro. The confusion matrix was calculated to provide the producer's accuracy, user's accuracy, overall accuracy, and kappa coefficient for each classification. The producer's accuracy is 96.30% and the user's accuracy is 92.86%. The overall accuracy is 85.19%. The kappa statistic is 0.828. The accuracy level met the suggested value by Lucas et al. (1994).

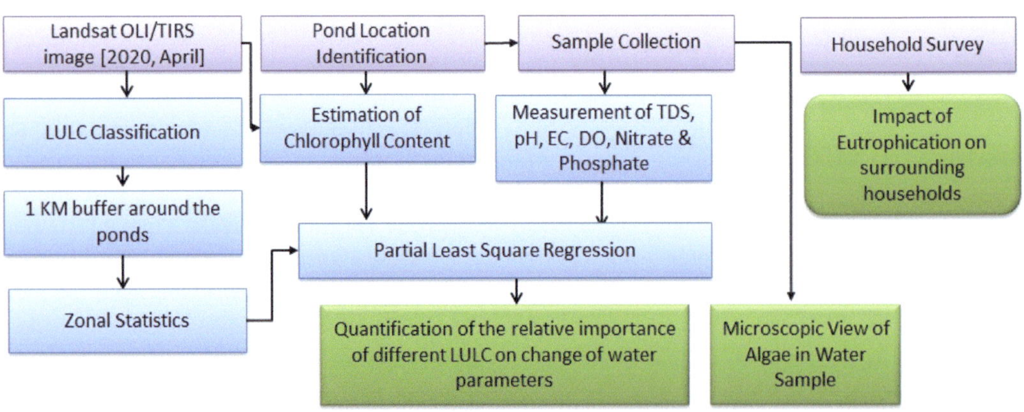

Fig. 12.5 Methodology

12.5.2 LULC Classification

By visually inspecting the classified maps (Fig. 12.6), it could be observed that the urban built-up is occupying nearly the entire city. It is covering approximately 64.52% of the city in 2020. Homestead with plantation which has the next largest share comprises of the semi dense built-up areas. Slum and squatter settlements of this type of floating migrant population along the peri-urban belt often give birth to these semi-dense built-up areas. The agricultural lands and vegetation areas are present in a scattered way near the periphery within the boundary. Only 1.76% of areas have been covered by water bodies and 0.41% is occupied by bare lands (Fig. 12.7). Almost all bare lands have been very rapid demolition of bare soil for converting it into residential property.

12.5.3 Conditions of the Ponds

Samples No. 1–7 are classified as Eutrophic, characterized by relatively moderate levels of Dissolved Oxygen (DO), Total Dissolved Solids (TDS), Electrical Conductivity (EC), pH, Nitrate, Phosphate, and Chlorophyll content (Fig. 12.8). These parameters collectively indicate a state of increased nutrient enrichment and biological productivity in these water samples (Fig. 12.10, Table 12.2).

Samples No. 8–10 exhibit characteristics of hyper-eutrophication (Fig. 12.9), with significantly higher levels of Nitrate, Phosphate, and Chlorophyll content compared to the Eutrophic samples. This indicates a more pronounced nutrient loading and algae proliferation, indicative of a highly eutrophic condition. Temperature ranges from 23.4 to 28.6 °C across the samples, with no

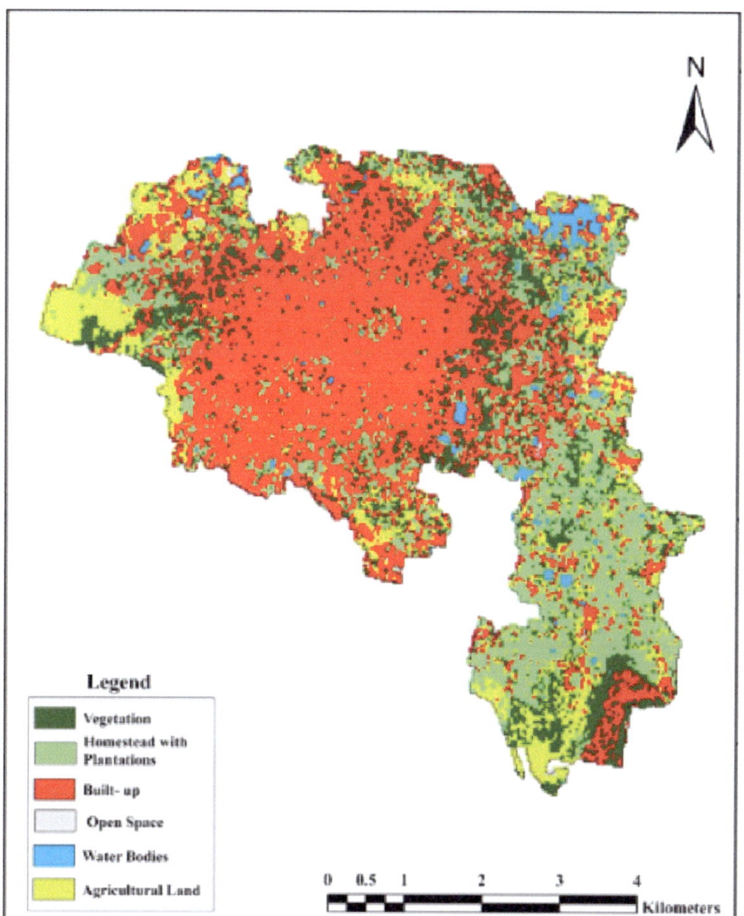

Fig. 12.6 LULC classification, 2020

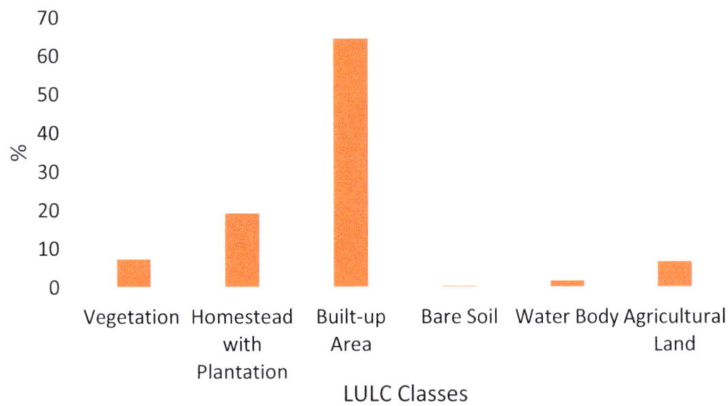

Fig. 12.7 Area of LULC classes (%)

Fig. 12.8 Ponds that are in eutrophic stage

Table 12.2 Result of water sample tests

Sample no.	Temperature (°C)	DO (mg/lit.)	TDS	EC	PH	Nitrate (mg/lit.)	Phosphate (mg/lit)	Chlorophyll content	Result
1.	28	4.5	370	358	7.38	0.81	3.01	0.493	Eutrophic
2.	24	4.3	183	276	7.13	0.75	2.8	0.642	Eutrophic
3.	23.4	5.4	408	637	7.29	0.81	2.8	0.474	Eutrophic
4.	24	3.4	351	522	6.58	0.76	3.06	0.327	Eutrophic
5.	25	4.5	272	415	6.74	0.87	3.22	0.427	Eutrophic
6.	27.6	0.2	429	651	7.2	0.86	4.06	0.408	Eutrophic
7.	28.6	4.1	361	558	6.97	0.8	1.23	0.345	Eutrophic
8.	28.5	4.1	360	546	7.17	1.03	0.6	0.528	Hyper eutrophic
9.	28.5	0.2	441	674	7.07	1.01	2.54	0.625	Hyper eutrophic
10.	26.1	0.3	314	472	6.77	1.18	1.6	0.395	Hyper eutrophic

Fig. 12.9 Ponds that are in hyper-eutrophic stage

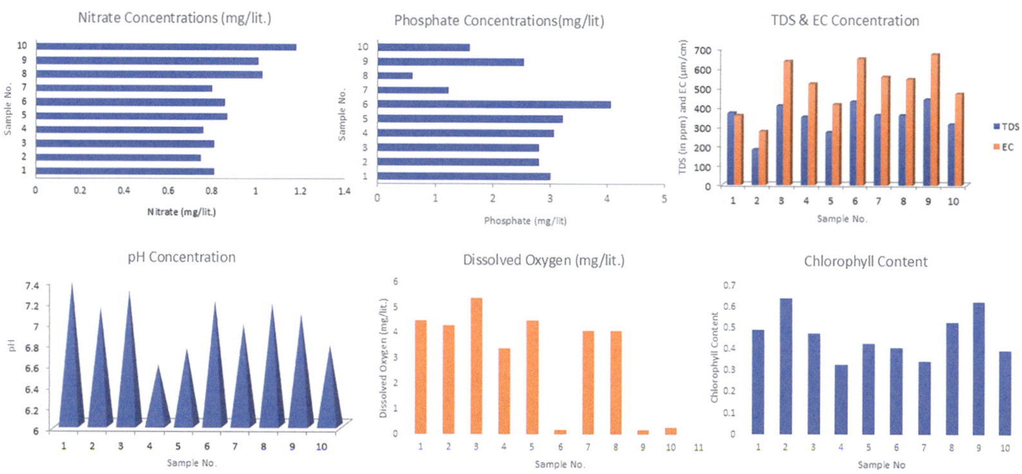

Fig. 12.10 Water parameters

clear correlation observed between temperature and eutrophication status. Dissolved Oxygen (DO) levels vary from 0.2 to 5.4 mg/lit., with lower DO concentrations observed in some Hyper-eutrophic samples (Sample Nos. 8 and 9). This can be attributed to increased organic matter decomposition and algal respiration, leading to oxygen depletion. Total Dissolved Solids (TDS) and Electrical Conductivity (EC) values are higher in the Hyper-eutrophic samples, reflecting the elevated nutrient and ion concentrations associated with eutrophication. pH values range from 6.58 to 7.38, indicating slightly acidic to neutral conditions across the samples, typical of freshwater ecosystems.

Nitrate and phosphate concentrations are notably elevated in the hyper-eutrophic samples, highlighting the excessive nutrient loading contributing to eutrophication. Chlorophyll content, a measure of algal biomass, increases significantly in the hyper-eutrophic samples, indicating intense algal growth and biomass accumulation due to nutrient enrichment. Overall, the data illustrates a gradient of eutrophication levels, with the hyper-eutrophic samples exhibiting more pronounced nutrient enrichment, algae proliferation, and associated water quality deterioration compared to the Eutrophic samples (Fig. 12.10).

12.6 Microscopic View of Algae in Water Sample

Microscopic view of the water sample showed the presence of blue-green and/or brownish-green cyanobacteria (pond No. 10) and spirogyra (pond No. 2) in abundance, and presence of other unicellular algae (Fig. 12.11). Further analysis of the algal population may help analyze the health hazards. As the biomass increases due to surplus nutrients and eventually dies, bacteria further decompose them, decreasing the oxygen content of the pond. Increase in nutrients causes an increase in HABs (harmful algal blooms) like cyanobacteria (Oh et al. 2023). Although cyanobacteria are technically bacteria, they act more like alga in aquatic ecosystems and are considered part of alga communities. A cyanobacterial bloom is evidenced by a discoloration of the water or a paint-like scum (typically blue-green) at the surface of the water that has a foul odor. Toxins produced by cyanobacteria may further affect human health when ingested through water or fish. For example, Cylindrospermopsin produced by freshwater cyanobacteria may cause vomiting, nausea, diarrhea, pain, abdominal tenderness, or acute liver failure (Diez-Quijada et al. 2021). Another freshwater cyanobacterial toxin, microcystin causes gastrointestinal pain and liver failure (Xu et al. 2020). Toxins produced by cyanobacteria may further affect human health when ingested through water or fish (where they may enter directly or through bioaccumulation).

12.6.1 Impact of Land Use and Land Cover on the Level of Eutrophication

The Variable Importance in the Projection (VIP) scores, derived through Partial Least Squares Regression (PLSR), highlight the significant influence of various land use and land cover (LULC) categories on eutrophication parameters (Fig. 12.12). Among these categories, homesteads with plantations, water bodies, bare soil, vegetation, and cropland emerge as major influencing factors contributing to eutrophication

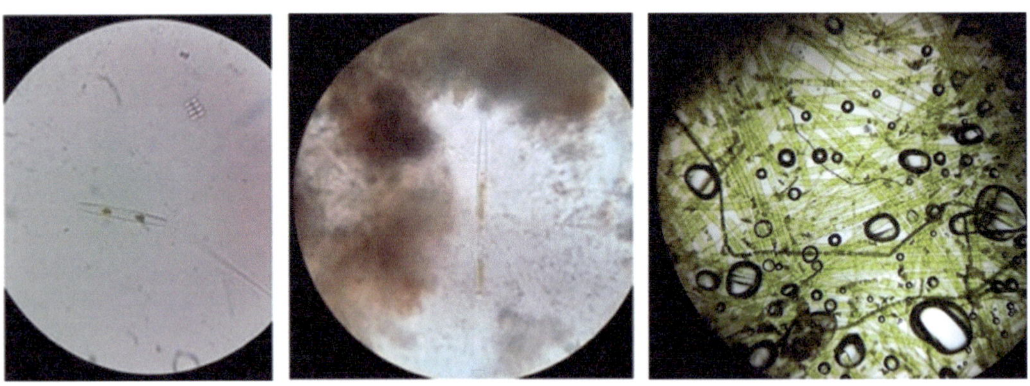

Microscopic view of Pond No. 2

Microscopic view of Pond No. 11

Fig. 12.11 Microscopic view of algae in water sample

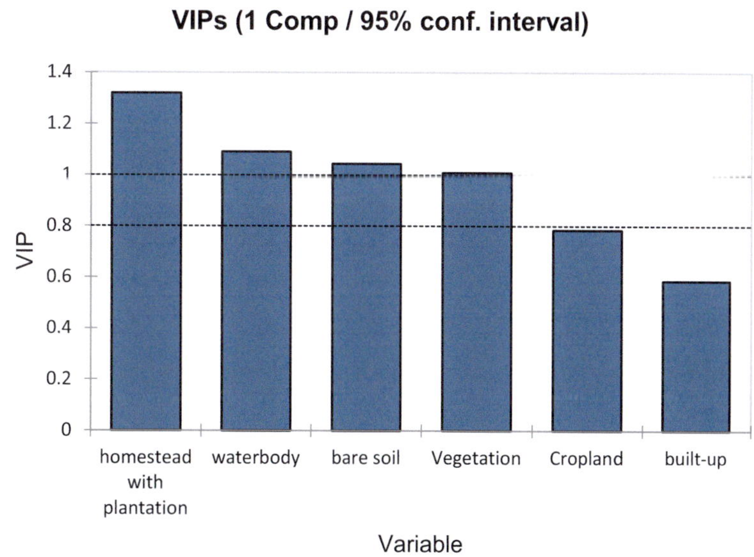

Fig. 12.12 VIP score of different LULC

dynamics. However, built-up areas show comparatively lower influence in this context, probably due to low contribution to surface runoff.

Homesteads with plantations exhibit a VIP score of 1.319, indicating a substantial impact on eutrophication processes. This influence may stem from factors such as agricultural runoff, nutrient leaching from plantation activities, or vegetation cover affecting nutrient retention and cycling. Water bodies play a pivotal role in eutrophication, as reflected in their VIP score of 1.092. Nutrient inputs from surrounding land areas, hydrological connectivity, and aquatic habitat characteristics contribute significantly to the eutrophication potential of water bodies. Bare soil areas, with a VIP score of 1.044, also contribute notably to eutrophication dynamics. Factors such as sediment and nutrient runoff from bare soil surfaces can elevate nutrient levels in water bodies, fueling eutrophication processes. Vegetated areas, represented by a VIP score of 1.009. Vegetation cover influences nutrient uptake, sediment retention, and overall ecological balance, impacting eutrophication levels in aquatic ecosystems. Cropland areas, with a VIP score of 0.784, demonstrate a considerable influence on eutrophication. Agricultural practices, including fertilizer use, irrigation, and land management, contribute to nutrient runoff and water pollution, affecting eutrophication dynamics in adjacent water bodies. In contrast, built-up areas exhibit a lower VIP score of 0.586, indicating a comparatively lesser impact on eutrophication processes. While urban activities can introduce pollutants into water bodies, the overall contribution of built-up areas to eutrophication appears to be less pronounced compared to other LULC categories.

12.6.2 Use of Ponds and Their Impacts

The socioeconomic survey findings reveal significant insights into the utilization and management of the sampled water bodies (Fig. 12.13). A substantial majority, comprising 73% of the sampled water bodies, is utilized primarily for domestic activities such as bathing, cleaning utensils, and washing clothes, highlighting the crucial role of these water bodies in fulfilling daily household needs. However, a notable portion (27%) of the water bodies remain unused for domestic purposes due to their poor condition, indicating environmental concerns and challenges in maintaining water quality. Additionally, activities like idol immersion, sewage disposal, and garbage dumping were observed, underscoring potential sources of pollution and environmental degradation. Pisciculture, or fish farming, is a prevalent activity observed in 73% of the sampled

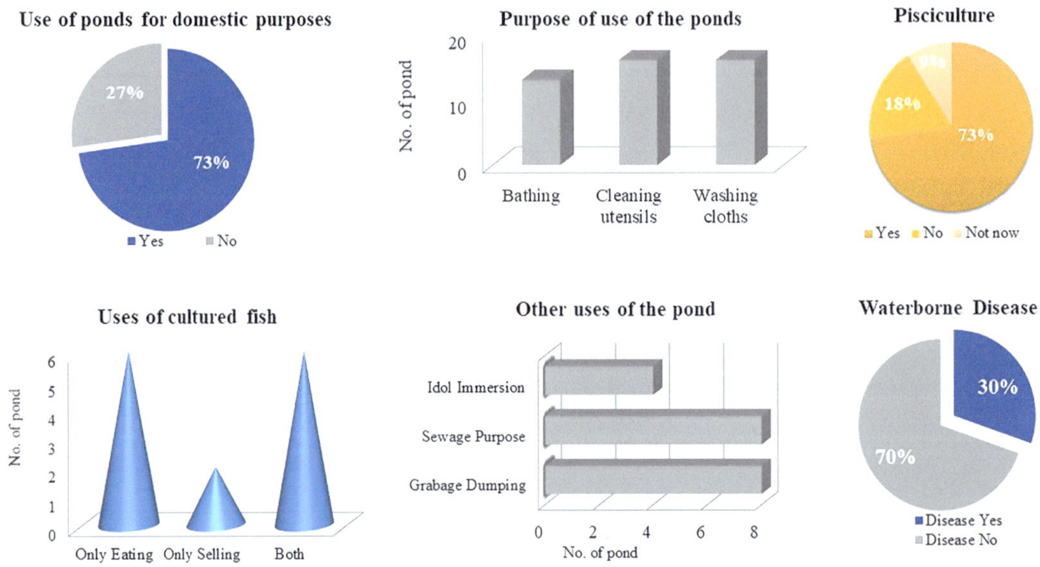

Fig. 12.13 Use of sampled ponds and the impact of water usage

water bodies. Local communities and pond owners engage in pisciculture, either for personal consumption or commercial purposes. Interestingly, some instances involve dual usage, where fish are utilized both for personal consumption and for sale. Approximately 9% of the sampled water bodies reported a cessation of pisciculture activities due to deteriorating conditions. This highlights the direct impact of environmental degradation on economic activities dependent on water resources, emphasizing the need for sustainable management practices.

30% surveyed population who use pond water for different purposes have reported suffering from waterborne diseases. The prevalent skin issues include rashes (59%), black spots (8%), itching (25%), and allergies (8%). Interestingly, females are more affected than males, constituting 57% of those suffering from skin diseases due to their higher contact with the water bodies. Despite the prevalence of waterborne diseases, many villagers do not seek medical treatment. Approximately 63% of the population consults allopathic doctors, while only 5% prefer homeopathic treatments. Surprisingly, 32% of the affected population does not seek medical help and chooses to ignore their health issues related to waterborne diseases.

These findings underscore the complex interplay between human activities, environmental conditions, and economic practices related to water resource utilization. Addressing challenges such as pollution, degradation, and sustainable resource management is crucial for maintaining the ecological integrity and socioeconomic viability of these water bodies.

12.7 Discussion

The present study aimed to explore the extent of nutrient enrichment and eutrophication in urban ponds, examine the influence of surrounding land use and land cover (LULC), and assess the associated health risks for local populations. The results clearly demonstrate the fulfillment of these objectives through a multi-pronged methodology that integrated water quality analysis, remote sensing-based chlorophyll assessment, and household health surveys. The analysis revealed a spatial gradient of eutrophication across Barasat's urban ponds, with seven classified as eutrophic and three as hyper-eutrophic. These categories were based on elevated levels of nitrates, phosphates, and chlorophyll content, coupled with declining dissolved oxygen levels.

These findings directly address the first objective by documenting the state of water degradation across different land use contexts.

Importantly, the study moves beyond descriptive categorization to critically examine the environmental and anthropogenic drivers behind these water quality conditions. Ponds located near areas dominated by homesteads with plantations, croplands, and bare soils showed significantly higher nutrient levels. This suggests that nutrient runoff from unregulated household effluents, localized agriculture, and exposed soil surfaces are key contributors to eutrophication. Built-up zones, interestingly, had lower influence, possibly due to better surface sealing and drainage systems, a nuance that highlights how different urban forms affect ecological processes differently. These results underscore the significance of land-water interactions and provide strong evidence in support of the second research objective.

The findings indicate that eutrophication in Barasat's urban ponds is primarily driven by elevated phosphate and nitrate concentrations, consistent with patterns reported by Prasad and Prasad (2019), who emphasize phosphate as the key nutrient triggering algal blooms and consequent hypoxia in Indian inland waters. The nutrient concentrations observed in this study were sufficient to support phytoplankton "bloomers"—species that thrive under low N:P ratios, a condition also identified in the Yamuna River study by Sharma et al. (2017), where the DIN/DIP ratio averaged around 33:1, indicating phosphorus limitation and potential for harmful algal blooms. Similarly, both studies report low dissolved oxygen levels in eutrophic stretches, especially during warmer months, which aligns with the hypoxic conditions observed in Barasat ponds with high nutrient loads. The high chlorophyll-a values and visible algal scum in the hypereutrophic ponds further underscore the severity of bloom activity, matching the ecological responses documented in both Yamuna and Ganga river systems. From a land use perspective, the influence of homesteads with plantations and croplands as significant contributors to nutrient runoff mirrors findings from other Indian studies, including the algal bloom incidents in Udaisagar Lake and Sankey Lake, where similar sources were implicated (Prasad and Prasad 2019).

Beyond environmental degradation, the study's health survey revealed that 30% of the surveyed population reported water-related health issues such as rashes, allergies, and skin infections. A gendered dimension was evident—57% of affected individuals were women, likely due to their more frequent exposure through bathing and household chores. This observation not only confirms the third research objective but also signals the need to integrate gender-sensitive strategies in water and health management. Alarmingly, 32% of affected individuals did not seek medical treatment, either due to a lack of awareness, financial constraints, or the normalization of symptoms, highlighting a hidden health burden that remains largely undocumented and under-addressed in policy frameworks.

The findings carry substantial implications for urban planning, public health, and environmental governance. Immediate policy responses could include establishing protective buffer zones around vulnerable water bodies, incorporating green infrastructure to manage runoff, and launching awareness campaigns on safe water practices—particularly targeting women and children. Urban zoning policies should be revised to prevent uncontrolled residential encroachments near ponds and promote integrated water resource management. Simultaneously, health systems must recognize the link between water quality and disease burden, and develop localized interventions accordingly.

This study's strength lies in its methodological integration. By combining remote sensing, in-situ water sampling, and household-level surveys, it presents a rare and comprehensive understanding of how land dynamics, ecological degradation, and human health intersect. This approach offers a replicable model for similar research in other fast-urbanizing regions and fills an important gap in current literature, which often treats water quality, land use, and public health as isolated domains.

Beyond its immediate context, the study also contributes to larger theoretical discussions in urban ecology, environmental justice, and sustainable development. The degradation of pond ecosystems in Barasat reflects deeper structural issues—unregulated urbanization, inadequate public services, and socioeconomic disparities—that resonate with global sustainability concerns. The results are directly aligned with several UN Sustainable Development Goals, particularly SDG 6 (Clean Water and Sanitation), SDG 3 (Good Health and Well-being), and SDG 11 (Sustainable Cities and Communities). The localized degradation of water quality in Barasat is thus a microcosm of broader urban environmental challenges, illustrating how ecological risks and social vulnerabilities are tightly interwoven.

12.8 Challenges and Limitations

Despite the comprehensive approach undertaken in this study, several limitations arose due to practical and resource-related constraints.

Firstly, the number of water sampling sites was limited to ten ponds. This was primarily due to limitations in funding and fieldwork resources, which restricted the scope of data collection and laboratory analysis. As a result, although the sites were selected to represent diverse land use types across Barasat, the sample size may not fully capture the broader spatial heterogeneity of eutrophication in the region.

Secondly, data collection was conducted only during the pre-monsoon season. This temporal limitation was influenced by both the academic calendar and financial constraints, which made year-round sampling unfeasible. Since eutrophication levels and runoff patterns can vary significantly across seasons—especially during the monsoon and post-monsoon periods—the findings reflect a seasonal snapshot rather than an annual trend.

Moreover, while the study included a household survey to assess perceived health impacts, it lacked access to formal longitudinal health records due to privacy regulations and institutional barriers. Furthermore, funding limitations prevented the incorporation of clinical testing or medical collaborations that could have strengthened the linkage between water quality and specific health outcomes. Thus, while self-reported health data provides valuable insights, the absence of clinical validation poses a limitation.

12.9 Recommendations

To build on the findings of this study and address its limitations, future research should consider expanding the number of water sampling sites and incorporating seasonal variations to better capture the dynamics of eutrophication throughout the year. Collaboration with public health departments and access to clinical health data would help establish stronger links between water quality and disease prevalence. Additionally, improved monitoring infrastructure and low-cost water testing solutions should be promoted at the community level. Sustainable land use practices, including the reduction of fertilizer runoff and the implementation of vegetative buffer zones around water bodies, can mitigate nutrient loading. Public awareness campaigns on safe water use and hygiene—especially targeting women who are more exposed—are essential. Lastly, securing institutional funding is recommended to support interdisciplinary research that integrates spatial analysis, water science, and public health for more robust and scalable insights.

12.10 Conclusion

The comprehensive analysis conducted in this study sheds light on the multifaceted dynamics influencing eutrophication in freshwater ecosystems and its implications for human health and socioeconomic activities. The findings reveal distinct eutrophication gradients, with Eutrophic and Hyper-eutrophic conditions observed in the sampled water bodies, driven by varying nutrient concentrations and algal biomass.

The investigation into land use and land cover (LULC) impacts on eutrophication highlights the significant role of different landscape elements. Homesteads with plantations, water bodies, bare soil, vegetation, and cropland emerge as key influencers, contributing to nutrient enrichment and biological productivity in aquatic environments. These insights underscore the need for integrated land and water management strategies to mitigate eutrophication risks.

The utilization patterns of the sampled water bodies reflect their vital role in fulfilling the domestic and economic needs of local communities. However, challenges such as pollution from anthropogenic activities, environmental degradation, and waterborne diseases pose significant concerns. The prevalence of skin ailments among individuals using pond water underscores the health impacts of deteriorating water quality, particularly affecting vulnerable groups like females.

The socioeconomic survey highlights gaps in healthcare access and awareness regarding waterborne diseases, emphasizing the importance of public health interventions and community engagement in water resource management. Sustainable practices, including pollution control measures, efficient waste management, and promotion of alternative healthcare options, are imperative for safeguarding human health and preserving ecosystem integrity.

This study aligns closely with the theme "*Wetlands and Urban Resilience: Ecological Studies—Analysis and Synthesis*" by highlighting the ecological degradation of urban ponds functioning as wetlands that are critical yet often overlooked in resilience planning. Through an integrated analysis of water quality, land use, and health outcomes, the research reveals how nutrient over-enrichment, driven by unregulated development and poor waste management, leads to eutrophication and public health risks. These findings underscore that urban wetlands are not just ecological assets but essential for regulating hydrology, supporting biodiversity, and protecting vulnerable communities. By demonstrating how their degradation undermines both ecological stability and human well-being, this study calls for embedding wetland conservation into urban resilience strategies to ensure long-term sustainability.

AI Disclosure The authors acknowledge the use of generative AI (ChatGPT by OpenAI) to assist in refining the language, structuring content, and formatting specific sections of this manuscript. The core ideas, data analysis, and interpretations are the original work of the authors. AI was used as a tool to enhance clarity and coherence in the presentation.

Potential Conflicts of Interest On behalf of all authors, the corresponding author states that there is no conflict of interest.

Informed Consent Informed consent was obtained from all individual participants included in the study.

Authors' Contribution All authors have contributed equally to this work.

Data Availability Statement The datasets during and/or analyzed during the current study available from the corresponding author on reasonable request.

Funding The research was not supported by any funding.

References

Aoun O, Benamara S, Aberkane L, Amrani N, Dahmoune F, Madani K (2013) Improvement of the diphenylamine-based spectrophotometric method for nitrate determination. Analytical Methods, 5(20):5830–5835

Bashkin VN, Howarth RW (2002) Modern biogeochemistry. Springer

Carmichael W (2008) A world overview—one-hundred-twenty-seven years of research on toxic cyanobacteria—where do we go from here? In: Cyanobacterial harmful algal blooms: state of the science and research needs. Springer, New York, pp 105–125

Constant KM, Sheldrick WF, Mundial B (1992) World nitrogen survey, No. 174. World Bank, Washington, DC, pp xxvi+−190pp

Diez-Quijada L, Benítez-González MDM, Puerto M, Jos A, Cameán AM (2021) Immunotoxic effects induced by microcystins and cylindrospermopsin: a review. Toxins 13(10):711. https://doi.org/10.3390/toxins13100711

Djodjic F, Bergström L, Grant C (2005) Phosphorus management in balanced agricultural systems. Soil Use Manag 21(1):94–101

Energy Information Administration (EIA) (2008) International energy outlook 2008. U.S. Department of Energy, Washington, DC. Available at www.eia.doe.gov/oiaf/ieo/pdf/0484(2007).pdf

Hunt ER, Daughtry CS, Eitel JU, Long DS (2011) Remote sensing leaf chlorophyll content using a visible band index. Agron J 103(4):1090–1099. https://doi.org/10.2134/agronj2010.0395

Iverson SJ (2009) Tracing aquatic food webs using fatty acids: from qualitative indicators to quantitative determination. In: Lipids in aquatic ecosystems. Springer, New York, pp 281–308

Jarvie HP, Smith DR, Norton LR, Edwards FK, Bowes MJ, King SM, Scarlett P, Davies S, Dils RM, Bachiller-Jareno N (2018) Phosphorus and nitrogen limitation and impairment of headwater streams relative to rivers in Great Britain: a national perspective on eutrophication. Sci Total Environ 621:849–862. https://doi.org/10.1016/j.scitotenv.2017.11.128

Jones RC (2020) Recovery of a tidal freshwater embayment from eutrophication: a multidecadal study. Estuar Coasts 43(6):1318–1334

Kadlec RH, Bevis FB (2009) Wastewater treatment at the Houghton Lake Wetland: vegetation response. Ecol Eng 35(9):1312–1332. https://doi.org/10.1016/j.ecoleng.2008.11.001

Khan FA, Ansari AA (2005) Eutrophication: an ecological vision. Bot Rev 71(4):449–482

Kuang QJ, Ma PM, Hu ZY, Zhou GJ (2005) Study on the evaluation and treatment of lake eutrophication by means of algae biology. J Saf Environ/Anquan Yu Huanjing Xuebao 5(2):87–91

Lucas I, Janssen F, van der Wel FJ (1994) Accuracy assessment of satellite derived landcover data: a review. Photogramm Eng Remote Sens 60:426–479

Lürling M, van Oosterhout F (2013) Controlling eutrophication by combined bloom precipitation and sediment phosphorus inactivation. Water Res 47(17):6527–6537

Mukherjee K, Das P (2018) Modelling the relationship between urban growth modes and the thermal environment—a case study of the Barasat Municipality, West Bengal. J Geogr Environ Earth Sci Int 17(2):1–19. https://doi.org/10.9734/jgeesi/2018/44047

Mukherjee K, Deb N, Roy AD, Dash P (2022) Impact of noise pollution on human health in Barasat Urban Area, West Bengal. In: Sustainable urbanism in developing countries, pp 375–393. https://doi.org/10.1201/9781003131922-25

Mukherjee K, Roy K, Pal B (2023) Gender parity in urbanizing city: a case study of Barasat City, India. Glob Soc Welf 10(4):313–333. https://doi.org/10.1007/s40609-023-00296-5

Mukherjee K, Saha S, Karmakar S, Dash P (2024) Uncovering spatial patterns of crime: a case study of Kolkata. Crime Prev Community Saf 26(1):47–90. https://doi.org/10.1057/s41300-024-00198-4

Oh JW, Pushparaj SSC, Muthu M, Gopal J (2023) Review of harmful algal blooms (HABs) causing marine fish kills: toxicity and mitigation. Plants (Basel, Switzerland) 12(23):3936. https://doi.org/10.3390/plants12233936

Ozimek T (1990) Aspects of the ecology of a filamentous alga in a eutrophicated lake. In: Trophic relationships in inland waters, pp 23–27. https://doi.org/10.1007/978-94-009-0467-5_4

Paerl HW, Otten TG, Kudela R (2018) Mitigating the expansion of harmful algal blooms across the freshwater-to-marine continuum. Environ Sci Technol 52(10):5519–5529

Petrucio MM, Esteves FA (2000) Uptake rates of nitrogen and phosphorus in the water by *Eichhornia crassipes* and *Salvinia auriculata*. Rev Bras Biol 60:229–236

Prasad R, Prasad S (2019) Algal blooms and phosphate eutrophication of inland water ecosystems with special reference to India. Int J Plant Environ 5(1):1–10. https://doi.org/10.18811/ijpen.v5i01.1

Ramesh HN, Krishnaiah MA, Supriya MM (2013) Role of moulding water content on the strength properties of red earth treated with mine tailings. In: Proceedings of the 1st International Conference on Innovations in Civil Engineering, 9th and 10th May, 2013, Emakulam, Kerala, India. Contributory papers, 45

Schindler DW (1977) Evolution of phosphorus limitation in lakes: natural mechanisms compensate for deficiencies of nitrogen and carbon in eutrophied lakes. Science 195(4275):260–262

Schroeder MJ, Shabanowitz J, Schwartz JC, Hunt DF, Coon JJ (2004) A neutral loss activation method for improved phosphopeptide sequence analysis by quadrupole ion trap mass spectrometry. Anal Chem 76(13):3590–3598

Sharma S, Jha PK, Ranjan MR, Singh UK, Kumar M, Jindal T (2017) Nutrient chemistry of River Yamuna, India. Asian J Water Environ Pollut 14(2):61–70. https://doi.org/10.3233/AJW-170016

Sharpley AN, Chapra SC, Wedepohl R, Sims JT, Daniel TC, Reddy KR (1994) Managing agricultural phosphorus for protection of surface waters: issues and options. J Environ Qual 23(3):437–451

THE 17 GOALS | Sustainable Development. https://sdgs.un.org/goals

Vitousek PM, Aber JD, Howarth RW, Likens GE, Matson PA, Schindler DW, Schlesinger WH, Tilman DG (1997) Human alteration of the global nitrogen cycle: sources and consequences. Ecol Appl 7(3):737–750. https://doi.org/10.1890/1051-0761(1997)007[0737:HAOTGN]2.0.CO;2

Wagner KJ (2004) The practical guide to lake management in Massachusetts: a companion to the final generic environmental impact report on eutrophication and aquatic plant management in Massachusetts. Commonwealth of Massachusetts, Executive Office of Environmental Affairs

Wetzel RG (1975) Limnology. W. B. Saunders Co, Philadelphia, p 734

Wold HOA (1985) Partial least squares. In: Kotz S, Johnson NL (eds) Encyclopedia of statistical sciences. Wiley, New York, pp 581–591

Xu S, Yi X, Liu W, Zhang C, Massey IY, Yang F, Tian L (2020) A review of nephrotoxicity of microcystins. Toxins 12(11):693. https://doi.org/10.3390/toxins12110693

13 Urban Wetland Unveiling the Socioeconomic Benefit and Sociocultural Significance: Reference to Five Different Case Studies

Roshan Baa

Abstract

Wetlands are green-blue infrastructure components that focus on various services. Their focus is on increasing the environmental and social sustainability of urban areas, which are generally considered the urban ecosystem of fundamental areas. To preserve ecosystems in urban areas, it is important to focus on protecting biodiversity while ensuring long-term survival for vulnerable species by contributing to the resilience of the urban ecosystem. There are five cases taken for the study: Tidal Wetlands Programme, New York State, USA, Eastern Kolkata Wetlands, Kolkata, India, Colombo Flood Detention Area, Colombo, Sri Lanka, Spatio-temporal Dynamics of Wetlands, Wuhan, China, Monavale and Surrounding Areas, Harare, Zimbabwe. The scope of the study is to help them understand the different dynamics of wetland and their effect on the local communities and other factors. Using chemical fertilisers and pesticides affects the global growth of different agricultural productivity. In the methodology part, the data analysis will be generated based on the code identified in the collected data.

Keywords

Habitat · Biodiversity · Environmental conservation · Ecology · Local communities · Natural habitat

13.1 Introduction

Wetlands, a critical part of our natural environment, offer a diverse array of services under the ecosystem. Urban wetlands, in particular, help to enhance the liveability of cities. They improve carbon sequestration and water quality and provide a habitat for a rich diversity of wildlife species, thereby reducing the effects of heat islands. However, the dynamic of wetlands in urban areas is under immediate threat due to challenges such as the loss of this crucial biodiversity, land-use change, and altered water regimes. This underscores the pressing need for immediate action to protect and restore these vital ecosystems. The growing dynamic of urbanisation in terms of density and area is affecting the disruption of natural infrastructure under the natural process of sustainability. It also affects the urban environment's ecological and social factors, leading to vegetation degradation. In most cases, urban areas have less water and less vegetation than the surrounding areas, and existing infrastructures also suffer from the adverse effects of climate change to preserve culture.

R. Baa (✉)
St. Xavier's College, Ranchi, Jharkhand, India

S. K. Chakraborty et al. (eds.), *Wetlands and Urban Resilience*, Sustainable Development Goals Series,
https://doi.org/10.1007/978-3-032-06881-1_13

The wetland is an important green-blue infrastructure component which focuses on a wide range of services. The focused optimal of the wetland is to increase the environmental and social sustainability of urban areas. The wetlands situated in urban areas are generally considered the urban ecosystem of fundamental areas. The wetlands of the urban areas basically provide suppliers and services to the human communities, which include leisure and recreation, carbon sequestration, and reduced air pollution. The urban wetland shows its own microclimate, which acts mainly cooler than the surrounding areas; it also helps in improving the environment and quality of life, which directly leads to sustainable and urban development (Alikhani et al. 2021). Importantly, the wetlands play a pivotal role in protecting global biodiversity and, significantly, in shaping the socioeconomic fundamentals of the environment. The wetland is considered a hot spot of biodiversity, as it is also considered an endangered ecosystem globally, and they are also ecologically sensitive under a compatible landscape (Fig. 13.1).

The unique characteristics of urban wetlands are diverse and fascinating. They can be classified based on their origin, water source, dominant species, sediment and soil features, geographical location, and dominant species. Understanding these characteristics is crucial for effective conservation and restoration efforts. The ecological benefit which influences the landscape depends on its structure, and several environmental consequences influence it, including habitat structure and function (Das et al. 2022). Urban wetlands act as naturally absorbing sponges that absorb excess water from rainfall and reduce the flooding risk in urban areas. It carries the ability to store and release water slowly, helps mitigate the danger of flood, and saves communities. Wetlands significantly help to improve water quality by filtering sediments and pollutants from runoff before they reach the larger bodies of water. The process of natural filtration not only helps enhance the health condition of living entities of aquatic ecosystems but also significantly helps provide clean drinking water to urban residents, which also helps reduce the need for costly water treatment.

Urban wetlands are trying to preserve the ecosystem's uniqueness under the urban landscape while overlooking the hustle. Biodiversity conservation also comes under the social and economic factors, which help support the rich diversity of species of animals and plants, which leads to a serving dynamic for the habitat migrant. Wetlands mainly represent ecosystems which are rich in biodiversity and function according to the essential buffering systems while contributing to flood mitigation and air quality regulation and serving as the development and breeding grounds of aquatic species (Syaban and Appiah-Opoku

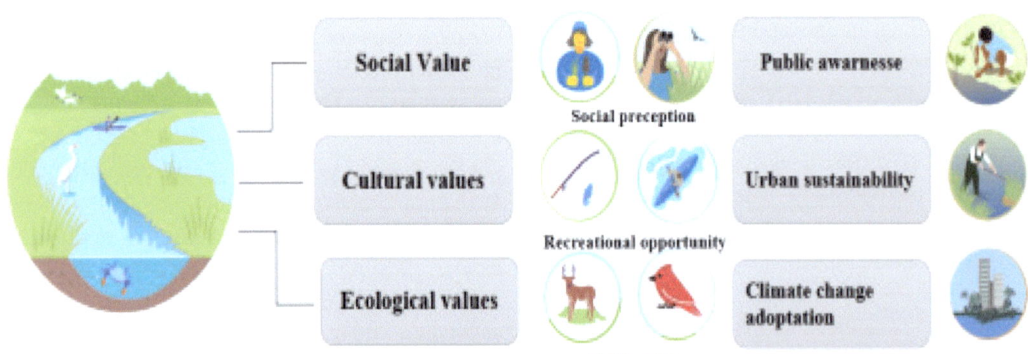

Fig. 13.1 Cultural and ecological values offered by the urban wetlands. (Source: Alikhani et al. 2021)

2024). In preserving ecosystems in urban areas, it is important to focus on protecting biodiversity while ensuring long-term survival for vulnerable species by contributing to the overall resilience of the urban ecosystem. These lands significantly offer valuable recreational opportunities for different urban residents, which works by providing spaces for activities like hiking, fishing, etc. The recreational activities help promote mental and physical well-being and foster a sense of connection to the community and nature of the urban dwellers.

Urban wetlands help enhance the attraction of tourism and property values while contributing to the local economy. The conservation and restoration of wetland projects stimulate economic growth and create job opportunities while demonstrating the socioeconomic importance of urban areas. This ecosystem helps in these ecosystems, repositories of biodiversity, and offers habitats for numerous endemic and endangered species. Wetlands are generally dubbed the "kidneys of ecosystems" because of their pivotal role in the ecosystem (Yang et al. 2024a, b). The role holds an important and irreplaceable position in the global ecological fabric, under its diversity and functionality, which greatly helps enrich the global biosphere. Beyond this role, they offer humanity invaluable ecological services while ranging between climate regulation and water purification and flood mitigation. Economically, several wetland areas had morphed into traditional medicine hubs, fisheries, and tourism, proffering resources of wealth such as medicine, food, and timber. The ecological shifts of wetlands have garnered considerable attention from policymakers.

Several urban wetlands hold significant historical and cultural value while representing indigenous communities' traditional fishing grounds, sacred sites, and ancestral lands. Preserving the wetlands is important for maintaining the heritage and honouring the cultural identity of past generations. The urban ecosystem includes cultivated land, wetlands, urban forests, parks, streams, sea, etc. (Mohtat and Khirfan 2023). Urban wetlands also serve as outdoor classrooms, which help to provide valuable educational resources for schools, community groups, and universities. Through hands-on learning, students can gain a deeper understanding of ecology, environmental stewardship, and conservation while being empowered towards sustainability. Poets, writers, and artists have mainly inspired the beauty of the wetland by serving as a source of inspiration and creativity. Due to submergence, the damming process destroys social habitats ecologically and upstream; downstream, it impacts the whole habitats by reducing the flow towards floodplains and wetlands (Ekka et al. 2020). Consequently, it directly leads to a direct decrease in the productivity of the floodplain. The urban wetlands are valuable ecological habitats and integral components of urban landscapes.

In the process of unveiling, they gain a deeper appreciation of the major role wetlands play in raising sustainable development and emerging urban resilience. The aim of the study is to unveil the socioeconomic benefits and sociocultural significance of wetlands, and the objectives of the study are:

- To evaluate the Socioeconomic Benefits of Urban Wetlands and discuss the Role of Urban Wetlands in Climate Change Mitigation and Adaptation
- To recommend the socioeconomic benefit and sociocultural significance of wetlands

This study investigates the different multifaceted contributing systems of urban wetlands under the urban environment's ecological and socioeconomic aspects. While focusing on the geographical dynamic, the study's findings bring out a scope that delimits the wetland into defined areas to facilitate the whole focus and create an understanding. The socioeconomic benefits help the research investigate the qualitative and quantitative impacts and benefits on the economy, infrastructure, and local communities. The sociocultural landscape helps the researcher to develop clear aspects of understanding the dynamic of the base, which contributes to collective memory, sense of place, and community identity. Moreover, if a researcher is researching the different socioeconomical and geographical factors of wetland, this study helps them to understand the different dynamics of wetland and its effect on the local communities and other factor.

13.2 Rationale of the Study

The study takes flight to study the local perceptions of the ecosystem services provided by wetlands and work to overlap them with different spatially explicit biodiversity indicators and socioeconomic indicators.

It also aims to evaluate the spatiotemporal characteristics of the cultural ecosystem services and their relationship to the factors of landscape, and to identify different missing cultural values and oversubscribed natural values by assessing the dynamic impacts on the sustainable changes. It also focuses on studying the land use and land cover changes that impact the ecosystems and other economic values of the wetlands of different adjacent communities and to identify the services and disservices of the ecosystem by associating it with the open space system and exploring the values under the socioeconomic spectrum.

The study evaluates the priorities that help protect the open space by enhancing the ecosystem services and gives a proper understanding of the worth of wetlands by accounting for regulating, cultural and provisioning services. It also characterises the assessment of the perceptions of other groups of residents and leaves with the discussion of the efficiency of natural treatment systems for managing wastewater in different developing countries. Finally, the findings integrate nature-based solutions, ecosystem services, well-being, and traditional ecological knowledge in urban planning to foster resilient and sustainable communities.

13.3 Materials and Methods

The research methodology describes how the research is conducted. This information mainly allows the reader to check whether the approach is accurate. Research methodology is an important part of research papers, as it is useful to show the necessary methods and requirements for data collection. This research will also use effective research methods, highlighting the point and indicating how the data provided is important, relevant, and appropriate. These methods will be useful for collecting data and conducting research properly. Systematic review includes the substantive findings and theoretical and methodological contributions (Newman and Gough 2020). These findings and discussion will help fulfil research objectives. Secondary sources of data are useful to conduct proper discussion and result analysis. The secondary sources will be inspected to collect data, and secondary qualitative data will be used to conduct this research and provide findings.

The research methodology provides an action plan and the grounding logic that help define the research problem and effectively analyse the collected data. Marketers and researchers must understand and choose the main performance metric for using the methodologies. The methodology allows the understanding and discussion of a constructive approach to a subject and gives a scientific sound of the findings. Proper research and data collection help to understand endogeneity better and improve communication between macro and micro researchers (Hill et al. 2021). The secondary qualitative data is the data that the other researcher collects, and the secondary data is already available and analysed in different places. The secondary data collection helps save time and resources and helps individuals build knowledge by following the company's expertise. The secondary data helps the data collector understand the data very well and also helps to know about the historical event and where the data is collected.

13.4 Result and Discussion with Case Studies

13.4.1 Case Study 1: Tidal Wetlands Programme, New York State, USA

The Tidal Wetlands Programme in New York State, established in 1973, aims to manage and protect tidal wetlands and the state's coastline. These wetlands are important factors for coastal resilience, biodiversity, and water quality, and this programme includes some efforts related to habitat restoration, public education and rigorous permitting. It focuses on supporting the measure of sustainable development which also contributes to preserving natural resources, ensuring different activities like pollution, construction and dredging are regulated carefully. The tidal wetlands helped to avoid $625 million in flood damage from Hurricane Sandy, and it is also estimated that the salt marshes helped in reducing annual flood losses by 16% (Weis et al. 2021). The initiative collaborates with the stakeholders and local communities to adapt to the climate change impacts and foster the stewardship of the environment, which ultimately helps in enhancing the sustainability and ecological health of New York's coastal areas.

13.4.2 Case Study 2: Eastern Kolkata Wetlands, Kolkata, India

The Eastern Kolkata Wetlands in India, a Ramsar site, is a serving point of economic and ecological resources. It is spanning near over 12,500 hectares, and the wetlands naturally treat the crucial dynamic of Kolkata's sewage and also very deeply support local fisheries and agriculture while contributing to thousands of livelihoods. This wetland mainly acted as a buffer which helped against the flood and maintained the microclimate of the city. The efforts of the conservation focus on preventing and maintaining urban encroachment and focus towards sustainable management. This wetland does not have any catchment area, which is the reason the hydrological condition of this wetland is quite different from others (Mondal et al. 2022). The initiatives based on the conservation focus also include policy advocacy, community engagement, and scientific research which help in balancing the ecological integrity and the development process. The multifunctional role of wetlands highlights their importance for environmental protection and urban sustainability.

13.4.3 Case Study 3: Colombo Flood Detention Area, Colombo, Sri Lanka

The Colombo Flood Detention Area in Sri Lanka has a critically designed infrastructure project which helps to mitigate the risk of urban flooding. While located under the region of Colombo metropolitan, this channel of the wetlands network and the absorption of retention basins help in regulating the stormwater after heavy rainfall. Through enhancing the flood resilience of the city, different detention area helps to reduce the economic losses and prevent damage to the infrastructure. The chances of the flood basin capacity in Colombo were reduced by 30%, and as a result, there are several infrastructure developments under which the wetland loss rate in Colombo city reached 1.2% per annum (Herath et al. 2023). The integrated project of ecological conservation under the urban planning aims to preserve the natural habitats while accommodating the plan to address the flood risks. The area which provides support to the biodiversity and maintains a recreational space. The ongoing management also included the procedure of community involvement, maintenance, and monitoring to ensure the effectiveness of the systems.

13.4.4 Case Study 4: Spatiotemporal Dynamics of Wetlands, Wuhan, China

The spatiotemporal dynamics of wetlands in Wuhan, China, reflects changes due to the rapid growth in climate variability and urbanisation. Over the last decades, infrastructure development and extensive land reclamation led to the fragmentation and reduction of wetland areas. The water levels due to the seasonal variations, which help influence the monsoonal patterns due to the further impact of wetland ecosystems. The city has a humid continental subtropical monsoon climate, which has an average annual temperature of about 16 °C, and its annual precipitation is 1200 mm (Wang et al. 2022). Research and different monitoring efforts help in employing the GIS tools and satellite imagery for tracking those changes and informing the conservation strategies. Restoration projects also aim towards the procedure of enhancing flood control, water management, and biodiversity, while these policies also focus on balancing urban growth under ecological sustainability, while emphasising the importance of the wetlands for maintaining environmental health in Wuhan.

13.4.5 Case Study 5: Monavale and Surrounding Areas, Harare, Zimbabwe

The Monavale Vlei in Harare, Zimbabwe, has a critical and diverse wetland area which plays an important role in biodiversity support, water purification and flood control. Efforts lead to Community-led conservation, which is mainly spearheaded by local communities and groups such as the "Conservation Society of Monavale (COSMO)", which focuses on the practices of habitat restoration and sustainable management. All these initiatives also include activities like promoting the measure of eco-friendly agriculture, removing several invasive species and replanting all the different native vegetation. During the wet season, the daily temperatures vary as the maximum is 29 °C and the minimum is 18 °C, the mean daily temperatures stand at 25 °C, and the daytime temperatures measure at 35 °C (Sharai et al. 2020). The wetland helps in raising awareness and serves as an educational site regarding environmental stewardship. Despite the several urbanisation pressures, the Monavale Vlei always focuses on exemplifying the collaborative measure of a successful community in a way to protect the natural resources and maintain the urban setting ecological balance.

The urban wetland helps enhance the role of the environment and creates a dynamic of suitability within the urban areas by providing different services to mitigate the impact of climate change. A wetland is an area of land that is distinguished by the occasional or continual presence of water, where the whole water table is very close to the soil surface or a place which has land and is shallowly covered by water (Zhang et al. 2024). The waterland serves as the sacred site that has been passed down through the whole generation (Fig. 13.2).

Though local classifiers are conducted among outperformed global classifiers, which are slightly changed, the whole global classifiers are also illustrated under the plausible accuracies in five different national wetland reserves with an overall mean accuracy of *86.40%* (Liu et al. 2023). These results also show the reliability of accurately mapping under the wetland categories in different regions (Fig. 13.3).

Urban wetlands are mainly considered prime real estate for several development projects due to their proximity to urban centres. Consequently, the wetlands also face relentless pressure from the process of urbanisation, which directly leads to degradation and habitat loss. Urbanisation results in a continuous loss, degradation, and homogeneity of wetlands (Yang et al. 2024a, b). Urbanisation threatens the wetland, and the lack of wetlands in or surrounding it directly affects nature and its habitat. Wetlands help to improve water quality by filtering sediments and pollutants, and the lack of wetlands affects the whole geographical landscape.

The Cultural Ecosystem Services (CES) help provide well-being and health through urban parks (Song et al. 2023). Urban sustainability is the enduring progression that aims to cultivate green spaces in urban areas by fostering the urban dwellers' well-being and providing facilities for nature's actual benefits (Fig. 13.4).

The benefits and values of the wetland are overlooked; the residing residents act as the vicinity, which has a great and direct impact, as it is also expedient for authorities to implement laws for protecting the wetlands during the process of development. The livelihood process includes activities such as agriculture and fishing.

Fig. 13.2 Number and locations of reviewed urban wetlands worldwide. (Source: Alikhani et al. 2021)

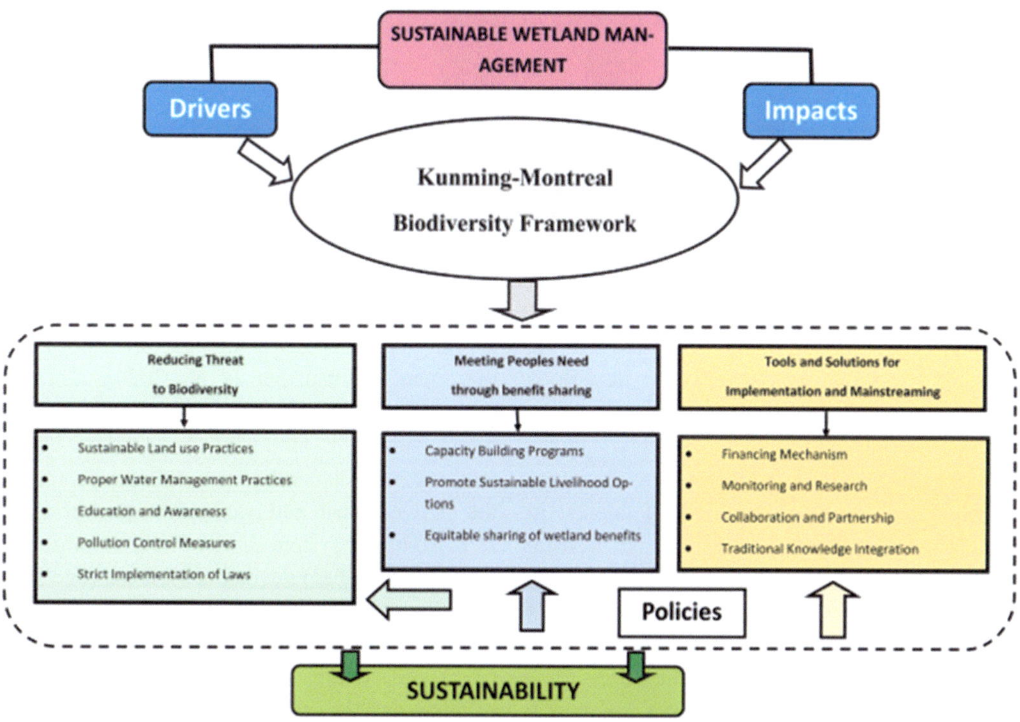

Fig. 13.3 Dynamics of changes in land cover and land use. (Source: Krivoguz 2024)

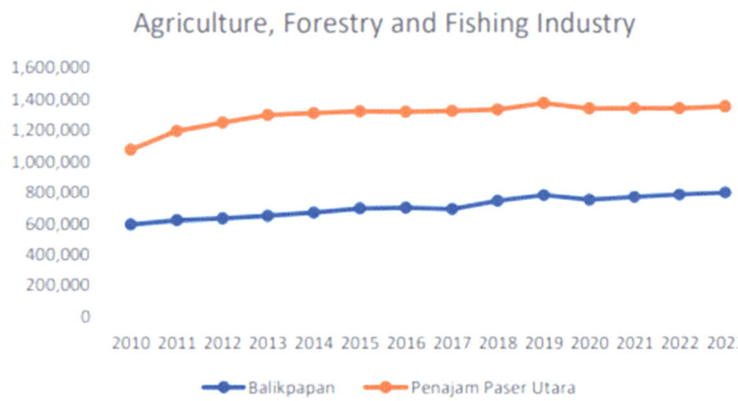

Fig. 13.4 Agriculture, fishing, and forestry industry in the IKN area. (Source: Syaban and Appiah-Opoku 2024)

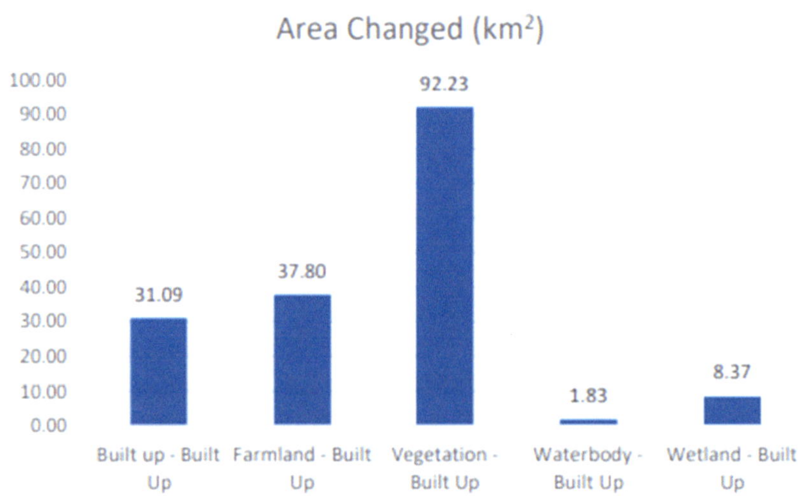

Fig. 13.5 Area changed as per different development processes. (Source: Syaban and Appiah-Opoku 2024)

Displacement is collectively identified as the forced residential movement due to hazardous changes, political or financial, in the different immediate surrounding environments (Miller 2020). Displacement pressure is due to the ongoing environmental clean-up and the interest in renewed development. Dismantling of coastal wetland fishing areas, ecosystems, and agricultural areas used by different local communities creates a bent on the development of capitalistic on both physical and social instability (Fig. 13.5).

The displacement causes social gaps, which impact people's well-being and mental health. It generally weakens social connections and creates loneliness, which is generated by the process of isolation and separates into the process of displacement (Matsushita et al. 2024). Through utilising collaborative initiatives, the potential transformation is gradually characterised by the low and high demand, which helps in protecting the native forest and wetland (Ghasemi et al. 2024). The ecosystem services help to value unit areas and also help develop a view on vegetation loss (Fig. 13.6).

Agriculture is the main cause of pollution in rivers and streams, which are the second main wetland, and the third is the lake source. Approximately 60–90% of phosphorus travels mainly with soil and affects the wetlands (Akhtar et al. 2021). Agriculture is an accountable phenomenon

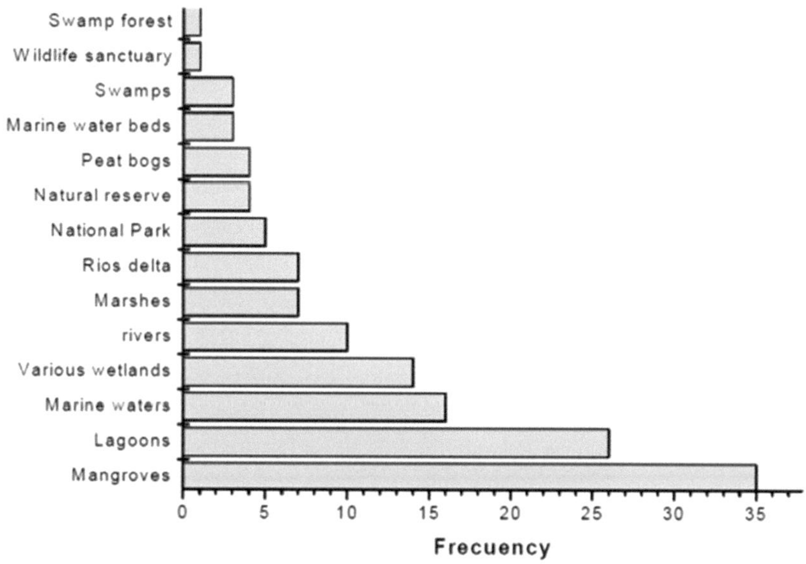

Fig. 13.6 Types of affected wetlands globally. (Source: Ballut-Dajud et al. 2022)

to the huge amount of groundwater and surface water pollution through nitrogen.

The global policies of ineffective development are creating an impact on the dynamic of wetlands due to their nature of uncontrollability or due to anthropogenic activities. Degraded wetlands and the installation of wetlands help in enhancing the evapotranspiration and attenuation of sulfur water at the time of heavy rainfalls (De Vito et al. 2022). The complete natural urban green spaces are wetlands, forests, and areas of natural origin (Pukowiec-Kurda 2022). The wetlands of the urban areas provide suppliers and services to the human communities, which include carbon sequestration and reduced air pollution. Urban wetlands can store and release water slowly, help mitigate the danger of flooding, and save communities.

Environmental education about its different phenomena helps create a conservation strategy that also impacts the synergistic space and another coverage system (Mahdianpari et al. 2020). Its work is the foreground of the value, practices, experience, and local knowledge under the place-based setting through encouraging the groups to be marginalised under the interferences of productivity.

Human perceptions and relations with wetlands have significantly changed over time and also vary from place to place (Maltby 2022). However, despite the different positive factors but still, it still gets overshadowed by socioeconomic pressures and negative perceptions.

Urban wetlands enhance the attraction of tourism and property values while contributing to the local economy. The global perspective of wetlands is based on the consequences of ecology that generate threats to the freshwater generated from wetlands (Bănăduc et al. 2022). Fragmentation of freshwater wetlands and ecological loss are problematic in nature in numerous ways.

The economic value of the floodplain wetlands generates billions of USD in ecosystem services for supporting the local inhabitants (Camacho-Valdez et al. 2020). This study discusses the urban wetlands' sociocultural significance and socioeconomic benefits and draws insights from case analyses and various studies. According to Guo et al. (2023), the dynamics of cultural ecosystem services provided by urban wetlands, such as recreation and aesthetic appreciation, are closely tied to the landscape, a fact often reflected in social media activity. The adjacent lake areas help reconstitute into natural riparian areas, which also include meadows and wetlands based on the different humidity levels (Viti et al. 2023). Different studies of wetlands observed that the trapping and hunting activities of wetlands basically range between freshwater lakes, aquaculture ponds, and coastal mudflats (T.D.W. et al. 2023). This observation is taken from places like Cambodia, Vietnam, Myanmar, etc. The conservation of wetlands helps in stimulating economic growth and creates job opportunities while demonstrating the socioeconomic importance of urban areas. The Okavango Delta World Heritage Site highlights the need to balance natural and cultural values to ensure the conservation of sustainable practices (Keitumetse et al. 2023) (Fig. 13.7).

Wetlands generally act as natural water filters, removing pollutants from water bodies and improving water quality. Wetland ecosystems are important in providing, regulating, and provisioning cultural services, and it is essential for these areas' sustainable conservation and management (Ramachandra et al. 2021). Wetlands help in mitigating climate change impacts by sequestering carbon dioxide which impacts the cycle of global carbon cycle significantly. The continental wetlands are more affected than the coastal and marine ones (Ballut-Dajud et al. 2022). Under the benefits of urban wetlands, policymakers adopt an integrated approach that considers the social, environmental, and economic dimensions. The wetlands also serve as educational opportunities, spiritual practices, and cultural activities sites. As per Wessels et al. (2021), the complex nature perceptions within the local communities found that wetlands are valuable for their social and cultural importance and ecological functions. The natural wetlands, open spaces, and different urban parks are used for illegal dumping and building rubble by the broader community, mainly in the urban areas. In the constructed wetlands, treatment is applied for the liquid sewage sludge, which results in the dried sludge production that can also be valorised (Tsatsou et al. 2023). That helps to gain the pressing need for immediate action to protect and restore these major landscapes and ecosystems.

The knowledge of traditional ecology helps encompass the practices and wisdom of local and indigenous communities, which also helps generate sustainable management in wetlands. As per Kiddle et al. (2021), the integration of the knowledge of traditional ecology under the modern design for the process of enhancing of services of the ecosystem also helps in promoting the well-being of the local communities. Wetlands are an effective factor for water purification, pollution reduction, and flood risk reduction (Cui et al. 2021). The coastal wetland also acts as a buffer for flood and sea level rise, where

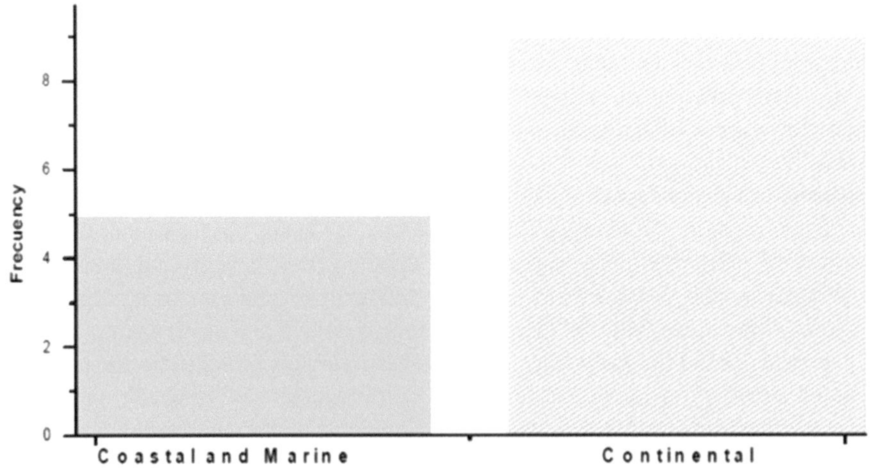

Fig. 13.7 Wetlands classification which is affected globally. (Source: Ballut-Dajud et al. 2022)

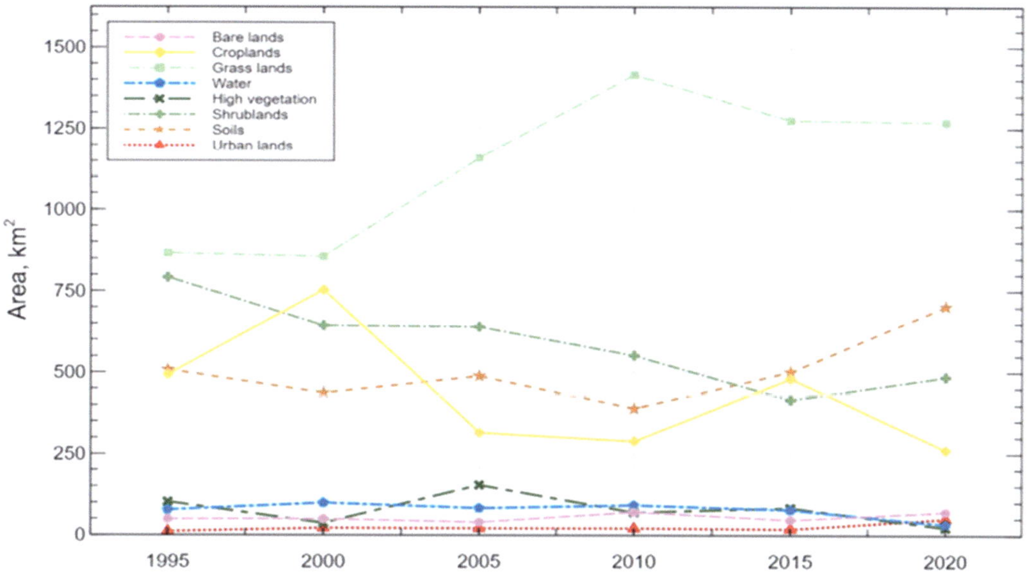

Fig. 13.8 Dynamics of changes in land cover and land use. (Source: Krivoguz 2024)

the urban wetlands are constructed to reduce the water pollution of urban areas. The constructed wetlands are generally used in urban areas to treat rainwater, sewer overflow, and outflow treatment for the existing plant's wastewater treatment, including contamination treatment through emerging concerns (Fig. 13.8).

Despite their various benefits, urban wetlands face challenges, including climate change, pollution, and encroachment. Strategic planning and effective management are essential for addressing these issues and ensuring the sustainability of the wetlands ecosystem. According to Odaro et al. (2023), the impact of land cover and land use changes the ecosystems of wetlands and creates stress on the need for comprehensive planning of the environmental and management strategies for enhancing the sustainability of wetlands. The increase in urbanised areas influences the decline of natural ecosystems like wetlands and forests (Krivoguz 2024). The most affected wetlands are marine waters, mangroves, and lagoons. Some drivers are political influence, population growth, economic opportunity, and public service accessibility (Fig. 13.9).

The wetland's ecosystem is vital for human life and biodiversity, and the community's involvement is important in managing urban wetlands. Tien et al. (2024) evaluate the role of community-based ecotourism (CBET) in promoting sustainable development. This study evaluates the significance of CBET, which can significantly enhance the efforts of environmental conservation and provide an economic benefit to the local communities. It also evaluates the importance of community participation and strategic planning in initiatives for sustainable development. Agriculture is the major cause of the loss of wetlands, followed by urbanisation and industry development in a lower percentage (Ballut-Dajud et al. 2022). Determination of the wetland condition and its evolution are among the important factors evaluated in the global context. The soil formation, food production, services of gas regulation, and climate change are directly connected with the health and abundance of ecological land, which also includes forests, wetlands, and grasslands. The processes of socio-economic proxies also include the dynamics of governmental policies, economic structure and demographic valuation of the society (Fig. 13.10).

The cultural heritage in the delta landscape explores the policies of existing heritage and its

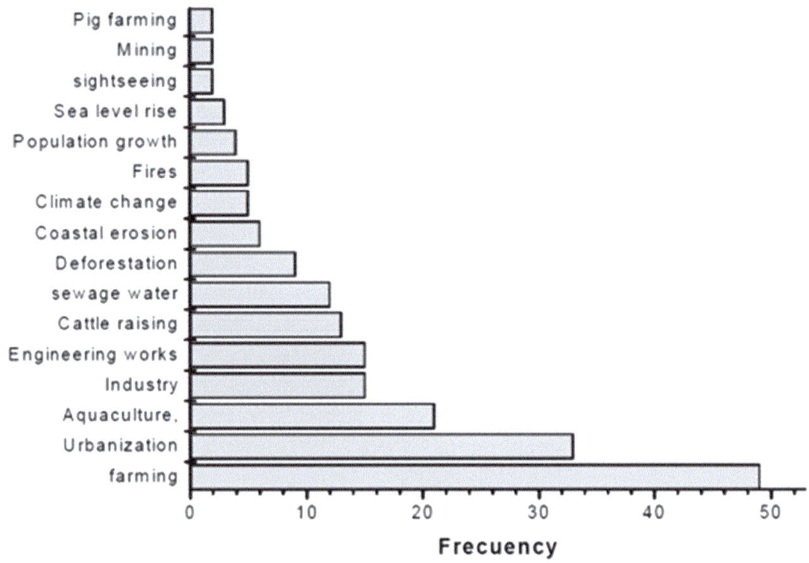

Fig. 13.9 Incident factor of loss of wetlands globally. (Source: Ballut-Dajud et al. 2022)

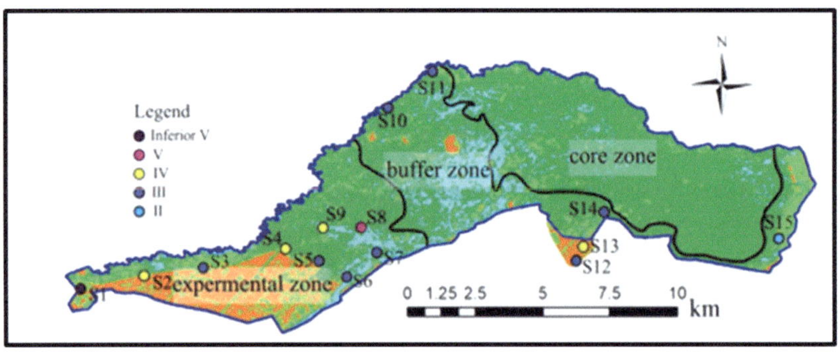

Fig. 13.10 Water quality grade of the River Qixing at the sampling sites of the wetland. (Source: Cui et al. 2020)

framework, which considers the values of the diverse sites. De et al. (2022) discuss that the natural treatment efficiency of systems manages the components of wastewater in developing countries. It also helps to emphasise the policymakers' need to take initiatives which support the solution of sustainable wastewater management. These initiatives extended into the broader context of the management of wetlands while ensuring the policies are inclusive and supportive of ecological and community needs. 58% of the loss of wetland habitat is mainly due to farmland expansion, along with the different factors of agronomic, which are mainly being and structuring under the influence of principal drivers of change in land use of wetland areas (Cui et al. 2020). The degradation and destruction of ecological land help spread the phenomenon in the world under a predominantly demand for urbanisation and the expansion of agriculture to meet the needs of the growing population, which also directly leads to substantial biodiversity loss.

The hydrological factor influences the quality of wetland reserves, whereas the landscape pattern affects the quality pattern of habitat (Lin et al. 2024). The cultural value of wetlands is gaining interest through the implementation goals of sustainable development, which demands a holistic approach towards resource conservation. The decentralised system, apocopate for the urban fringes and suburban zones, is based on the factor of inflow concentration, which is also multifaceted in the irrigation process of wetland discharge. Wetlands are perceived as not constituting a benefit under a good portion (Targetti et al. 2021). These landscapes are shaped by the interaction dynamics among the sociocultural and natural forces under which the population perceives the assessment of representative, which also contributes to social well-being. It is categorised among the broad and encompasses natural elements like wetlands, sociocultural components like human settlement, and other forms (Tan et al. 2024). These angles of landscape and their combination have emerged to shape the different forms of landscape. In the process of adopting the approach of integration, the consideration of diverse values of different wetlands helps in ensuring their sustainable use and conservation for future generations.

The utilisation process of technologies for decentralised treatment, which includes artificial wetlands and bioreactors, is the major dynamic for wastewater treatment (Félix-López et al. 2023). Wetlands are temporarily or permanently inundated with artificial and natural waters, which also consist of peatland, swampy, or brackish areas used for agriculture (Noortyani et al. 2023). From an economic perspective, urban wetlands provide invaluable services to the ecosystem. Ecological connectivity is increasingly recognised in the process of resiliency of long-term coastal systems under diverse flora and disturbance benefits of biodiversity (Pittman et al. 2022). The wetlands also provide an extended range of ecosystem services and goods, but they are also at risk an unsustainable use (Azadi et al. 2021). The wetlands are situated around or in urban areas, and they are considered important for the species as they include roosting, breeding, and feeding (Aticho et al. 2023). The process of urbanisation threatens the whole biodiversity very badly in the planning of the rural to urban conservation; the species are affected more badly as the wetlands are used in the process of urbanisation.

The wetland helps reduce the event volume peak flow and eliminates the flooding surfaces, which benefits informal housing areas (Irvine et al. 2021). The wetlands are examples of the production of the socioecological landscape, which is mainly lifted into the area of wetland ecosystem; it also exposes the different disturbances of activities like flooding and agricultural activities (Salas 2023). Urban wetlands have the potential to contribute to sustainable urban development and green infrastructure initiatives (Fig. 13.11).

Through recognising the different valuable and adaptable benefits of those urban wetlands and prioritising their process of restoration and conservation, urban areas like cities use valuable ecosystems' full potential by creating a healthier, more resilient and more sustainable urban environment for future and current generations. The

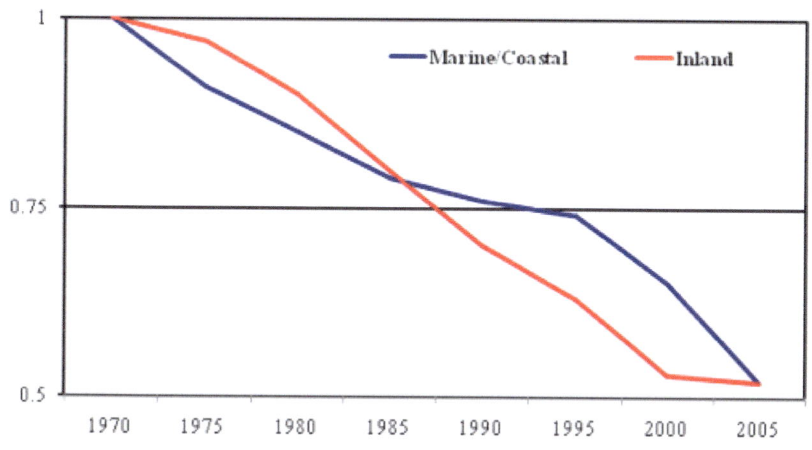

Fig. 13.11 Index of global wetland extent. (Source: Dar et al. 2020)

process of sustainable management of wetlands is important for ecosystems that offer an ecological array of functions that help sustain the livelihoods of people worldwide. Index of Wetland Extent estimated a *40%* decline in the coverage of coastal/marine and inland wetland ecosystems in the last 40 years, and this is mainly due to degradation and fragmentation (Dar et al. 2020). Accordingly, wetlands are the hot spots for the entire biodiversity, and they are also the most endangered ecosystems globally and ecologically sensitive to a compatible system. Its characteristics have offered wetlands great diversity due to their developing chemical and water regime, sediment and soil features, origin, geographical location, and dominant species (Fig. 13.12).

Throughout the process of preserving ecosystems in urban areas, it is important to focus on protecting biodiversity while ensuring long-term survival for vulnerable species by contributing to the resilience of the urban ecosystem. The dataset of Global annual wetland_FCS30 splits the wetlands into seven different wetland subcategories, which are three coastal and four inland subcategories (Zhang et al. 2023). Wetlands are distinguished by the occasional and continual presence of water, where the water table is close to the soil surface, and the land is entirely covered with water (Zhang et al. 2022). Through purifying water, the wetlands tend to reduce the need for expensive water treatment, which directly saves the regulatory government money while saving water resources for ecological health and human consumption. Wetland is the boundary of aquatic and terrestrial ecosystems, which is strongly reliant but also has distinct characteristics (Yaseen and Chalotra 2022). Owing towards the increased rate of anthropogenic and anthropogenic activities driven by the changes in the natural processes, the wetlands are starting to degrade at a rate of alarming rate, which affects human health and the quality of water.

The function of the generated blue-green structure is important to alleviate the pressure of the environment of urbanisation and enhance the well-being of urban residents (Lourdes et al. 2021). Urban residents use natural wetlands as dumping grounds for household chores and building rubble for the broader community, mainly in urban areas. The natural ecosystem is the set of interactions between species and their nonbiological and local environment, which also function together to sustain life (Khirfan et al. 2021). Several inevitable ecosystem losses result in the direct disappearance of the nonbiological and biological natural environment, like wetlands, soil, trees, forests, rivers, etc. Your social-cultural valuation helps in moving beyond the monetary technique, which relies on direct, live experience with the surrounding environment (García Llorente et al. 2020). Moreover, wetlands in urban areas contribute to community social cohesion and well-being by providing space for relaxation, social interaction and recreation. The measure of ecosystem-based adaptation includes growing and protecting urban green cover to improve the local climate and maintaining the services of an ecosystem like water supply and recharge (Singh et al. 2020). Through the community's involvement in conservation efforts

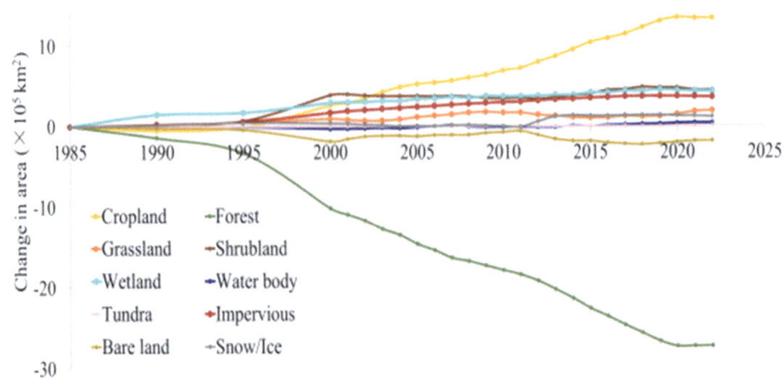

Fig. 13.12 Distribution of global intensity of land-cover-change from 1985 to 2022. (Source: Zhang et al. 2023)

and wetlands restoration, cities can promote a sense of responsibility towards the environment and empower citizens to take action to protect their ecosystems. Habitat destruction and invasive species further aggravate the degradation of wetlands in urban areas, leading to ecological imbalance and biodiversity loss. The schemes of afforestation practise the ecosystem for easy access to restoration areas (Sigman and Elias 2021). The ability to recreate the supplement and degraded ecosystem services challenges the wetland ecologically for lost wetlands. This cumulative impact improves the integrity and flexibility of the wetland ecosystem and benefits communities and cities.

13.5 Limitations of the Study

The study consists of a secondary systematic review, which is not based on the method of assembling the information from existing data. This research only contains the information already discovered, and as a result, this study lacks of having new data. The method of primary data collection includes interviews, observations, and queries from the database that help gather new data and insights, but this research does not include such type of information. Apart from that, there is a lack of information on urban wetlands changing dynamic globally, and the majority of articles are country- and region-based and contain similar findings, which have also limited this study.

13.6 Recommendation

13.6.1 Integrated Management and Planning Approaches

Implementing these approaches might help integrate urban planning and wetland management systems to ensure the long-term sustainability and conservation of wetlands. The incorporation of the objective of the conservation of wetlands is important in land-use policies, zoning regulations, and development plans. Integrating the wetland consideration in a border aspect can help this framework of urban planning balance the need for development with the goal of environmental conservation. Taking approaches and planning about the conservation of wetlands through the process of direct management should enable the retention of species diversity and enable the actual use of the wetlands and their resources (Tomaškinová et al. 2021). These approaches also help in ensuring the wetland's ecosystem protection while meeting growing urban population needs.

13.6.2 Education and Community Engagement

Engaging the local communities in the effort of wetland conservation through decision-making, outreach to the communities, and an education system can be helpful and also important for fostering support. Implementing community-based programmes, educational work programmes, and volunteer initiatives might help empower the local residents to take an active role in the wetland management system. Sociocultural values concepts try to access the ecosystem services group and community to engage in social-cultural, economic, and ecological services (Yao et al. 2023). Prioritising wetland sustainable management programmes should help to ensure the continued existence of the ecosystem and its benefits for future generations.

13.6.3 Ecological Rehabilitation and Restoration

Investing in ecological rehabilitation and restoration programmes and projects can help restore the degraded habitat of wetlands and improve the ecosystem and biodiversity. This process should involve restringing the hydrological connectivity, reintroducing the native vegetation and removing the invasive species of degraded wetland areas. Rehabilitation and management strategies, which are based on the principle of scientific knowledge, help in enforcing effective measures for regularising development activities (Ragavan et al. 2021). Restoring urban wetlands' ecological integrity might help the place provide needed ecosystem services like carbon sequestration, water purification, and flood mitigation, integrated and long-term ecosystem protection.

13.6.4 Strengthened Regulatory and Legal Frameworks

Enhancing the regulatory and legal framework for the conservation of wetlands can help to provide the necessary incentives and protection under an effective management system. This also might include enforcing and enacting laws that designate the wetlands as protected areas and establishing regulated activities and buffer zones that also impact the wetlands' ecosystem. Integrating specific rights for environmental defenders helps to create accountability and enhance environmental rights, helping to protect the ecosystem (Sam et al. 2023). Providing tax breaks, financial initiatives, and other beneficial breaks for landowners might help those who are fiercely engaged in the conservation process of wetlands encourage voluntary compliance.

13.7 Conclusion

The urban wetlands are an essential ecosystem component that benefits communities and cities worldwide. The dynamic of water purification and flood mitigation to cultural enrichment and biodiversity purification, these areas help enhance sustainability. The cities are struggling with the challenges of habitat destruction, urbanisation and climate change, and the urban wetlands are getting under the most critical dynamic. This paper explored the different dimensions of urban wetlands and their sociocultural significance and socioeconomic contribution. The study also highlights the natural buffer of the urban wetlands against flooding, and by absorbing the extra water, the wetlands help reduce the urban inundation risks. The role of the wetlands is to improve the water quality by filtering the contamination and pollutants and making it better for ecological health and human consumption. The natural spaces also help to provide repose in the urban landscape, environmental stewardship, and social cohesion and foster the dynamic of community engagement.

The expansion of cities leads to a conversion of wetlands into a built-up environment, resulting in degradation, fragmentation, and habitat loss. Urban runoff pollution, agricultural practices, and industrial activities impair the ecological health of the population and pose risks to wildlife and human health. The changes in climate provoke challenges and lead to a rise in sea level, flooding, and habitat loss, which have lowered the resilience and integrity of urban wetlands. The adoption of integrated practices and planning helps strengthen the regulatory and legal framework, invest in ecological restoration, engage the local communities, and help to protect the wetlands for future generations. Water scarcity and climate change are the major issues that are faced by the global population. Urban wetlands also have interpretive and educational sites that offer insights into biodiversity, environmental conservation and local ecology. Through attracting natural enthusiasts and tourists, urban wetlands restore economic activity in local areas, creating revenue for their communities and supporting their local businesses.

Acknowledgement We would like to thank all the stakeholders of the college and officials related for their constant support, understanding, for solving our every query and problem. They have helped us every time we were in any doubt. We also want to thank our friends for their constant help during our research. We would like to thank our team and close people; they have been our mental support during our hard times and encouraged us.

References

Akhtar N, SyakirIshak MI, Bhawani SA, Umar K (2021) Various natural and anthropogenic factors responsible for water quality degradation: a review. Water 13(19): 2660

Alikhani S, Nummi P, Ojala A (2021) Urban wetlands: a review on ecological and cultural values. Water 13(22):3301

Aticho A, Beyene A, Zelelew SA, Nigussie A, Gutema TM, Bakkestuen V, Chala D (2023) Observed and projected changes in urbanization and nature conservation in a typical fast growing city of Ethiopia, Jimma. Biodivers Conserv 32(12):4007–4028

Azadi H, Van Passel S, Cools J (2021) Rapid economic valuation of ecosystem services in man and biosphere reserves in Africa: a review. Glob Ecol Conserv 28: e01697

Ballut-Dajud GA, Sandoval Herazo LC, Fernández-Lambert G, Marín-Muñiz JL, López Méndez MC, Betanzo-Torres EA (2022) Factors affecting wetland loss: a review. Land 11(3):434

Bănăduc D, Simić V, Cianfaglione K, Barinova S, Afanasyev S, Öktener A, McCall G, Simić S, Curtean-Bănăduc A (2022) Freshwater as a sustainable resource and generator of secondary resources in the 21st century: stressors, threats, risks, management and protection strategies, and conservation approaches. Int J Environ Res Public Health 19(24):16570

Camacho-Valdez V, Saenz-Arroyo A, Ghermandi A, Navarrete-Gutiérrez DA, Rodiles-Hernández R (2020) Spatial analysis, local people's perception and economic valuation of wetland ecosystem services in the Usumacinta floodplain, Southern Mexico. PeerJ 8: e8395

Cui S, Yu T, Zhang F, Fu Q, Hough R, An L, Gao S, Zhang Z, Hu P, Zhu Q, Pei Z (2020) Understanding the risks from diffuse pollution on wetland eco-systems: the effectiveness of water quality classification schemes. Ecol Eng 155:105929

Cui M, Ferreira F, Fung TK, Matos JS (2021) Tale of two cities: how nature-based solutions help create adaptive and resilient urban water management practices in Singapore and Lisbon. Sustainability 13(18):10427

Dar et al (2020) Current status of wetlands in Srinagar city: threats, management strategies, and future perspectives. Front Environ Sci 7. https://doi.org/10.3389/fenvs.2019.00199

Das S, Adhikary PP, Shit PK, Bera B (2022) Urban wetland fragmentation and ecosystem service assessment using integrated machine learning algorithm and spatial landscape analysis. Geocarto Int 37(25): 7800–7818

De Vito L, Staddon C, Zuniga-Teran AA, Gerlak AK, Schoeman Y, Hart A, Booth G (2022) Aligning green infrastructure to sustainable development: a geographical contribution to an ongoing debate. Area 54(2): 242–251

De I, Hasan R, Iqbal M (2022) Natural treatment systems and importance of social cost benefit analysis in developing countries: a critical review. Sustainability 14(7): 3913

Ekka A, Pande S, Jiang Y, der Zaag PV (2020) Anthropogenic modifications and river ecosystem services: a landscape perspective. Water 12(10):2706

Félix-López M, Ormaza-Murillo M, Álvarez-Santana C, Banchon C (2023) Exploring the impact of reclaimed water on Latin America's development. J Ecol Eng 24(10):157–173

García-Llorente M, Castro AJ, Quintas-Soriano C, Oteros-Rozas E, Iniesta-Arandia I, González J, García del Amo D, Hernández-Arroyo M, Casado-Arzuaga I, Palomo I et al (2020) Local perceptions of ecosystem services across multiple ecosystem types in Spain. Land 9(9):330. https://doi.org/10.3390/land9090330

Ghasemi M, González-García A, Charrahy Z, Serrao-Neumann S (2024) Utilizing supply-demand bundles in nature-based recreation offers insights into specific strategies for sustainable tourism management. Sci Total Environ 922(2024):171185

Guo RZ, Lin L, Xu JF, Dai WH, Song YB, Dong M (2023) Spatio-temporal characteristics of cultural ecosystem services and their relations to landscape factors in Hangzhou Xixi National Wetland Park, China. Ecol Indic 154:110910

Herath M, Jayathilaka T, Azamathulla HM, Mandala V, Rathnayake N, Rathnayake U (2023) Sensitivity analysis of parameters affecting wetland water levels: a study of flood detention basin, Colombo, Sri Lanka. Sensors 23(7):3680

Hill AD, Johnson SG, Greco LM, O'Boyle EH, Walter SL (2021) Endogeneity: a review and agenda for the methodology-practice divide affecting micro and macro research. J Manag 47(1):105–143

Irvine K, Loc HH, Sovann C, Suwanarit A, Likitswat F, Jindal R, Koottatep T, Gaut J, Chua L, Qi LW, De Wandeler K (2021) Bridging the form and function gap in urban green space design through environmental systems modeling. J Water Manag Model 29:1–19

Keitumetse SO, Mwale KP, Satau G, Velempini K, Baitsiseng VO, Ntema OPBM, Manga J, Mogotsi ST (2023) Exploring cultural values of African wetlands for sustainable conservation: Okavango Delta world heritage site, Botswana. J Cult Herit Manag Sustain Dev 13(3):501–516

Khirfan L et al (2021) Reading an urban palimpsest: how the gradual loss of an urban stream impacts urban form's connections and ecosystem functions. Front Water 3. https://doi.org/10.3389/frwa.2021.754679

Kiddle GL, Pedersen Zari M, Blaschke P, Chanse V, Kiddle R (2021) An Oceania urban design agenda linking ecosystem services, nature-based solutions, traditional ecological knowledge and wellbeing. Sustainability 13(22):12660

Krivoguz D (2024) Geo-spatial analysis of urbanization and environmental changes with deep neural networks:

insights from a three-decade study in Kerch peninsula. Eco Inform 80:102513
Lin Y, Zhang X, Zhu H, Li R (2024) Spatiotemporal evolution and mechanisms of habitat quality in nature reserve land: a case study of 18 nature reserves in Hubei Province. Land 13(3):363
Liu H, Liao T, Wang Y, Qian X, Liu X, Li C, Li S, Guan Z, Zhu L, Zhou X, Liu C (2023) Fine-grained wetland classification for national wetland reserves using multi-source remote sensing data and pixel information expert engine (PIE-engine). GIScience Remote Sens 60(1):2286746
Lourdes KT, Gibbins CN, Hamel P, Sanusi R, Azhar B, Lechner A (2021) A review of urban ecosystem services research in Southeast Asia. Land 10:40. https://doi.org/10.3390/land10010040
Mahdianpari M, Granger JE, Mohammadimanesh F, Salehi B, Brisco B, Homayouni S, Gill E, Huberty B, Lang M (2020) Meta-analysis of wetland classification using remote sensing: a systematic review of a 40-year trend in North America. Remote Sens 12(11):1882
Maltby E (2022) The wetlands paradigm shift in response to changing societal priorities: a reflective review. Land 11(9):1526
Matsushita T, Ghezelloo Y, Maly E, Kondo T, Meyer M, Newman G (2024) Placemaking mediating dilemmas by addressing the gaps in post-disaster recovery process: long-term citizen-driven place-nurturing in New Orleans after hurricane Katrina. Int J Disaster Risk Reduct 106:104457
Miller JT (2020) Temporal analysis of displacement: racial capitalism and settler colonial urban space. Geoforum 116:180–192
Mohtat N, Khirfan L (2023) Epistemic justice in flood-adaptive green infrastructure planning: the recognition of local experiential knowledge in Thorncliffe Park, Toronto. Landsc Urban Plann 238:104834
Mondal BK, Kumari S, Ghosh A, Mishra PK (2022) Transformation and risk assessment of the East Kolkata wetlands (India) using fuzzy MCDM method and geospatial technology. Geogr Sustain 3(3): 191–203
Newman M, Gough D (2020) Systematic reviews in educational research: methodology, perspectives and application. Syst Rev Educ Res 64(3):3–22
Noortyani R, Mu'in F, Munawwarah R, Normelani E (2023) Ethnolinguistic study of the traditional Indonesian parenting song "DindangMaayunAnak": cultural reflection in the socio-cultural life of the Banjar in South Kalimantan. SAGE Open 13(2): 21582440231175127
Odaro DO, Ucakuwun EK, Daudi F (2023) Analysis of land use land cover changes of Okana wetland ecosystems in lower Nyando River basin, Kisumu County, Kenya. J Res Innov Implic Educ 7(4):493–507
Pittman SJ, Stamoulis KA, Antonopoulou M, Das HS, Shahid M, Delevaux J, Wedding LM, Mateos-Molina D (2022) Rapid site selection to prioritize coastal seascapes for nature-based solutions with multiple benefits. Front Mar Sci 9:832480
Pukowiec-Kurda K (2022) The urban ecosystem services index as a new indicator for sustainable urban planning and human well-being in cities. Ecol Indic 144:109532
Ragavan MI, Risser L, Duplessis V, DeGue S, Villaveces A, Hurley T et al (2021) The impact of the COVID-19 pandemic on the needs and lived experiences of intimate partner violence survivors in the United States: advocate perspectives. Violence Against Women 28(12–13):3114–3134
Ramachandra TV, Sincy V, Asulabha KS (2021) Accounting of ecosystem services of wetlands in Karnataka state, India. J Resour Energy Dev 18(1–2):1–26
Salas GM (2023) An indicator framework for assessing socio-ecological resilience of Candaba wetlands, Philippines. Curr J Appl Sci Technol 42(23):1–19
Sam SB, Morgenroth E, Strande L (2023) Changes in organic fractions, cations, and stabilization from feces to fecal sludge: implications for dewatering performance and management solutions. J Water Sanit Hyg Dev 13(9):699–710
Sharai S, Tawanda T, Gladman C (2020) Threats and conservation strategies on urban wetlands: a case of Monavale and surrounding areas in Harare, Zimbabwe. Am J Environ Protect 9(2):36–43
Sigman E, Elias M (2021) Three approaches to restoration and their implications for social inclusion. Ecol Restor 39:27–35. https://doi.org/10.3368/er.39.1-2.27
Singh A, Sharma J, Paichha M, Chakrabarti R (2020) Achyranthes aspera (prickly chaff flower) leaves- and seeds-supplemented diets regulate growth, innate immunity, and oxidative stress in Aeromonas hydrophila-challenged Labeo rohita. J Appl Aquac 32(3):250–267
Song L, Wu M, Wu Y, Xu X, Xie C (2023) Research on the evaluation of cultural ecosystem services in Zhengzhou Urban Parks based on public perceptions. Sustainability 15(15):11964
Syaban ASN, Appiah-Opoku S (2024) Unveiling the complexities of land use transition in Indonesia's New Capital City IKN Nusantara: a multidimensional conflict analysis. Land 13(5):606
Tan J, Gu K, Zheng Y (2024) Peri-urban planning: a landscape perspective. Plan Theory 23(1):42–63
Targetti S, Raggi M, Zavalloni M, Viaggi D (2021) Perceived benefits from reclaimed rural landscapes: evidence from the lowlands of the Po River Delta, Italy. Ecosyst Serv 49:101288
Tien ND, Duyen TNL, Huyen NTT, Anh PQ, Oanh NT, Van Tich V, Dat DT, Hanh NTH, Trang VH (2024) Community-based ecotourism for sustainability: an evaluative analysis of Binh Son district, Quang Ngai province in Vietnam. Soc Sci Humanit Open 9:100807
Tomaškinová J, Tomaškin J, Theuma H, Alcántara Valero AF, Attard V (2021) Addressing present challenges in the life-cycle of wetlands management to successfully integrate sustainability and good governance. J

Environ Eng Landsc Manag 29(1):48–60. https://doi.org/10.3846/jeelm.2021.14142
Tsatsou A, Frantzeskaki N, Malamis S (2023) Nature-based solutions for circular urban water systems: a scoping literature review and a proposal for urban design and planning. J Clean Prod 394:136325
Viti M, Löwe R, Sørup HJ, Ladenburg J, Gebhardt O, Iversen S, McKnight US, Arnbjerg-Nielsen K (2023) Holistic valuation of nature-based solutions accounting for human perceptions and nature benefits. J Environ Manag 334:117498
Wang C, Ma L, Zhang Y, Chen N, Wang W (2022) Spatiotemporal dynamics of wetlands and their driving factors based on PLS-SEM: a case study in Wuhan. Sci Total Environ 806:151310
Weis JS, Watson EB, Ravit B, Harman C, Yepsen M (2021) The status and future of tidal marshes in New Jersey faced with sea level rise. Anthr Coasts 4(1): 168–192
Wessels N, Sitas N, Esler KJ, O'Farrell P (2021) Understanding community perceptions of a natural open space system for urban conservation and stewardship in a metropolitan city in Africa. Environ Conserv 48(4):244–254
Yang D, Yang Q, Tong Z, Du W, Zhang J (2024a) Coupling coordination analysis of production, living, and ecological spaces in wetlands: a case study of Xianghai wetland nature reserve, China. Ecol Indic 158:111578
Yang C, Suleiman AK, Lang Y (2024b) Sustainable wetland management using the Kunming-Montreal global biodiversity framework as a guide in the Sierra Leone case. J Geosci Environ Protect 12(3):1–15
Yao Y et al (2023) Deletion of MEC1 suppresses the replicative senescence of the cdc13-2 mutant in Saccharomyces cerevisiae. G3 13(5):jkad065
Yaseen, Chalotra (2022) Sustainable management strategies urban water bodies and wetlands in the Indian context. Int J Innov Res Eng Manage 9(5): 257–265. ISSN 2347–5552
Zhang Y, Rossow WB, Lacis AA, Oinas V (2022) Calculation, evaluation and application of long-term, global radiative flux datasets at ISCCP: past and present. In: Luo ZJ, Tselioudis G, Rossow WB (eds) Studies of cloud, convection and precipitation processes using satellite observations, Lectures in climate change, vol 3. World Scientific, pp 151–171. https://doi.org/10.1142/9789811256912_0009
Zhang X, Zhao T, Xu H, Liu W, Wang J, Chen X, Liu L (2023) GLC_FCS30D: the first global 30-m land-cover dynamic monitoring product with a fine classification system from 1985 to 2022 using dense time-series Landsat imagery and continuous change-detection method. Earth Syst Sci Data Discuss 2023:1–32. https://doi.org/10.5194/essd-2023-320
Zhang X, Liu L, Zhao T, Wang J, Liu W, Chen X (2024) Global annual wetland dataset at 30 m with a fine classification system from 2000 to 2022. Sci Data 11(1):310

Social and Cultural Significance of Urban Wetlands 14

Ruslan Halifu, La Ode Abdul Munafi, Kuswinton, Andy Arya Maulana Wijaya, Ulfa Sevia Azni, Restiyana Agustine, Uchaimid Biridlo'i Robby, Prameshwara Anggahegari, and Hari Harjanto Setiawan

Abstract

The wise use of wetlands is currently a major concern in developing urban settlements. This is influenced by the lack of commitment to maintaining the existence of wetlands in urban areas. The emergence of these problems can impact long-term ecological and community sustainability, especially in urban areas. The lack of commitment to maintaining the existence of wetlands can be influenced by sociocultural factors (aspects of local wisdom support, government policies, and social changes that occur). Therefore, this study aims to determine the effects of local wisdom (cultural), government policies, and social changes that occur on the resilience of urban wetlands. This study used a qualitative design with a case study approach and was located in Baubau City, Southeast Sulawesi Province (Indonesia). As a result of research and discussion, the researcher found that the people of Baubau City still uphold their cultural values. The local government always strives to maintain community behavior so as not to disturb the existence of urban wetlands through various programs that support wetlands. The state of social change that has occurred is also very supportive of the existence of wetlands. This study concludes that support for local wisdom (culture), government policies, and social change is quite influential in efforts to maintain the resilience of urban wetlands. Recommendations from this study include the need to strengthen local wisdom-based urban wetland management, making environmentally friendly infrastructure development a top priority for the government, realizing wetlands-based economic development, and increasing commitment and collaboration from various parties.

Keywords

Coastal communities · Cultural · Policy · Social changes · Urban · Wetlands

R. Halifu (✉) · U. S. Azni · R. Agustine · U. B. Robby · P. Anggahegari · H. H. Setiawan
National Research and Innovation Agency (BRIN), Jakarta, Indonesia
e-mail: ruslan.halifu@brin.go.id; ulfa007@brin.go.id; restiyana.agustine@brin.go.id; uchaimid.biridloi.robby@brin.go.id; prameshwara.anggahegari@brin.go.id; hari.harjanto.setiawan@brin.go.id

L. O. A. Munafi
University of Dayanu Ikhsanuddin, Baubau, Indonesia
e-mail: laodeabdulmunafi@unidayan.ac.id

Kuswinton
Politeknik Baubau, Baubau, Indonesia
e-mail: kuswinton@poltekbaubau.ac.id

A. A. M. Wijaya
University Muhammadiyah of Buton, Baubau, Indonesia
e-mail: andy.arya@umbuton.ac.id

S. K. Chakraborty et al. (eds.), *Wetlands and Urban Resilience*, Sustainable Development Goals Series,
https://doi.org/10.1007/978-3-032-06881-1_14

14.1 Introduction

Preserving the environment is very important because a clean or healthy environment can improve human health. An environment polluted by pollution can cause various diseases (Almetwally et al. 2020). Taking care of the environment can help preserve various species of flora and fauna (Zhou et al. 2024). The existence of these various species is very important for the ecosystem to be maintained. A well-maintained ecosystem will ensure that human nutritional and food needs are met because human nutritional needs come from flora and fauna (Penuelas et al. 2020).

A well-preserved environment can ensure that natural resources such as clean water, forests, and soil are preserved. A damaged environment can widen the negative impacts of climate change.[1] A healthy environment can bring economic value to society, for example, tourism destinations, agricultural land, and fishing grounds (Mejjad et al. 2022). Natural resources must be used wisely or not excessively to maintain environmental balance (Erdoğan et al. 2021). This principle is intended to ensure that natural resources can be renewed and used by future generations. Rehabilitation efforts are essential when the environment or biodiversity has been damaged (Mentis 2020). These efforts can be made by raising awareness among individuals or communities about the importance of the environment (Liu et al. 2023a, b).

In addition to raising awareness, efforts to recycle waste, reduce waste, and so on are needed to resolve environmental pollution quickly (Kurniawan et al. 2022; Skirtun et al. 2022). Implementing environmentally friendly technology can also reduce the potential for environmental damage. These various things can be done in line with strengthening innovation in products/goods/technologies directly related to everyday life (China et al. 2020; Zhao et al. 2021a, b). Until now, various innovations aimed at preserving the environment have been produced/discovered by humans. One of them is raising awareness to preserve wetlands (Wada et al. 2020).

Wetlands are ecosystems with sufficient amounts/volumes of water on the surface/underground (Espinosa et al. 2021; Zhao et al. 2021a, b). Wetlands are often considered useless land, so they need to be drained or converted. Since 1900, more than 64% of the world's wetlands have been destroyed or converted due to drainage or other conversion efforts (Dajud et al. 2022). At present, the function of wetlands has generally been degraded. The number of environments with sufficient water volume has drastically decreased. With the increasing human population and massive climate change, the existence of wetlands is decreasing, raising concerns about a water, soil, and food crisis in the future (Wang and Gu 2021; Zhong et al. 2022).

Wetlands are being damaged in various areas, including urban areas (Yang et al. 2021). Various factors cause this damage. Urban areas are often the target of urbanization, making them increasingly densely populated. The high population density increases housing areas and infrastructure development, resulting in a decrease in wetlands (Askari 2022; Rojas et al. 2022). Housing and infrastructure development will trigger deforestation, which will ultimately disrupt wetlands. This development will also increase the risk of water pollution and the increase of waste and sewage, exacerbating the damage to wetlands (Danso et al. 2021).

Various global efforts have been made to preserve urban wetlands. In 1971, countries worldwide made a joint agreement focused on protecting wetlands around the world, which later became known as the Ramsar Convention.[2] In 2015, the countries belonging to the United Nations held a scientific meeting and officially endorsed the Sustainable Development Goals (SDGs) agenda as a global development agreement (Li et al. 2022). The agenda, specifically in the sixth program, reinforces the need to maintain water quality sustainably, including wetlands

[1] Abbass et al. (2022).

[2] Bridgewater and Kim (2021).

protection (Weise et al. 2020). In addition, the 15th program emphasizes the preservation of land ecosystems, including the restoration of wetlands in urban areas (Zhang et al. 2023).

The government and relevant stakeholders in Indonesia have also launched various programs to support various policies for the maintenance of urban wetlands at the global level. The government and the legislature have issued Law No. 32 of 2009 on environmental protection and management (Nirwana 2023; Suhendra et al. 2020).

Various problems arise when wetland protection policies are implemented in urban areas. Difficulty in coordination between sectors (governments) in handling the damage and maintenance of wetlands, so the policies created by the various sectors are often not integrated (Dixon et al. 2021). Lack of funding/budget allocated to finance various wetland protection programs, thus disrupting the program's effectiveness (Dai et al. 2021). The limited number of environmental experts and human resources with the capacity to manage wetland protection programs, especially at the local level, means that the program is not implemented effectively (Clare and Creed 2022; Taylor et al. 2021).

Implementing wetlands policies in urban areas is often also confronted with the complexity of community behavior that does not support wetlands protection (Hu et al. 2022). Many urban communities are more concerned with economic value and stability than with the urgency of wetlands. They consider the economic aspect directly related to daily life (short-term needs), while the benefits of wetlands are long-term, so they consider them less important (Adhya and Banerjee 2022; Aguilar et al. 2022). This lack of awareness is due to the lack of available information, supporting policies, and community involvement (community empowerment) in efforts to protect urban wetlands (Imdad et al. 2023; Sinthumule 2021). To support information about wetlands for urban communities, various studies have been conducted by researchers. These studies include research by Assefa et al. (2021), Cobbinah et al. (2022), Ekumah et al. (2020), Maltby (2022), Shah and Atisa (2021), and Hu et al. (2022).

Various studies have stated that public awareness in urban areas is important for protecting urban wetlands. The question is how sociocultural factors influence the protection of urban wetlands, especially when these factors are broken down into aspects of local wisdom (culture), social change, and government policy.

The above question arises because aspects of local wisdom (culture), social change, and government policy are components of socioculture that can influence the formation of social awareness of a situation (Lin and Kaewkhunok 2021; Pasya Ashsyafa et al. 2024), and coastal areas are areas with unique community characteristics.[3,] This question is important to answer because it is directly related to the commitment to maintaining the resilience of urban wetlands. This study aims to determine the effect of local wisdom (cultural) support, government policies and social changes on the resilience of urban wetlands. This is a new study because it examines the simultaneous influence of sociocultural factors on the resilience of urban wetlands in coastal areas.

14.2 Rationale of the Study

Efforts to maintain the resilience of urban wetlands are necessary, given that they are closely related to environmental balance and social welfare now and in the future. Research on urban wetlands must be carried out as a serious commitment to these efforts. The substance of the study will be very important if the context includes how efforts to protect urban wetlands are related to aspects of local wisdom (culture), social change, and government policy. These various aspects, as important components in the sociocultural dimension, contribute to efforts to maintain the resilience of urban wetlands. The results of this study will also integrate the importance of maintaining wetland resilience into

[3] Phillips and Murphy (2021),

Fig. 14.1 Baubau River (wetlands in Baubau City). (Source: Primary Data 2025)

efforts to maintain the city's environmental resilience system (Fig. 14.1).

14.3 Materials and Methods

This is a qualitative study with a case study approach. Using a case study approach, this research explores social phenomena in urban communities, especially in coastal areas. These phenomena relate to how local wisdom (culture), social change, and government policies support the resilience of urban wetlands. The results of this exploration produce qualitative data.

This study's population includes urban policymakers (local government) and the communities living in the area. The informants are representatives of the local government and real community leaders. The selection and determination of informants in this study were based on capacity, competence, or academic expertise in understanding local wisdom (culture), social change, and government policies related to wetlands.

This research was conducted in Baubau City, Southeast Sulawesi Province. The location was chosen because of its position/location on the coastal area. The data in this study were collected through in-depth interviews, observations, and documentation. The data processing in this study was carried out carefully. The research data collected, especially from the interview results, was compiled into a record. The data collected was qualitative. After inputting it, the researcher carried out the data analysis process.

14.4 Results and Discussion

This section will discuss how culture (local wisdom), government policy, and social change affect the resilience of wetlands in urban areas based on field research and reinforced by the views of various previous theories.

14.4.1 The Influence of Local Wisdom on the Resilience of Urban Wetland

Most of the people of Baubau City are part of the Buton Tribe. In the Buton culture, there is a system related to the pattern of utilization/conservation of river and coastal environmental resources. Coastal communities have unique traditions or behaviors compared to other communities (Almutairi et al. 2020; Phillips and Murphy 2021). All their behavior is often associated with the philosophy of the marine or river environment. They generally have rituals that are a legacy from their ancestors. They also preserve these rituals by carrying them out consistently.[4] This effort is their responsibility to the mandate given to them by their ancestors.

This cultural institution is reflected in the *kaombo*. *Kaombo* is a cultural institution that reflects a traditional conservation system for environmental resources (forests, rivers, and coastal and marine resources). *Kaombo* comes from the word *ombo*, which means "to protect." Essentially, *kaombo* refers to the community's "protecting" (natural resources) behavior. Because it is protected, customary provisions apply in the form of prohibitions on exploitation so that the resources (on land and at sea) that are under protection can be preserved. *Kaombo*, thus, is a cultural institution that provides behavioral guidance (orientation) to humans in managing and utilizing environmental resources (on land and at sea).

Today's *kaombo*-based environmental resource conservation system maintained in Buton society is a wise practice inherited from the Buton Sultanate era. The Buton constitution (the Seven Dignities Law) states that every kadie (village) is obliged to maintain and protect an area (on land or the coast). *Kaombo*-based area protection is aimed at preserving all area resources (plants and animals). Maintaining the sustainability of environmental resources in the *kaombo* area is a collective obligation of the community in each *kadie*. The traditional *kaombo*-based conservation system functionally enables the sustainability of the carrying capacity of environmental resources.

In conserving natural environmental resources, *kaombo* refers to a "covered area" in terrestrial and coastal aquatic environments. Therefore, *kaombo* is divided into land *kaombo* and sea *kaombo*. Land *kaombo* is a covered area (conservation area) located in a terrestrial environment, including forests, rivers, and springs. Meanwhile, sea *kaombo* is a coverage area located in coastal and marine waters. Within the *kaombo* area (coverage/conservation area), customary regulations apply in the form of prohibitions on the exploitation of all the resources contained therein, both plant and animal resources. Prohibitions on exploiting resources in *kaombo* apply for a certain period (open-close system) or permanently.

Land *kaombo* is a coverage area (conservation area) in the land environment. The Buton people deeply appreciate the importance of preserving the environment (on land), especially the forest, river, and spring environments. Based on this awareness, the (specific) forest environment, rivers, and springs are designated as *kaombo* areas close to utilizing environmental resources. The designation of *kaombo* for a forest area has consequences, namely enforcing a (customary) ban on logging or exploiting resources (wood) in the area.

Apart from the forest environment, *kaombo* (a conservation area) is also applied to the river environment, especially in the upstream area. For the people who live in the interior of Buton Island and depend on the dryland agricultural sector (peasant farmers), rivers are one of the supports of their lives. In addition to the utilization of river water (as a source of drinking water and household needs), various river biota (such as fish, shrimp, eels, and river shellfish) are a source of food security for the local community, in addition to the plantation crops they produce.

In the sea *kaombo* area, there is a customary prohibition on exploiting the resources contained therein, including the biota and ecosystem environment. Thus, the sea *kaombo* is a marine protected area established to ensure the sustainability of marine life and the ecosystem environment. The *kaombo*-based coastal environmental resource utilization pattern is also intended to avoid the sea as an open access area.

In practice, the designation of a coastal area as a *kaombo* begins with a *Sara Kadie* consultation. The following matters are discussed in the consultation: (1) the determination of the boundaries of the waters to be temporarily closed from exploitation activities, (2) the determination of the types of biota that may not be exploited while the area is closed, and (3) sanctions against destructive actions in the closed zone. Once these provisions have been agreed upon, local traditional authorities will install boundary markers

[4] Zhao et al. (2021a).

for the closed zone. Traditionally, the boundary mark of the *kaombo* zone, which is commonly used, is a coconut tree trunk stuck into the seabed.

The installation of the *kaombo* zone boundary mark is carried out in a traditional ritual. In the ritual, a warning is also given to the community not to take seafood during the *kaombo* zone. Violations of the ban are subject to strict traditional sanctions. The local community supervises the *kaombo* zone. Supervision is very important to prevent theft of marine resources and other destructive actions in the *kaombo* zone, both by local fishermen and fishermen from outside.

Based on the description above, the people of Baubau City have a unique culture. Culture can be defined as the totality of ways or patterns of life shaped by aspects of social interaction, history, and social environment (Dupuis 2022). Culture is not only in the form of material aspects but also includes perspectives, attitudes, and behaviors, which then become the identity of a person/community (Glăveanu 2020; Liu et al. 2023a, b). Some examples of material culture are traditional buildings, historical monuments, fine arts (sculptures, paintings, and so on), traditional clothing, and traditional food and drinks. Meanwhile, examples of culture in the form of perspectives are the behavior of cooperation in social services and maintaining environmental sustainability with local wisdom values.

The culture can be formed by two things: the result of someone's discovery in the past and its historical value.[5] The person's reverence and admiration for the discovery are high enough that they are spread to others by various methods, with the intention that others will also admire them. After others admire it and the number increases, the admirers will make it an object of shared pride and a social fact.

In addition, culture can also be formed by coercion (Nyirenda et al. 2020; Wang 2020). Initially, the object/behavior was carried out or adapted by someone in a powerful position in a community. With their power, this person forces others to follow their behavior. Because it was coercive, everyone adopted the behavior, and it gradually became a social norm that had to be followed.

Based on the research results and the description above, it can be said that the local wisdom in the community of Baubau City strongly supports the resilience of the wetlands. This local wisdom is demonstrated by the cultural institutions in the community that strongly support the preservation of the wetlands. The resilience of the wetlands in Baubau City has been well maintained until now. This provides strong evidence that local wisdom has a significant effect on the resilience of wetlands, especially in urban areas.

14.4.2 The Influence of Government Policies on the Resilience of Urban Wetlands

Baubau City is located in Southeast Sulawesi Province and has an area of 294.99 km^2. With a coastline length of 42 km and a sea area of 30 km^2, Baubau City also has the unique shape of an area divided into two parts by the Baubau River, which has a length of 15,704 km and an area including the Baubau watershed, which has an area of 76,117 km^2 (BPS or Baubau City Statistics Agency 2024).

The position of the Baubau River as the largest river in Baubau City has played a very important role in the development of Baubau City until now. In Baubau City Regional Regulation Number 4 of 2014 concerning the Baubau City Spatial Plan 2014–2034, it mentions the function of the Baubau River as a provider of raw water networks, communal water supply systems, and primary drainage network systems. Meanwhile, the 2018–2023 Baubau City RPJMD Document only mentions the Baubau River as part of the function and carrying capacity of the Baubau City environment. These various regulations are to protect the people of Baubau City as a coastal community.

The management of the Baubau River is currently not a priority for regional development, even though the river has the potential to be a source of electricity, agriculture, fisheries,

[5] Paszkowska et al. (2022).

industrial needs, household needs, and tourism. The 2024–2026 Baubau City Regional Development Plan (RPD) only mentions the hydrological role of the river, but no plans have been found in the form of regional governance policies or infrastructure development plans that can support the optimization of the economic function of other rivers.

Based on the above description, the Baubau City government believes strong policies are needed to preserve the environment and protect coastal communities, especially those in Baubau City. Government policy can be defined as a series of decisions or actions formulated by the government as the leader of the people and the state to achieve a specific goal.[6] Policies generally relate to various fields, such as the economy, society, politics, the environment, health, and education. Policies are made for the welfare of the community. In formulating policies, the government always involves various stakeholders, including the community, to obtain input or support for the policies being formulated (Masuda et al. 2022; Van der Wal 2020).

Based on this stage, policies are formulated through problem identification, analysis, and the creation/development of policy alternatives (Capano et al. 2020). In this process, policymakers usually compile academic papers related to the policy. Once formulated, the policy is implemented based on its designation and subject or target (Bibri et al. 2020). The implementation process usually involves many parties or stakeholders. When implemented, instruments are prepared to assess effectiveness, efficiency, and impact. These instruments will be used when the policy has been implemented. The process is a policy evaluation step (Capano and Howlett 2020).

The strength of a government policy is influenced by various factors, including active participation from various stakeholders (government, nongovernmental organizations, civil society and the private sector) (Roziqin et al. 2021), integration with other policies (Petridou and Mintrom 2021), policy consistency (Weible et al. 2020), clarity in the aspects of sanctions and supervision (Felbermayr et al. 2021; Wang et al. 2023), the strength of policy monitoring and evaluation (Capano and Howlett 2020), the level of understanding and awareness of the public/policy implementers (Capano et al. 2020) and the availability of budget for policy implementation (Razumovskaia et al. 2020). The resulting/implemented policy will feel strong and binding when these factors are considered and fulfilled. On the other hand, when the formulation and implementation of policies do not pay attention to these matters, the resulting/implemented policy will be weak, and the community will neglect its implementation.

In several studies, it is explained that the function of rivers is constantly changing. Namely, land cover in upstream areas continues to increase, coupled with the reduced capacity of river channels, especially in downstream areas, due to sedimentation and changes in the topography of river basins due to settlements (Nuzul and Achmad 2023). On the other hand, critical land is increasing every year. This phenomenon has caused a decrease in the watershed's ability to store water in the dry season, so that the frequency and magnitude of flooding will increase, as well as higher sedimentation resulting in river siltation, thus reducing its carrying capacity and flow.

In the historical landscape of Baubau City, rivers have been an important element in its development since the founding of the Buton Sultanate. In the past, rivers were the community's lifeline, and in some historical traces, rivers are also mentioned as busy lines of city activity. This trace is still recorded by naming several village areas through which the Baubau River passes, such as Nganganaumala (mouth of the river).

In addition, some watersheds are currently icons of the productive sector of the city of Baubau, where the upstream area of the river from Batu Poopi to Kali Ambon is used as a source of drinking water, agricultural and community plantation areas, and tourist areas. Meanwhile, the estuary area of the river from the Beli

[6] Mansoor (2021).

Bridge to Kotamara is used as a residential area for fishing boat mooring activities, and Kotamara has become a popular public space in the city.

Based on the research results and the discussion above, the policy formulated by the Baubau City Government is strong enough to support the conservation of wetlands in Baubau City. The resilience of the wetlands in Baubau City has been well maintained until now. The government strives to ensure that the formulated wetlands area resilience policy remains relevant to the times and the dynamics of community life. These results provide strong evidence that the structure/posture of policy significantly influences the resilience of wetlands in urban areas.

14.4.3 The Influence of Social Changes That Occur on the Resilience of Urban Wetlands

The people of Baubau City can be considered a coastal community. A coastal community is a group of people who live in an area directly adjacent to the sea or river (Andrews et al. 2021). Because they live in these areas, they generally depend on fishing, seaweed farming, etc. In urban areas, the government often turns coastal areas into tourist areas, so the community also becomes involved in micro, small, and medium enterprises (MSMEs) in the tourism sector (Ji and Wang 2022). To add to the exotic value of the area, the government builds various supporting public facilities, such as locations for trading and vehicle parking. Nevertheless, coastal communities are often faced with the problem of poverty.[7] Their working patterns, which are generally still traditional, result in relatively low incomes.

The people of Baubau City have a unique and diverse approach to river conservation, reflecting their local wisdom and social solidarity. The informant mentioned that one of the traditions carried out by the community is cooperation to clean the river periodically. This activity involves various community groups, both young and old, who voluntarily work together to ensure the river remains clean of waste and pollution-free.

The community has initiatives that include river ecosystem rehabilitation, such as planting mangroves around streams. This program prevents erosion and improves natural habitats for various species. Meanwhile, fishermen in Baubau see the preservation of rivers as an integral part of their livelihoods. They work with the government to keep rivers clean and functioning as safe passageways to the sea. The younger generation in Baubau is also very supportive of this.

The people mentioned that the river in Baubau City is a source of life and plays an important role in supporting the local community's economy. Many residents use the river for various economic activities that have been going on for generations. One community group that is highly dependent on the river is the traditional fishermen. The river functions as a fishing ground for catching fish. The river has the potential to generate electricity, tourism, and agriculture, as well as meet the daily needs of the local community.

Then, the informant mentioned that people who live near rivers are often involved in processing natural products. An example is the selective use of river sand and stones as building materials, without damaging the ecosystem. On the other hand, local women's communities use the river as a water source for traditional crafts such as bamboo weaving. Indirectly, the river is also a place for ecotourism-based activities. Some community groups are beginning to realize the river's potential as a tourist attraction and offer boat services along the river while introducing Baubau's natural and cultural wealth to visitors. This opens up new opportunities for residents to earn additional income. However, the sustainability of this use of the river depends heavily on its preservation.

Coastal communities generally have strong social bonds or solidarity.[8] They are quite obedient in carrying out the rules initiated by the

[7] Dube et al. (2022).

[8] Van Assche et al. (2020).

government (Hamid et al. 2021). Dependence on river/sea natural resources can make them vulnerable to dynamic environmental changes (Dube et al. 2022; Sinay and Carter 2020). They are greatly affected by climate change, such as rising sea/river levels. As a result of climate change, their homes or environments are often threatened by seawater abrasion or flooding (river water). In addition, the livelihoods of those who rely heavily on marine/river resources are often disrupted.

With the increasing population and economic activity in the river's coastal area, there is increased pressure on the river, such as pollution and the narrowing of the river channel. Domestic and household waste is a major challenge, especially in densely populated areas with limited waste management facilities. Nevertheless, there are joint efforts from local communities to keep the river clean through cooperation and environmental education programs. For people living in riverine areas, rivers are a natural resource and part of their identity and way of life. Despite the challenges of modernization and urbanization, they strive to preserve the function of the river so that it can continue to be utilized by future generations. This pattern indicates that they can adapt to social change.

Social change is a transformation process of structures, norms, values, and patterns of social interaction or communication in a community or society (de Haas 2021; Hölscher et al. 2021; Lage 2022). This transformation can occur in various fields, such as economics, culture, and politics. Social change is not a momentum that occurs only once but can occur continuously or sustainably. Social change can cause changes in people's behavior, patterns of communication, and social interaction in society (Bhukya and Paul 2023; Saefudin et al. 2023).

Social change can be influenced by various factors, including industrialization, technological innovation or modernization, policy (regulation), the media, the desire for a better life, and the culture of society (Beno et al. 2022; Liu et al. 2020; Marijuán and Navarro 2020). From these factors, it can be said that various internal and external factors of a community can influence social change. Social change can have positive impacts, such as increased social welfare and respect for human rights (Castellacci 2023).

The description of the results of this study shows that social changes in the lives of the people of Baubau City play a significant role in the efforts to preserve the wetlands. Community behavior is always in line with efforts to preserve wetlands. The community realizes that wetlands are a place to live and earn a living. For them, damaging wetlands can disrupt their current and future well-being. The resilience of the wetlands in Baubau City has been maintained to this day. These results provide evidence that social change significantly affects the resilience of wetlands in urban areas.

14.5 Limitations of the Study

The limitations of this research are that it is only a limited study of the influence of social and cultural factors on urban wetlands, while there are still other factors that can affect urban wetlands. The sociocultural aspects in question include local wisdom factors in the community, government policies that have been formulated, and the social conditions/changes that occur. Based on these limitations, there are several constructive solutions for further research.

This research has not examined various other factors, such as (1) mitigation efforts carried out by the community/government in dealing with environmental disasters. This needs to be studied, considering the potential for environmental disasters, such as industrial and mining activities. These activities are quite prevalent at this time and can affect the balance of the ecosystem in the wetland area. (2) Soil structure and flora/fauna resilience. This needs to be studied because both will always experience dynamics that align with the times. These dynamics can certainly affect the resilience of wetlands, especially in urban areas. (3) The volume of waste the community or industry generates each year. This study needs to be carried out because the amount of waste will increase along with the increase in population, which can affect the ecosystem's resilience in urban wetlands. These various factors have an important role in the resilience of urban wetlands.

14.6 Recommendations

Wetlands in Baubau City have significant economic, cultural, and ecological value. As a result, planned actions are necessary to ensure the preservation and sustainable use of the area. There are several recommendations from this research, including: *First*, urban wetland management should be based on integrating local wisdom, such as the *kaombo* system, which has long been used by the people of Baubau City and demonstrates high principles of conservation and respect for nature. *Second*, the community must actively participate in preserving wetlands. Through cooperation and implementing customary rules prohibiting pollution, the people of Baubau City have committed to preserving the river's sustainability. *Third*, developing environmentally friendly infrastructure must be a top priority for the government. Creating a sustainable drainage and waste management system can reduce pollution and maintain the health of wetland ecosystems.

Fourth, the development of wetlands-based economies can also be a good option. Ecotourism programs, environmentally friendly fish farming, and traditional handicrafts can preserve wetlands' sustainability while improving the community's welfare. *Fifth*, the government, academics, nongovernmental organizations, and local communities must collaborate more. Forming a platform involving various stakeholders will produce more accurate and effective wetlands conservation formulations and strategies.

Sixth, continuous monitoring and evaluation are necessary to ensure conservation efforts succeed. The use of contemporary technology, such as community-based monitoring and geographic information systems (GIS), can help monitor the condition of wetlands and evaluate the policies implemented. *Seventh*, the government/stakeholders need to create contemporary policies by considering local wisdom, community participation, evaluation of previous policies, and community adaptation to social changes.

14.7 Conclusions

The conclusions of this study are: *First*, the local wisdom (culture) of coastal communities significantly influences the resilience of urban wetlands. The people of Baubau City have a unique culture of maintaining the resilience of wetlands. This culture is reflected in *kaombo*. *Kaombo* is a cultural institution that reflects a traditional conservation system for environmental resources (forests, rivers, and coastal and marine resources). *Second*, policies that have been announced or planned by the urban government have a significant effect on the resilience of urban wetlands. Baubau City Regulation Number 4 of 2014, concerning the Baubau City Spatial Plan for 2014–2034, mentions the function of the Baubau River as a provider of raw water networks, communal water supply systems, and primary drainage systems.

Third, with all their circumstances and social changes, urban coastal communities have a significant impact on the resilience of urban wetlands. The people of Baubau have a unique and diverse approach to preserving coastal/river areas, reflecting their local wisdom and social solidarity. They uphold customary rules prohibiting household or business waste disposal into rivers. This study has concluded that the support of local wisdom (culture), government policies, and social change is quite influential in efforts to maintain the resilience of urban wetlands. However, this study still has limitations. The limitations of this research are that it is only a limited study of the support of social and cultural factors on urban wetlands, while there are still other factors that can affect urban wetlands. Therefore, it is hoped that future research can examine the influence of various other factors that can affect the environmental resilience of wetlands in urban areas, such as soil structure and the resilience of flora/fauna around wetlands, mitigation efforts during natural disasters, the volume of waste produced by the community or industry each year, and so on. These various factors have an important role in the resilience of urban wetlands. The presence of this study will complement this research so that, in the future, a complete study of the resilience of urban wetlands will be created.

Acknowledgments We would like to express our deepest gratitude to the community leaders of Baubau City in Southeast Sulawesi Province, Indonesia, for their willingness to share insights and experiences that are very important for this research. We would also like to thank the Baubau City Government for providing us with support and assistance by providing various information relevant to this research. Finally, we are grateful to the peer reviewers and editors for their constructive feedback, which has provided various inputs to improve the quality of this chapter.

Potential Conflicts of Interest On behalf of all authors, the corresponding author states that there is no conflict of interest.

Informed Consent Informed consent was obtained from all individual participants included in the study.

Authors' Contribution All authors have contributed equally to this work.

Data Availability Statement The datasets during and/or analyzed during the current study are available from the corresponding author on reasonable request.

AI Discloser We always maintain the accuracy and reliability of the information we present in this chapter. We are also committed to ensuring that the use of AI in the preparation of this chapter is carried out ethically and responsibly.

Funding The research was not supported by any funding.

References

Abbass K, Qasim MZ, Song H, Murshed M, Mahmood H, Younis I (2022) A review of the global climate change impacts, adaptation, and sustainable mitigation measures. Environ Sci Pollut Res 29(28): 42539–42559. https://doi.org/10.1007/s11356-022-19718-6

Adhya T, Banerjee S (2022) Impact of wetland development and degradation on the livelihoods of wetland-dependent communities: a case study from the lower Gangetic floodplains. Wetlands 42(7):1–12. https://doi.org/10.1007/s13157-022-01588-w

Aguilar AG, Flores MA, Lara LF (2022) Peri-urbanization and land use fragmentation in Mexico City. Informality, environmental deterioration, and ineffective urban policy. Front Sustain Cities 4:1–19. https://doi.org/10.3389/frsc.2022.790474

Almetwally AA, Bin-Jumah M, Allam AA (2020) Ambient air pollution and its influence on human health and welfare: an overview. Environ Sci Pollut Res 27(20): 24815–24830. https://doi.org/10.1007/s11356-020-09042-2

Almutairi A, Mourshed M, Ameen RFM (2020) Coastal community resilience frameworks for disaster risk management. Nat Hazards 101(2). https://doi.org/10.1007/s11069-020-03875-3

Andrews N, Bennett NJ, Le Billon P, Green SJ, Cisneros-Montemayor AM, Amongin S, Gray NJ, Sumaila UR (2021) Oil, fisheries and coastal communities: a review of impacts on the environment, livelihoods, space and governance. Energy Res Soc Sci 75:1–15. https://doi.org/10.1016/j.erss.2021.102009

Askari KOA (2022) Review of the effects of the anthropogenic on the wetland environment. Appl Water Sci 12(12):1–18. https://doi.org/10.1007/s13201-022-01767-4

Assefa WW, Eneyew BG, Wondie A (2021) The impacts of land-use and land-cover change on wetland ecosystem service values in peri-urban and urban area of Bahir Dar City, upper Blue Nile Basin, Northwestern Ethiopia. Ecol Process 10(1):1–18. https://doi.org/10.1186/s13717-021-00310-8

Beno J, Silen A, Yanti M (2022) Globalisation, industrialisation, urbanisation, modernisation and I.C.T as correlates to developing social structure in Nigeria. Braz Dent J 33(1):1–12

Bhukya R, Paul J (2023) Social influence research in consumer behavior: what we learned and what we need to learn?—a hybrid systematic literature review. J Bus Res 162:1–21. https://doi.org/10.1016/j.jbusres.2023.113870

Bibri SE, Krogstie J, Kärrholm M (2020) Compact city planning and development: emerging practices and strategies for achieving the goals of sustainability. Dev Built Environ 4:1–20. https://doi.org/10.1016/j.dibe.2020.100021

BPS (2024) Kota Baubau dalam Angka Tahun 2024

Bridgewater P, Kim RE (2021) 50 years on, w(h)ither the Ramsar convention? A case of institutional drift. Biodivers Conserv 30(13):3919–3937. https://doi.org/10.1007/s10531-021-02281-w

Capano G, Howlett M (2020) The knowns and unknowns of policy instrument analysis: policy tools and the current research agenda on policy mixes. SAGE Open 10(1):1–13. https://doi.org/10.1177/2158244019900568

Capano G, Howlett M, Jarvis DSL, Ramesh M, Goyal N (2020) Mobilizing policy (in)capacity to fight COVID-19: understanding variations in state responses. Polic Soc 39(3):285–308. https://doi.org/10.1080/14494035.2020.1787628

Castellacci F (2023) Innovation and social welfare: a new research agenda. J Econ Surv 37(4):1156–1191. https://doi.org/10.1111/joes.12537

China CR, Maguta MM, Nyandoro SS, Hilonga A, Kanth SV, Njau KN (2020) Alternative tanning technologies and their suitability in curbing environmental pollution from the leather industry: a comprehensive review. Chemosphere 254:1–18. https://doi.org/10.1016/j.chemosphere.2020.126804

Clare S, Creed IF (2022) The essential role of wetland restoration practitioners in the science-policy-practice process. Front Ecol Evol 10:1–13. https://doi.org/10.3389/fevo.2022.838502

Cobbinah PB, Korah PI, Bardoe JB, Darkwah RM, Nunbogu AM (2022) Contested urban spaces in unplanned urbanization: wetlands under siege. Cities 121:1–14. https://doi.org/10.1016/j.cities.2021.103489

Dai W, Goodale E, He R, Mammides C, Liu S, Zhou L, Tang S, Su B, Lao X, Jiang A (2021) An eco-compensation policy increases shorebird diversity during the non-farming period for aquaculture. Wetlands 41(1):1–9. https://doi.org/10.1007/s13157-021-01397-7

Dajud GAB, Herazo LCS, Fernández-Lambert G, Marín-Muñiz JL, Méndez MCL, Betanzo-Torres EA (2022) Factors affecting wetland loss: a review. Land 11(3): 1–43. https://doi.org/10.3390/land11030434

Danso GK, Takyi SA, Amponsah O, Yeboah AS, Owusu RO (2021) Exploring the effects of rapid urbanization on wetlands: insights from the Greater Accra metropolitan area, Ghana. SN Soc Sci 1(8):1–21. https://doi.org/10.1007/s43545-021-00218-2

de Haas H (2021) A theory of migration: the aspirations-capabilities framework. Comp Migr Stud 9(1):8. https://doi.org/10.1186/s40878-020-00210-4

Dixon A, Wood A, Hailu A (2021) Wetlands in Ethiopia: lessons from 20 years of research, policy and practice. Wetlands 41(2):1–14. https://doi.org/10.1007/s13157-021-01420-x

Dube K, Nhamo G, Chikodzi D (2022) Flooding trends and their impacts on coastal communities of Western Cape Province, South Africa. GeoJournal 87(s4): 453–468. https://doi.org/10.1007/s10708-021-10460-z

Dupuis D (2022) The socialization of hallucinations: cultural priors, social interactions, and contextual factors in the use of psychedelics. Transcult Psychiatry 59(5): 625–637. https://doi.org/10.1177/13634615211036388

Ekumah B, Armah FA, Afrifa EKA, Aheto DW, Odoi JO, Afitiri AR (2020) Geospatial assessment of ecosystem health of coastal urban wetlands in Ghana. Ocean Coast Manag 193:1–10. https://doi.org/10.1016/j.ocecoaman.2020.105226

Erdoğan S, Çakar ND, Ulucak R, Danish, & Kassouri, Y. (2021) The role of natural resources abundance and dependence in achieving environmental sustainability: evidence from resource-based economies. Sustain Dev 29(1):143–154. https://doi.org/10.1002/sd.2137

Espinosa CM, Sauvage S, Al Bitar A, Green PA, Vörösmarty CJ, Sánchez-Pérez JM (2021) Denitrification in wetlands: a review towards a quantification at global scale. Sci Total Environ 754:1–16. https://doi.org/10.1016/j.scitotenv.2020.142398

Felbermayr G, Morgan TC, Syropoulos C, Yotov YV (2021) Understanding economic sanctions: interdisciplinary perspectives on theory and evidence. Eur Econ Rev 135:1–15. https://doi.org/10.1016/j.euroecorev.2021.103720

Glăveanu VP (2020) A sociocultural theory of creativity: bridging the social, the material, and the psychological. Rev Gen Psychol 24(4):335–354. https://doi.org/10.1177/1089268020961763

Hamid SK, Teniwut WA, Renhoran M, Teniwut RMK (2021) A novel framework for marine protected areas in small Island regions using integrated local wisdom. Reg Stud Mar Sci 45:1–10. https://doi.org/10.1016/j.rsma.2021.101819

Hölscher K, Wittmayer JM, Hirschnitz-Garbers M, Olfert A, Walther J, Schiller G, Brunnow B (2021) Transforming science and society? Methodological lessons from and for transformation research. Res Eval 30(1):73–89. https://doi.org/10.1093/reseval/rvaa034

Hu C, Wright AL, He S (2022) Public perception and willingness to pay for urban wetland ecosystem services: evidence from China. Wetlands 42(2):1–10. https://doi.org/10.1007/s13157-022-01538-6

Imdad K, Rihan M, Sahana M, Parween S, Ahmed R, Costache R, Chaudhary A, Tripathi R (2023) Wetland health, water quality, and resident perceptions of declining ecosystem services: a case study of Mount Abu, Rajasthan, India. Environ Sci Pollut Res 30(55): 116617–116643. https://doi.org/10.1007/s11356-022-21902-7

Ji J, Wang D (2022) Regional differences, dynamic evolution, and driving factors of tourism development in Chinese coastal cities. Ocean Coast Manag 226:1–11. https://doi.org/10.1016/j.ocecoaman.2022.106262

Kurniawan TA, Liang X, O'Callaghan E, Goh H, Othman MHD, Avtar R, Kusworo TD (2022) Transformation of solid waste Management in China: moving towards sustainability through digitalization-based circular economy. Sustainability 14(4):1–15. https://doi.org/10.3390/su14042374

Lage J (2022) Sufficiency and transformation—a semi-systematic literature review of notions of social change in different concepts of sufficiency. Front Sustain 3:1–21. https://doi.org/10.3389/frsus.2022.95466

Li H, Mao D, Wang Z, Huang X, Li L, Jia M (2022) Invasion of Spartina alterniflora in the coastal zone of mainland China: control achievements from

2015 to 2020 towards the sustainable development goals. J Environ Manag 323:1–11. https://doi.org/10.1016/j.jenvman.2022.116242
Lin B, Kaewkhunok S (2021) The role of socio-culture in the solar power adoption: the inability to reach government policies of marginalized groups. Renew Sust Energ Rev 144:1–9. https://doi.org/10.1016/j.rser.2021.111035
Liu Y, Zhu J, Li EY, Meng Z, Song Y (2020) Environmental regulation, green technological innovation, and eco-efficiency: the case of Yangtze river economic belt in China. Technol Forecast Soc Chang 155:1–21. https://doi.org/10.1016/j.techfore.2020.119993
Liu T, Yu L, Chen X, Wu H, Lin H, Li C, Hou J (2023a) Environmental laws and ecological restoration projects enhancing ecosystem services in China: a meta-analysis. J Environ Manag 327:1–13. https://doi.org/10.1016/j.jenvman.2022.116810
Liu X, Zhang J, Zhou Z, Feng Z (2023b) Investigating the impact of environmental graphics on local culture in sustainable rural cultural tourism spaces. Sustainability (Switzerland) 15(13):1–15. https://doi.org/10.3390/su151310207
Maltby E (2022) The wetlands paradigm shift in response to changing societal priorities: a reflective review. Land 11(9):1–42. https://doi.org/10.3390/land11091526
Mansoor M (2021) Citizens' trust in government as a function of good governance and government agency's provision of quality information on social media during COVID-19. Gov Inf Q 38(4):1–14. https://doi.org/10.1016/j.giq.2021.101597
Marijuán PC, Navarro J (2020) Sociotype and cultural evolution the acceleration of cultural change alongside industrial revolutions. Biosystems 195:1–12. https://doi.org/10.1016/j.biosystems.2020.104170
Masuda H, Kawakubo S, Okitasari M, Morita K (2022) Exploring the role of local governments as intermediaries to facilitate partnerships for the sustainable development goals. Sustain Cities Soc 82:1–12. https://doi.org/10.1016/j.scs.2022.103883
Mejjad N, Rossi A, Pavel AB (2022) The coastal tourism industry in the Mediterranean: a critical review of the socio-economic and environmental pressures & impacts. Tour Manag Perspect 44:1–15. https://doi.org/10.1016/j.tmp.2022.101007
Mentis M (2020) Environmental rehabilitation of damaged land. Forest Ecosyst 7(1):1–16. https://doi.org/10.1186/s40663-020-00233-4
Nirwana MA (2023) Regulatory approaches and strategies for household waste pollution management in the East River Flood Canal of Semarang City. Walisongo Law Review (Walrev) 5(1):55–72. https://doi.org/10.21580/walrev.2023.5.1.14770
Nuzul M, Achmad M (2023) Pengaruh Penggunaan Lahan Terhadap Debit Banjir Rancangan Di DAS Baubau. Dampak: Jurnal Teknik Lingkungan Universitas Andalas 19(1):22–29. https://doi.org/10.25077/dampak.19.1.22-29.2022
Nyirenda D, Sariola S, Kingori P, Squire B, Bandawe C, Parker M, Desmond N (2020) Structural coercion in the context of community engagement in global health research conducted in a low resource setting in Africa. BMC Med Ethics 21(1):1–10. https://doi.org/10.1186/s12910-020-00530-1
Pasya Ashsyafa Z, Sari W, Amalya Rahayu P, Dikrurahman D, Yusup Permana D (2024) Dynamics of interaction between law and socio-culture in the context of a multicultural society. Int J Soc Serv Res 4(4):1171–1178. https://doi.org/10.46799/ijssr.v4i04.773
Paszkowska MS, Milosz M, Powroznik P, Lukasik E (2022) 3D technologies for intangible cultural heritage preservation—literature review for selected databases. Herit Sci 10(1):1–24. https://doi.org/10.1186/s40494-021-00633-x
Penuelas J, Janssens IA, Ciais P, Obersteiner M, Sardans J (2020) Anthropogenic global shifts in biospheric N and P concentrations and ratios and their impacts on biodiversity, ecosystem productivity, food security, and human health. Glob Chang Biol 26(4):1962–1985. https://doi.org/10.1111/gcb.14981
Petridou E, Mintrom M (2021) A research agenda for the study of policy entrepreneurs. Policy Stud J 49(4): 943–967. https://doi.org/10.1111/psj.12405
Phillips C, Murphy C (2021) Solastalgia, place attachment and disruption: insights from a coastal community on the front line. Reg Environ Chang 21(2):1–14. https://doi.org/10.1007/s10113-021-01778-y
Razumovskaia E, Yuzvovich L, Kniazeva E, Klimenko M, Shelyakin V (2020) The effectiveness of Russian government policy to support smes in the COVID-19 pandemic. J Open Innov: Technol Mark Complex 6(4):1–20. https://doi.org/10.3390/joitmc6040160
Rojas O, Soto E, Rojas C, López JJ (2022) Assessment of the flood mitigation ecosystem service in a coastal wetland and potential impact of future urban development in Chile. Habitat Int 123:1–10. https://doi.org/10.1016/j.habitatint.2022.102554
Roziqin A, Mas'udi SYF, Sihidi IT (2021) An analysis of Indonesian government policies against. Public Adm Policy 24(1):92–107. https://doi.org/10.1108/PAP-08-2020-0039
Saefudin A, Santyaningtyas AC, Lubis AF, Mokodenseho S (2023) History, cultural shifts, and adaptation in social change: An ethnographic study in the Aboge Islamic community. J Innov Educ Cult Res 4(2): 303–310. https://doi.org/10.46843/jiecr.v4i2.596
Shah P, Atisa G (2021) Environmental education and awareness: the present and future key to the sustainable management of Ramsar convention sites in Kenya. Int Environ Agreem: Politics Law Econ 21(4):611–630. https://doi.org/10.1007/s10784-021-09534-7

Sinay L, Carter RWB (2020) Communities and local governments. Climate 8(7):1–15. https://doi.org/10.3390/cli8010007
Sinthumule NI (2021) An analysis of communities' attitudes towards wetlands and implications for sustainability. Glob Ecol Conserv 27:1–10. https://doi.org/10.1016/j.gecco.2021.e01604
Skirtun M, Sandra M, Strietman WJ, van den Burg SWK, De Raedemaecker F, Devriese LI (2022) Plastic pollution pathways from marine aquaculture practices and potential solutions for the North-East Atlantic region. Mar Pollut Bull 174:1–11. https://doi.org/10.1016/j.marpolbul.2021.113178
Suhendra D, Yahya A, Faisal S, Syarifuddin. (2020) How effective is the environmental law for the conservation of the Leuser ecosystem area in Indonesia? Adv Soc Sci Educ Humanit Res 413(276):61–67. https://doi.org/10.2991/assehr.k.200306.182
Taylor NG, Grillas P, Al Hreisha H, Balkız Ö, Borie M, Boutron O, Catita A, Champagnon J, Cherif S, Çiçek K, Costa LT, Dakki M, Fois M, Galewski T, Galli A, Georgiadis NM, Green AJ, Hermoso V, Kapedani R, Sutherland WJ (2021) The future for Mediterranean wetlands: 50 key issues and 50 important conservation research questions. Reg Environ Chang 21(2):1–17. https://doi.org/10.1007/s10113-020-01743-1
Van Assche K, Hornidge AK, Schlüter A, Vaidianu N (2020) Governance and the coastal condition: towards new modes of observation, adaptation and integration. Mar Policy 112:1–10. https://doi.org/10.1016/j.marpol.2019.01.002
Van der Wal Z (2020) Being a public manager in times of crisis: the art of managing stakeholders, political masters, and collaborative networks. Public Adm Rev 80(5):759–764. https://doi.org/10.1111/puar.13245
Wada JE, Kettenring KM, Sutton-Grier A (2020) Protecting wetlands for people: strategic policy action can help wetlands mitigate risks and enhance resilience. Environ Sci Policy 108:37–44. https://doi.org/10.1016/j.envsci.2020.01.016
Wang JJ (2020) How managers use culture and controls to impose a '996' work regime in China that constitutes modern slavery. Account Finance 60(4):4331–4359. https://doi.org/10.1111/acfi.12682
Wang YS, Gu JD (2021) Ecological responses, adaptation and mechanisms of mangrove wetland ecosystem to global climate change and anthropogenic activities. Int Biodeterior Biodegrad 162:1–14. https://doi.org/10.1016/j.ibiod.2021.105248
Wang J, Zhang H, Jin X, Ma L, Chen Y, Wang C, Zhao J, An T (2023) Subsidy policy with punishment mechanism can promote voluntary vaccination behaviors in structured populations. Chaos Solitons Fractals 174:1–10. https://doi.org/10.1016/j.chaos.2023.113863
Weible CM, Nohrstedt D, Cairney P, Carter DP, Crow DA, Durnová AP, Heikkila T, Ingold K, McConnell A, Stone D (2020) COVID-19 and the policy sciences: initial reactions and perspectives. Policy Sci 53(2):225–241. https://doi.org/10.1007/s11077-020-09381-4
Weise, K., Höfer R, Franke J, Guelmami A, Simonson W, Muro J, O'Connor B, Strauch A, Flink S, Eberle J, Mino E, Thulin S, Philipson P, van Valkengoed E, Truckenbrodt J, Zander F, Sánchez A, Schröder C, Thonfeld F et al (2020) Wetland extent tools for SDG 6.6.1 reporting from the satellite-based wetland observation service (SWOS). Remote Sens Environ 247:1–12. https://doi.org/10.1016/j.rse.2020.1118921
Yang X, Liu S, Jia C, Liu Y, Yu C (2021) Vulnerability assessment and management planning for the ecological environment in urban wetlands. J Environ Manag 298(72):1–14. https://doi.org/10.1016/j.jenvman.2021.113540
Zhang Z, Jiang W, Peng K, Wu Z, Ling Z, Li Z (2023) Assessment of the impact of wetland changes on carbon storage in coastal urban agglomerations from 1990 to 2035 in support of SDG15.1. Sci Total Environ 877:1–15. https://doi.org/10.1016/j.scitotenv.2023.162824
Zhao L, Li T, Przybysz A, Guan Y, Ji P, Ren B, Zhu C (2021a) Effect of urban lake wetlands and neighboring urban greenery on air PM10 and PM2.5 mitigation. Build Environ 206:1–14. https://doi.org/10.1016/j.buildenv.2021.108291
Zhao J, Xi X, Na Q, Wang S, Kadry SN, Kumar PM (2021b) The technological innovation of hybrid and plug-in electric vehicles for environment carbon pollution control. Environ Impact Assess Rev 86:1–13. https://doi.org/10.1016/j.eiar.2020.106506
Zhong Y, Xue Z, Davis CC, Moreno-Mateos D, Jiang M, Liu B, Wang G (2022) Shrinking habitats and native species loss under climate change: a multifactorial risk assessment of China's inland wetlands. Earth's Future 10(6):1–27. https://doi.org/10.1029/2021EF002630
Zhou W, Li M, Achal V (2024) A comprehensive review on environmental and human health impacts of chemical pesticide usage. Emerg Contam 11(1):1–13. https://doi.org/10.1016/j.emcon.2024.100410

15 Environment and Community Conflict Management: Exploring the Scope of Community Engagement as an Efficient Measure to Localized Common-Pool Resource (CPR) Management in Halti Beel, Bangladesh

Anutosh Das, Most. Mousumi Akhter Putul, and Borsha Ghosh

Abstract

This chapter examines the mismanagement of Common-Pool Resources (CPRs) in the Halti Beel area, focusing on environmental degradation, resource depletion, and conflicts. Employing Participatory Rural Appraisal (PRA) methodology, the research aims to assess current CPR extraction, engage the community, and empower locals for sustainable resource management. Various participatory tools such as social and resource mapping, transect walks, mobility mapping, dream mapping, and SWOT analysis are utilized to collect data and engage the community in resource management. The study examines conflicts within the Halti Beel community that stem from resource extraction. Using resource mapping and historical analysis, it uncovers contradictions among community members. It highlights Halti Beel's ecological importance, emphasizing its diverse biodiversity, and addresses the challenges of managing shared resources, particularly land, water, and fisheries. The study reveals conflicts between farmers and fishermen, issues with land acquisition, wetland management, and implications of over-exploitation. Factors contributing to degradation include corruption, bird-killing, improper fishing methods, and a lack of employment opportunities. Recommendations include community engagement, negotiation, localized management, and better regulation implementation. Policy suggestions involve environmental and participatory governance and time-bound resource use agreements. The study underscores the importance of managing resource consumption for sustainable development, especially in addressing over-exploitation in developing countries like Bangladesh. It serves as a baseline for future conflict management efforts related to Common-Pool Resource degradation.

Keywords

Common-Pool Resources (CPR) · Participatory Rural Appraisal (PRA) · Community conflict · Resource management · Environmental degradation · Sustainable development

15.1 Introduction

The environment is a critical life support system, yet the Earth's reservoir of ecological resources is finite. Humans' relentless pursuit of growth and

A. Das (✉) · M. M. A. Putul · B. Ghosh
Department of Urban & Regional Planning, Rajshahi University of Engineering & Technology, Rajshahi, Bangladesh
e-mail: anutosh@urp.ruet.ac.bd

S. K. Chakraborty et al. (eds.), *Wetlands and Urban Resilience*, Sustainable Development Goals Series,
https://doi.org/10.1007/978-3-032-06881-1_15

comfort pushes the environment to its limits. In developing nations such as Bangladesh, green spaces[1] have become scarce due to rapid urbanization and the disappearance of wetlands, exacerbating environmental issues and increasing vulnerability to urban flooding and natural disasters. Bangladesh is identified as the most vulnerable country to extreme climate events by the Global Climate Risk Index 2021 and the Sixth IPCC Assessment Report (AR6, 2023).[2] The Global Goals and New Urban Agenda challenge developing nations to create inclusive, safe, and resilient human settlements. The extravagant use and free access to common-pool characteristics of these resources have made them even scarcer for future generations to consume. This eventually leads to the tragedy of the commons.[3] G. Hardin, in his famous book "Tragedy of the Commons," argued that the communities that share resources inexorably pave the way for their destruction. In a reverse way, the tragedy of the commons reappears in problems of environmental degradation itself. In this way, the economy and the environment are inextricably linked. As resources are scarce, choices must be made about how to use them optimally and effectively (Hardin1968; Ostrom 1999; Gordon 1954).

Widespread individual excesses and inadequate regulatory measures exacerbate environmental degradation in Bangladesh. Despite global case studies, there is a significant research gap, highlighting the need for localized, field-level investigations to understand the socioeconomic and geopolitical context. Such studies are essential to devise effective policy measures for optimal management of common-pool natural resources in the country (Das et al. 2021). The management of Common-Pool Resources often leads to disputes and conflicts of interest among different groups, emphasizing the urgent need for dispute resolution measures in environmental preservation, particularly in countries like Bangladesh.

Common-Pool Resources make important contributions to the livelihoods of many rural populations in developing countries. Degradation and loss of Common-Pool Resources are caused by increasing population pressure and high levels of extraction, such as the cultivation of high-yielding varieties of crops, fishing, and extraction of water, grazing, construction of roads, dams, bridges, urban sprawl, and new settlements. The increasing pressure on CPRs tends to increase the likelihood of conflicts over access to them, so conflict issues can be expected to grow in importance.

The primary objectives of this study are two-fold. First, the study aims to assess the current state of Common-Pool Resource (CPR) extraction and identify issues related to the Tragedy of the Commons within the selected study area. This involves examining patterns of resource use, levels of exploitation, and the socio-environmental impacts arising from unregulated or poorly managed CPR utilization. The identification of these issues will be based on direct observation, stakeholder input, and analysis of relevant environmental and socioeconomic indicators. Second, the study seeks to evaluate the potential of community participation as a viable solution for the effective management of CPRs. This includes exploring the current level of community engagement, local governance structures, and existing practices that involve community-led conservation or resource management. The analysis will incorporate qualitative and quantitative methods to assess the feasibility and impact of participatory approaches in resource governance.

Based on these two core objectives, the study will propose context-specific strategies to strengthen sustainable CPR management through

[1] Green spaces are areas of natural vegetation within urban or rural environments that support ecological balance, provide recreational areas, and help improve air quality and overall well-being.

[2] The Global Climate Risk Index (GCRI) ranks countries based on their vulnerability to extreme weather events, highlighting regions most affected by climate change impacts to guide urgent adaptation efforts.

[3] The "tragedy of the commons" is a concept describing how individuals, acting in their self-interest, can overuse and deplete shared resources (like air, water, or fisheries), ultimately harming the entire community that depends on them.

enhanced community involvement. These recommendations aim to mitigate resource degradation, reduce user conflicts, and promote long-term ecological and socioeconomic resilience in the study area.

15.2 Study Area

The primary focus of the research is centered on the Common-Pool Resources, namely Wetland services, water, and fisheries, within the Halti Beel region in Natore (Fig. 15.1). Halti Beel, situated in the Natore Sadar Upazila (sub-district) of Natore district, holds significant importance as a wetland in north-western Bangladesh. It is characterized as a semi-closed perennial, irregularly shaped water body. This particular wetland is renowned for its abundant fish production throughout the year, upon which the livelihoods of numerous fishermen from neighboring villages heavily rely. Positioned between two rivers, namely the Atrai and the Barnai, Halti Beel receives regular floodwater during the monsoon season from these two rivers. The total expanse of the water body measures approximately 1012.5 hectares during the monsoon season and 15.95 hectares during the dry season. The depth of the water fluctuates between 1.5 and 6 meters, contingent upon the season.

Halti Beel encompasses the regions of Madhnogor, Brahmapur, Khajura, and Piprul union. This wetland harbors a complex ecosystem and is well-known for its natural fish breeding, as it is connected to the Atrai River. Additionally, Halti Beel is categorized as a Jal mohal (fishery estate) and operates under a leasing system of management, which is administered by the Ministry of Land. The District and Upazila administration, acting on behalf of the Ministry of Land,

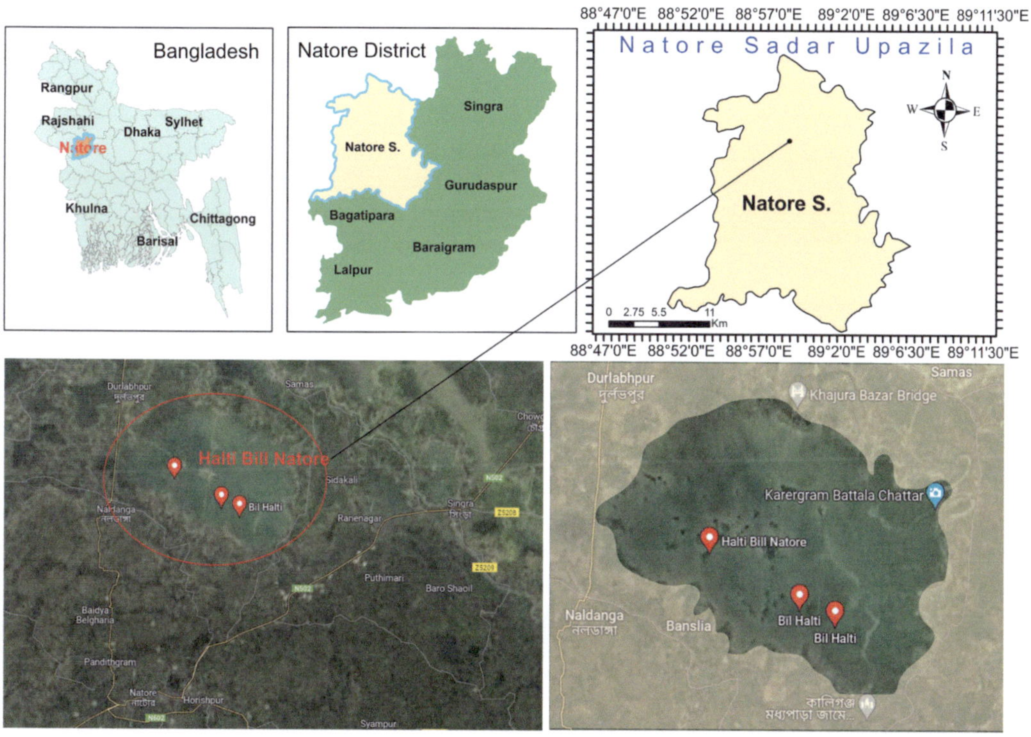

Fig. 15.1 Case study area of Halti Beel, Natore. This figure shows the study area location. (Source: prepared by the authors 2023)

possess the executive authorities for the collection of revenue from the leasing of Jal mahal.[4]

The community in Halti Beel comprises two main groups: Farmers, cultivating land seasonally and facing unemployment in the rainy season, and Fishermen, who are engaged in year-round fishing. Resource competition between the two groups is evident, driven by concerns about depletion. Two types of fishermen, Halder (Hindu) and Jioni (Muslim), reside in specific areas. The Beel hosts 63 fish species, with some critically endangered, endangered, and vulnerable (Galib and Imteazzaman 2013, Alamgir 1988). Conservation efforts, like "tanki"[5] have improved fish diversity. Halti Beel is a crucial wetland habitat for migratory and local birds, though wildlife hunting, particularly waterfowl, persists, driven by economic necessity. Legal measures against hunting exist, but lack robust enforcement.

15.3 Materials and Methods

The investigation utilizes the case study approach employing Participatory Rural Assessment (PRA) methodology, leveraging its cost-effectiveness and time-saving features to access and analyze information on rural conditions. Various PRA tools are employed for data collection, emphasizing community engagement as a crucial tool for development practitioners to gain a comprehensive understanding of rural issues, potentials, resources, and solutions within a specified timeframe (Chambers 1992). The literature on natural resources and communally owned property for resource economics and resource planning provides numerous instances of how community engagement is currently becoming a key strategy in rural development (Baginski et al. 2003, Kasch 2023). Participatory procedures locally open up the prospect of establishing connections between sustainable livelihoods, awareness webs, consciousness schemes, and development techniques (Thomson and Gray 2009, Tough 1987, Walker et al. 2000).

The study utilizes participatory rural assessment to explore the impact of land policy and structural adjustment on the poor, emphasizing the significance of community involvement in providing a realistic understanding of current resource management within the Common-Pool Resources. The implementation of participatory rural assessment instruments for our specific aims is shown in Table 15.1. The study employs participatory rural assessment tools, including resource and social network mapping, transect walks for land resource evaluation, and mobility mapping to identify key sources for information on the management problems related to land in the research area.

The study prioritizes community expectations in Common-Pool Resource management, utilizing dream mapping to grasp community aspirations. Employing participatory research methods, including Participatory Rural Appraisal, focus group talks, and key informant interviews, the study integrates a seasonal diagram (Figs. 15.8, 15.9) to assess conflict likelihood and employs SWOT analysis to identify strategies for mitigating resource deterioration.

Semi-structured questions were designed to prompt essential conversations and gather data in group discussions, with secondary databases from various administrative offices, such as local agencies, union parishad, AC land office, and BMO, used as a guide, and their information verified for accuracy.

The study emphasizes the crucial role of focus group discussions in Participatory Rural Appraisal (PRA), employing an expert knowledgeable in various ideologies. It extensively gathers data on watersheds, soil and water conservation, agroforestry, fisheries, aquaculture, biodiversity, and wildlife reserves. The focus is on understanding resource extraction and conflicts and providing policy recommendations for the management system, specifically in crops and animal husbandry, utilizing methods like

[4] A **Jal mahal** is a government-leased water body in Bangladesh designated for fishing, managed to support local livelihoods and generate revenue.

[5] "Tanki" refers to a conservation effort or practice aimed at enhancing fish diversity and preserving fish populations within the wetland ecosystem.

Table 15.1 List of PRA tools to be applied in the study area

Issue(s) purposes	Tools used
Assessment of community (physical/natural) assets	Social and resource mapping
Exploration of local resources and development conditions, and observation of physical, natural, and capital assets	Transect walk
Exploration of mobility pattern of the community	Mobility map
Exploration of the aspirations of community people	Dream map
Exploration of the degradation of resources and conflict	Cause-effect diagram
Exploration of resources and how they are extracted and the degradation of the resource, and their conflict of extraction	Focus group discussion
Gaining in-depth knowledge of specific issues, structures, and organizations	Key information discussion
Income generation present, future and trend, and constraints	Seasonal diagram
Strategies to reduce the degradation of resources	SWOT analysis

Source: Prepared by the authors 2023

farmer participatory research, farming systems research, and problem identification in irrigation and marketing. Various technologies, including social and resource mapping, transect mapping, seasonal diagrams, mobility mapping, and dream maps, were employed in the study field to understand the relationships between residents and shared resources in Halti Beel, determining the current state and tragedy of common resource extraction. Techniques such as social and resource mapping (Figs. 15.3, 15.7) were crucial in identifying the Common-Pool Resources, assessing their deterioration, and evaluating the economic standing of the Halti Beel community, while studies on shared resource reliance utilized seasonal calendars and bio-resource flow diagrams to enhance problem-solving discussions and generate farmer-driven solutions (Atieno et al. 2014, Pomeroy 1999). With the use of a transect walk (Fig. 15.4), local resources, physical and natural assets, and their development pattern were explored. To understand the seasonal variations in terms of migration, cropping patterns, and income production, seasonal diagrams (Figs. 15.8 and 15.9) were employed.

A mobility map (Figs. 15.5 and 15.6) was employed to identify daily challenges, such as disputes and issues related to land management and livelihood, while data on land-based Common-Pool Resources (CPRs) contribution was gathered through various instruments. The survey analysis followed sequences such as mobility patterns, distance, trip purpose, frequency, and the gender of trip makers, supplemented by key informant surveys after group discussions and case studies detailing household history, crisis management techniques, and conflict resolution strategies. The study's topic has been chosen with the aid of secondary data sources. The study's goal and aim were taken into consideration when choosing the study region. The study's characteristics and those of Halti Beel are identical. Sources, both primary and secondary, were used to gather the data. Ultimately, the data were interpreted to provide recommendations.

15.3.1 Literature Review

Common resources are goods that cannot be sufficiently excluded from private use and are often shared by a community or society. They are characterized by collective ownership and have the potential to be neglected or overused if not properly regulated, as mentioned by Garrett Hardin in his classic article "The Tragedy of the Commons" (1968). Therefore, it is crucial to engage in cautious and sustainable administration of these resources since numerous households rely on them for their livelihood and sustenance characterize commons as resources that can prevent users from accessing them. Common-Pool

Resources(CPRs) encompass the natural resources consumed by a group of individuals or a community.

It is crucial to differentiate between open-access Common-Pool Resources and common property resources, with the key distinction lying in property rights; open-access resources lack ownership and governance, making them susceptible to over-exploitation and degradation, sharing characteristics with both private and public goods according to Ostrom. In contrast, Bromley (1990:20) refutes the existence of such common resources. He posits that only resources controlled and managed as common property, state property, or private property are present. The latter, common property regimes, postulate that there is a degree of human involvement through formal or informal institutions to manage and regulate it (Bromley 1990; Osterm et al. 1994; Singh et al. 1997, Smith 1981). It operates and functions as property regimes both in the private and state spheres (Bromley 1990: 3). This encompasses a set of rules that delineate access, use, exclusion, management, monitoring, sanctioning, and arbitration behaviors of users about specific resources (Schlager and Ostrom 1992; Blackburn and Holland 1998,Kabir et al. 2011). The present study deems the Common-Pool Resources of Halti Beel as common property resources/regimes, as the common resources consumed by villagers have been autonomously managed and controlled by themselves since ancient times.

The concept of Common-Pool Resources (CPR) emerged in response to Hardin's "Tragedy of the Commons" in 1968, where he advocated for privatization or state management to prevent degradation; however, his oversimplified approach faced criticism, prompting later writings to provide corrections and refine the understanding of CPR. Berkes et al. (1990) criticized Hardin's insightful but incomplete article, prompting a correction in 1998. Hardin acknowledged the mistake of not using the term "unmanaged commons" and highlighted that unmanaged commons would face issues like overgrazing. Hardin's model significantly influenced global development and environmental policies, often misguiding them with the assumption that common property resources are open access. Scholars noted two key factors driving changes: the increasing demand for natural resources and environmental services, and the techniques applied by humans to manage these resources (Tang 1991, Thompson et al. 1999, Thompson et al. 2003).

CPR theorists, like Ostrom (1999) and Agrawal (2003), have developed conditions for sustainable management, emphasizing small user groups, proximity to resources, group homogeneity, effective enforcement mechanisms, and past experiences of cooperation. They identified regularities for successful CPR management, considering characteristics of resources, the nature of user groups, institutional regimes, and the relationship with external forces. The Community-Based Natural Resource Management (CBNRM) approach draws theoretical insights from CPR theories. While initially designed based on conditions postulated by earlier theorists, contemporary CPR theorists have evolved their thinking beyond conventional understandings.

Past research on the local communities in Manang Valley of Manang and Dolpo of Dolpa in Nepal has acknowledged that villagers demonstrate a willingness to adhere to societal rules and regulations when utilizing Common-Pool Resources (CPRs). This compliance may be attributed to the longstanding embedded rules and regulations that have been followed for generations in managing, controlling, and distributing such resources. These local regulations contribute to the "established structure of power and authority" within the community (Dasgupta 1995:208; cited in Sethi & Somanathan, 1996: 768). The communities also possess the capacity to enforce specific behavioral rules and bear the costs associated with their implementation (Ibid).

For example, in the Manang Valley, farmers adhere to fixed planting and harvesting dates determined by the village headman, who consults with a lama (Buddhist priest). This practice, rather than spreading the risk by extending the planting and harvesting seasons, is followed to prevent potential damage to all crops. Similarly, in Dolpo, individuals come to agreements with their neighbors regarding a set of enforceable rules and regulations that govern households' access to and use of community resources (Siddique et al. 2016,2020)

CPR theorists also recognize the pivotal role of local institutions in shaping human behavior. Local norms play a crucial role in influencing the management and control of common resources, as they both facilitate and constrain the actions and property rights of individuals (Agrawal 2003:244). When people realize their dependence on CPRs, they establish their institutional arrangements to allocate and distribute resources and benefits equitably. Participatory Rural Appraisal (PRA) is a community-focused approach involving various techniques, emphasizing active engagement, cross-verification, and "optimal ignorance" to gather accurate information. PRA integrates rural communities' knowledge through methods like mapping, transect walks, and mobility maps, facilitating community involvement in project planning. Analyzing urban agriculture and gender trends, PRA reveals changes in cropping patterns, land access, and gender participation. The concept of community involves a social unit with shared norms, values, and identity, while the Biodiversity Management Organization (BMO) in Halti Beel manages natural resources with a focus on fisheries, agriculture, and animal resources, employing legal aid provisions and membership categories.

15.4 Results and Discussions

15.4.1 Existing Conditions of the Study Area

15.4.1.1 The Existing Social Status of Communities

Farmers are generally wealthier than fishermen in the described area (Fig. 15.2). They reside in well-constructed two-story houses, primarily of bricks, while some have semi-permanent structures. Farmers take the initiative to build barricades to protect their homes from floodwaters, particularly after shifting to high-yield crops. Despite economic prosperity from seasonal crop cultivation, they often migrate or engage in various activities during the off-season. Landownership is significant, with an average of 4–5 bighas per individual, and some owning up to 100 bighas.

Fishermen, divided into Hindu ("Halder") and Muslim ("Jioni") communities, live in specific areas. A fishermen's association, "Motsho Somobai Somity," with 2309 members contributing BDT 20 monthly, promotes water body development ("Jal mahal"). Only somity members can fish during the dry season, while non-members can do so by making a monetary deposit. Fish are collectively harvested, and proceeds are evenly distributed. Unaffiliated fishermen may face unemployment during the dry season, leading to migration. Overall, there's a clear distinction in economic status and housing between farmers and fishermen in the community.

Communities' Basic Services and Facilities

Different types of facilities are available there (Fig. 15.2). There are six primary schools, one high school, one madrasa, and one community clinic. There are also seven mosques and three bazaars. Recently, there has been a restaurant under construction. There is a memorial as a tourist spot (Table 15.2).

15.4.1.2 Observation of Physical, Natural, and Capital Assets

The transect map (Fig. 15.3) covers locations in Halti Beel, Natore district, Rajshahi division, along the main street from south to north, encompassing three villages connected by the primary road. Data collection focused on key variables: land type, crops, trees, livestock, fish, and soil type, including pitch-covered streets. However, some sections have gaps in pitch coverage, affecting street conditions. The locals predominantly cultivate catfish, hornfish, silver carp, tilapia, etc. The landscape is adorned with various trees, including neem, mango, guava, jackfruit, tamarind, litchi, date palm, coconut, betel nuts, papaya, and banyan, providing seasonal fruits. Additionally, households along the street raise diverse livestock. The collected data has been organized into a tabular format (Fig. 15.3) for analysis.

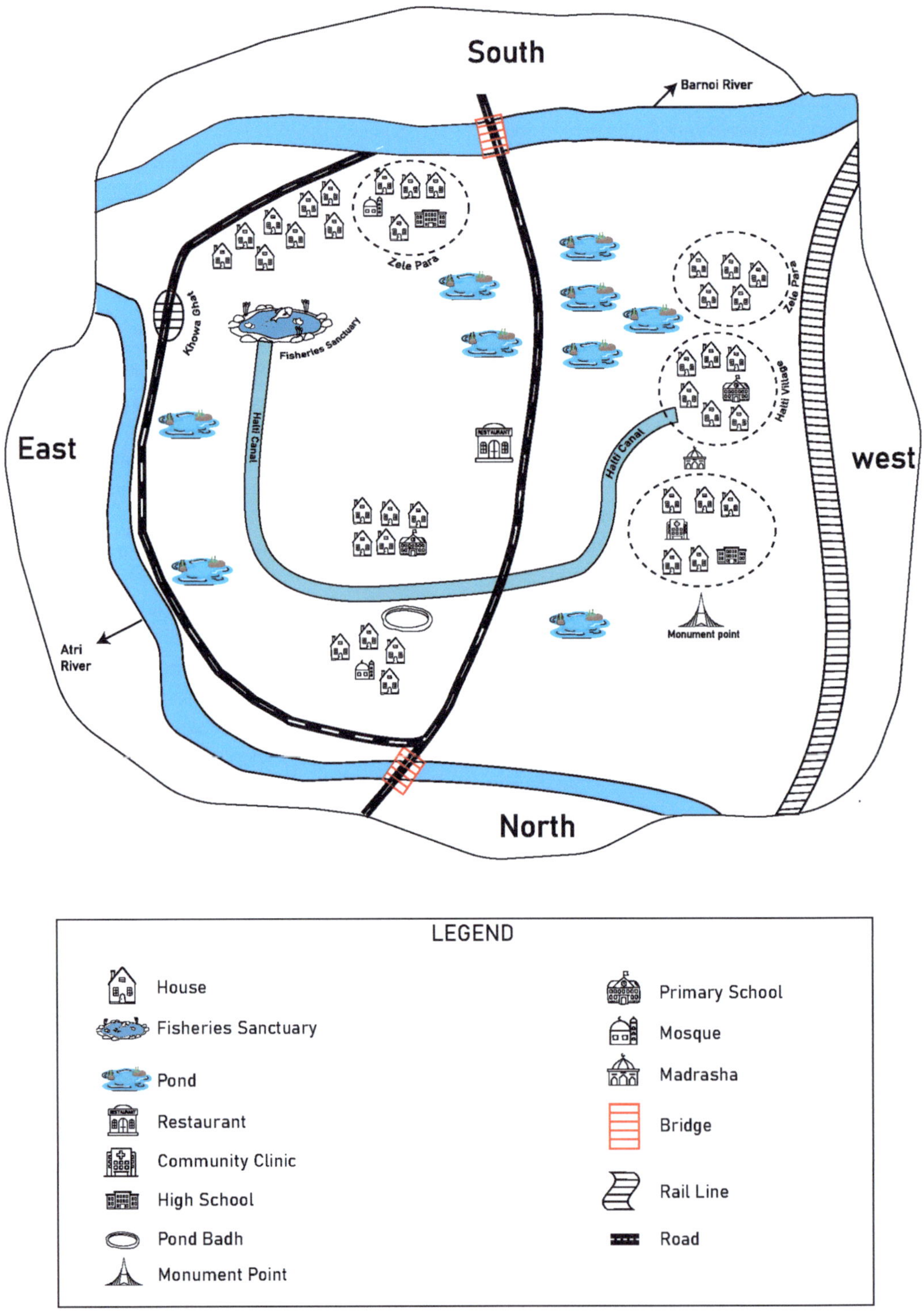

Fig. 15.2 Social map. Community-based social mapping of the study area. (Source: Field Survey, 2023)

Livestock in the area include goats, buffaloes, ducks, hens, and swans. During the dry season, paddy cultivation occurs in the flat land areas. However, there has been a noticeable decline in both the value and quantity of resources over time. This decline is attributed to a significant increase in consumption and changes in geographic conditions. Consumption patterns are closely tied to ownership. According to local villagers, individuals who invest capital in crop cultivation tend to benefit from fixed land. Those involved in cultivating crops are the primary consumers of the harvested produce. A consensus has emerged that individuals consuming a particular crop should refrain from cultivating it to maintain a balance. The crops harvested include paddy, corn, pulses, cabbage, cauliflower, ladies' finger, and "Momordica Charantia."

Table 15.2 Basic services and facilities

Basic services	Number
Primary school	6
High school	1
Madasa	1
Mosque	7
Restaurant	1
Bazar	3
Community clinic	1

Source: Field Survey 2023

The soil types comprise atel, bele-do-ash, do ash, and bele, with the most common being atel bele-do-ash and bele. The soil along roads consists of bele and atel, while cultivated land primarily features bele-do-ash soil. The soil in the house premises is of the bele type, allowing for effective drainage and keeping the area dry. Houses are constructed using a combination of brick, tin, and straw materials. Moving along the street from south to north, some areas serve as agricultural land within the village premises. Occasionally, these lands contain low canals meticulously used for fish cultivation.

The transect map (Fig. 15.3) delineates resource boundaries, identifying individuals with the right to harvest at night, and land ownership privileges are confined within these boundaries, allowing for the establishment of rules. The area, rich in wild resources, serves as a habitat for migratory birds in winter, considered a significant tourist destination by villagers who propose government involvement in establishing tourist motels. The Common-Pool Resources, providing

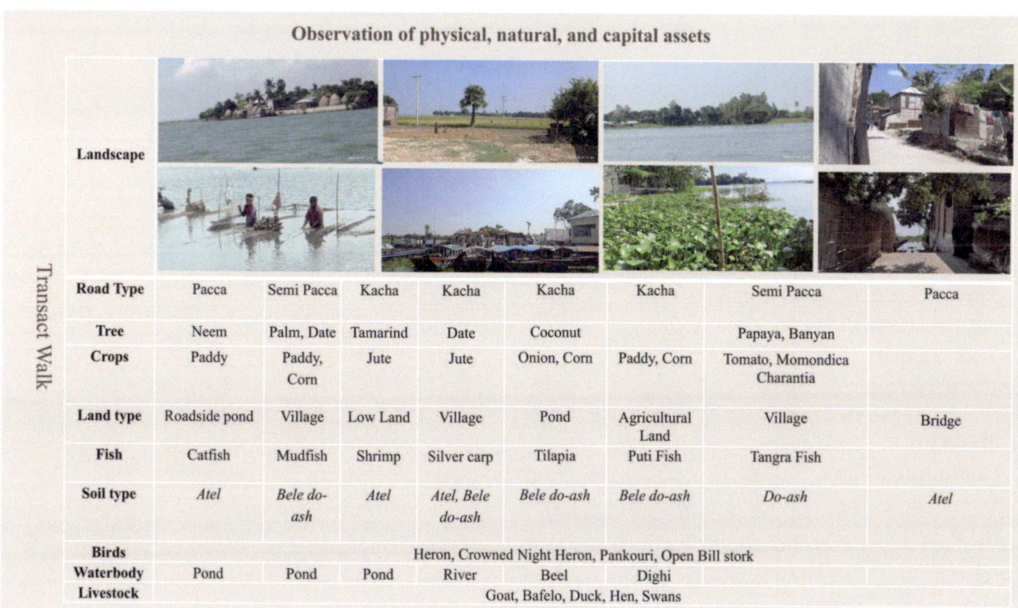

Fig. 15.3 Transect map. Transect mapping of physical, natural, and capital assets across different land use zones. (Produced by the focus group participants)

forage for livestock, play a crucial role in supporting their livelihoods. The moderate pressure on Common-Pool Resources at Halti Beel impacts the living environment and potentially increases benefits for specific groups. Intensifying pressure leads to conflicts, particularly between fishermen and farmers, stemming from one villager consuming resources to the detriment of others. Economic indicators such as income, land ownership, and livestock ownership measure household economic conditions in this context, where the government owns the majority of land and redistributes it to economically vulnerable individuals without property.

15.4.1.3 Mobility Pattern of the Community

The creation of a mobility map (Fig. 15.4) for villagers is essential for understanding their interactions with Common-Pool Resources and

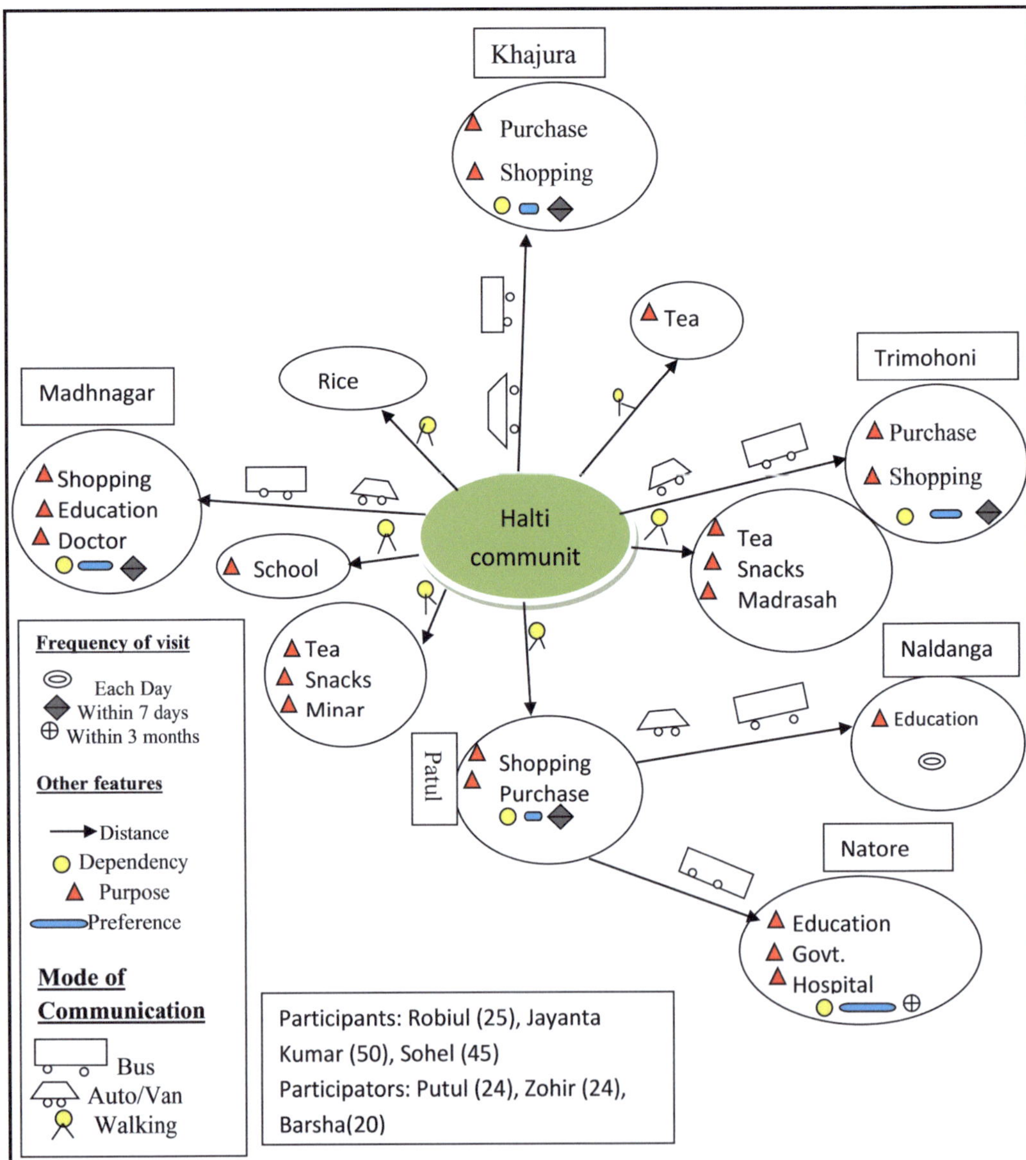

Fig. 15.4 Mobility map for male. (Source: Field Survey, 2023)

related problems. To achieve this, several steps must be followed. Identifying the target group is crucial, as it helps elucidate individuals closely connected to the identified issues. Communicating the map's purpose to the community allows them to provide insights into activities related to Common-Pool Resource problems. Once the target individuals are identified, they are asked to draw places they frequent, using alphabetical letters and symbols to denote locations and distances. In Halti Beel, a mobility map (Fig. 15.4) was conducted to establish significant relationships between government bodies, institutional facilities, and the targeted people. The selection of target individuals involved understanding their activities and conversations, with careful evaluation of their input. Clear communication about the objectives ensured accurate representation on paper.

The Halti community comprises five major localities, with people visiting locations such as rice mills, schools, and tea stalls daily. Patul is a frequently used area, serving as a hub for meals, education, and shopping, with its stalls remaining open during the dry season. To access land-related services, people visit the AC land office near Patul Union Parishad. The mobility pattern (Fig. 15.4) also extends to Natore Upazila, where community members go for medical check-ups, clothing purchases, and better educational facilities. Fishermen travel to Madhna Gorstation to sell fish to other parts of the country, leveraging the area's reputation for quality fish. Additionally, children attend educational facilities in Noldanga. The mobility map (Fig. 15.4) effectively summarizes the community's movement patterns and interactions with various facilities.

It is a well-established fact that, in the social context of Bangladesh, women are less engaged in the daily activities of rural villages compared to men. The women residing there obtain essential individual services, such as emergency assistance, for themselves. Girls residing in Noldanga, which is the closest vicinity to Patul, have access to educational opportunities. Rawson Ara, a resident of our research area, expressed, "In Patul, we never receive readily available healthcare services; we have to rely on the hospital in Natore." Women receive healthcare from community clinics, which is not sufficient for them. During the dry season, they utilize auto vans and rickshaws to access their desired locations. Regrettably, the community lacks access to bus services. To reach Natore or Rajshahi, they travel by auto to the Natore-Rajshahi road and then switch to buses or CNG vehicles. The mobility behavior of women (Fig. 15.5) has been effectively depicted through a mobility map (Fig. 15.5) specifically designed for women.

15.4.2 Resource Extraction and the Tragedy of the Commons

15.4.2.1 Assessing Resource Extraction and Conflict Dynamics Within the Community

Halti Beel is renowned for its intricate ecosystem and serves as a natural breeding ground for fish, connecting to the Atria River. Within Halti Beel, there are 63 fish species, including 3 critically endangered, 11 endangered, and 8 vulnerable species (Imteazzaman and Galib 2013, Hanna 1998, Stevenson 1991). Among the endangered species are Bata, calbaus, pabda, chitol, bou, lohachata, tangra, boali pabda, gojar, and baim, while the vulnerable ones include foli, raikhor, titoputi, gulsha tangra, chanda, ranga chanda, and chang (Adekola and Olumati 2023, Ni Nyoman Clara Listya 2023, Participatory Forestry 1996). Halti Beel experiences water levels ranging from 5 to 8 feet underwater during the Bengali months of Ashar to Kartik. A designated portion of Halti Beel has been declared a fishery sanctuary, serving as a habitat and natural breeding ground for fish. During the pre-monsoon period, fishing is prohibited, and the monsoon period becomes a vital source of livelihood as people engage in fishing activities.

Water stands out as a critical resource for Halti Beel, characterized as a semi-closed perennial water body with an irregular shape. This wetland is particularly known for its consistent fish production, supporting the livelihoods of numerous fishermen in nearby villages. Positioned between

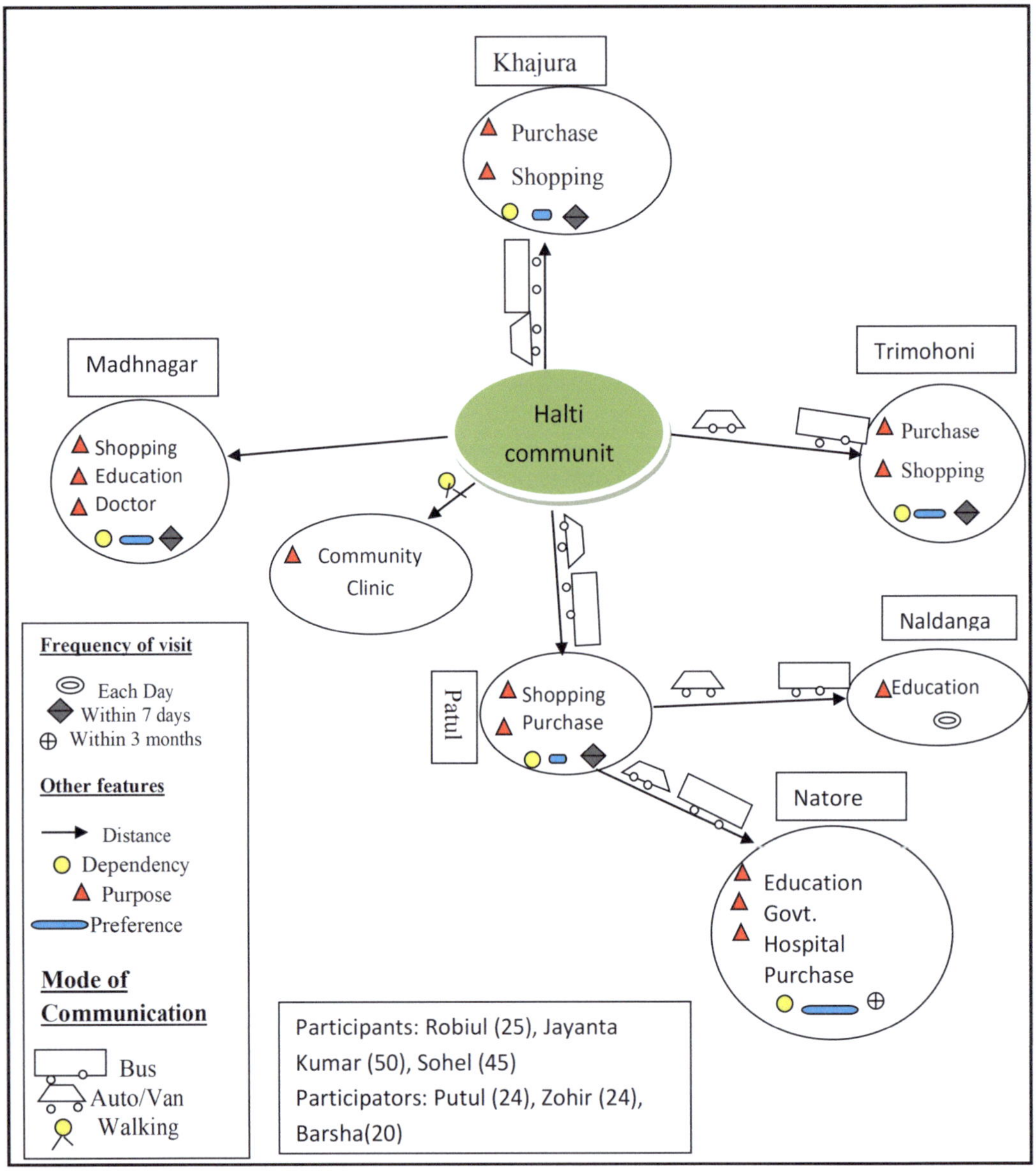

Fig. 15.5 Mobility map for female. (Source: Field Survey, 2023)

the Atrai and Barnai rivers, Halti Beel receives regular floodwaters during the monsoon season. There is an association known as "Motsho Somobai Somity" that collects contributions from its members to fund the development of "Jal mahal." "Jalmahal" is also known as "Tanki" locally. This somity also organizes various training sessions for fishermen. Before the establishment of the fishery sanctuary, there was a shortage of fish, particularly larger ones. The construction of "tanki" resulted in the presence of a diverse range of fish, including larger specimens weighing around 10–15 kg, in the beel (Fig. 15.6).

During the dry season, only somity members collaborate to catch and sell fish in the markets. In contrast, during the rainy season, all individuals are allowed to catch fish, but those using large fishing nets are typically fishermen.

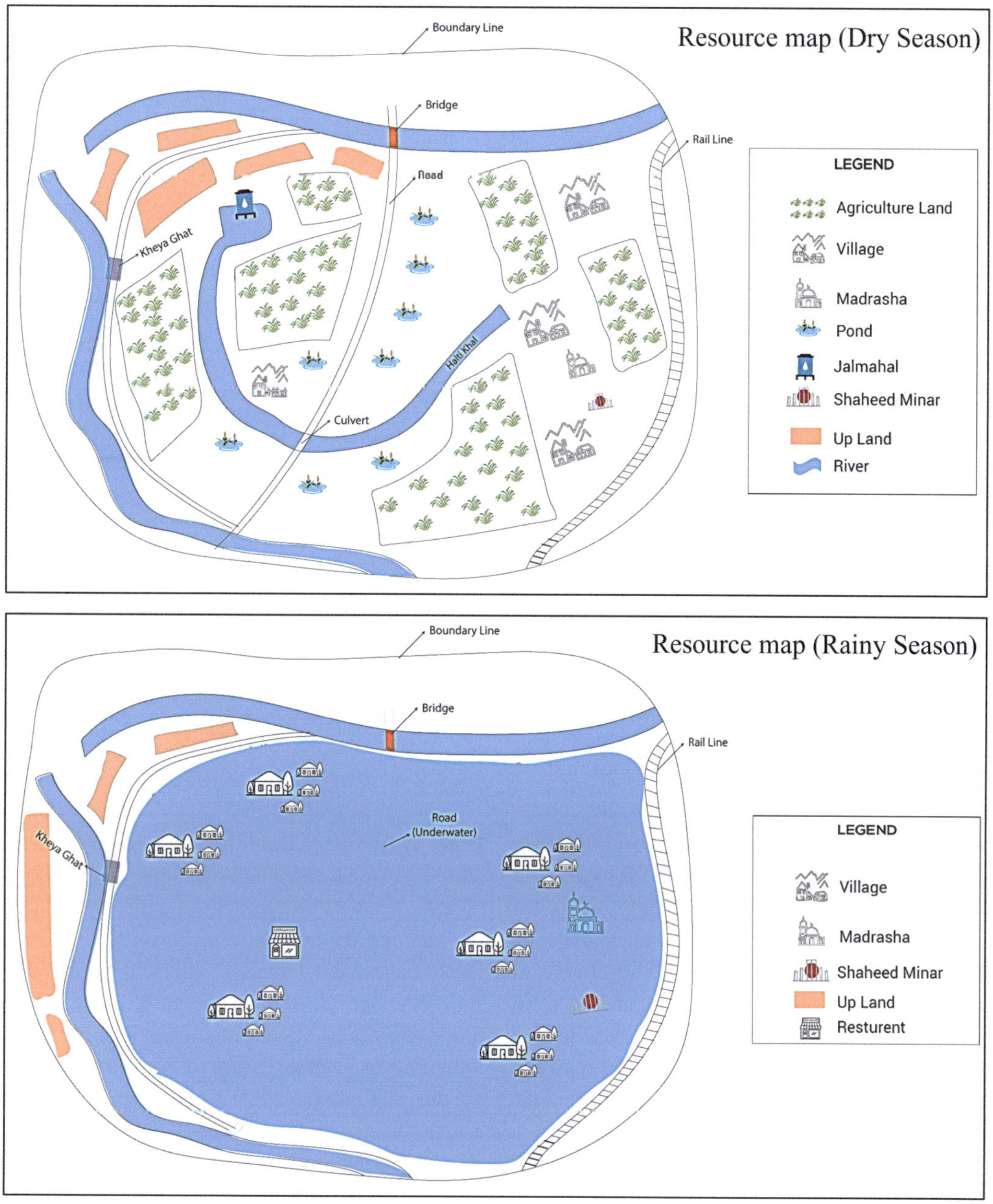

Fig. 15.6 Resource map of the dry season and rainy season. (Source: Field Survey, 2023)

Land serves as a crucial resource in Halti Beel, especially during the dry season when the water recedes. People utilize two main types of land: infield and upland. Farmers own the majority of the land, including some "Khas" land. The primary crops cultivated by farmers are paddy and maize, with various types of Paddy such as hybrid, unotirish, miniket, and atash. Initially, only "Amon" was cultivated, but technological advancements and increased production led to the adoption of hybrid and miniket varieties, with the cultivation of "Akfosla" crops.

Farmers in the area are relatively prosperous, owning an average of 4–5 bigha of land, while some possess up to 100 bigha. Besides maize, they cultivate a variety of vegetables, including radish, tomato, pea, pumpkin, bottle gourd, wax gourd, cucumber, ribbed gourd, bitter gourd, palwal, and brinjal. Various trees, such as mango, jackfruit, coconut, palm, guava, litchi, custard-apple, and betel nuts, are also present.

Halti Beel boasts rich biodiversity, featuring diverse fish species, trees, herbs, and birds during the rainy season. Notable bird species include "parijai duck" like "chokhachokhi," "Rajshorali," "goradmatha," "dholabele," "ghughu," and "Machraga," "Heron," "Pankouri," "Crowned Night Heron," and "Open Bill Stork."

In the winter months, hunters from different parts of the country engage in hunting activities in the Halti Beel area, often facilitated by lessees and local elites. While hunting wildlife is illegal and subject to punishment, the enforcement of legal measures is not robust enough to deter such activities effectively.

15.4.2.2 Challenges in Managing Common-Pool Resources: Conflicts Unveiled

The primary conflict in Halti Beel arises from restrictions on farmers using big fishing nets during flooding, intensifying clashes with fishermen. Disparities in fishing rights, especially the use of large nets for personal consumption, and tensions over "Khas" land allocation contribute to conflicts, exacerbated by unfair practices in obtaining land. These issues escalate during the rainy season, disrupting livelihoods and sparking disagreements over fishing rights and property ownership (Ahmed et al. 1995).

During the rainy season, uplands become submerged in water, creating an abundance of water and fish. This period brings joy and financial stability to fishermen, as they can freely engage in fishing activities, yielding ample fish for consumption and sale. Various fish species, including "katala," "Rui," "shol," "koi," "Baim," and "Jiol," are readily available during this time. Farmers also participate in fishing, but they confine their activities to the submerged areas of their land. However, the rainy season poses challenges for the general population, as the excess water restricts movement, causing inconvenience. People, especially farmers, experience a sense of monotony and purposelessness during this time, as their usual agricultural activities come to a standstill.

During the rainy season, the Beel experiences extensive flooding, causing difficulties for people to sustain their livelihoods. Mobility becomes a challenge, resulting in widespread unemployment. Farmers adopt various means of earning a living, such as fishing, pulling rickshaws in nearby cities, crafting diverse handicrafts for fishing, and operating boats for communication with nearby villages or cities. To protect their homes from flooding, farmers construct barriers and embankments. During the rainy season, cultivable lands are submerged, leading farmers to engage in fishing on their portions. However, local authorities, including magistrates and community leaders, prohibit them from using large nets for fishing. Despite the restrictions, farmers express a desire to employ big nets, citing their property rights as a compelling reason for doing so.

15.4.2.3 Unraveling Challenges in Land Acquisition

In the early days of Halti Beel, the absence of land ownership and the non-designation of land as "Khas" resulted in unrestricted cultivation, leading to conflicts over land resources. To address these issues, the Landlord Acquisition and the Republic Act of 1950 were enacted, leading to the government acquiring these lands, now termed "Khas" land. The intention was to distribute this land among landless poor farmers through a general procedure. However, the implementation faced challenges as political and financial influences allowed many wealthy farmers to acquire ownership of "Khas" land. A testimony from farmer Rashid Ali highlighted the plight of genuinely landless individuals who, despite cultivating leased land, struggled to gain ownership of "Khas" land. The absence of effective monitoring and management systems in the land acquisition process perpetuated landlessness among vulnerable farmers, contributing to the increasing concentration of land in the hands of wealthier individuals over time.

15.4.2.4 Wetland Acquisition

In the early days, fishing in Halti Beel was unregulated, with fishermen, farmers, and anyone catching fish without restrictions. With a growing population and increased fishing pressure, the "Jaal Jar Jola Tar" Act was enacted, granting exclusive fishing rights to fishermen. The Halti Beel Biodiversity Management Organization (BMO) was established, and a fishing sanctuary, "Tanki," was created to protect fish habitats and breeding grounds. Fishing during the pre-monsoon period was forbidden, supervised by the Upazila chairman, and the "Chalan Beel Prokolpo" initiative led to the construction of "Jal mahal."

In the dry season, only BMO members could fish from "Jal mahal," contributing funds for its development. The somity, with 400 members paying BDT 20 monthly, organizes training for fishermen. Non-members can fish by depositing money, and the proceeds are equally distributed. In 2014, BMO collaborated with WBMO-DOF, a German institute, to construct "Jal mahal" for the benefit of fishermen. The community receives institutional support from WBRP, including microcredit for impoverished fishermen, reflecting a comprehensive effort toward sustainable fishing practices and community development.

15.4.2.5 Grasping the Variability in Crop Yields Across Seasons

December-January marks the winter season, characterized by dry conditions conducive to cultivation. Toward the end of Magh, farmers commence the sowing of paddy, with their primary focus on the cultivation of rice and maize. Initially producing "Amon," they later shifted to "Boro," "Untrish," and maize due to higher production yields. They also cultivate "Akfosla" crops, including maize and various vegetables such as cauliflower, cabbage, tomatoes, and carrots. Given the dry season, fish are not widespread and can only be found in "Jal mahal," though farmers are unable to catch them. Despite initial idleness, farmers find fulfillment as they engage in work during this season, leading to overall contentment (Fig. 15.7).

During this dry season, individuals are actively involved in cultivating paddy, maize, and vegetables. The cultivable lands thrive with abundant crops, bringing joy to farmers. However, fish are not present during this season. During the crop-cutting period, farmers are occupied with harvesting activities. Many possess ample land, with some owning up to a maximum of 100 bighas. Despite cultivating "Akfosla" crops, farmers find satisfaction in their production. Following the sale of crops, most farmers spend the

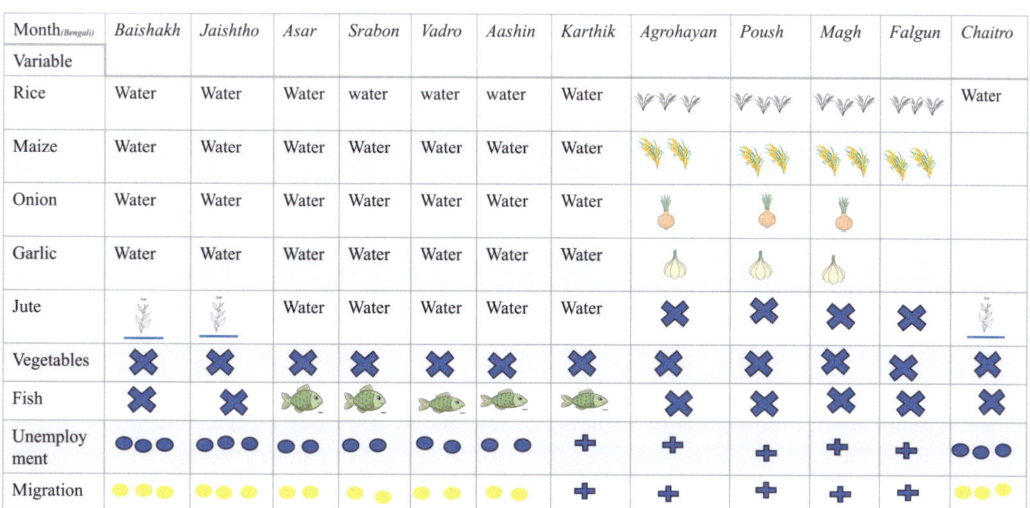

Fig. 15.7 Seasonal diagram for farmers (produced by the focus group participants)

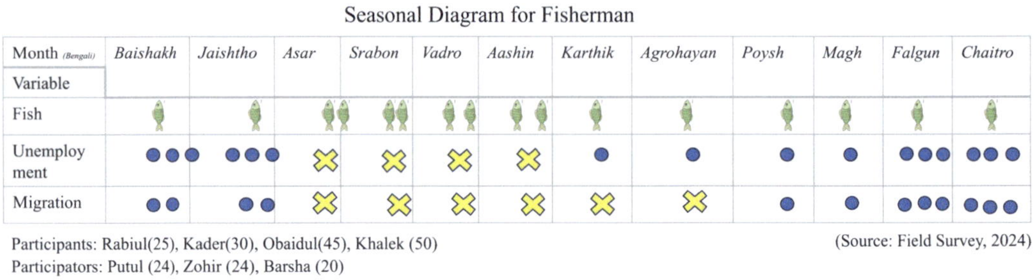

Fig. 15.8 Seasonal diagram for fishermen

remainder of the month without engaging in further activities. They also cultivate various vegetables like brinjal, ladies' fingers, pumpkin, radish, etc.

With the onset of the rainy season, lands gradually submerge in water, reaching full submersion by the end of Ashar. People resort to boat transportation as their primary mode of movement. During this time, farmers face a lack of work in agriculture, leading some to pursue alternative activities such as van pulling and shopkeeping. In the absence of viable work options, farmers may migrate to cities like "nature" and "Rajshahi," where they often take on labor-intensive tasks. Fish become abundant during the rainy season, and farmers, despite restrictions on large fishing nets, catch fish for consumption. The constraints on movement during this season can lead to boredom among farmers.

As Ashin concludes, water levels begin to recede, signaling a period when farmers abstain from cultivation. During this time, there is a noticeable decline in the availability of certain fish species. By the end of Kartik, water sources are completely dry, and with the onset of Agrahayan, farmers commence the sowing of maize. Additionally, they cultivate various seasonal vegetables like cauliflower, cabbage, peas, carrots, and more. Individuals who migrated during the off-season return home and actively participate in cultivation, bringing joy to farmers as they embark on the new agricultural season.

15.4.2.6 Fish Extraction by Fishermen

In the rainy season, the village becomes inundated with water, offering a variety of fish like "shol," "Jiol," "koi," "Magur," and more. Fishermen capitalize on this abundance, catching fish and generating income through their sales. This period brings happiness to fishermen due to the ample availability of fish (Fig. 15.8).

During the initial stages of the drying period at the end of Ashin, fish remain accessible, allowing fishermen to earn a livelihood without the need for migration. As the season progresses and the water dries up, fishermen face unemployment, except for those affiliated with the "Motsho Somobai Somity." Members of this association can still catch fish from "Jal mahal" by contributing a fee known as "Joma." Despite the dry conditions, fishermen affiliated with the "Motsho Somobai Somity" continue to find employment by fishing in "Jolmohol." However, those fishermen who are not members of the association encounter unemployment and resort to various activities. Some fishermen opt to migrate during this dry season, seeking opportunities in places like "Natore" and "Rajshahi," where they often engage in labor-intensive work. The absence of a stable income source during this season leaves fishermen discontent (Salan et al. 2018).

15.4.2.7 Excessive Exploitation of Natural Resources

Through an examination of the trend (Fig. 15.9), alterations in various variables, such as fish size, fish species, fish quantity, birds, trees, rice cultivation, and maize cultivation, to water levels have been observed. Post-independence, there was a discernible decline in both the size and quantity of fish, accompanied by a reduction in fish species

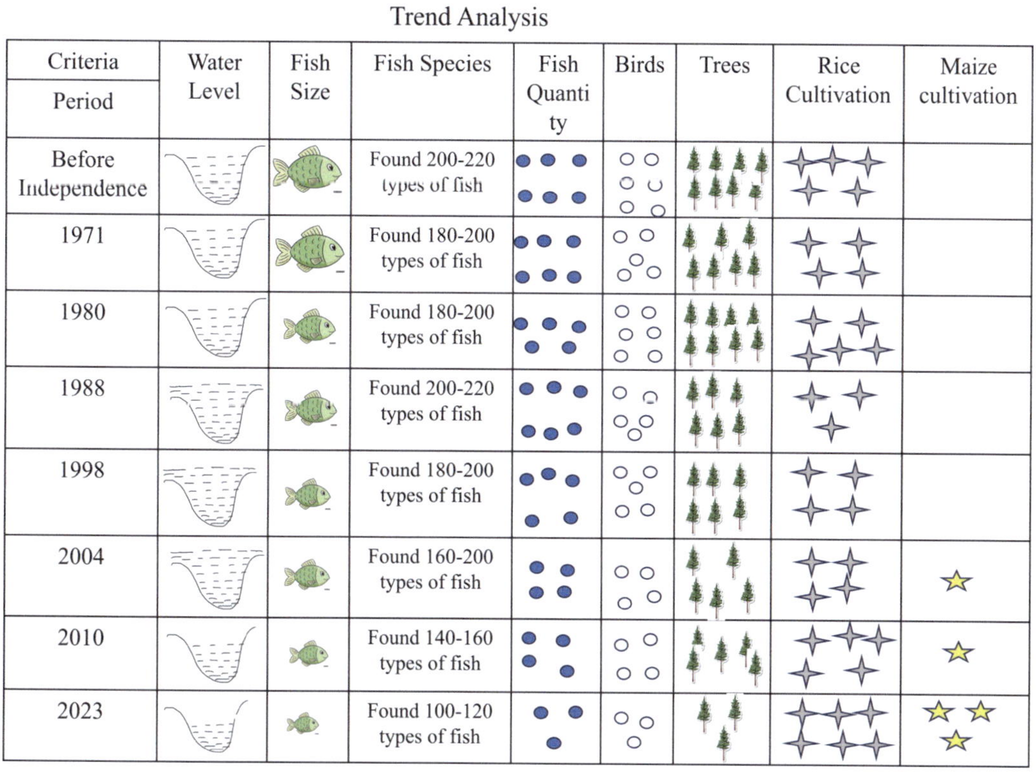

Fig. 15.9 Trend analysis. (Source: Field Survey, 2023)

from 200–220 to 180–200. Simultaneously, the number of birds and trees experienced a decrease. Rice cultivation remained consistent during this period, with no cultivation of maize.

The pivotal year of 1988 marked a significant shift when the water level of Halti Beel surged dramatically due to a devastating flood that ravaged numerous villages, agricultural lands, and homes. This calamitous event led to widespread suffering among the populace, forcing many to seek refuge in safer locations. One individual, Asraf Ali (76), vividly recalled the hardships faced during the 1988 flood, emphasizing the severity of the situation.

Subsequently, from 1988 to 2004, the water level remained relatively stable. However, after 2004, a noticeable decline in water levels commenced. Correspondingly, as water levels rose, there was an augmentation in fish size and quantity, and the number of fish species increased from 180–200 to 200–220. Unfortunately, this positive trend was accompanied by a decline in the number of trees and a reduction in rice cultivation.

Post-2004, as water levels receded, a reversal in trends occurred. Fish size and quantity diminished, and the number of fish species decreased to the current range of 100–120. Additionally, this period witnessed the initiation of maize cultivation, which started gaining popularity after the 1988 flood. People started to cultivate maize after 1988. As the water level reduced, Jute cultivation also started.

15.4.2.8 Degradation of Resources in Halti Beel

Over time, various transformations have occurred in Halti Beel, impacting factors like water levels, fish size and quantity, fish species, bird population, tree numbers, and rice cultivation. The majority of these changes are attributed to resource degradation, leading to a loss of biodiversity in Halti Beel.

Participatory Rural Appraisal (PRA) tools serve as effective instruments to analyze the shifts in these aspects. This chapter utilizes trend analysis (Fig. 15.9) to illustrate both the degradation of resources and the ensuing consequences. The size and quantity of fish in Halti Beel are diminishing progressively, with the reduction attributed to the over-extraction of Common-Pool Resources. Before 1971, Halti housed 200–220 fish species, including substantial ones like "Gojar," "Title," "Boal," "Baus," and "Mrigel," weighing between 2 and 20 kg. However, due to conflicts over resources, the number of fish species has dwindled to 120–140. Fishermen like "Shreepodo Halder" highlight the extinction of flavorful species like "Dhoda," "Baitka," and "Guipoi," attributing it to over-extraction.

In the past, Halti Beel attracted various winter migratory birds like "Balihash," "Komontil," "Sovelor," "Melard," and "Orchard," contributing to tourism and income sources. Unfortunately, the bird population has steadily declined due to ongoing conflicts.

Similarly, the number and types of trees have diminished over time, impacting herbal plants that once served medicinal purposes. Rice, the primary crop in the Halti Beel area, has changed both quantity and types of cultivation. Initially, "Amon" rice was prevalent, but due to low production and increased population, farmers transitioned to "Boro" production. The varieties of "Boro" paddy include "Atash," "Minikate," and "Hybrid." Additionally, some farmers shifted to maize production, experiencing increased yields after 2000.

15.4.3 Management of the Conflict

15.4.3.1 Exploring the Reasons for Common-Pool Resource Degradation and Its Implications

Numerous factors contribute to the degradation of Common-Pool Resources, and an exploration of these causes and effects (Fig. 15.10) has been facilitated through a cause-and-effect diagram. Among the various reasons, corruption in land acquisition stands out as a significant factor. Individuals often manipulate land ownership, falsely representing their sons as landless to acquire "Khas land." However, true landless farmers, relying on leased land, and fishermen dependent solely on fishing find themselves without land, relying heavily on wetlands like Halti Beel for their livelihoods.

Another notable cause is the indiscriminate killing of birds at Halti Beel. Despite the area's rich biodiversity, the population of migratory birds declines steadily due to rampant bird

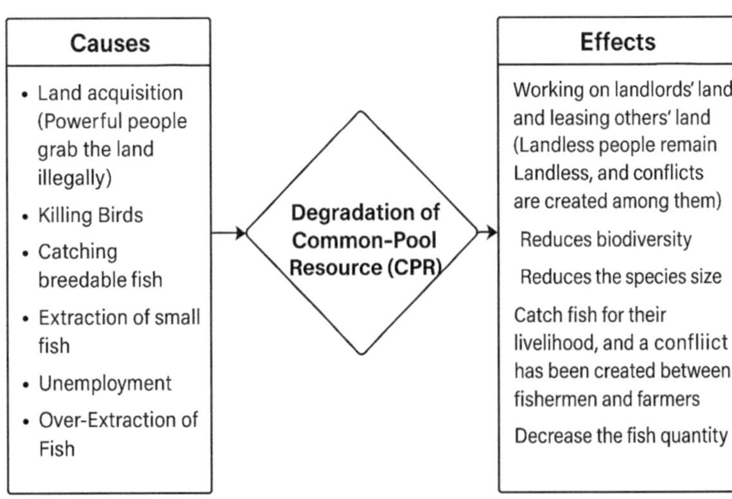

Fig. 15.10 Cause-effect diagram of degradation of CPR in Halti Beel. (Source: Field Survey, 2023)

hunting during the winter season. The practice of breeding fish using current nets is a detrimental factor leading to the degradation of fish species in Halti Beel. To curb this, the government has designated a portion of Halti Beel as a fishing sanctuary, prohibiting fishing during the breeding season.

Furthermore, the tendency to catch small fish, known as "Jatka," using current nets has led to a decline in the size of fish over time. To address this issue, the government has imposed restrictions on using the current nets for fish catching. The scarcity of employment opportunities in the Halti Beel area has compelled a significant portion of the population to rely on fishing for their livelihoods. Lastly, over-extraction of fish emerges as a predominant cause contributing to the degradation of fishery resources in Halti Beel.

15.4.3.2 Involving Community Members in Conflict Management

In managing conflicts related to Common-Pool Resources, the costly endeavor faces challenges due to complexities in administration, influenced government regulations, and potential lack of transparency, hindering fair access to land and water resources for the general populace, including the impoverished and landless individuals.

In the management of Common-Pool Resources, challenges such as distortions and biases in regulations necessitate a conflict management system, highlighting the crucial importance of establishing a strong sense of belonging among resource users and clear delineation of ownership for effective conflict resolution.

Ostrom's principle suggests that individuals affected by regulations should have the opportunity to be involved in modifying them, yet it doesn't specify the integration process. The Land Acquisition Act, designed to benefit landless farmers, has failed due to corruption and political influence, highlighting the need to modify regulations actively involving landless farmers to address the issue. Ostrom asserts that Common-Pool Resources should be localized, yet he fails to establish a definitive limit and duration for resource extraction. Consequently, individuals can degrade resources according to their whims and exclude others from accessing these resources. In his set of principles, Ostrom suggests the development of a system that is implemented by community members to monitor the behavior of its members. The absence of such a monitoring system has resulted in numerous problems and conflicts in Halti Beel, particularly in the realm of land acquisition, where there is no mechanism in place to identify landless farmers.

Furthermore, Ostrom advocates for the implementation of sanctions for those who violate regulations. However, in Halti Beel, there is a lack of such sanctions, which contributes to degradation and conflicts. Policymakers must exercise caution when formulating policies. Ostrom emphasizes the importance of identifying the specific users of resources who possess the ability to consume said resources. In addition, inclusive policies should be implemented instead of singular rules. Merely establishing rules without proper enforcement and regulation will perpetuate management issues related to Common-Pool Resources.

The government must ensure efficient operations and adherence to conflict management regulations, with departments responding to victims' claims based on their consumption patterns. If farmers face difficulties, they should report to the nearest administration, prompting the administrative body to take appropriate measures to address and mitigate the problem.

The administrative body, such as the assistant land commissioner, must conduct field research to observe and understand phenomena, presenting findings with transparent evidence to higher authorities like the district commissioner. Each step should adhere to a government-established timeframe, and accountability is crucial for any delays. Conflict management doesn't conclude at the initial stage; victims can appeal unfavorable verdicts within a specified period. After land allocation management, both the authority and the community should oversee the procedure.

In the management of Common-Pool Resources (CPRs), the local government faces administrative challenges, and central government empowerment of various departments, including the matsho office, the agricultural office, and the water resource department, becomes crucial to overcome internal obstacles and enable effective fulfillment of their roles in designated areas. These offices are essential for providing necessary initiatives to assist consumers during conflicting times.

In many countries, Common-Pool Resources are managed under a complex set of regulations involving both state and traditional laws, with defined access rights granted to specific individuals or communities. However, in some cases, resources operate under free and open-access conditions, subject to localized restrictions by local governing bodies. Examples from Cambodia and Indonesia highlight instances where valuable fish resources are controlled by private individuals or traditional tenure systems, often excluding poor rural households. Community participation is crucial in enhancing the active management of common-pool resources in many countries.

During Participatory Rural Appraisal (PRA) in Halti Beel, rural expectations highlight the need for effective land acquisition management and prevention of fish over-exploitation to mitigate conflicts and reduce degradation. In the dream map exercise, villagers emphasize the importance of enforcing land acquisition and the Republic Act of 1950.

Providing landless individuals with diara property, known as Khas land, can sustain their livelihoods year-round, reducing the need to exploit resources from other sources and mitigating conflicts between fishermen and farmers. Encouraging community members' engagement in non-farm activities through educational initiatives can reduce reliance on crop cultivation and mitigate the depletion of resources due to competitive consumption. The community desires a reformation of geographical and structural conditions to establish a designated area for tourism, including a park, with the expectation of significant economic profit and a shift in business focus during the rainy season. Another faction of the fisherman community in Halti Beel desires an additional fish sanctuary, specifically near Jal mahal or Tanki, to address and resolve conflicts related to resource consumption.

Agriculturalists in Halti expressed a desire for the construction of a reservoir to enable the year-round collection of aquatic fauna, with the village chief highlighting its potential to generate monetary resources consistently throughout the year.

15.4.3.3 The Consequences of Their Wants

If farmers have access to a marketplace beside the park, they can sustain their livelihood throughout the year, reducing the need for conflict-inducing fish collection. This would lead to decreased consumption compared to previous levels. If the restaurant sources its ingredients from the halti community, it will increase employment opportunities for farmers and fishermen, which can help reduce the over-exploitation of fish (Fig. 15.11).

The proposed reservoir for the farmer community is expected to provide a consistent source of fish throughout all seasons. This, in turn, could contribute to a reduction in conflicts related to land acquisition, fostering biodiversity growth and increasing overall production value.

Fishermen are advocating for an additional sanctuary beside Jal mahal, anticipating that it would eliminate conflicts in fish consumption. A year-round supply of fish would support their livelihoods, contributing to sustained and reduced conflict associated with unrestrained consumption. These positive outcomes hinge on adherence to rules and regulations governing the proposed program as per the community's preference.

15.4.3.4 Strategies to Reduce Conflict and Degradation

Community engagement through focus group discussions can offer effective strategies to mitigate conflict and degradation (G. Adekola, 01 Jan 2023). Negotiation is the most effective strategy for reducing conflict and degradation in the management of common-pool resources (Dewi, 16 Feb 2023). In the dry season, there should be good management of ponds so that farmers and fishermen can catch fish equally.

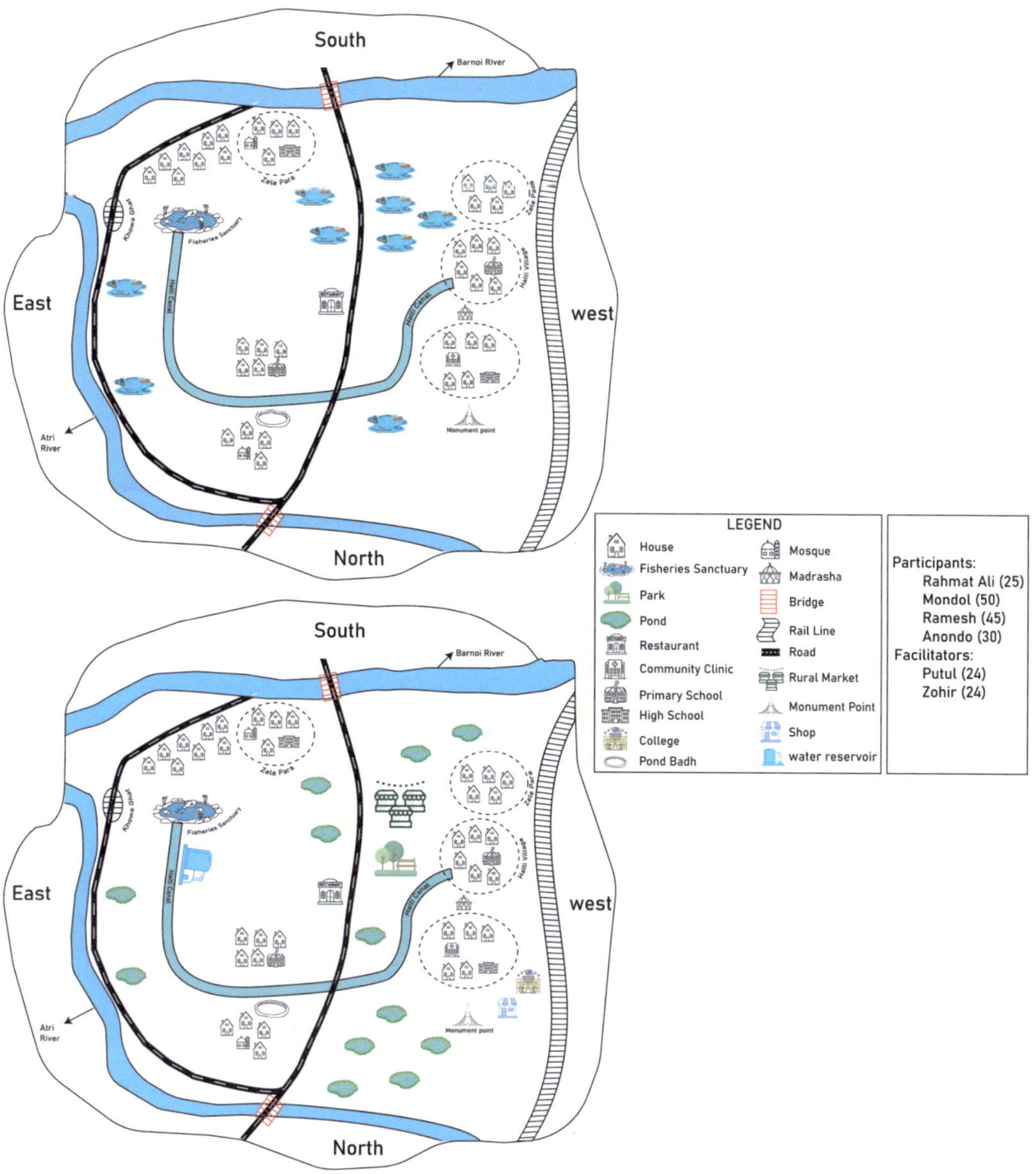

Fig. 15.11 Social and dream map. (Source: Field Survey, 2023)

In the community's view, addressing the structural weaknesses of community-based organizations like the BMO involves granting independence to the BMO, enabling it to make decisions autonomously based on the prevailing circumstances. The current lack of organization and political influence in the BMO necessitates corrective measures.

Given the seasonal submersion of the entire Halti Beel area during the rainy season, localizing fishery resources becomes challenging. The non-localization of fishery resources complicates their management and contributes to conflicts. To address this, the community advocates for the localization of fishery resources through active community involvement.

The improper implementation of the Land Acquisition Act results in landless individuals being deprived of ownership rights to "Khas land," while wealthy individuals accumulate more wealth through land ownership. To rectify this, a proper monitoring system is deemed necessary before the distribution of "Khas land." Additionally, the application of environmental governance is suggested to ensure equitable benefits for all resource users. Identifying local norms to convey to them not to over-exploit the breeding fish (Fig. 15.12).

15.4.3.5 Policy Recommendation

The rural populace must unite for effective oversight of fish and crop resources, recognizing the potential degradation of assets when resources become scarce. Privatization can mitigate conflicts by incentivizing private entities to maximize profits, but in a developing nation with limited resources, environmental governance is crucial for equitable decision-making in natural resource management.

Participatory governance is pivotal in fostering relationships between the state administration, civil society, and the private sector to confront emerging challenges (Kooiman et al. 2005; Pimbert 2004; UNDP 2000). By ensuring the engagement of community members, conflicts can potentially be minimized, thereby simultaneously mitigating resource degradation. The administration can designate specific areas of land and water based on their respective uses, thus enabling effective resource allocation. The state bears a crucial role in dispute resolution, monitoring the implementation of resource use agreements, providing technical and managerial support to local organizations, and conducting environmental assessments.

Additionally, the government can establish time durations through the assistance of informal and formal institutions. For instance, one group may be granted the right to collect fish from a reservoir for the initial 15 days, while another group assumes this responsibility for the subsequent 15 days. One of the primary goals of management is to delineate the group of individuals who lack land and have no alternative but to engage in agriculture and fishing. The susceptibility of a group of landless farmers is not greater than that of other groups, as an individual farmer is unable to withstand the influence exerted by dominant factions.

Consequently, the collective territory may be fragmented into various smaller groups. These smaller groups can effectively facilitate the

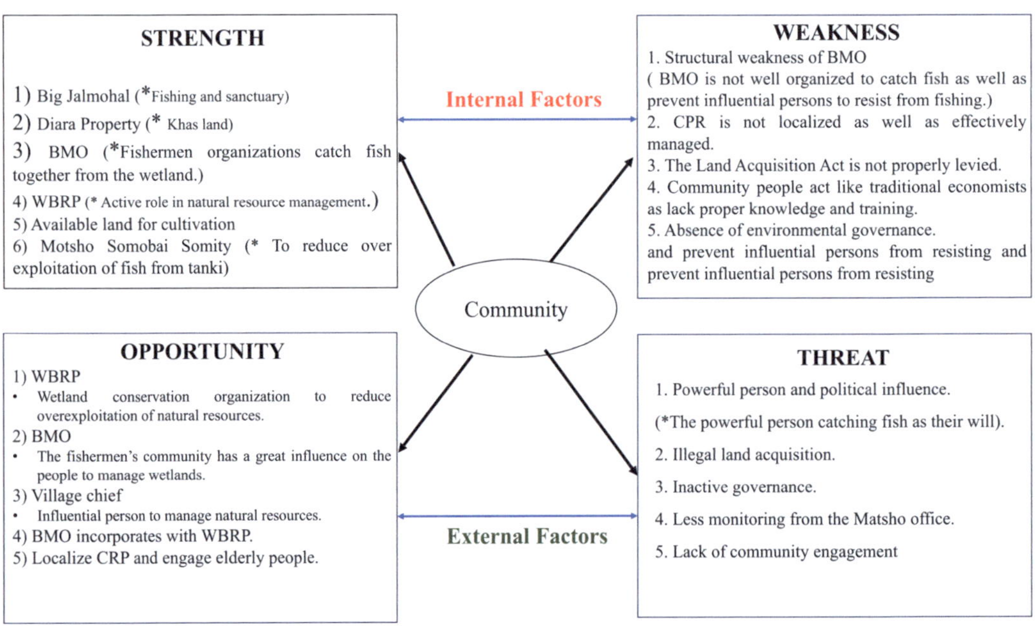

Fig. 15.12 SWOT analysis community of Halti Beel. (Source: Field Survey, 2023)

exchange of information within the management process using regular meetings focused on resource extraction. Consequently, they will establish a distinctive force capable of confronting conflicts with influential individuals or entities. The formation and discussions of such groups have proven to be effective in conflict management. In all instances, governments have endeavored to clarify issues related to land tenure and reinforce the rights of local communities to autonomously manage their resources. This has been achieved through the bestowal of legal recognition and decision-making authority.

15.5 Conclusions

Community participation is crucial for managing conflicts and resource degradation, helping identify the tragedy of the commons and key issues. Involving the local populace ensures the transparent execution of resource extraction, mitigating conflicts by understanding factors contributing to overfishing and uncontrolled land usage. The Participatory Rural Appraisal method facilitates the acquisition of in-depth information from local individuals, providing a platform for minority groups to connect with the government and hold it accountable for addressing conflicts and resource management issues.

While the study encompasses a broad scope, it does have some limitations. The primary outcome is the ongoing degradation of resources in Halti Beel. Currently, the community is unable to obtain the expected number of resources, such as large fish and valuable crops. The conflict surrounding resource management is prevalent in the Beel due to unrestricted competition for resources. Therefore, community engagement in the management system, coupled with an awareness of resource degradation, will help mitigate conflicts and find a solution. The findings of this chapter can be valuable in developing effective strategies, policies, and plans for the existing organizational procedures related to Common-Pool Resource management.

This study on the management of Common-Pool Resources (CPRs) in Halti Beel offers a nuanced understanding of how participatory, community-led approaches can mitigate the degradation of wetland ecosystems while addressing socio-environmental conflicts. By assessing patterns of over-extraction, resource-related disputes, and institutional gaps, the research illustrates the direct link between ecological vulnerability and weak governance structures within wetland systems.

Importantly, the findings underscore the centrality of wetlands like Halti Beel to broader urban and peri-urban resilience. As urban sprawl increasingly encroaches upon ecologically sensitive areas, the mismanagement of CPRs poses significant threats not only to local livelihoods but also to urban sustainability, particularly in terms of flood regulation, food security, and biodiversity conservation. This positions wetlands as critical assets in climate adaptation and urban ecological planning.

By emphasizing community engagement, localized governance, and inclusive resource management, the study contributes a scalable framework that bridges ecological science with participatory development. These insights are highly relevant to the synthesis of integrated strategies for wetland conservation and urban resilience, reinforcing the need to incorporate community-based wetland stewardship into national and international sustainability agendas.

Thus, this chapter not only enhances the scientific understanding of CPR dynamics in wetland environments but also advocates for transformative governance models that align ecological integrity with long-term societal resilience.

Acknowledgments We are grateful to the local community of Halti Beel, Bangladesh, whose participation and insights were invaluable to this research. Their collaboration and enthusiastic participation in the Participatory Rural Appraisal (PRA) process significantly enhanced our understanding of the region's Common-Pool Resource management issues and potential solutions.

Potential Conflicts of Interest The corresponding author declares that there is no conflict of interest on behalf of all authors.

Informed Consent Every individual participant in the study gave their informed consent.

Authors' Contribution All authors have contributed equally to this work.

Data Availability Statement The corresponding author will provide the datasets used and/or analyzed in the current study on reasonable request.

Funding This research was funded by the **International Centre for Climate Change and Development (ICCCAD)** as part of the **Capacity Strengthening of Multi-Actors to Limit Climate Change Impacts and Enhance Resilience (CAP-RES) Project** originally funded by the Embassy of Sweden. The authors gratefully acknowledge the support and guidance provided by ICCCAD throughout the project.

References

Adekola G, Olumati ES (2023) Focus group discussion: a research method in community development. Int J Res Innov Soc Sci 7(5):392–399

Ahmed M, Capistrano AD, Hossain M (1995) Fisheries co-management in Bangladesh; experiences with GO-NGO- fisher partnership models. Presented at "reinventing the commons", the fifth annual conference of the International Association for the Study of common property (IASCP), Bodoe, Norway, May 24–28. https://dlc.dlib.indiana.edu/dlc/items/167958d2-f838-4949-b8d4-a71f6f34a8d8/full

Alamgir M (1988) Poverty alleviation through participatory development. Development 2(3):97–102. https://scholar.google.com/scholar?hl=en&as_sdt=0%2C5&q=Alamgir%2C+Mohiuddin+%281988%29.+Poverty+Alleviation+through+Participatory+Development.+Development%3A+Journal+of+the+Society+for+International+Development.+2%283%29%3A97102.&btnG=#:~:text=%5BCITATION%5D%20Poverty-,alleviation, through%20participatory%20development

Atieno MPO, Mwangi SW, Ngetich K (2014) Mechanisms of conflict management over common pool resources among the residents of lower river Nyando floodplains, Kisumu County, Kenya. Res Humanit Soc Sci 4:23. https://www.academia.edu/download/69478282/17240.pdf

Baginski O, Dev OP, Yadav NP, Soussan J (2003) Community forest management in the middle hills of Nepal: the changing context. J For Livelihood 3(1):5–20. https://www.researchgate.net/publication/265740977_Community_Forest_Management_in_the_Middle_Hills_of_Nepal_The_Changing_Context

Berkes F, Colding J, Folke C (2000) Rediscovery of traditional ecological knowledge as adaptive management. Ecological Applications, 10(5): 1251–1262.https://doi.org/10.2307/2641280

Blackburn J, Holland J (1998) Who changes? Institutionalizing participation in development. Intermediate Technology Books, London. https://www.taylorfrancis.com/chapters/edit/10.4324/9781351324 7627/mainstreamingparticipation-development-james-blackburn-robert-chambers-john-gaventa

Bromley DW (1990) The ideology of efficiency: Searching for a theory of policy analysis. Journal of Environmental Economics and Management, 19(1): 86–107.https://doi.org/10.1016/0095-0696(90)90062-4

Chambers R (1992) Rural appraisal: rapid, relaxed and participatory, IDS discussion paper no. 331. Institute of Development Studies, U.K. (1994). The Origins and Practice of Participatory Rural Appraisal. World Development. 22(7): 953–69. (1997) Whose Reality Counts. Intermediate Technology Publications, London (2004). Notes for participants in PRA-PLA familiarization workshops in 2004.Institute of Development Studies, U.K. https://www.academia.edu/download/58193656/Participatory_Rural_Appriasal_by_Robert_Chambers-1992.pdf

Das A, Kashem S, Hasan M (2021) Using market mechanism to stimulate sustainable use of the non-renewable environmental resource (groundwater) in Barind tract of Bangladesh. Trans Hong Kong Inst Eng 28:39–48. https://doi.org/10.33430/V28N1THIE-2020-0036. https://www.bip.org.bd/admin/uploads/memberpublication/2021/MP-ff7de954f9.pdf

Dasgupta P (1995) The population problem: Theory and evidence. Journal of Economic Literature, 33(4): 1879–1902. https://www.jstor.org/stable/2729316

Gordon S (1954) The economic theory of a common–property resource: the fishery. J Polit Econ 62:124–142. https://ideas.repec.org/a/bcp/journl/v7y2023i5p392-399.html, https://www.taylorfrancis.com/chapters/edit/10.4324/9781315193182-1/economic-theory-common-property-resource-fishery-1-scott-gordon

Hanna S (1998) Co-management in small-scale fisheries: creating effective links among stakeholders. In: Paper presented at international workshop on community-based natural resource management. The World Bank, Washington, DC. May 10-14

Hardin G (1968) The tragedy of the commons: the population problem has no technical solution; it requires a fundamental extension in morality. Science 162(3859): 1243–1248. https://www.science.org/doi/abs/10.1126/science.162.3859.1243

Imteazzaman AM, Galib SM (2013) Fish fauna of Halti beel, Bangladesh. Int J Curr Res 5(1):187–190. https://www.researchgate.net/publication/263661605_Fish_Fauna_of_Halti_Beel_Bangladesh

Kabir GS, Yew TS, Noh KM, Hook LS (2011) Assessing fishers' empowerment in inland openwater fisheries in Bangladesh. Ecol Econ 70(11):2114–2123

Kasch G (2023) Common-pool resource management and conflict resolution: a case study of two self-governed irrigation schemes in Ntcheu, Malawi. Capstone Collection, p 3295. https://digitalcollections.sit.edu/capstones/3295/

Kooiman J, Jentoft S, Bavinck M, Pullin R (2005) Fish for life: interactive governance for fisheries (p 432). Amsterdam university press.

Ni Nyoman Clara Listya D (2023) Collective actions and challenges analysis on management of the Mekong River as common pool resources. Glob South Rev. https://doi.org/10.22146/globalsouth.68422. https://www.researchgate.net/publication/347926905_Fish_sanctuary_as_a_sustainable_management_tool_for_recovering_fish_biodiversity_production_and_livelihood_A_case_study_on_Halti_Beel_tank_sanctuary_Bangladesh

Ostrom E (1999) Coping with tragedies of the commons. Annu Rev Polit Sci 2(1):493535. https://www.annualreviews.org/content/journals/10.1146/annurev.polisci.2.1.493

Participatory Forestry: The process of change in India and Nepal By: Mary Hobley,1996. https://odi.org/documents/3232/8144.pdf

Paudel S (2006) Community forestry in Nepal. Himal J Sci 1. https://doi.org/10.3126/hjs.v1i1.190. https://www.researchgate.net/publication/250191655_Community_forestry_in_Nepal

Pimbert MP (2004) Institutionalising participation and people-centered processes in natural resource management. IIED.

Pomeroy RS (1999) Institutional analysis, from participatory methods in community-based coastal resource management, Vol. II. In: Langill S (ed) Institutional analysis: readings and resources for researchers, vol V. International Development Research Centre, Ottawa. https://idl-bnc-idrc.dspacedirect.org/bitstream/10625/26556/5/114510.pdf

Salan MSA, Rahman S, Mustari R, Wakil MA (2018) Rural resource management: a study to promote the sustainable use of land resources. Int J Rural Reg Plann Dev 4(2):36–45. https://www.academia.edu/download/58860448/3645_Rural_Resource_Management_A_Study_to_Promote_the_Sustainable_Use_of_Land_Resource20190411-120702-1ch8bdb.pdf

Sethi R, Somanathan E (1996) The evolution of social norms in common property resource use. The American Economic Review, 86(4): 766–788. https://www.jstor.org/stable/2118304

Schlager E, Ostrom E (1992) Property-rights regimes and natural resources: A conceptual analysis. Land Economics, 68(3): 249–262. https://www.jstor.org/stable/3146375

Siddique MAB, Hussain MA, Flowra FA, Alam MM (2016) Assessment of fish fauna about biodiversity indices of Chalan Beel, Bangladesh. Int J Aquat Biol 4(5):345–352. https://doi.org/10.22034/IJAB.V4I5.234. https://www.sciencedirect.com/science/article/pii/S0921800911002412

Siddique MAB, Hussain MA, Alam MM, Haque MA, Flowra FA (2020) Fish sanctuary as a sustainable management tool for recovering fish biodiversity, production and livelihood: a case study on Halti Beel tank sanctuary, Bangladesh. Arch Agric Environ Sci 5(4): 567–575

Singh GS, Vishvakarma, Subhash CR (1997) Prospects of sustainable development of Kullu Valley in North Western Himalaya. J Rural Dev 12:359–368 https://www.researchgate.net/publication/327106915_Prospects_of_Sustainable_Development_of_Kullu_Valley_in_North_Western_Himalaya

Smith R (1981) Resolving the tragedy of the commons by creating private property rights in wildlife. Cato J 1: 439–468. https://www.researchgate.net/publication/239806443_Resolving_the_Tragedy_of_the_Commons_by_Creating_Private_Property_Rights_in_Wildlife

Stevenson GG (1991) Common property economics: a general theory and land use applications (pp. xiv+–256pp). https://www.cabidigitallibrary.org/doi/full/10.5555/19921801532

Tang SY (1991) Institutional arrangements and the management of common-pool resources. Public Adm Rev 51(1):42–51. https://doi.org/10.2307/976635

Thompson PM, Sultana P, Islam MN, Kabir MM, Hossain MM, Kabir MS (1999) An assessment of co-management arrangements developed by the community based fisheries management project in Bangladesh. In: International workshop on fisheries co-management, pp 23–28

Thompson PM, Sultana P, Islam N (2003) Lessons from community-based management of floodplain fisheries in Bangladesh. https://www.academia.edu/download/76896333/21_20Thompson.pdf

Thomson K, Gray T (2009) From community-based to co-management: improvement or deterioration in fisheries governance in the Cherai Poyil fishery in the Cochin Estuary, Kerala, India? Mar Policy 33(4): 537–543

Tough F (1987) Fisheries economics and the tragedy of the commons: the case of Manitoba's inland commercial fisheries, Discussion paper no.33. Department of Geography, York University, Toronto. https://www.sciencedirect.com/science/article/pii/S0308597X08001899

UNDP (2000) Consolidating conservation through people's participation: parks and people programme (DNPWC/UNDP). His Majesty's Government of Nepal, Kathmandu. https://doi.org/10.1111/1468-0297.00497

Walker JM, Gardner R, Herr A, Ostrom E (2000) Collective choice in the commons: experimental results on proposed allocation rules and votes. Econ J 200:212–234

Part V

Challenges and Future Perspectives of Urban Wetlands

Integration of Wetland and Urban Resilience to Conserve Fireflies (Coleoptera: Lampyridae: Luciolinae): A Potential Tool for Ecotourism Development and Environmental Monitoring at the Urban Wetland-Terrestrial Interfaces

16

Srinjana Ghosh and Susanta Kumar Chakraborty

Abstract

Recent encroachment of urbanisation leads to the transformation of different simple ecological setups, e.g. forest, rural, coastal, and riverine zones, to complex systems like towns, cities, mostly consisting of built environments with high human density. In urban setups by prioritising human needs, the natural carrying capacity of that area gets exhausted, leading to the decline in biodiversity and a hike in the level of pollution. Urban resilience, on the other hand, paves the way for returning to ecological harmony following the perturbation caused by environmental alterations due to urban sprawling. However, specific segments of ecosystems, e.g., wetlands, representing the interface between terrestrial and aquatic habitats, facilitate the development of urban resilience and serve as conservation hubs for biodiversity. Interestingly, throughout different stages of their life cycle, fireflies (Coleoptera: Lampyridae: Luciolinae), the bioluminescent entomofauna with an aesthetic appeal, are intricately associated with the hydrologic environments prevailing in wetlands surrounding the urban habitats, for securing vital resources like shelter and nutrition. In the present chapter, the scope of utilisation of fireflies as bioindicators of deteriorating environmental conditions caused by urbanisation has been highlighted. Moreover, firefly-based ecotourism approaches centring wetlands near the urban localities not only favour the urban economy but also, by promoting urban resilience, safeguard the fireflies, their habitats and associated biotic organisms.

Keywords

Conservation · Ecological integrity · Fireflies · Habitats · India · Lampyridae · Wetlands

16.1 Introduction

In recent days, urbanisation has been sprawling globally, leading to significant alterations in the environmental integrity of any locality. Various urban drivers, through the dynamic process of environmental alterations, result in differential impacts to natural resources (Fig. 16.1). Extremely modified urban environments (i.e. the intra-urban or urban core areas) are separated from comparatively less drastically transformed

S. Ghosh (✉)
Department of Zoology, Bethune College, Kolkata, West Bengal, India

S. K. Chakraborty
Department of Zoology, Vidyasagar University, West Midnapore, West Bengal, India

S. K. Chakraborty et al. (eds.), *Wetlands and Urban Resilience*, Sustainable Development Goals Series,
https://doi.org/10.1007/978-3-032-06881-1_16

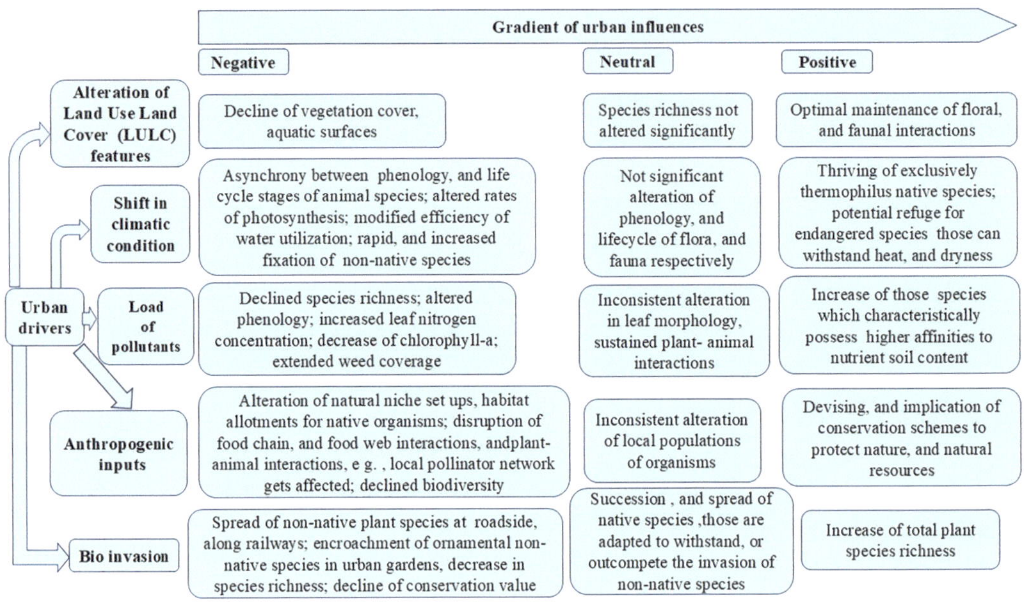

Fig. 16.1 Schematic representation of the gradient of urban influences resulting from dynamic interactions (either with positive, negative, or neutral outcomes) of different urban drivers

suburban, or extra-urban zones (e.g., rural, agricultural, and wild areas) by an ecotone, i.e. peri-urban zones acting as a continuum, blurring the distinctive boundaries between urban and non-urban areas (Gupta et al. 2017, Pérez-Hernández et al. 2023).

In India, like many other Asian countries, the extensive human agglomerations often exceed the carrying capacity of the urban, as well as peri-urban region and demand the availability of huge natural resources (food, water, shelter) to support those populations. Remote sensing data, evidently have shown in various Indian cities like Delhi (Yadav et al. 2017), Mumbai (Dwivedi et al. 2019), Ahmedabad (Mohammad et al. 2019), Kolkata (Gupta et al. 2017), Pune (Deosthali 2000), Lucknow (Singh et al. 2017), Chennai (Jeganathan et al. 2016), Nagpur (Kotharkar and Bagade 2018), Jaipur (Chandra et al. 2018), etc., that unplanned as well as haphazard urban development involves substantial alterations in Land Use Land Cover (LULC) patterns and release of heat from automobiles and industries. Such factors remain responsible for the increase in land surface temperatures, evaporation rates, changes in local wind patterns, hydrological cycles, and emergence of urban heat islands (Gupta et al. 2017). Thus, functionally, urban areas act as geographical barriers for biodiversity (Pérez-Hernández et al. 2023).

In this scenario, legitimate utilisation of natural resources can only effectively maintain the coherence of urban ecology and economy. Urban resilience can be described as the process to integrate ecosystem services like provision of those resources (direct extraction of food, water, forest products, etc.), providing supports to biogeochemical processes (formation of soil, photosynthesis, and nutrient recycling), regulation of natural energy cycles (modulation of climatic conditions, soil, water qualities, regulation of floods, drought, etc.), maintenance of cultural values (establishment of recreational e.g., tourism, aesthetic, and spiritual benefits, etc.). It also helps recovery against all kinds of eco degradation (pollution, soil erosion, groundwater depletion, irrational consumption of fossil fuel, global warming, desertification, deforestation, and so on). Such resiliencies, in the long term, help the urban ecosystems along with their biotic communities in combating, resisting, absorbing, accommodating, and recovering the hazards or

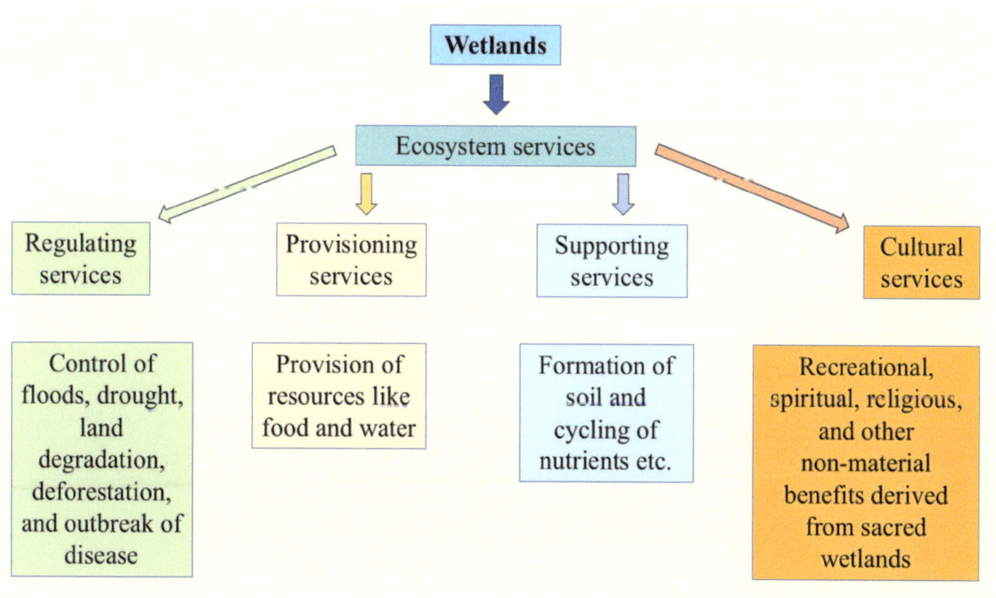

Fig. 16.2 Ecosystem services provided by wetlands surrounding urban habitats adding to the urban resilience

stresses imposed by natural factors or anthropogenic issues in a sustainable way (Gupta et al. 2017).

However, natural ecosystems in the periphery of urban and peri-urban regions like wetlands provide the "life-support system" to the urban centres by offering essential ecological services (Fig. 16.2) and securing the existence of biotic communities of those ecosystems. Decline in the ecological health and hygiene of wetlands influences the resources availability to biotic communities, ultimately leading to increased levels of vulnerability towards environmental hazards and a decrease in the intrinsic property of resilience to disaster (Chakraborty et al. 2023). Therefore, the relationship between wetland ecosystems and urban setups could be considered as a multi-layered eco technology (Figs. 16.3 and 16.4), both in terms of resource harvesting and buffering the ecosystem provisions. Management of the transient ecosystems delineating the interface of urban-wetland ecozones is essential to meet up the challenges thrown by the risk factors associated with rapid urbanisation, e.g., modification of land use land cover patterns, chemical and particulate pollution of urban air, soil and water, attack of heat waves, deforestation, carbon sequestration, etc. (Gupta et al. 2017; Chakraborty et al. 2023).

An urban gradient of entomodiversity is prominent as insect communities in urban core zones or urban centres tend to be more homogeneous, rather less diversified with respect to their ecological functionalities, as they are prone to increased susceptibility towards extinction. Contrastingly, in suburban, as well as in peri-urban areas, insect diversity is often represented with quite higher levels. Such distinctive entomodiversity is the result of their intrinsic ability to persist within the more diverse, rather less modified residual vegetation patches remaining through different urban sectors (Piano et al. 2017).

In this chapter, fireflies (Coleoptera: Lampyridae: Luciolinae), the ecologically as well as aesthetically valuable entomofauna have been selected as the focal tool to highlight the vital resilience offered by wetlands and associated macrophytic vegetations to act as the "conservation hubs" for this insect in the urban, suburban, and as well as at the peri-urban habitats (Pérez-Hernández et al. 2023). Fireflies act as a potentially efficient bioindicator of the environmental

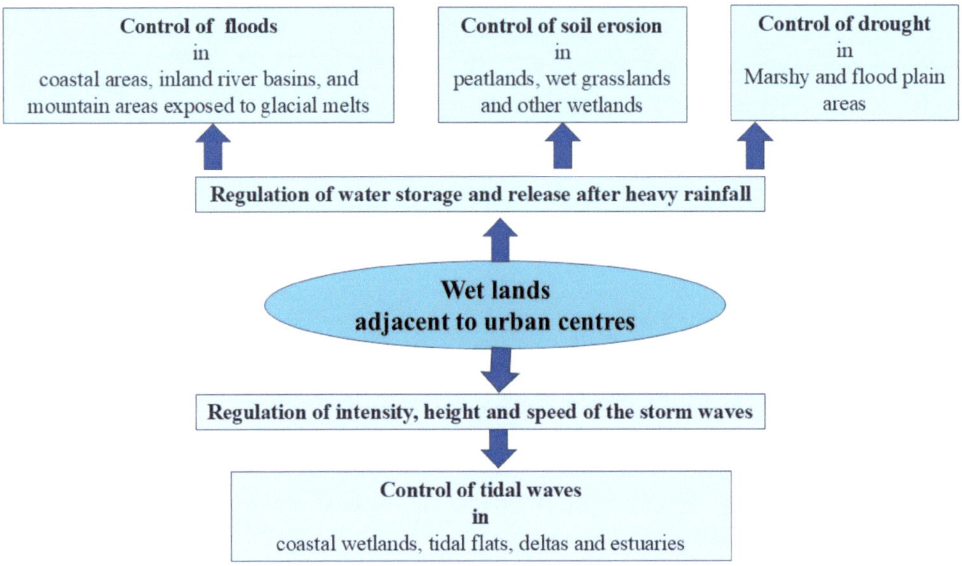

Fig. 16.3 Ecotechnology offered by wetlands adjacent to urban centres to provide urban resilience and mitigate the environmental hazards

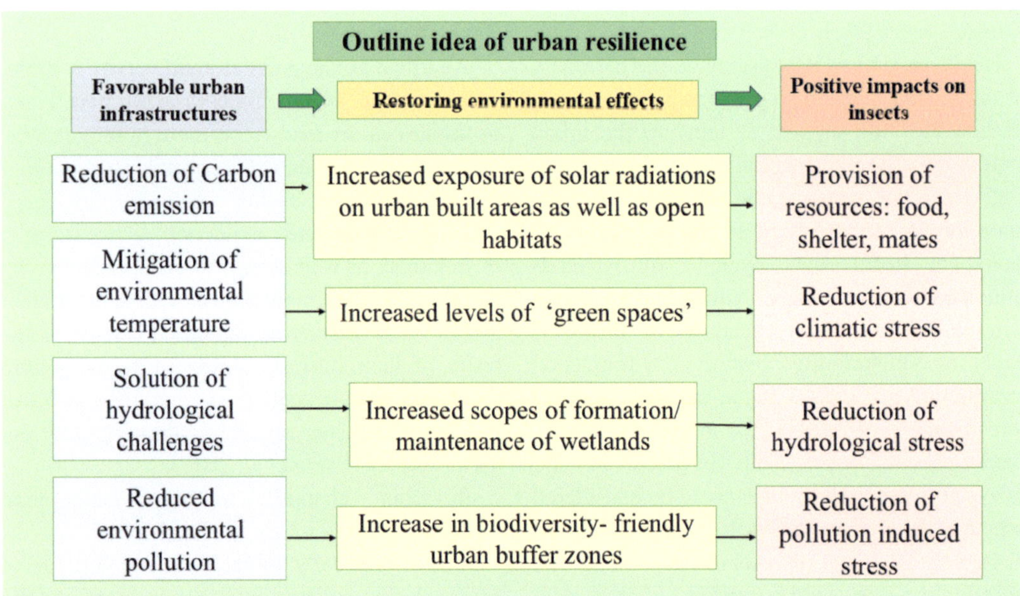

Fig. 16.4 Positive influences of urban factors leading to urban resilience

conditions prevailing in the urban setups exposed to different hazardous factors (Takeda et al. 2006; Ghosh and Chakraborty 2025).

Fireflies, one of Earth's most charismatic insects, consist of about 2200 species worldwide. In India, there are 35 species of Luciolinae fireflies (of those, two species are newly reported to science, *Triangulara frontoflava*, and *Medeopteryx bengalensis*) belonging to 11 different genera. Throughout the two eastern Indian states of India, Odisha, and West Bengal, different species of fireflies have been found to inhabit contrasting ecozones, and utilise 56 different plant species under 25 families (consisting of Associated Plants, APs, and Display Plants, DPs) (Ballantyne et al. 2019; Ghosh et al. 2021, 2023a, b, d, 2025a, b).

Fireflies, depicting species-specific resource utilisation patterns, have been found to depend on the ecosystems prevailing with moist conditions around the streams, wetlands, rivers, lakes, tributaries, even paddy fields with static storage of water, etc., to prevent desiccation of their eggs, larvae, pupae, along with their soft-bodied prey. The riparian vegetations consisting of their APs and DPs provide a platform for their perching, bioluminescent courtship displays, mating, and oviposition. Their carnivorous larvae show a predatory nature by voraciously feeding on soft-bodied invertebrates including snails, slugs, and worms, often inhabiting the wetland areas. Whereas, adults are often described to be nonfeeding and somctimcs adults are reported to nectar on flowers, mouthing leaves, and consuming stem sap of macrophytes (Fallon et al. 2021; Ghosh et al. 2023b, c). Researchers in Japan have found that hydroenvironments consisting of watershed areas with flat plains, non-urbanised areas with urban ratios less than 0.1, deciduous forests, cultivated lands, particularly paddy fields, and shallow and undisturbed aquatic bodies provide ideal geophysical conditions for firefly habitat (Kazama et al. 2007). In a nutshell, wetland habitats worldwide add urban resilience qualities to furnish favourable habitats for firefly populations in and around cities, towns, and other urban setups (Figs. 16.5, 16.6 and 16.7) (Fallon et al. 2021). At the same time, larvae of fireflies have the potential to act as a good indicator of the water quality of wetlands situated in the vicinity of cities and towns (Fig. 16.8). In Japan, firefly larvae are used for the biological testing of treated wastewater before the water is discharged into the source water body (Takeda et al. 2006).

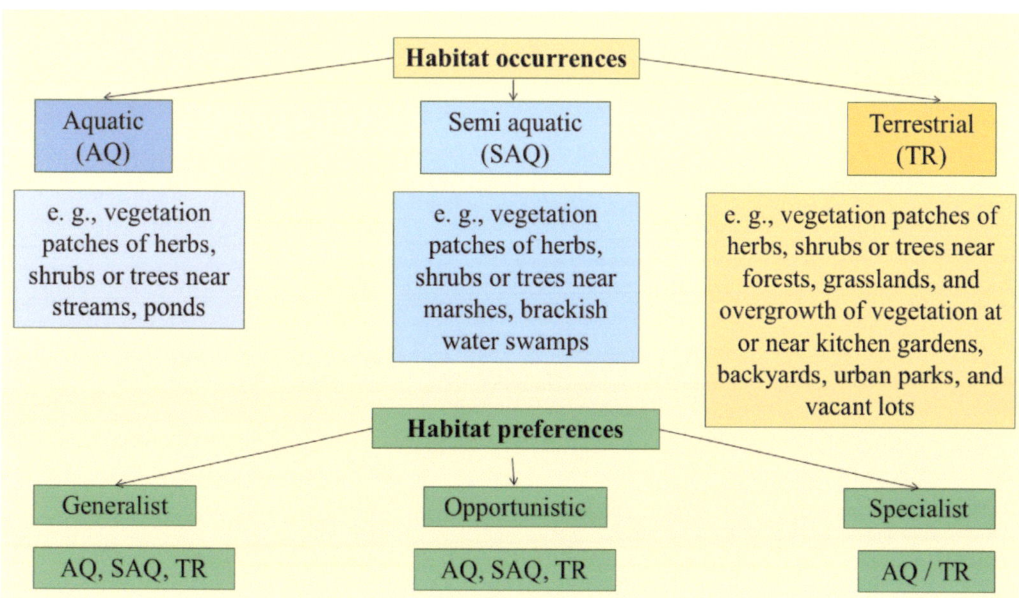

Fig. 16.5 Overall trend of habitat utilisation and preferences displayed by fireflies

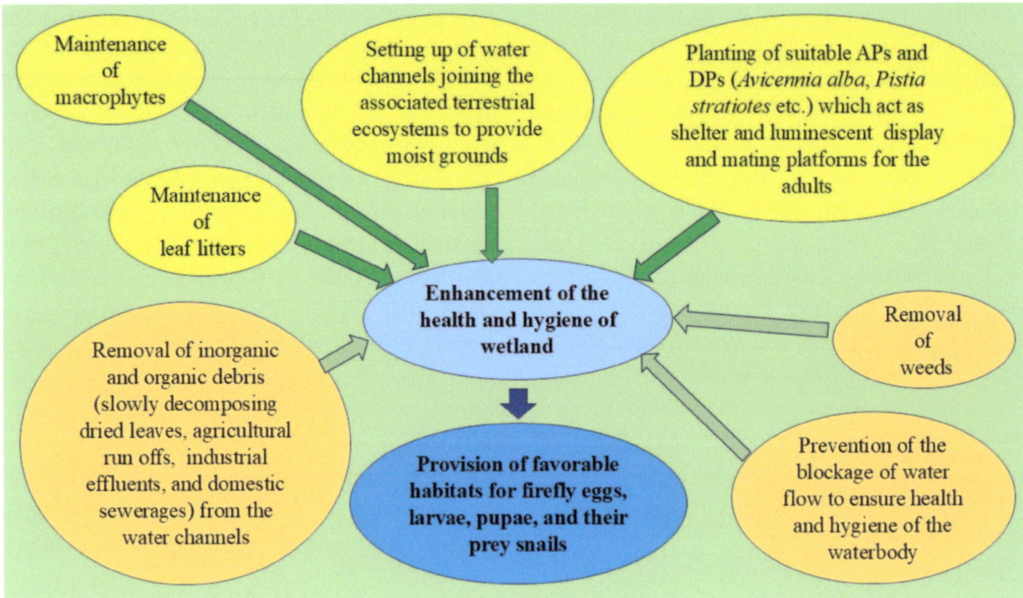

Fig. 16.6 Habitat enhancement properties of wetlands to promote favourable habitat qualities for fireflies

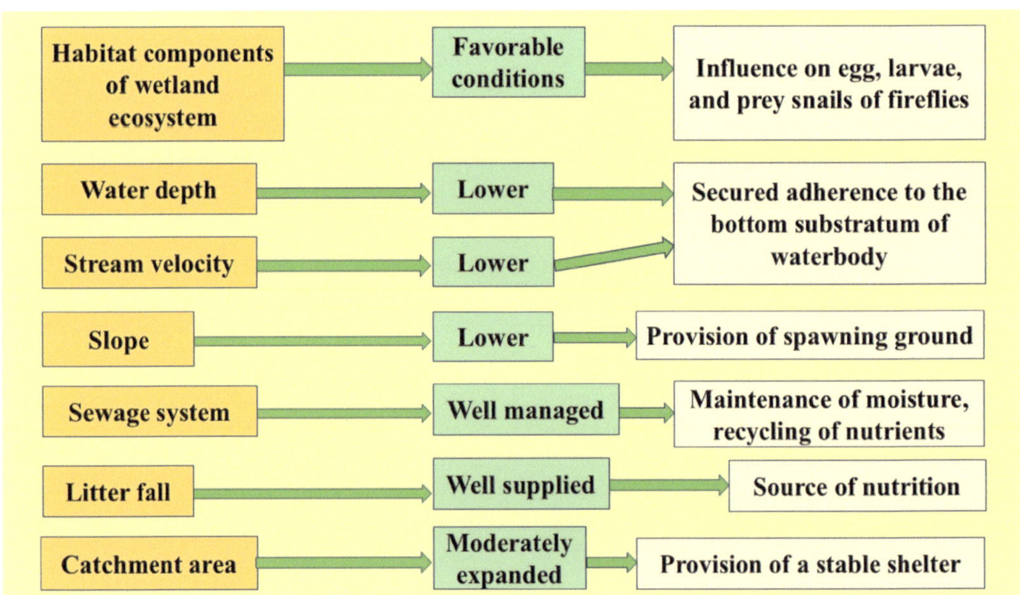

Fig. 16.7 Influence of the hydroenvironment of wetlands on the resilience of the adjacent urban habitat to enhance habitat qualities for fireflies

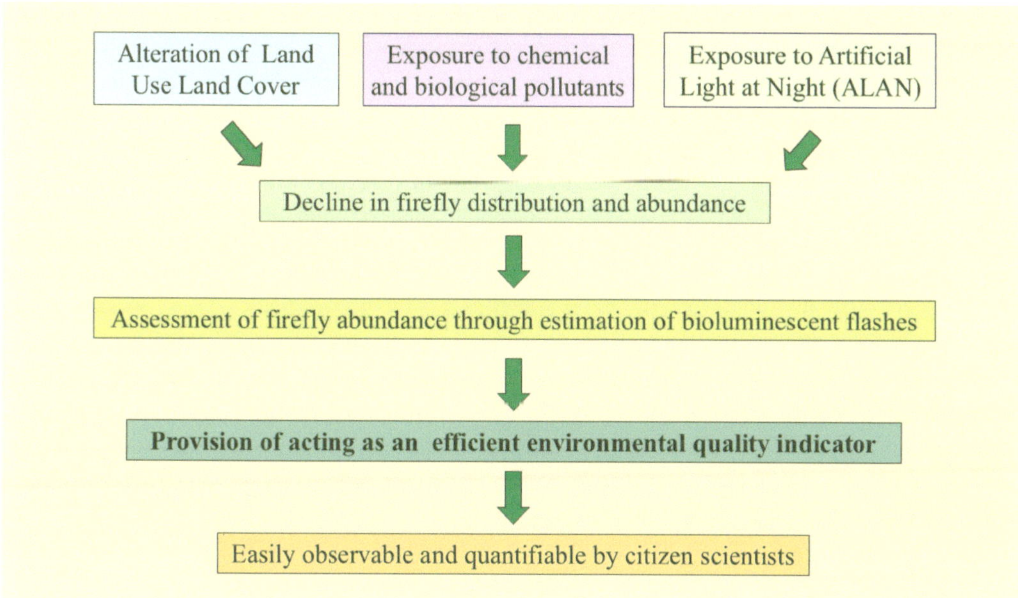

Fig. 16.8 Fireflies—the potential bioindicator of urban ecosystem

Upshots of socioeconomic ventures by utilising the association of fireflies with wetlands, firefly themed ecotourism activities have been sprung in different countries in Europe (Belgium, Portugal, Italy, Switzerland—*Lampyris* spp., *Lamprohiza* spp., *Luciola* spp.), USA (Great Smoky Mountain—*Photinus* sp., *Photuris* sp., Mexico—*Photinus* sp.), SE Asian countries (Malaysia—*Pteroptyx* spp., *Asymmetricata* sp., Thailand—*Pteroptyx* spp., *Asymmetricata* sp., Taiwan—*Abscondita* spp., *Aquatica* sp., *Pyrocoelia* sp., Japan—*Luciola* spp., South Korea—*Pyrocoelia* sp., Hong Kong—*Abscondita* sp., *Pygoluciola* sp., *Pyrocoelia* sp., *Diaphanes* sp) (Idris et al. 2021). Such ecotourism activities related to fireflies pave the way for the upliftment of the urban economy. Well-managed strategies could assure the importance of ecotourism activities to be continued without causing extirpations of ecological integrity of natural habitats surrounding urban areas (Fallon et al. 2021).

However, in recent days, globally wetlands are threatened with the risks of decline due to natural (climate change, drought, storms, dropping of water tables, drying up of ephemeral water bodies, alteration of river flows, and sea level rise, etc.) and anthropogenic causes (misuse of water resources, unmanaged drainage system, manipulation of coastal regions, poor water quality, uncontrolled exposure of agrochemicals, and light pollution) (Gupta et al. 2017). Presently, fireflies are threatened by light pollution imposed by exposure to ALAN as it masks their intrinsic property of bioluminescence and negatively affects the life processes of feeding, defence against predators, and reproduction (MacGregor-Fors 2011; Ghosh et al. 2024). The disappearance of wetlands imposes negative impacts on fireflies, particularly on the aquatic and semiaquatic species and their prey also (Fallon et al. 2021). They are also faced with challenges like habitat loss and degradation, infestation of heavy metals, and pesticides encroachment of invasive species (Fallon et al. 2021, Ghosh et al. 2024, 2025). However, despite the simultaneous presence of these threat factors, a few firefly populations have still been found to persist in urbanised areas (Viviani et al. 2010).

Under such scenarios, the present chapter aims at documenting Firefly Observation Sites (FOSs) (spotted, and selected by the observability of their presence, guided by their luminescent flashes) distributed throughout the suburban, peri-urban, and urban zones of study, situated at the urbanised areas of Southern Gangetic Plains (SGP), adjacent to Kolkata metropolis, West Bengal, India, in one hand, and the estuarine mangrove vegetation rich wetlands of Sundarbans delta (Sundarban Biosphere Reserve, SBR), in the district of South 24 Parganas, West Bengal, India, on the other. It also includes characterisation of those sites along different urbanisation parameters consisting of human population load, prominent and detailed LULC types, proportion of vegetated areas, and types of associated wetlands. The occurrence of different firefly species throughout these urbanisation gradients was chronicled. The contrasting trends of habitat preferences and microhabitat utilisations were specified through the discriminatory urban courses. Divergent existence of FOSs at sites with variable levels of exposure to potential threat factors like pesticides and Artificial Light At Night (ALAN) were delineated. A proposed scheme of planning for successful ventures of firefly-themed urban ecotourism has also been represented.

All the data had been gathered with the aim of introspecting how the urban wetlands, and associated ecological components remain instrumental in furnishing a conducive environmental compartment supporting the local populations of fireflies, the potential focal indicator of the urban ecological setup. Moreover, the findings also suggested that implementing suitable conservation strategies and planning for sustainable urban development could favour the conservation of fireflies and other vital resources for their existence (Pérez-Hernández et al. 2023).

16.2 Materials and Methods

16.2.1 Study Sites

Altogether 12 sites were selected from the southern Gangetic plains (SGP), and reclaimed mangrove habitats from the Sundarban Biosphere Reserve (SBR) (Table 16.1, Fig. 16.9) situated in the districts of North and South 24 Parganas, surrounding the metropolis of Kolkata [SGP (Southern Gangetic Plains): The site has evergreen vegetation in the deltaic alluvial plains; SBR (Sundarban Biosphere Reserve): The site represents islands having typical mangrove vegetation and reclaimed areas having normal mesophytic vegetation]. Average elevation was noted to be 8–10 m. The seasons are conveniently recognised as pre-monsoon from March to June (daily sun hours ranged between 7.8 and 9.8 h, temperature ranged from 10° to 41 °C and average rainfall up to 95 cm and occasionally accompanied by thunder-storms), monsoon comprising of July to October (daily sun hours 5.6–8.8 h, temperature ranged between 22° and 36 °C and with an average rainfall up to 540 cm), and post-monsoon during November to February (characterised by daily sun hours 6.8–9 h, temperature range between 10° and 15 °C, average rainfall up to 380 cm). Land Use Land Cover (LULC) profiles of all those sites (as observed and the major types as prevailed there) were documented (Ghosh et al. 2025a). The distribution of larvae and adults of fireflies through different habitats, microhabitats, and their association with other entomofauna in different habitat conditions were determined based on field observation at the same study site.

16.2.2 Taxonomic and Distributional Data on the Organisms of Study

Distributional data of the firefly populations throughout the study sites were obtained from field observation-based entomological sampling performed from 2017 to 2023 (Fig. 16.10, Table 16.1). In their native habitats, the fireflies often could be located prominently by production of reciprocal bioluminescent flashes to attract mates (in case of adult individuals) or prey (in case of larvae) (Takeda et al. 2006, Ghosh et al. 2023b, Pérez-Hernández et al. 2023). The feedback of the residents about the visibility of fireflies was also taken into consideration (MacGregor-Fors 2011). Based on the specimens available from collections and samplings, and

Table 16.1 Details of study areas and Firefly Observation Sites (FOSs) representing their urban status based on prominent Land Use Land Cover (LULC) features

Ecoregions of study and study areas (abbreviations)	Lat—long of the study areas	Geographical locations	Description of sampling sites	Major Land Use Land Cover (LULC) types	Categories of urban surfaces					Urban status
					R	S	CB	MV	IS	
SBR Sajnekhali (SJNKHLI)	22° 7′ 12″ N, 88° 46′ 49″ E	South 24 Parganas district, West Bengal	Marsh land and other aquatic bodies, riverside mangrove vegetation patches and display trees, herbs, and shrubs along dykes, abandoned lands, crop fields	CON, GRN, RES, COM	++	++	++	++	+	SU
SBR Kalash (KLSH)	21° 35′ 24″ N 88° 34′ 50″ E			CON, GRN	++	+	++	+	+	PU
SBR Bonnie (BNNIE)	21° 49′ 50″ N 88° 37′ 26″ E			CON, GRN	+	+	+	+		PU
SBR Dhainchi (DHNCHI)	21° 42′ 05″ N 88° 26′ 00″ E			CON, GRN	+	+	+	+		PU
SBR Indrapore village (INDRPR)	21°39′ 00″ N, 88°24′ 36″E			RES, GRN, COM	+	+	+	+	+	PU
SGP Pailan (PLN)	22° 25′ 12″ N 88° 18′ E			RES, COM, GRN	++	+	+	+	+	U
SGP Kulpi (KLPI)	22° 4′ 44″ N 88° 14′ 42″ E			RES, GRN	++	++	++	++	++	U
SGP Naihati (NHTI)	22° 52′ 48″ N 88° 25′ 48″ E	North 24 Parganas, West Bengal	Macrophytic vegetations near aquatic sites, cultivated lands, grassland and abandoned land, herbs, shrubs and display trees and roadside vegetation patches	INDUS, RES	++	+	+	+	+	U
SGP Kasthodanga (KSTDNG)	22° 58′ 16″ N 88° 38′ 01″ E	North 24 Parganas, West Bengal		GRN	+++	+++	+++	+++	+++	SU
SGP Sonarpur (SNRPR)	22° 26′ 57″ N, 88° 23′ 25″ E	South 24 Parganas district, West Bengal		RES, COM	++	+	+	+	+	SU
SGP Aamtala (AMTLA)	22° 21′ 59″ N, 88° 16′ 37″ E	South 24 Parganas district, West Bengal		RES, COM	++	++	++	++	+	SU
SGP Behala (BHLA)	22°30′ 21″ N, 88°19′ 23″ E	Kolkata and South 24 Parganas district, West Bengal		RES, COM	++	++	++	++	+	U

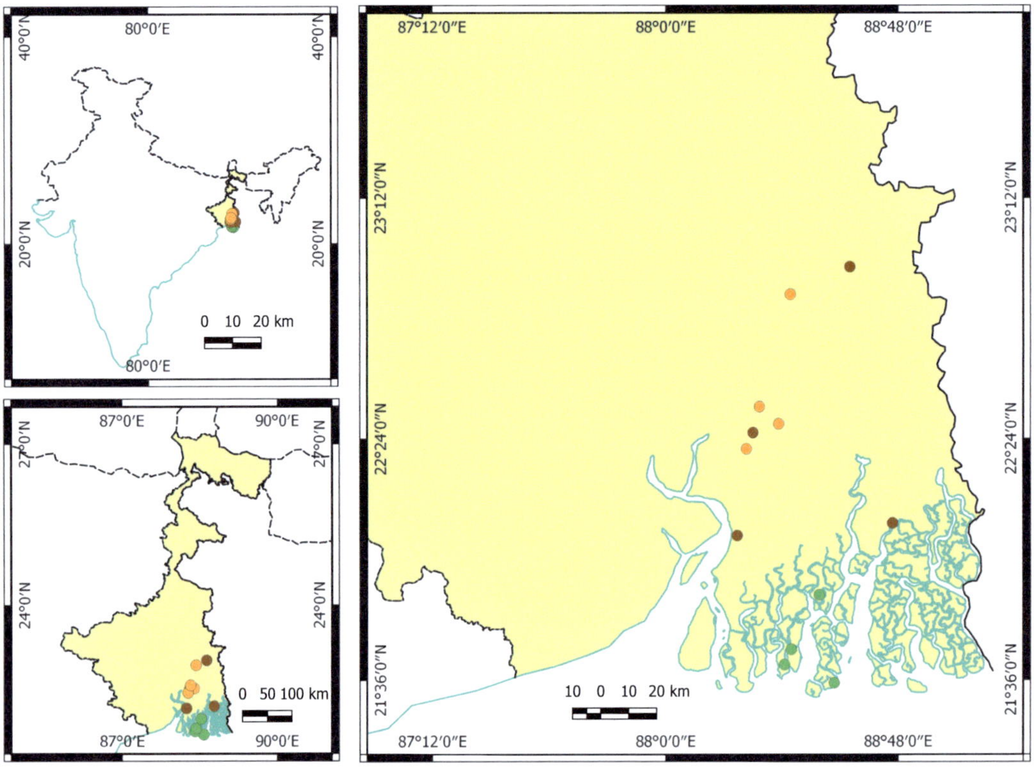

Fig. 16.9 Study areas and FOSs representing their differential urban status [differential urban status has been depicted by different colours; Green: peri-urban, light brown: suburban, and dark brown: urban]

Fig. 16.10 Ratio (%) of FOSs according to their differential urban status

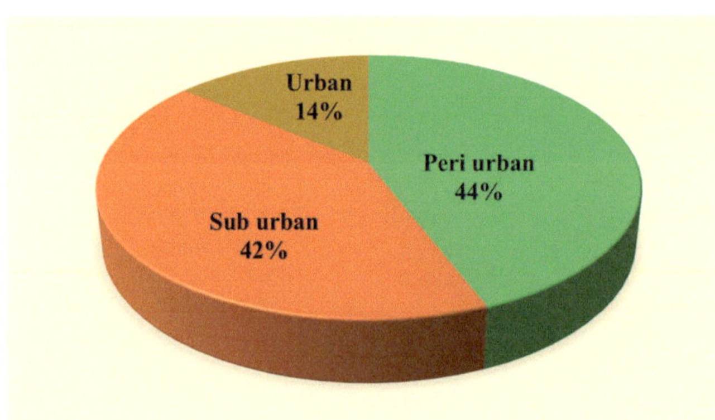

utilising their photographs, the genera and species were identified (Ballantyne et al. 2013, 2016, 2019, Ghosh et al. 2023a, d). Sites where fireflies were available or could be observed were denoted as the Firefly Observation Sites or FOSs.

Information regarding different habitat variables of each of those sites was documented to characterise fireflies' habitats (Fallon et al. 2021, Pérez-Hernández et al. 2023). Five main categories were considered: (i) geographic location (decimal coordinates), (ii) characteristics of the sampling habitats, (iii) major LULC classes, (iv) categories of urban surfaces ("commercial," "conservation," "green space," "residential," or "industrial"), (v) urban status (residential areas, streets, concrete buildings, exposure to motor vehicles, industrial setups) of each sites, (Table 16.1) [vegetation types were documented according to Champion and Seth 1968 classification; urban land use categories were characterised as follows: commercial—COM, conservation—CON, green space—GRN, residential—RES, industrial—INDUS; categorisation of urban surfaces were labelled as: residential areas—R; streets—S; concrete buildings—CB; exposure to motor vehicles—MV; industrial setups—IS ($\leq$ 30% lower—+, 30–60% moderate—++, $\geq$ 60% higher—+++)].

16.2.3 Characterisation of Firefly Habitats in the Scale of Urban Parameters

Firefly habitat size (the area of the specific site where fireflies were observed or recorded to exist) was measured using satellite imagery from Google Earth (2023). The habitat size included the area within a 500-m radius circular buffer (0.79 km^2) around the exact geographical site of occurrence of fireflies (either observed or reported to occur), for homogenous treatment for all sites comparable (MacGregor-Fors 2011).

Areas covered with vegetation were categorised into forests (woods possessing a tree layer), shrubs (bush layer), open grassland (meadows), cultivation land (crop land), backyards or domestic gardens, kitchen gardens, city parks, sports ground (urban green spaces), parking lot, garage space (vacant habitats with overgrowth of vegetation), roadside vegetation patches (e.g. median strips, green areas along rivers or between residential areas), and conservation areas (sanctuaries, etc.). Categories of microhabitat utilisations by fireflies were also documented (Figs. 16.13 and 16.14).

The eye estimation combining with Google Earth image-based percentage of urban surfaces (within a 500-m radius circular buffer zone around the centre of each site of firefly observation) were categorised based on population status [categories of urban population status: $\geq$ 10 inhabitants per hectare—highly populated (HP); 6–10 inhabitants per hectare—moderately populated (MP); $\leq$ 6 inhabitants per hectare—less populated (LP); categories were based on eye estimation within a 500-m radius circular buffer zone around the site of firefly observation at each study site], vegetation cover [categories of vegetated area: 67–100% vegetated area—highly vegetated (HV); 34–66% vegetated area—moderately vegetated (MV), and 0–33% vegetated area—sparsely vegetated (SV); categories were based on eye estimation within a 500-m radius circular buffer zone around the site of firefly observation at each study site], and concrete-built setups [categories of concrete-built setups: 0–33% built cover—sparsely developed (SD); 34–66% built cover—moderately developed (MD); 67–100% built cover—highly developed (HD); categories were based on eye estimation within a 500-m radius circular buffer zone around the site of firefly observation at each study site] (Fig. 16.16a–c) (MacGregor-Fors 2010, 2011, Pérez-Hernández et al. 2023).

16.2.4 Characterisation of Firefly Microhabitats in Relation to the Association with Wetlands

Microhabitat utilisation by fireflies had been documented into three distinctive categories, e.g., Exposed (above water surface, upper surfaces of leaves of macrophytes, and other plants), Transitional (root layer of larger trees, surface of stem, inside flower petals), Hidden (crevices under muddy substratum, within leaf litter beds; upper surfaces of leaves of macrophytes, and other plants). We had documented the presence of different stages of the lifecycle of fireflies at or near seasonal or permanent watercourses (e.g. lakes, creeks, rivers, ponds, channels, and wetlands) in the vicinity of their habitats (Fig. 16.15).

16.2.5 Estimation of Potential Disturbance Factors

FOSs were categorised according to the urban gradients based on their exposure to two of the most hazardous potential disturbance factors, agrochemicals, and ALAN. Such categorisation represented an estimation of the level of the direct or indirect influences of urban development on the city's ecosystems (Fig. 16.17a, b). The degree of light pollution to which fireflies in the study sites are exposed, data were compiled from the nighttime light maps produced by the Earth Observation Group (2021). The radiance values (nW/cm^2 /sr) dataset provided by the Visible and Infrared Imaging Suite (VIIRS) and Day Night Band (DNB) satellites represent an indirect measure of light pollution and acts as a proxy for urbanisation for each site where fireflies were observed (Vaz et al. 2022). Each pixel in the light pollution map shows radiance values (in selected year 2024) ranging from (0.3–0.5) to (70–71) via 5–7. Where (0.3–0.5) represents darkness (often associated with rural zones), (70–71) represents very bright sites associated with urban sites near Kolkata metropolis, and (5–7) denotes peri-urban areas. Pesticidal exposures were categorised into three sites—low (<30 g/ ha) at urban, moderate (30–60 g/ ha) at peri-urban, and high (> 60 g/ ha) suburban, based on the trend of annual application of pesticides on local crop fields by the farmers.

16.2.6 Schematic Framing of Ecotourism Involving Local Firefly Populations

The scheme was proposed based on literature reviews, field surveys, and feedback obtained from periodic interactions with community stakeholders (Fig. 16.18) (Ghosh et al. 2025c).

16.2.7 Statistical Analysis

Kruskal-Wallis tests had been exercised to check whether significant differences exist among the abundance and distribution of firefly observation sites among the three different urban setups with reference to site specific species wise distribution, tendency of habitat and microhabitat utilisation, responding to various urban resilience factors like vicinity to wetlands, population status, built setup conditions, vegetation covers and impacts of potentially hazardous factors like exposures to agrochemicals and ALAN. All statistical analyses were performed with PAST version 4.17.

16.3 Results and Discussion

Altogether, 12 FOSs were selected where periodic samplings were performed to find out the diversity and distribution of fireflies, as an indicator organism to detect the influence of urbanisation on these entomofaunal populations. By minute observations of their habitat distribution, microhabitat utilisation, and resource utilisation patterns, an idea could be obtained about the impact of urban resilience on the survival potential of these insects. Also, the present research unearthed the importance of wetlands as a pro-conservative tool for urban ecosystems, as it has also been highlighted by many other researchers (Takeda et al. 2006; Kazama et al. 2007; Idris et al. 2021).

16.3.1 Distribution and Urban Status of the Study Areas

The FOSs were categorised as peri-urban (PU), suburban (SU), and urban (U) based on major land use land cover (LULC) features and urban structures. Following these criteria, there were four sites belonging to each of these three categories (Table 16.1).

16.3.2 Firefly Observation Sites (FOSs): Distribution Along Urban Setups

As depicted in Fig. 16.10, the maximum number of firefly observation sites were found to be situated at PU (44%), followed by those at SU (42%) and at U, least number (14%) of these sites were located.

16.3.3 Differential Occurrence of Firefly Populations Through Urban Setups

As it is reflected in Fig. 16.11, periodic samplings at firefly observation sites throughout all the study areas resulted in a total of seven species of Luciolinae fireflies belonging to four different genera. Of those, *Abscondita chinensis* was found to be present at all three urban setups, and *Triangulara sunderbanensis* had been restricted only to the SU zones of the mangrove-populated part of the Sunderban biosphere reserve, at Sajnekhali (SJNKHLI) area. *Sclerotia substriata* had also been represented as a SU species, prevalently observed at the periphery (AMTLA, Aamtala, South 24 Parganas) of the Kolkata metropolis. *Pteroptyx malaccae* is the mangrove species, venturing both SU (SJNKHLI, Sajnekhali, Sunderban Biosphere Reserve, South 24 Parganas) and PU (DHNCHI, Dhainchi and BNNIE, Bonnie islands, Sunderban Biosphere Reserve, South 24 Parganas) localities. On the other hand, *Abscondita perplexa*, *Abscondita terminalis*, and *Sclerotia aquatilis* populations could have been made themselves compatible both for U as well as SU habitat setups.

16.3.4 Differential Habitat Preferences Manifested by Larvae and Adults of Fireflies at Various Urban Setups

Present researchers had registered the trend of habitat preferences manifested by larvae and adults of the study insects, fireflies throughout aquatic, semiaquatic, and terrestrial habitat conditions under the background of PU, SU, and U scenarios.

Figure 16.12 depicts that, at PU localities, adults were found to prefer the semiaquatic habitats most, especially to conduct their courtship and breeding activities. There, the larvae also found the semiaquatic habitats suitable, perhaps due to the presence of their prey, gastropod snails. The moist ambience facilitated their instar switching and the macrophytic surfaces provided refuge from their predators.

At the SU setups, the aquatic habitats were mostly preferred by larvae and adults. There at semiaquatic and terrestrial parts, where sprawling human populations and multifactorial anthropogenic activities might have compromised the habitat qualities of semiaquatic bodies, at least to certain extents. That is why the larvae and adults

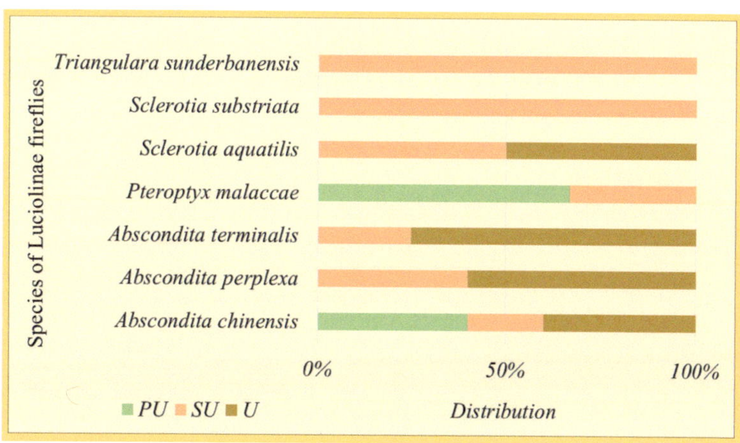

Fig. 16.11 Differential occurrences of firefly species populations through various urban setups

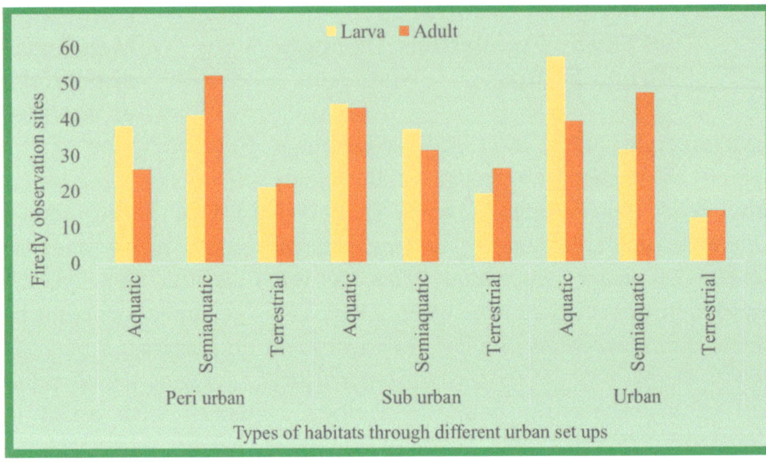

Fig. 16.12 Preferences for habitats in different stages of the life cycle of fireflies as observed in peri-urban, suburban, and urban study sites

of fireflies restricted themselves to aquatic sites for effective resource utilisation.

At U localities, the aquatic sites had offered the most suitable habitat for larval development by maintaining an intact supply of prey species, providing shelter and moisture content. Adults in urban zones have been found to cope with the environmental hazards and stress by finding safe shelter at semiaquatic sites.

16.3.5 Differential Microhabitat Utilisation by Fireflies Along the Distinct Urban Gradients

As observed from Fig. 16.13, at PU sites, fireflies had been found to occur significantly at exposed microhabitats, e.g., roaming above the water surfaces near aquatic bodies, particularly while emitting bioluminescent courtship flashes and flights.

At the SU backgrounds, the tendency to utilise all the types of microhabitats, exposed, transient, and hidden, was quite noticeable.

At U backdrops, they were found mostly on exposed surfaces. Different prevailing stress factors, acting as stimulants or external cues, perhaps drove them to come out of their hidden refuges and get exposed. Such a tendency, in turn, adds to their vulnerability to different threat factors.

16.3.6 Terrestrial Components Promoting Suitable Habitat Characteristics and Urban Resilience Along Urban Gradients

In Fig. 16.14, it is represented that the firefly habitats significantly vary in their characteristics along PU, SU, and U zones throughout the study areas (Kruskal-Wallis test had represented $p \leq 0.05$, $H = 12.97$). Throughout PU, SU, and U, backyard vegetations and kitchen gardens significantly supported firefly populations. At the SU and U sites, despite dwindling features of firefly populations, green patches of domestic gardens and roadside shrublands had contributed as "urban firefly hubs." At PU, selected patches of mangrove forests were inhabited by particular species of firefly, e.g., *Pteroptyx malaccae*. The city parks of U sites, despite possessing lots of "green components," had not been able to act as efficient shelters due to their repeated exposures to different agrochemicals, constructive designing processes, and many other anthropogenic modifiers.

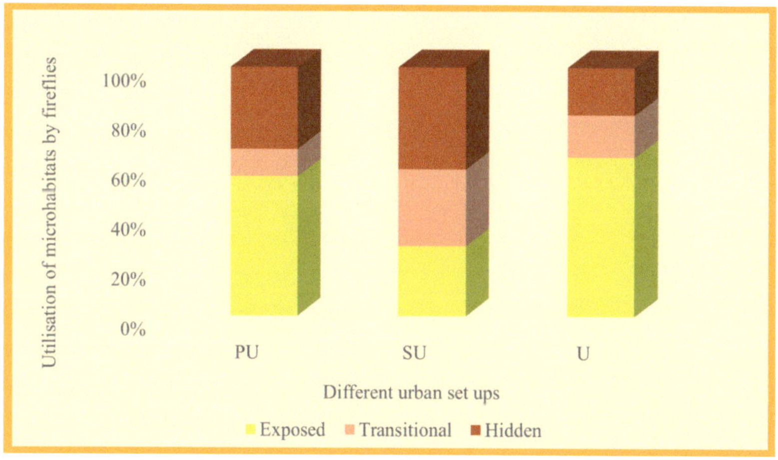

Fig. 16.13 Differential utilisation of microhabitats by fireflies as observed in peri-urban, suburban, and urban study sites

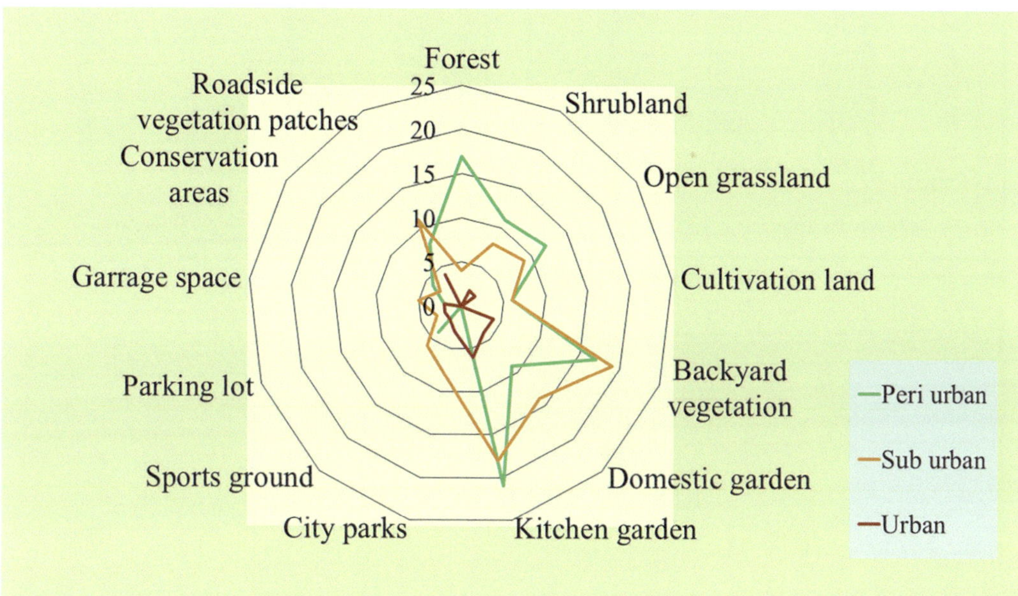

Fig. 16.14 Variations in characteristics of terrestrial firefly habitats among peri-urban, suburban, and urban zones

16.3.7 Wetland Components Promoting Suitable Habitat Characteristics and Urban Resilience Along Urban Gradients

From Fig. 16.15, it could be observed that, in PU, SU, and U localities, the aquatic sites like ponds offered the most favourable habitat conditions to local firefly populations by providing suitable courtship platforms (macrophytic surfaces consisting of APs and DPs), sites for breeding, oviposition, larval maturation, and pupal development. The larval prey items (gastropod molluscs, leeches, earthworms, larvae of dragonflies, etc.) were also abundant. Moreover, the macrophytes also provided them protection against predatory exposures.

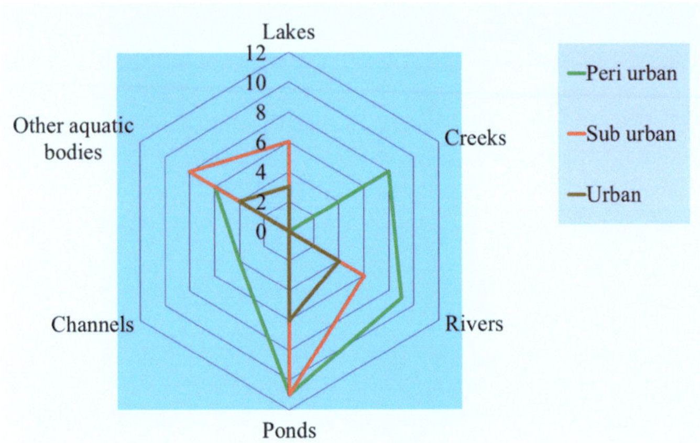

Fig. 16.15 Variations in characteristics of watercourses near firefly habitats among peri-urban, suburban, and urban zones

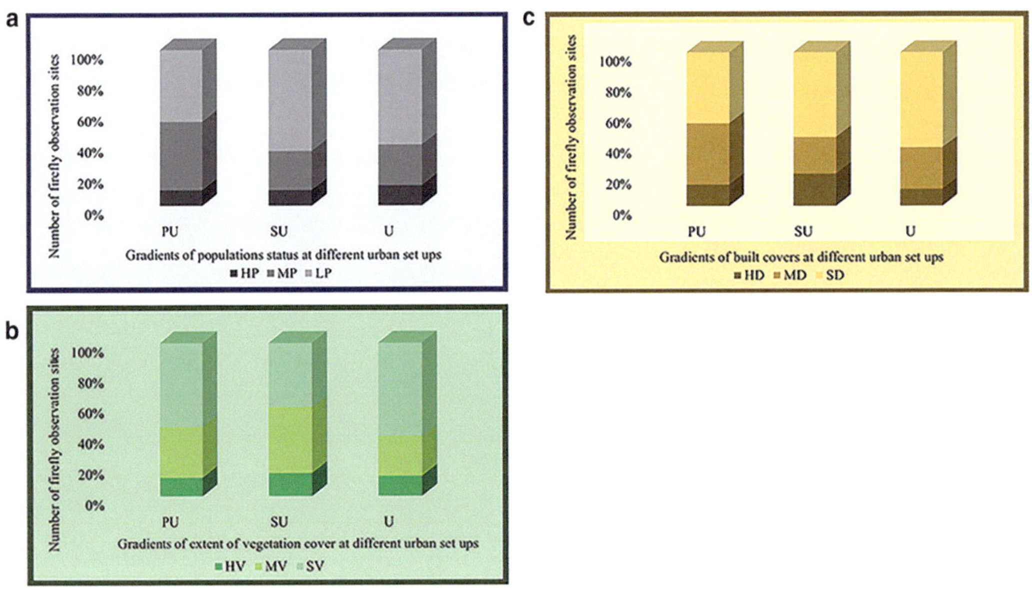

Fig. 16.16 (**a**) Differential distribution of FOSs along the intensity gradients of urban parameter: load of human populations. (**b**) Differential distribution of FOSs along the intensity gradients of urban parameter: extent of vegetation coverage. (**c**) Differential distribution of FOSs along the intensity gradients of urban parameter: concrete-built setup

16.3.8 Urban Resilience Corresponding to Gradients of Intensities of Urban Parameters

Three different urban parameters had been considered, e.g., load of human populations, extent of vegetation coverages, and categories of build setups. As evident from Fig. 16.16a, at PU, the maximum number of FOSs had been fallen under the least populated sites (46%), followed by those at moderately populated sites (44%). Likewise, at SU, and U, the highest numbers of FOSs were situated at least populated sites (65% at SU, and 61% at U, respectively).

Figure 16.16b shows that the highest numbers of FOSs were found to be restricted to sparsely vegetated sites, both at PU (55%), and U (61%).

Contrastingly, at SU, at moderately vegetated patches, the highest numbers of FOSs were found to be located (43%), followed by those at sparsely vegetated sites (42%). Throughout the gradient of vegetation cover along the urban setups, the number of FOSs was found to vary significantly (Kruskal-Wallis test resulted in $p \leq 0.05$, $H = 6.49$). These findings suggest that, at dense vegetation sites, the transmission of luminescent flashes gets affected or obstructed, thus the efficiency of courtship may be challenged. So, at all the sites, fireflies avoided highly vegetated parts. At the same time, the existence of their APs and DPs is essential, so that they can utilise those as perching sites and platforms for luminescent courtship displays, breeding grounds, oviposition sites and for larval attachment and pupal settlement at sparsely and moderately vegetated sites.

Figure 16.16c delineates that, at PU, SU, and U, sparsely developed built setup areas contained the highest numbers of FOSs (46% at PU, 55% at SU, and 62% at U, respectively). Thus, it can be deciphered that concretisation leads to loss of firefly habitats at all three types of urban setups.

16.3.9 Influence of Potential Threats to Urban Populations of Fireflies: Exposure to Pesticides and ALAN

At PU and SU, the lowest ALAN exposure levels (radiance 0.3–0.5 nW/ cm^2/ sr) were associated with maximum numbers of FOSs (88 at PU, 90 at SU) (Fig. 16.17a). Figure 16.17b shows the least to moderate levels of pesticidal pollution allowed fireflies to occur prominently. On the other hand, sites exposed to higher levels (> 100 g/ ha/ cropping year) of pesticides (Khatun and Jamal 2021) were documented with the least number of firefly observation sites. These findings indicated that deterioration of habitat qualities led to declining effects on the distribution of fireflies. At the same time, the tiny creatures, at least to some extent, could have made themselves capable of withstanding the selection load imposed by the hazardous factors, as evident from the occurrence of the highest observation sites (31) belonged to the moderate ALAN exposure range (radiance 5–7 nW/ cm^2/ sr) at U (Fig. 16.17a).

16.3.10 Firefly Themed Urban Ecotourism Centres Adjacent to Wetlands Utilising Urban Resilience

Figure 16.18 briefs the idea of firefly-based ecotourism development focusing on the long-term conservation approaches. The inherent property of bioluminescence of fireflies is often considered an attribute of tourist attractions. Centring this public interest and query, the firefly-themed ecotourism could be developed and promoted by surveying species diversity, distribution, and the influence of physical factors upon natural populations of fireflies, planning sustainable maintenance of firefly habitats and harmonising the natural balance of the ecosystem, including the derivation of guidelines for promoting the learning resources. In this perspective, academicians provide academic data to frame a handbook of knowledge regarding fireflies and their life cycles in relation to the ecoscape shared by them. Maps to be prepared to fix tourism routes (tour map) for firefly sightseeing, following the trend of abundance of fireflies in nature. Training schedules could be organised for local community members regarding the presentation of natural tourist attractions following the life cycle, habitat occupancy, and behavioural manifestations of fireflies. Co-management in parts of community members, academicians and policy makers could lead the firefly themed ecotourism attempts to an ultimate success (Chaiwongsaen et al. 2020).

16.4 Conclusions

In conclusion, the major implications of the research could be mentioned. First, there exist a few urban resilience-promoting environmental drivers, e.g., wetlands, and vegetation patches

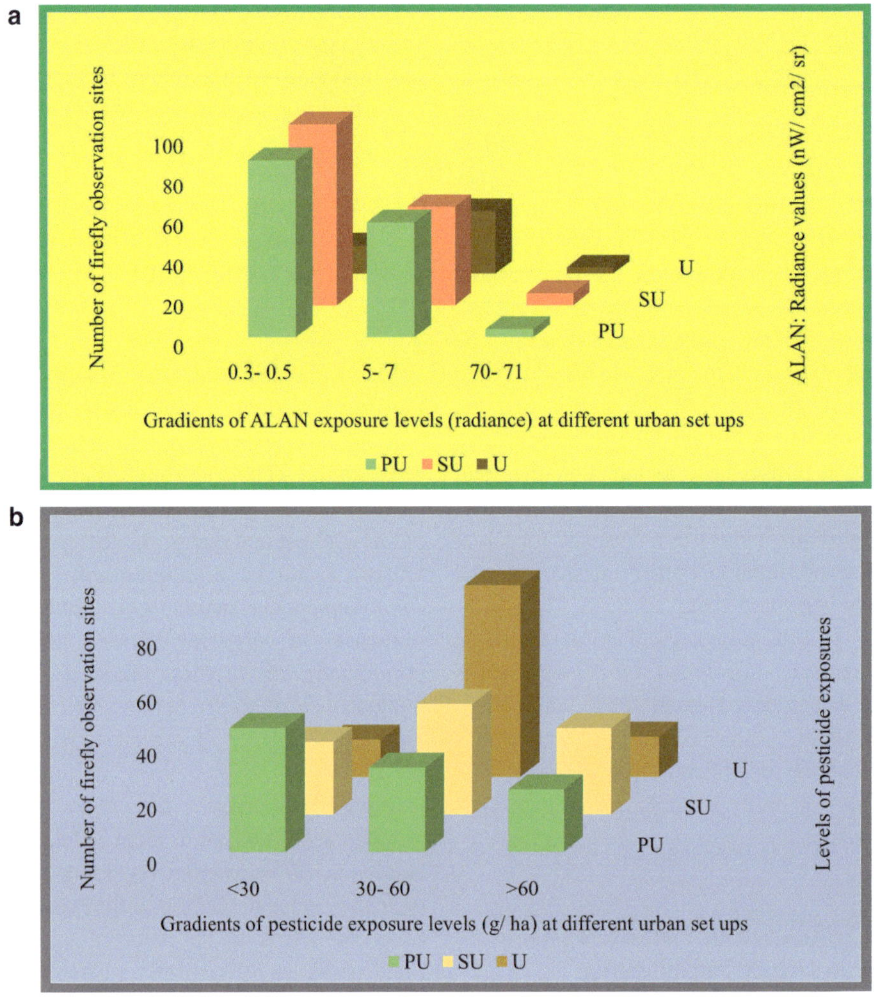

Fig. 16.17 (**a**) Differential distribution of FOSs along the varying exposure of different levels (radiance) of ALAN. (**b**) Differential distribution of FOSs along the varying exposure of different levels (g/ ha) of pesticides

which act as conservation hubs for urban populations of fireflies. Second, a few factors play a role as environmental stressors which compromise urban resilience. Those are population load, concrete setups, exposure to agrochemicals, and sources of ALAN. These findings provide a footprint in scientific research related to the beneficial role of urban wetlands to provide urban resilience and pave the way for long-term conservation of urban biodiversity and the ecosystem as well. These research outcomes also satisfy the global interest that how the tiny jewel of nature, the fireflies could be sustained in urban localities utilising the intrinsic balancing attributes of the urbanisation process itself. Last but not least, as dependence of fireflies on aquatic sites was found to be greater corresponding to the gradual intensification of urbanising parameters, the significance of wetlands for sustenance of urban biodiversity became evident.

From the outcomes of the above-mentioned research and review, a few hypotheses could be consolidated as follows:

First, fireflies could be considered as potentially effective bioindicators of integrity of urban ecosystems as those were found to be associated to specific, as well as diversified habitats, exhibited preference for proximity to water

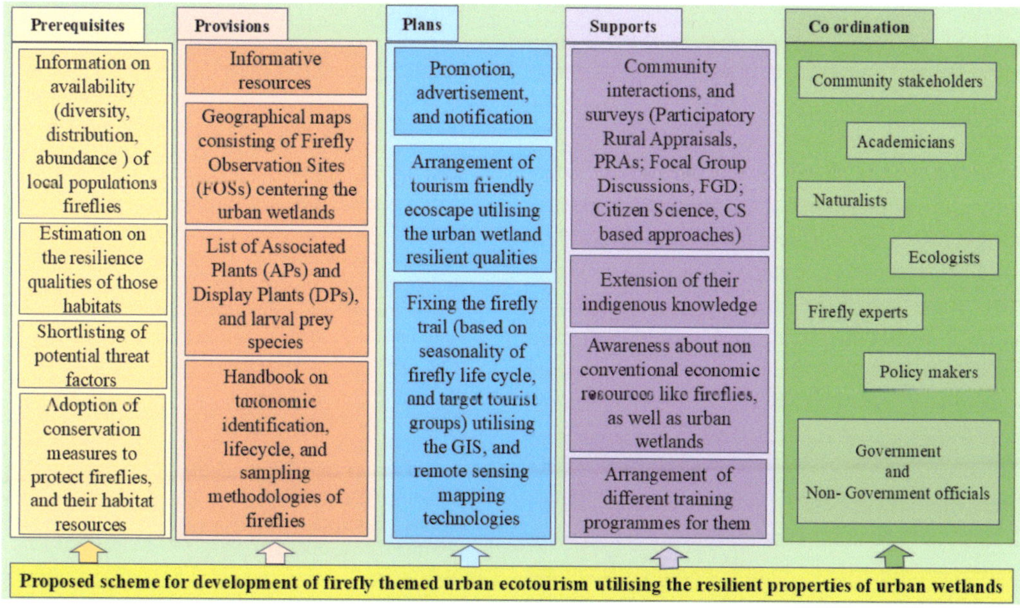

Fig. 16.18 Planning and prerequisites for firefly-themed ecotourism in the background of an urban ecosystem: a proposed scheme

bodies, distinct vegetations (APs and DPs) in different stages of lifecycle and sensitivity towards fluctuations of environmental parameters, e.g., exposure to pesticides and light pollution (Fallon et al. 2021; Ghosh et al. 2024).

Second, gaining an idea about the habitat qualities of the target bio-object, here the fireflies, would facilitate deriving the habitat suitability index to assess the carrying capacity of natural habitat compartments to support the biotic potentiality for community members inhabiting the habitat. Digital maps extracted by the application of advanced technologies like geographic information systems (GIS) and hydrological simulation, the levels of urban resilience and wetland conditions could be estimated. The information obtained helps to investigate the habitat qualities for fireflies in the urban backdrop in relation to geological conditions, hydrological conditions, and land use types. After validation by field-based observations, such suitability criteria for firefly habitats could be utilised to derive a habitat suitability index (HSI) for the estimation of habitat qualities (Pérez-Hernández et al. 2023).

Third proper land use planning and ecosystem management assist the process of urban resilience, which promotes the harmony between ecological resource utilisation and sustenance of species populations in their natural habitats. By examining the distribution patterns of firefly observation sites throughout the gradients of urban parameters, it is quite discernible that urban resilience gets manifested with the dual axes of robustness and flexibility at the same time to assure the recovery and restoration of the stressed biotic communities (Gupta et al. 2017). Even, protection and restoration of both the occupied (e.g., PU, SU, or U habitats) and adjacent (e.g., different wetlands, forests, shrublands, marsh, swamps) habitats of the concerned species can extend support to those species (Fallon et al. 2021). In case of fireflies as the research object, it is indicative that restoration and maintenance of green areas and nighttime environments are essential for maintenance of their overall diversity and distribution (Pérez-Hernández et al. 2023).

Fourth, merging the indigenous knowledge procured from community stakeholders with expertise from academicians, naturalists, scientists, field biologists, and experts is an essential prerequisite for the purpose of conservation of target organisms in the urban backdrop. To promote the conservation of aesthetic as well as ecologically significant entomofauna like fireflies in the urban setups, in different countries, various approaches related to citizen science have been applied to build up sustainable ecotourism activities (Wong 2014; Rana et al. 2022). It can turn into a successful event once those schemes get attestation from policy makers following the review of concerned government and non-government organisations (Chaiwongsaen et al. 2020). With the spread of ecotourism, fireflies can potentially enjoy the direct or indirect benefit of conservation. Threat factors could be omitted or minimised, habitats for fireflies and related community members like their larval prey species, APs and DPs as well could be protected (Chaiwongsaen et al. 2020).

Acknowledgments The authors express their gratitude to the authorities of the Vidyasagar University, Midnapore, West Bengal, India and to the Principal of Bethune College, Kolkata, West Bengal, India, for providing necessary research facilities.

Disclaimer (Artificial Intelligence) Authors hereby declare that no generative AI technologies were used during the writing or editing of this manuscript.

Potential Conflicts of Interest On behalf of the authors, the corresponding author states that there is no conflict of interest.

Informed Consent Informed consent was obtained from all individual participants included in the study.

Authors' Contributions Conceptualisation, SG, and SKC; methodology, SG; field sampling, SG; primary data collection, SG; map preparation, SG; resources, SG; data analysis, SG, and SKC; writing and original draft preparation, SG; review and editing, SG, and SKC; final review, SKC. Both authors checked and approved the last version of the manuscript.

Data Availability Statement The datasets during and/or analysed during the current study are available from the corresponding author on reasonable request.

Funding The research was not supported by any funding.

References

Ballantyne LA, Fu XH, Lambkin CL et al (2013) Studies on south-east Asian fireflies: *Abscondita*, a new genus with details of life history, flashing patterns and behaviour of *Abs. chinensis* (L.) And *Abs. terminalis* (Olivier) (Coleoptera: Lampyridae: Luciolinae). Zootaxa 3721(1):1–48. https://doi.org/10.11646/zootaxa.3721.1.1

Ballantyne LA, Lambkin CL, Luan X et al (2016) Further studies on south eastern Asian Luciolinae: 1. *Sclerotia* Ballantyne, a new genus of fireflies with back swimming larvae. 2. *Triangulara* Pimpasalee, a new genus from Thailand (Coleoptera: Lampyridae). Zootaxa 4170(2):201–249

Ballantyne LA, Lambkin CL, Ho JZ et al (2019) The Luciolinae of S. E. Asia and the Australopacific region: a revisionary checklist (Coleoptera: Lampyridae) including description of three new genera and 13 new species. Zootaxa 4687(1):1–174. https://doi.org/10.11646/zootaxa.4687.1.1

Chaiwongsaen P, Pinmongkhonkul S, Madhayamapurush W (2020) Guidelines for promoting tourism from biodiversity of fireflies in agricultural areas of ban-buo, Phayao Province, Thailand. Acad Strateg Manag J 21 (S2):1–13

Chakraborty SK, Sanyal P, Ray R (2023) Wetland ecology. Springer, Kolkata. 709pp

Champion HG, Seth SK (1968) A revised survey of the forest types of India. The Manager of Publications, Delhi

Chandra S, Sharma D, Dubey SK (2018) Linkage of urban expansion and land surface temperature using geospatial techniques for Jaipur City, India. Arab J Geosci 11(2):31. https://doi.org/10.1007/s12517-017-3357-6

Deosthali V (2000) Impact of rapid urban growth on heat and moisture islands in Pune City, India. Atmos Environ 34(17):2745–2754. https://doi.org/10.1016/S1352-2310(99)00370-2

Dwivedi A, Khire MV, Mohan BK, Shah S (2019) The role of structure cooling to reduce the effect of urban heat Island in Mumbai. Adv Build Energy Resour 13(2):174–192. https://doi.org/10.1080/17512549.2018.1488611

Earth Observation Group (2021) Visible and infrared imaging suite (VIIRS) day night band (DNB). Available at https://eogdata.mines.edu/products/vnl/

Fallon CE, Walker AC, Lewis S, Cicero J, Faust L, Heckscher CM et al (2021) Evaluating firefly extinction risk: initial red list assessments for North America. PLoS One 16(11):e0259379. https://doi.org/10.1371/journal.pone.0259379

Ghosh S, Chakraborty S (2025) Bioaccumulation and biotransportation of heavy metals along a water-soil-plant-firefly (Coleoptera: Lampyridae: Luciolinae) food chain: a case study from the Southern Gangetic Plains, West Bengal, India. Nature Environ Pollut Technol 24(4) Paper no. B-4300. Prepublication online version published on 28.03.2025. https://doi.org/10.46488/NEPT.2025.v24i04.B4300

Ghosh S, Sarkar SK, Chakraborty S (2021) Two new records of the subfamily Luciolinae Lacordaire, 1857 (Coleoptera: Lampyridae) with a checklist of genus *Abscondita* from India. J Asia Pac Biodivers 14(2021):53–59. https://doi.org/10.1016/j.japb.2020.10.004

Ghosh S, Sarkar SK, Chakraborty S (2023a) New distributional records of fireflies (Coleoptera, Lampyridae, Luciolinae) from two Eastern States of India with notes on their biology and an updated Indian checklist. Biodivers Data J 11:e98948. https://doi.org/10.3897/BDJ.11.e98948

Ghosh S, Saha S, Chakraborty S (2023b) The floral associates of fireflies (Coleoptera: Lampyridae: Luciolinae) as recorded in two eastern Indian states with reference to their display plants. J Asia Pac Biodivers 16(2):174–183. https://doi.org/10.1016/j.japb.2023.03.008

Ghosh S, Bera D, Chakraborty SK (2023c) Synchronisation of gut enzymes with differential foraging spectrum of two species of firefly (Coleoptera: Lampyridae: Luciolinae) from India with special reference on prey preferences of their larvae. J Entomol Res 47(2):266–272. https://doi.org/10.5958/0974-4576.2023.00048.8

Ghosh S, Sarkar SK, Chakraborty SK (2023d) *Triangulara sunderbanensis* sp. nov. (Coleoptera: Lampyridae)—a new firefly species from the Sunderban biosphere reserve in India. Orient Insects 58(3):393–404. https://doi.org/10.1080/00305316.2023.2289968

Ghosh S, Chattoraj A, Chakraborty SK (2024) The effect of ALAN on the ecobiological attributes of fireflies (Coleoptera: Lampyridae: Luciolinae): a study from a tropical habitat in India. Int J Trop Insect Sci 44:1159–1184. https://doi.org/10.1007/s42690-024-01202-y

Ghosh S, Sarkar SK, Paria S, Chakraborty S (2025a) Impact of changing landscapes on diversity and distribution of firefly (Coleoptera: Lampyridae: Luciolinae) assemblages: an approach for conservation. Trop Ecol. https://doi.org/10.1007/s42965-025-00370-3

Ghosh S, Sarkar SK, Chakraborty S (2025b) *Medeopteryx bengalensis* sp. nov. (Coleoptera: Lampyridae: Luciolinae)—a new firefly species from the urban city of Kolkata, West Bengal, India. Oriental Insects (accepted, under process of publication). https://doi.org/10.1080/00305316.2025.2491466

Ghosh S, Acharya S, Chakraborty S (2025c) Participatory rural appraisal (PRA) methods reveal the perceived status of firefly (Coleoptera: Lampyridae: Luciolinae) conservation: case studies in West Bengal, India. Trop Ecol (communicated)

Gupta AK, Singh S, Wajih SA, Mani N, Singh AK (2017) Urban resilience and sustainability. Gorakhpur Environmental Action Group (GEAG), Gorakhpur, p 107

Idris NS, Mustapha MA, Sulaiman N (2021) The dynamics of landscape changes surrounding a firefly ecotourism area. Glob Ecol Conserv 29(17):e01741. https://doi.org/10.1016/j.gecco.2021.e01741

Jeganathan A, Andimuthu R, Prasanna Venkatesh R, Kumar DS (2016) Spatial variation of temperature and indicative of the urban heat Island in Chennai metropolitan area, India. Theor Appl Climatol 123(1–2):83–95

Kazama S, Matsumoto S, Ranjan SP, Hamamoto H, Sawamoto M (2007) Characterization of firefly habitat using a geographical information system with hydrological simulation. Ecol Model 209(2):392–400

Khatun H, Jamal A (2021) Uses of chemical pesticides in agricultural fields in West Bengal and their effects on non-target species—a review study. Int J Innov Res Technol 8(6):64–69

Kotharkar R, Bagade A (2018) Evaluating urban heat Island in the critical local climate zones of an Indian city. Landsc Urban Plan 169:92–104. https://doi.org/10.1016/j.landurbplan.2017.08.009

MacGregor-Fors I (2010) How to measure the urban-wildland ecotone: redefining 'peri-urban' areas. Ecol Res 25(4):883–887. https://doi.org/10.1007/s11284-010-0717-z

MacGregor-Fors I (2011) Misconceptions or misunderstandings? On the standardization of basic terms and definitions in urban ecology. Landsc Urban Plan 100(4):347–349. https://doi.org/10.1016/j.landurbplan.2011.01.013

Mohammad P, Goswami A, Bonafoni S (2019) The impact of the land cover dynamics on surface urban heat Island variations in semi-arid cities: a case study in Ahmedabad City, India, using multi-sensor/source data. Sensors 19(17):3701. https://doi.org/10.3390/s19173701

Pérez-Hernández CX, Gutiérrez Mancillas AM, del-Val E, Mendoza-Cuenca L (2023) Living on the edge: urban fireflies (Coleoptera, Lampyridae) in Morelia, Michoacán, Mexico. Peer J 11:e16622. https://doi.org/10.7717/peerj.16622

Piano E, De Wolf K, Bona F, Bonte D, Bowler DE, Isaia M, Lens L, Merckx T, Mertens D, van Kerckvoorde M, De Meester L, Hendrickx F (2017) Urbanization drives community shifts towards thermophilic and dispersive species at local and landscape scales. Glob Chang Biol 23(7):2554–2564. https://doi.org/10.1111/gcb.13606

Rana N, Rayal R, Uniyal VP (2022) Firefly survey: adopting citizen science approach to record the status of flashing beetles. J Threat Taxa 14(10):22016–22020. https://doi.org/10.11609/jott.7939.14.10.22016-22020

Singh P, Kikon N, Verma P (2017) Impact of land use change and urbanization on urban heat Island in Lucknow city, Central India. A remote sensing-based estimate. Sustain Cities Soc 32:100–114. https://doi.org/10.1016/j.scs.2017.02.018

Takeda M, Amano T, Katoh K et al (2006) The habitat requirement of the Genji-firefly Luciola cruciata (Coleoptera: Lampyridae), a representative endemic species of Japanese rural landscapes. Biodivers Conserv 15: 191–203

Vaz S, Manes S, Gama-Maia D, Silveira L, Paiva P, Figueiredo M et al (2022) Light pollution is the fastest growing potential threat to firefly conservation in the Atlantic Forest hotspot. Insect Conserv Divers 14:211–224

Viviani VR, Rocha MY, Hagen O (2010) Bioluminescent beetles (Coleoptera: Elateroidea: Lampyridae, Phengodidae, Elateridae) in the municipalities of Campinas, sorocabavotorantim and Rio Claro-Limeira (SP, Brazil): biodiversity and influence of urban sprawl. Biota Neotropica 10:103–116

Wong CH (2014) Piloting and using citizen science to monitor the fireflies in Kuala Selangor Nature Park, Selangor State, Malaysia. In: International Firefly Symposium 2014, Florida, USA (Power Point Press). http://www.conference.ifas.ufl.edu/firefly/Presentations/3%20-%20Thursday/Session%208/1100%20Wong.pdf. Accessed 17 Nov 2023

Yadav N, Sharma C, Peshin SK, Masiwal R (2017) Study of intra-city urban heat Island intensity and its influence on atmospheric chemistry and energy consumption in Delhi. Sustain Cities Soc 32:202–211. https://doi.org/10.1016/j.scs.2017.04.003

Challenges and Opportunities in Managing Urban Wetlands for Resilience in Harare Metropolitan Province, Zimbabwe

17

Isheanesu Daniel Mangwaya, Terence Makumbe, and Thomas Karakadzai

Abstract

Current urban development patterns in Zimbabwe have led to fragmented spatial growth that follows an unsustainable trajectory. The encroachment on wetlands due to reckless urban expansion has resulted in the degradation of these ecologically sensitive areas, exacerbating risks associated with natural disasters such as flooding and biodiversity loss particularly in the Harare Metropolitan Province. As a signatory to the Ramsar Convention on Wetlands, Zimbabwe has committed to designating wetlands of international importance (Ramsar Sites). Ongoing urban development has largely overlooked the critical functions that these ecologically sensitive areas serve in urban environments. This research employed a mixed-method approach, utilizing document reviews, focus group discussions (FDGs), key informant interviews with urban development practitioners, GIS and remote sensing techniques. The findings reveal that wetlands are integral to nature-based solutions (NBS) that promote environmental sustainability, human health and wildlife conservation. This underscores the need for city managers to consider wetlands as Urban Living Labs (ULLs), where innovative strategies for urban resilience can be developed and tested. Consequently, this chapter advocates for a paradigm shift in policy implementation, emphasizing the necessity of collaboration between state and non-state actors. Furthermore, it promotes fostering a sense of community ownership over wetlands, recognizing that local engagement is vital for the sustainable management and preservation of these critical ecosystems. This synthesis of ecological studies and urban policy underscores the importance of wetlands in creating resilient urban environments capable of withstanding future challenges.

Keywords

Wetlands · Urban · Wetland management · Ecologically sensitive areas · Nature-based solutions (NBS)

I. D. Mangwaya (✉) · T. Makumbe
Department of Community and Social Development, University of Zimbabwe, Harare, Zimbabwe
e-mail: imangwaya@sociol.uz.ac.zw; tmakumbe@sociol.uz.ac.zw

T. Karakadzai
Department of Construction Economics and Management, University of Cape Town, Cape Town, South Africa

Dialogue on Shelter for the Homeless Trust, Harare, Zimbabwe
e-mail: tkarakadzai@sociol.uz.ac.zw

S. K. Chakraborty et al. (eds.), *Wetlands and Urban Resilience*, Sustainable Development Goals Series,
https://doi.org/10.1007/978-3-032-06881-1_17

17.1 Introduction

A myriad of definitions have been proposed by scholars regarding wetlands (Ferreira et al. 2023; Afonin et al. 2024; Abdullah et al. 2024). Wetlands are broadly defined as 'any area of marsh, fen, peatland or water, whether natural or artificial, permanent or temporary, with water that is static or flowing, fresh, brackish or salt, and includes riparian land adjacent to the wetland' (Ramsar Convention Secretariat 2016:9). This multifaceted concept has led to a substantial body of literature that identifies various inherent attributes of wetlands (Dixon et al. 2021; Eyvaz and Albahnasawi 2023; Bernard et al. 2024). Increasingly, researchers emphasize the role of wetlands as a key component of nature-based solutions (NBS), to address urban challenges and enhance urban resilience (World Bank 2021; Tsiko 2022; Oliver and Marsters 2024). Wetlands are also regarded as a form of infrastructure, specifically urban ecological infrastructure (UEI), defined as, 'an organic integration of blue (water-based), green (vegetated), gray non-living landscapes, combined with exits and arteries, at an ecosystem level' (Mantel et al. 2021:10). Wetlands can be broadly categorized into two types: inland and coastal wetlands. Inland wetlands are further divided into natural and man-made categories (Fig. 17.1).

Natural inland wetlands encompass lakes, oxbow lakes, riverine systems, waterlogged areas, rivers and streams, *dambos* or *vleis* and flood plains (Kumar and Kanaujia 2014). In contrast, man-made inland wetlands include reservoirs, tanks, constructed waterlogged areas and salt pans (ibid.). Natural coastal wetlands are characterized by lagoons, creeks, beaches, salt marshes, mangroves and coral reefs, while coastal man-made wetlands consist of aquaculture ponds and salt pans (Fig. 17.1). In Zimbabwe, the landscape features diverse wetlands, including flood plains, dambos/vleis, pans, swamps, and both natural and man-made dams and lakes (GoZ 2020a, b).

Globally, wetlands cover approximately 1.2 billion hectares, and yet, there have been varying perspectives on the extent of wetland loss (Xu et al. 2019; Convention on Wetlands 2021; Fluet-Chouinard et al. 2023). The urgent call for action reflected in the Sustainable Development Goals (SDGs)—particularly goals 11, 13 and 15—aligns closely with effective wetland management (Waly et al. 2022; Wang 2023; Mohanty et al. 2024). Goal 15.1 emphasizes the conservation, restoration and sustainable use of terrestrial and inland fresh water ecosystems, including wetlands, in accordance with international agreements. Additionally, goal 6.6 aims to protect and restore water-related ecosystems, further highlighting the role of wetlands in fostering sustainable urban environments. These SDGs collectively strive to enhance the sustainability of cities and communities while safeguarding environmental integrity, particularly regarding terrestrial life.

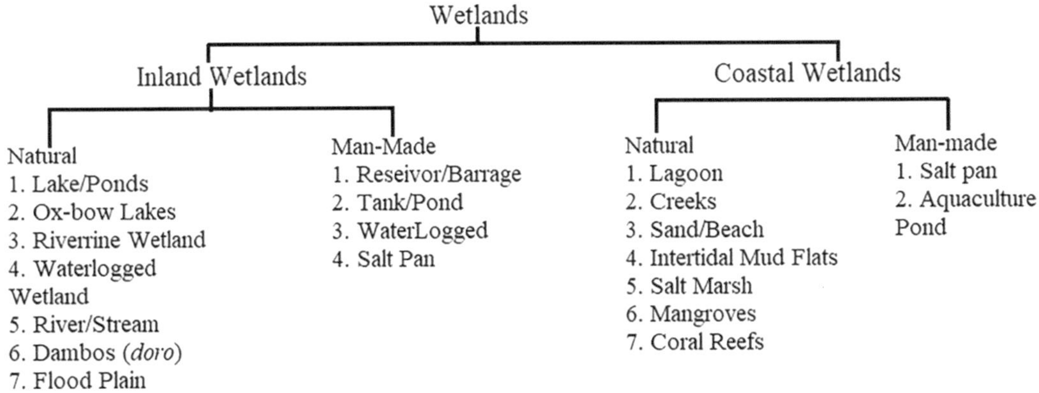

Fig. 17.1 Wetland typologies. (Source: Adapted from Kumar and Kanaujia 2014; GoZ 2020a, b)

Wetlands and their ecosystems are central to three global conventions, namely the Ramsar Convention of 1971, Convention on Migratory Species of 1979 and the Convention on Biological Diversity of 1992 (McInnes and Davidson 2018; Kingsford et al. 2021; Bulkeley et al. 2023). The Ramsar Convention promotes sustainable management efforts for wetlands deemed of national importance, while the 1979 Convention on Migratory Species recognizes the necessity of conserving migratory species, including those dependent on wetlands. The Convention on Biological Diversity underscores the intrinsic value of biodiversity and the anthropological threats facing these ecosystems, particularly in urban contexts (Ansari et al. 2011; Spataro and Rojas 2020). Despite a global consensus on the invaluable benefits that wetlands provide, challenges persist in managing these 'biological supermarkets' (Krauss et al. 2021; Gojendro et al. 2023; Donde et al. 2023). Poor management of wetlands has quantifiable impacts on urban environments and biodiversity, underscoring the need for effective governance and sustainable practices.

This study aims to examine the challenges and opportunities associated with managing urban wetlands for resilience in Zimbabwe, with a particular emphasis on the Harare Metropolitan Province. The novelty of this chapter is grounded in its threefold objectives: first, to identify key actors responsible for wetland management in Zimbabwe; second, to explore the impacts of urbanization on land use change within Harare Metropolitan Province; and third, to uncover the environmental challenges witnessed in wetland management while highlighting opportunities for conservation and resilience strategies. Following the contextualization of the study area in Sect. 17.2, Sect. 17.3 details the materials and methods employed for data collection and analysis. Sect. 17.4 details context-specific challenges and opportunities related to effective wetland management in Harare Metropolitan Province. This culminates in a conclusion in Sect. 17.5 that synthesizes findings and insights. Through integrating ecological studies with resilient pathways, this research seeks to inform strategies that enhance urban resilience through effective wetland management. Ultimately, the findings aim to contribute to a broader understanding of how urban wetlands can be leveraged as vital components of sustainable urban planning and climate adaptation efforts in Zimbabwe.

17.2 Study Area

17.2.1 Geographical Characteristics

The study sites are located at coordinates 17°49′39.79″ south latitude and 31°03′12.13″ east longitude, encompassing the Harare Metropolitan Province. This province is divided into four distinct local authorities: Harare City Council (HCC), Chitungwiza Municipality, Epworth Local Board (ELB) and Ruwa Local Board (RLB) (Fig. 17.2). As the capital city of Zimbabwe, Harare is experiencing rapid urbanization driven by rural-to-urban migration, with individuals seeking better livelihoods and job opportunities.

Geologically, the Harare Metropolitan Province features a diverse bedrock composition. Granites dominate the eastern and southwestern regions, while gabbro and dolerite are prevalent in the north. The central core is primarily composed of phyllite and metagreywacke. The terrain in the south-western part is gently undulating, contrasting with the rugged northern landscape characterized by U-shaped incised valleys. The bedrock is often hidden by layers of saprolite, several decimetres thick, which are subject to cyclic erosion, periodically exposing the underlying bedrock.

This region represents a dynamic landscape where geological variations intersect with urbanization trends, influencing the environment and human settlement patterns. The interplay between geological features, land cover dynamics and human activities highlights the complex relationship between natural processes and anthropogenic impacts within the Harare Metropolitan Province. The climate for Harare Metropolitan Province is subtropical, exhibiting four distinct seasons: a cool-dry period stretching from mid-May to August, a hot-dry season from September to mid-November, a rainy-wet season

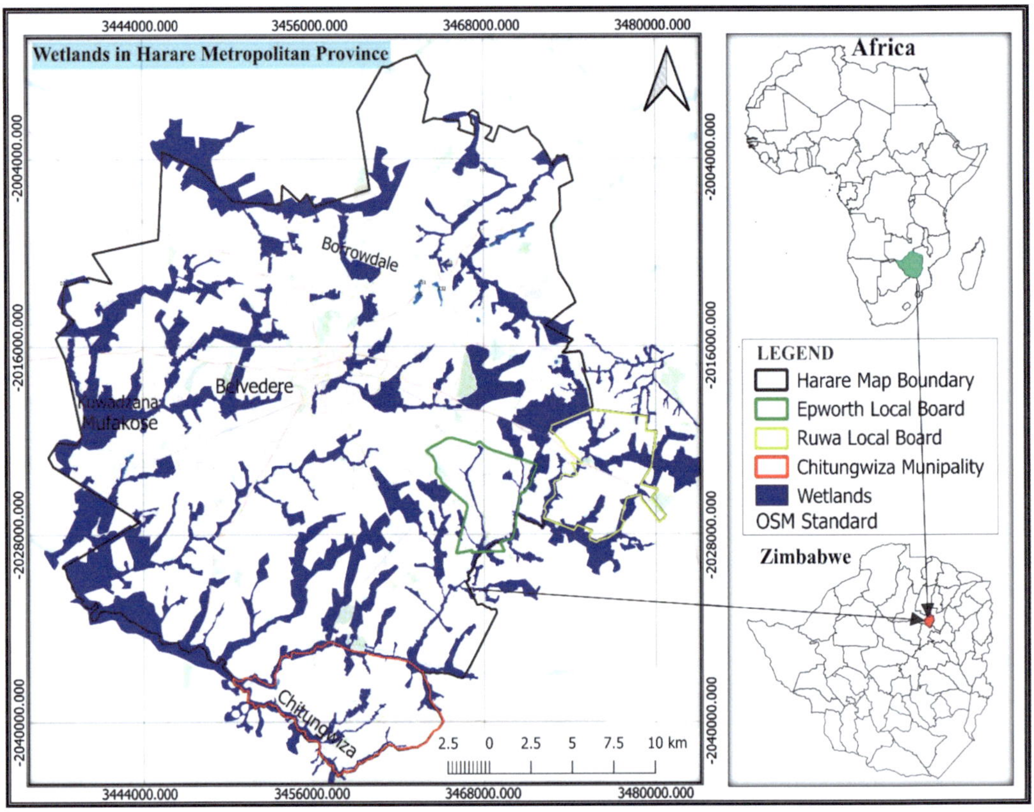

Fig. 17.2 Study area: Harare Metropolitan Province

extending from mid-November to mid-March and a post-rainy season extending from mid-March to mid-May. During the cool-dry season, temperatures typically range from 7 °C and 20 °C, while in the hot-dry season, temperatures fluctuate between 13 °C and 28 °C. The region receives an average annual rainfall of 470 mm to 1350 mm, with the majority of precipitation occurring during the rainy season.

17.2.2 Demographic Characteristics

Demographic shifts have been significant within Harare Metropolitan Province. The urban district of Harare recorded a population of 1,435,784 in 2002, which increased to 1,485,231 by 2012 and 1,849,617 by 2022 (ZimStat 2002, 2012, 2022). Similarly, the Epworth district witnessed a population surge from 114,047 in 2002 to 167,462 in 2012 and 206,368 in 2022 (ibid.). Currently, Chitungwiza has a population of approximately 371,246 (ibid.). The total area of the Harare Metropolitan Province, encompassing the study area, spans over 940 km^2, reflecting the dynamic interplay between climate patterns, population growth and land use transformations within this region.

17.3 Material and Methods

17.3.1 Focus Group Discussion (FGDs) and Key Informant Interviews (KIIs)

This study is based on data from five focus groups conducted in selected study sites: Monavale, Borrowdale West, Mufakose-Kuwadzana, Belvedere and Chitungwiza under Harare Metropolitan Province. One focus group was held in each of these communities. Additionally, eight key informant interviews were conducted with

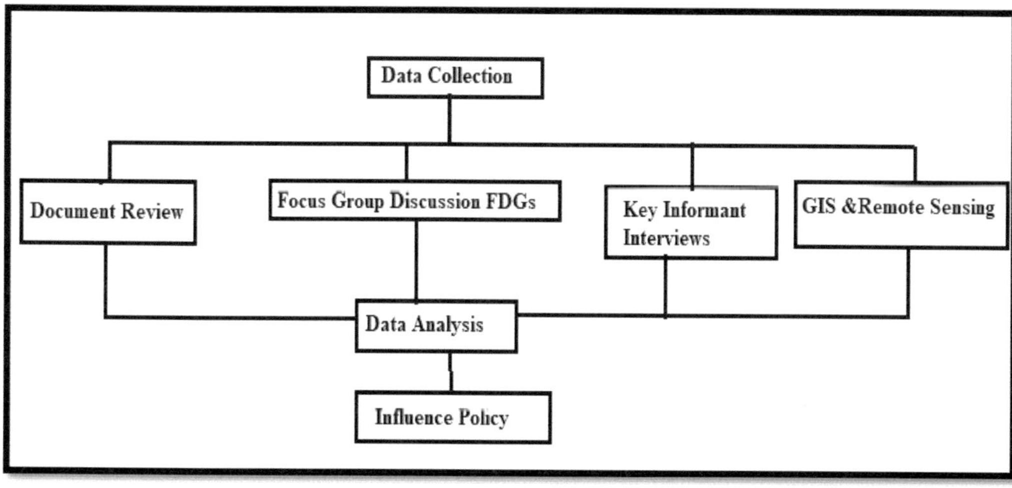

Fig. 17.3 Methodological flowchart

representatives from the following ministries and government departments: Ministry of Environment, Climate, Tourism and Hospitality (MoECTH), Ministry of Local Government and Public Works (MoLGPW), Ministry of Lands, Agriculture, Water and Rural Resettlement (MoLAWRR), local authorities (City of Harare and Chitungwiza Municipality), Environmental Management Agency (EMA), Zimbabwe National Water Authority (ZINWA) and the Department of Civil Protection Unit. Data collection utilized a focus group discussion guide and a key informant interview guide (Fig. 17.3). The scope of this research is summarized in Table 17.1.

17.3.2 Remote Sensing and Data Processing

Geographic Information System (GIS) and Remote Sensing were employed in the data collecting process to monitor changes in land use across selected wetlands in Monavale, Borrowdale West, Mufakose-Kuwadzana, Belvedere and Chitungwiza within the Harare Metropolitan Province from 2013 to 2023 (Fig. 17.3). Freely available Landsat 8 imagery was downloaded from Earth Explorer (USGS) and processed to identify land cover changes. The process of land use and land cover classification was conducted using ENVI software, which offers a range of geo-processing algorithms for image analysis. Prior to performing supervised classification, the images underwent processing to enhance quality and minimize noise through radiometric calibration and atmospheric correction. Radiometric calibration was essential for improving the accuracy of remote sensing data, facilitating the interpretation of images by providing true representations of surface features.

Additionally, atmospheric correction was applied in order to eliminate atmospheric substances, such as gases, aerosols and water vapour, that can affect reflectance values captured by sensors. This correction enhances the reliability of the image data. After image correction, Regions of Interest (ROIs) were defined based on land cover types, including built-up, cultivated and vegetated areas. Training samples for each land cover class within the ROIs were selected to guide the classification process. Supervised classification method was employed because it allows for the identification of pixels that represent specific land use features, informed by other ground truth data (Iqbal and Khan 2014). The support vector machine (SVM) algorithm was subsequently employed to execute the classification process, grouping areas with similar reflectance values into distinct land use categories. The

Table 17.1 FGDs and key informant interviews

Region	Settlement	FGD session count and reach		Key informant interviews
		Count	Participants	Count
Harare Metropolitan Province	Monavale, Borrowdale West, Mufakose-Kuwadzana, belvedere and Chitungwiza	5	12	8
Total number of research participants			60	8

Source: Field Work 2024

classification results were validated by comparing them with ground truth and high-resolution imagery, with Google Earth serving as a tool for this validation.

By comparing Landsat 8 images from 2013 and 2023, changes in land cover over the 10-year period were assessed. Cross-tabulation analysis (Eastman 2012) was utilized to quantify Land use/Land cover (LU/LC) changes over this decade. This analysis facilitated the identification of areas transitioning from one land use class to another, both spatially and quantitatively (Ozturk 2015). Furthermore, statistics extracted for the land use change analysis provided insights into the magnitude of change for each class and the transitions occurring between classes. The study ultimately elucidates the direction and rate at which human activities are transforming or modifying wetlands in the area of study.

17.4 Results and Discussion

17.4.1 Actors Managing Wetlands in Zimbabwe

A diverse array of stakeholders is involved in the management of ecologically sensitive areas in Zimbabwe. The country operates under a three-tier government system comprising central, provincial and local governments (GoZ 2013). The central government plays a critical oversight role concerning wetlands, coordinating efforts through key ministries, including the Ministry of Environment, Climate, Tourism and Hospitality (MoECTH), the Ministry of Lands Agriculture and Rural Resettlement (MoLARR), and the Ministry of Local Government and Public Works (MoLGPW) (Table. 17.2). Day-to-day management of urban development is delegated to local authorities, which collaborate with the Department of Spatial Planning and Development, a government parastatal. Legally, each local authority is required to draft and adopt a Local Environmental Action Plan (LEAP), which outlines strategies for managing and protecting within their jurisdiction, as mandated by the Environmental Management Act [20:27]. While local authorities are responsible for developing LEAPs, the Environmental Management Agency (EMA) is tasked with the review, evaluation and approval of these plans.

A representative from the Ministry of Local Government and Public Works emphasized that:

> Wetlands are a critical component of our environment. That is why the central government, through several pieces of legislation, has established different ministries and parastatals, with an oversight role on these ecologically sensitive areas. Managing these wetlands has its challenges; however, the government remains committed to ensuring that all stakeholders—including unconventional actors, are all hands on deck, to integrate this vital ecosystem into urban resilient frameworks.

Identified actors mandated with the oversight role of wetland management in Zimbabwe represent a positive step towards incorporating wetlands into urban resilience discussions. The ongoing infill and sprawling development patterns across Zimbabwean cities threaten the integrity of these essential ecosystem, often referred to as the kidneys of urban environments. These findings align with the perspectives of Kabiri et al. (2022), Musasa et al. (2023) and Mumuni et al. (2025), who concur that collaborative urban governance on wetland management is a panacea to unsustainable utilization of these

Table 17.2 Actors and their roles in managing wetlands in Zimbabwe

Actor(s)	Key responsibility
Ministry of Environment, Climate, Tourism and Hospitality (MoECTH)	Localizes and implements the global and regional agreements Zimbabwe adopts
Ministry of Local Government and Public Works (MoLGPW)	Oversees spatial development undertaken by all local authorities in Zimbabwe
Ministry of Lands Agriculture Water and Rural Resettlement (MoLAWRR)	MoLGPW liaises with the Ministry of Lands, which issues a 'certificate of no present interest' in amalgamating rural farm(s) into urban boundaries
Local Authorities (LA)	Regulates and approves all forms of development undertaken under their area of jurisdiction
Environmental Management Agency (EMA)	Custodian of the environment, and works in liaison with all local authorities
Zimbabwe National Water Authority (ZINWA)	It conducts hydrological studies for surface and underground water
Department of Spatial Planning and Development (DSPD)	Approves master plans that direct spatial development under the jurisdiction of all local authorities in Zimbabwe

Source: Adopted from GoZ (2020a, b); OAG (2021)

areas. Through fostering cooperation among diverse stakeholders, cities can enhance their resilience, which is crucial in maintaining ecological balance.

17.4.2 Urbanization Impacts (Land Use Changes)

The study sites encompass land classified into various land uses, including built-up areas, wetland, vegetation and open spaces. Figures 17.4, 17.5, 17.6, 17.7 and 17.8 present classified images from study sites for 2013 and 2023. The land use and land cover change analysis for all wetlands reveal significant alterations in wetland size over this decade (Figs. 17.4, 17.6 and 17.8). In the 2013 analysis of the Mufakose-Kuwadzana wetland, it covered 47.1% of the area, with built-up areas closely following at 46.9%. Open spaces accounted for a modest 6%. However, by 2023, the spatial extent of wetlands declined to 38.7%, while built-up areas surged to 59% of the total wetland area (Fig. 17.4). This rapid transformation indicates an 8.3% reduction in wetland areas and a substantial 12.2% increase in built-up areas.

Land use class transitions have been prominently observed in the Mufakose-Kuwadzana wetland, where areas classified as wetland have been significantly encroached by built-up developments (Fig. 17.5). The Harare City Council town planner emphasized that:

> Individuals have fallen prey to land barons who have sold land in ecologically sensitive areas that is *ulta vires* the confines of the Regional Town and Country Planning Act [29:12] and the Environmental Management Act [20:27]. These instruments are intended to guide urban development, and yet, if you consider developments along the Mufakose-Kuwadzana Wetland, proper planning procedures were ignored. The local authority appreciates efforts made through the Ministry of Environment, Climate, Tourism and Hospitality (MoECTH) that saw the promulgation of the Government Gazette Declaration for the Protection of Wetlands in 2022 under Harare Metropolitan Province.

Anthropogenic activities in these ecologically sensitive areas have also resulted in some tracts of land transitioning from wetland to open spaces (Fig. 17.5).

Table 17.3 illustrates land use transitions in the Mufakose-Kuwadzana wetland, highlighting significant transformations identified through cross-tabulation analysis. Key findings reveal that approximately 10.99% of wetland transitioned to a built-up area, underscoring the impact of urbanization on wetland areas (Table. 17.3). Another notable transition includes a 2.4% change from open spaces to built-up areas, indicating the conversion of previously undeveloped land for urban use. Remarkably, about 35.2% of wetland areas remained unchanged during this period, reflecting

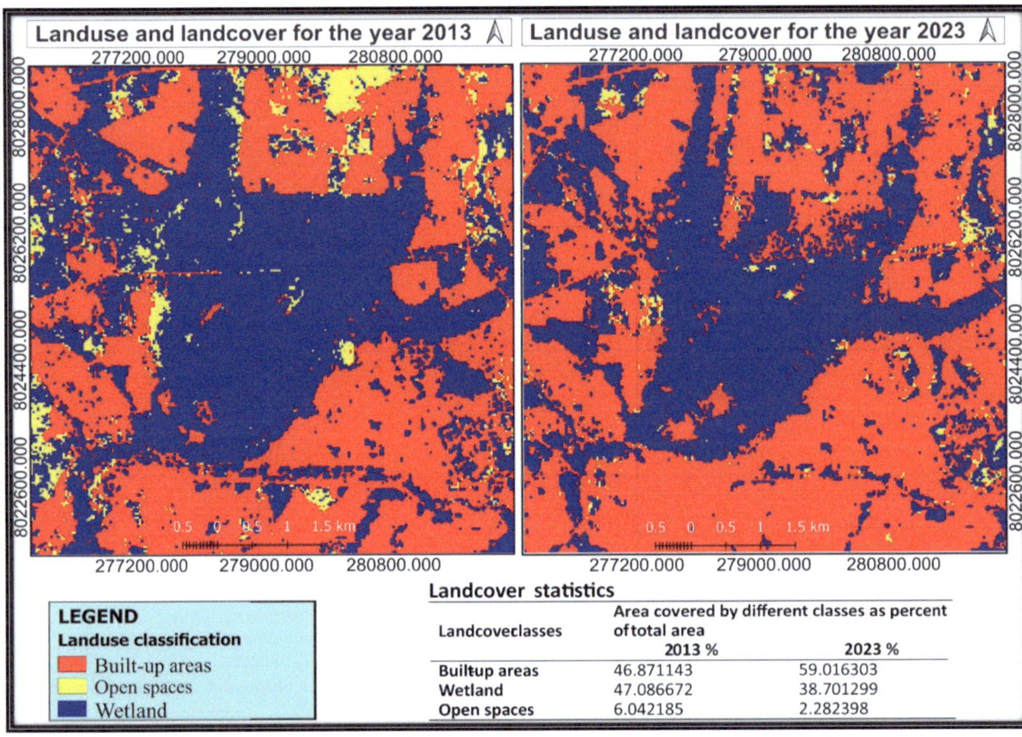

Fig. 17.4 Changes in the size of Mufakose-Kuwadzana Wetland between 2013 and 2023. (Source: Field Survey 2024)

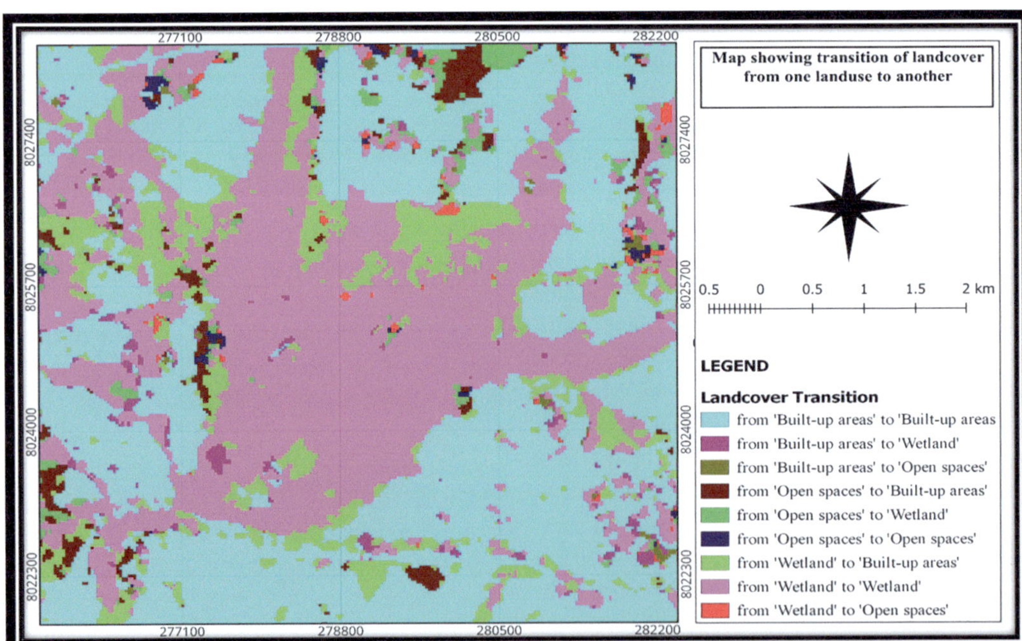

Fig. 17.5 Land use class transitions for Mufakose-Kuwadzana wetland between 2023 and 2023. (Source: Field Survey 2024)

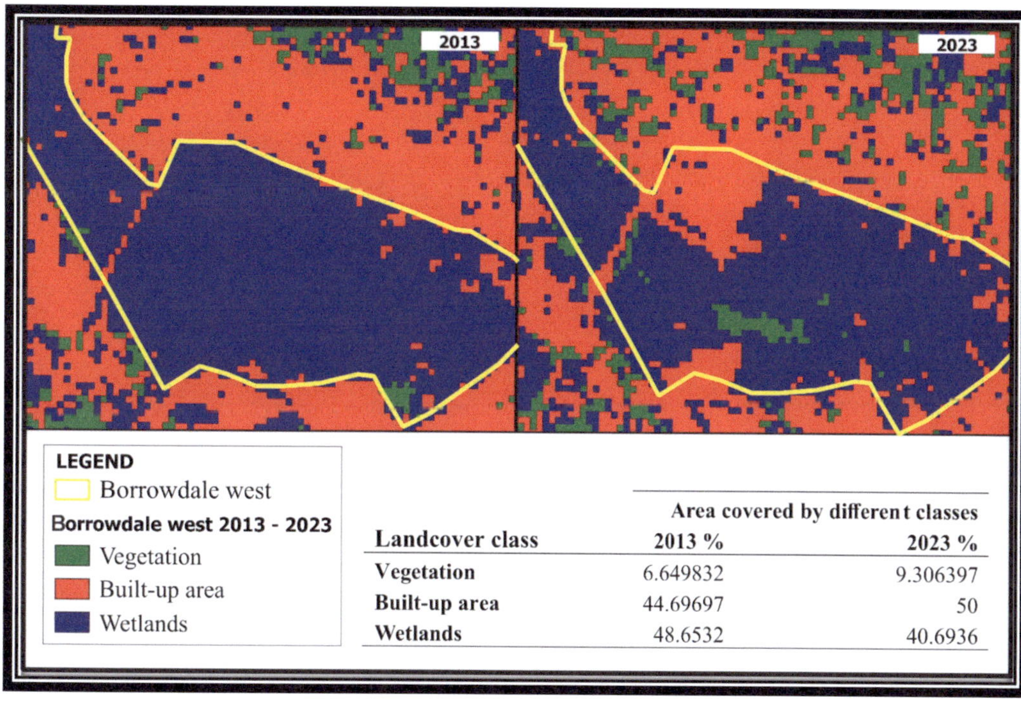

Landcover class	Area covered by different classes 2013 %	2023 %
Vegetation	6.649832	9.306397
Built-up area	44.69697	50
Wetlands	48.6532	40.6936

Fig. 17.6 Changes in the size of Borrowdale West Wetland between 2023 and 2023. (Source: Field Survey 2024)

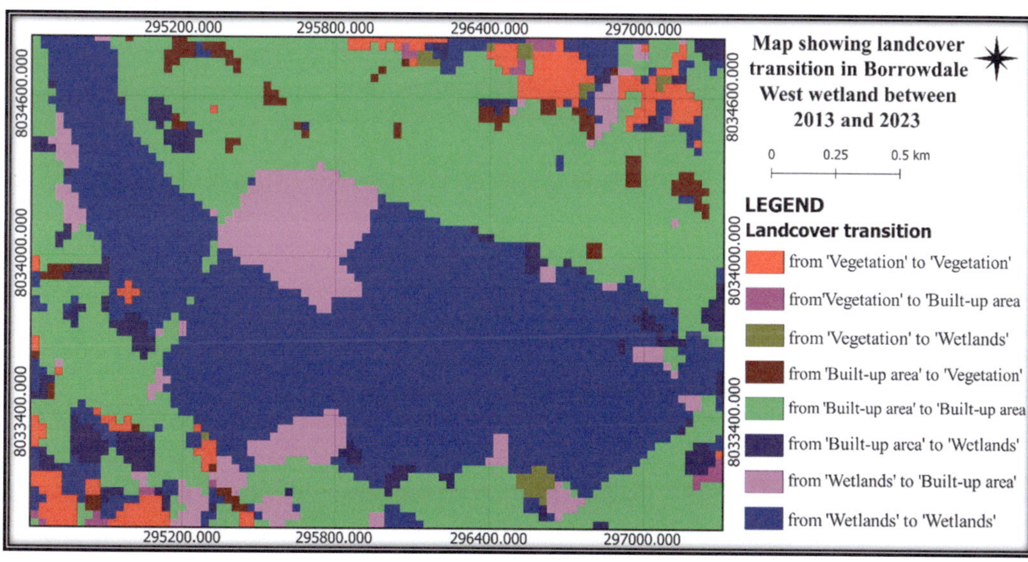

Fig. 17.7 Land use class transitions of Borrowdale West Wetland between 2013 and 2023. (Source: Field survey 2024)

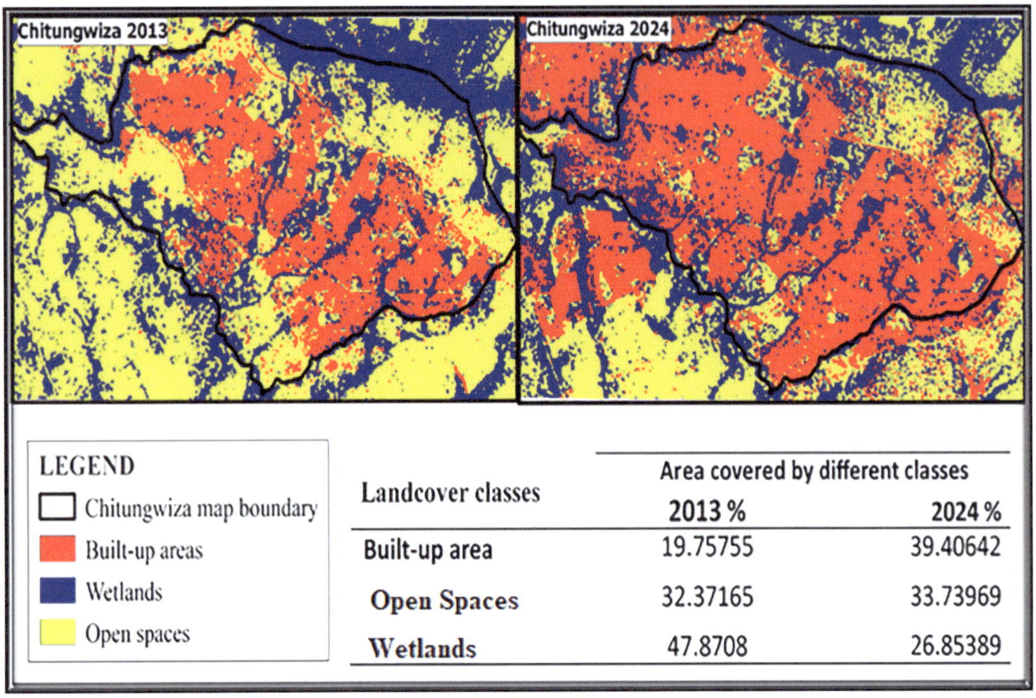

Fig. 17.8 Changes in the size of Chitungwiza Wetlands between 2013 and 2023. (Source: Field survey 2024)

Table 17.3 Land use class transition matrix for Mufakose-Kuwadzana wetland (2013–2023)

Land cover transition	Number of pixels	Percent of land area converted to another class
Built-up area to built-up area	25,386	46.764300
Built-up areas to wetland	766	1.411071
Open spaces to built-up area	1301	2.396610
Wetland to built-up areas	5967	10.991987
Wetland to wetland	19,125	35.230727

their resilience against encroachment. Additionally, 46.8% of built-up areas did not undergo any change, providing insights into the evolving spatial dynamics of land use within the study area. This transformation underscores the urgent need for effective management and protection strategies to preserve the ecological integrity of wetlands amidst increasing urban pressures.

A similar trend is evident at Borrowdale West, where built-up areas rose from approximately 44.7% in 2013 to 50% in 2023, while wetland size decreased from 48.6% to 40.1%, marking a 9.6% loss (Fig. 17.6). These trends demonstrate a negative functional relationship, where the expansion of built-up areas corresponds with the decline of wetlands. The decline in wetland area is largely driven by growing demands for housing stemming from rapid population growth, corruption and inadequate urban planning. Encroachment on wetlands for urban housing exacerbated urban sprawl, undermining both ecological integrity and urban resilience. One of the study respondents noted that:

> As Borrowdale residents, we have objected to several development projects, particularly those that threaten our wetland conservation efforts. We are often surprised when we see land developers subdividing and selling parcels of these wetlands. At one time, we even took the Harare City Council to court over the illegal sale of wetland areas. Sometimes, we suspect that local governments are corrupt.

The Borrowdale Wetland development case in northern Harare illustrates how political and financial interests can compromise urban development procedures, ultimately threatening environmental sustainability.

Land use transition on Borrowdale West Wetland (Fig. 17.7, detailed in Table. 17.4) shows that approximately 8.2% of the wetland area was converted into a built-up area. This shift aligns with Harare city council's efforts to operationalize densification as proposed in the Zimbabwe National Human Settlements Policy of 2020. While such developments aim to meet housing demands, they have attracted criticism from environmentalists who argue that these approaches are unsustainable (Mhlanga et al. 2014; Musasa and Marambanyika 2020; Sinthumule 2021).

A large parcel of land transitioned from a wetland into a built-up area on Borrowdale West Wetland (Fig. 17.7). Similar trends have been documented in various studies across the Harare Metropolitan Province (Cunliffe 2020; OAG 2021; Kudumba 2022). For instance, Olawole et al. (2011) noted an increase in Harare's built-up area from 24.13% in 1989 to 30.58% in 2000, coinciding with a decline in bare surfaces from 39.22% to 34.18%. This growth correlates with urbanization and population increases, with Harare expanding at an average rate of 5.65% from 1982 to 2002. The annual population growth rate surpasses the 1.27% growth rate of built-up areas, emphasizing the urgent need for more urban infrastructure, including housing, roads and utilities. The trends observed in the Borrowdale West Wetland contrast with the findings of Cobbinah et al. (2022) in Kumasi, Ghana, where wetlands under siege were characterized by unplanned urban growth. This divergence underscores the unique challenges faced by different regions in managing wetlands amidst urban development pressures.

In Chitungwiza, wetlands decreased from 40.4% coverage in 2013 to 32.7% in 2023, while built-up areas expanded from 19.8% to 39.4% during the same period, as shown in Fig. 17.8. Wetlands also declined significantly, from 47.9% to 26.9%. The increasing demand for housing and commercial spaces fuels a cycle of encroachment, which threatens wetland ecosystems while undermining urban resilience. An overall 9% wetland loss to built-up development from 2013 to 2023 was recorded under the jurisdiction of Chitungwiza Municipality (Table 17.5). Chitungwiza Municipality town planner highlighted that:

> Some of the development seen under our jurisdiction is highly informal. While we have authority

Table 17.4 Land use class transition matrix for Borrowdale West Wetland (2013–2023)

Land cover transition	Number of pixels	Per cent of land area converted to another class
Vegetation to vegetation	304	5.117845
Built-up area to vegetation	134	2.255892
Built-up area to built-up area	2439	41.060606
Wetland to built-up area	488	8.215488
Wetland to wetland	2131	35.875421

Table 17.5 Land use class transition matrix for wetlands under Chitungwiza Municipality (2013–2023)

Landcover transition	Number of pixels	Percentage of land area converted to another class
Built-up area to built-up area	27,061	20.566977
Wetlands to built-up area	4948	8.760593
Wetland to wetland	31,559	23.985560
Wetland to open spaces	5160	3.921718
Open spaces to built-up area	21,269	16.164925
Open spaces to wetland	10,776	8.190006
Open spaces to open spaces	30,320	23.043891

over the nature and form of development in Chitungwiza, enforcing relevant legislation can be cumbersome. By law, we are required to obtain a court order before any demolition can take place. The rising number of people on the housing waiting list has created a sense of desperation, leading many to purchase and build their homes on undesignated land. Considering the investments made, we often weigh the pros and cons and opt to regularize some developments. However, if development encroaches ecologically sensitive areas, we recommend demolition.

Depicted land use class transitions (Fig. 17.9 and Table. 17.5) clearly highlight the role land barons play in owning land under the jurisdiction of Chitungwiza Municipality. Corrupt council officials are also embroiled in such shoddy land deals (Munhende 2020; Civacts 2021; Pachawo 2023). This state of affairs clearly depicts the complex web of actors responsible for wetland degradation across Harare Metropolitan province. The demand for homeownership and commercial stands has left home seekers and informal traders desperate to own urban land, creating and sustaining this vicious cycle of wetland degradation across various localities in Zimbabwe.

Anthropogenic activities, such as deforestation, sand mining and dredging, significantly contribute to wetland degradation. These findings align with those of Adedeji et al. (2020), who observed similar issues in Lagos and Ogun, Nigeria, where deforestation and illegal sand mining impede effective wetland management. Govender-Ragubeer et al. (2014) further support these conclusions, highlighting the different roles played by government officials and experts in wetland exploitation, as illustrated by the Libradene Wetland in Boksburg, Gauteng, South Africa. Such changes adversely impact

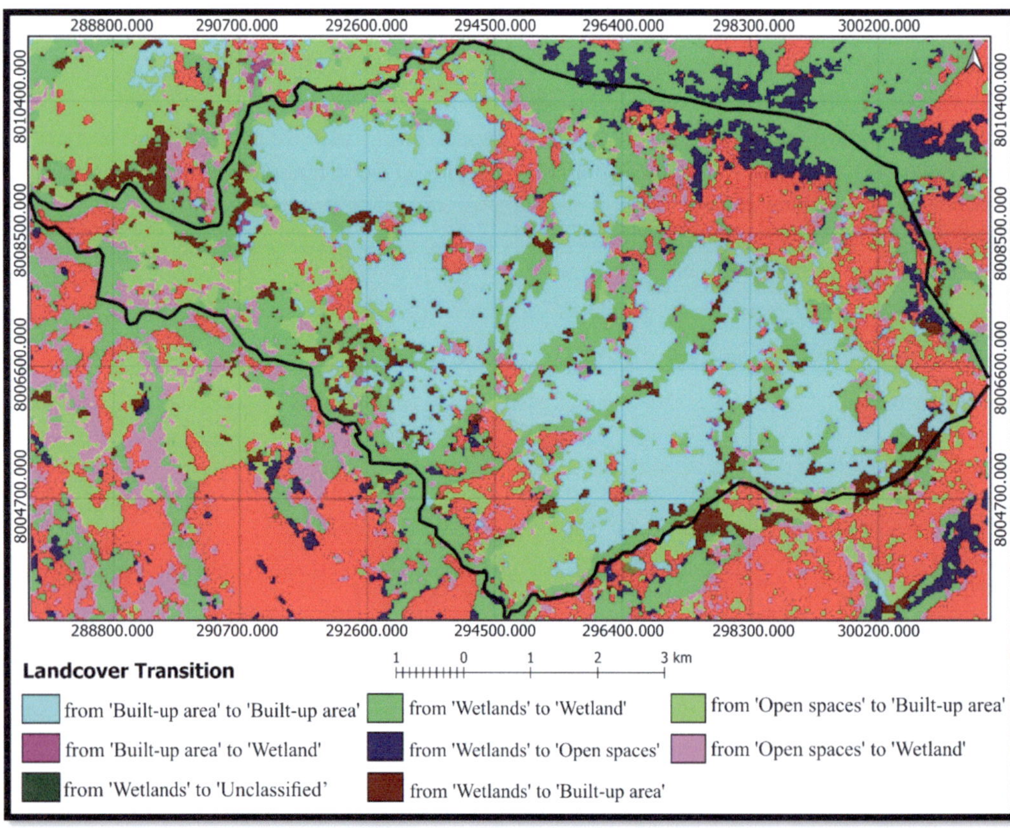

Fig. 17.9 Land use class transitions of Chitungwiza Wetlands between 2013 and 2023. (Source: Field survey 2024)

both human populations and biodiversity within urban ecosystems. In Chitungwiza, urban development has largely occurred without adequate planning, resulting in haphazard growth—a widespread challenge among several local governments in Zimbabwe. The rapid urbanization and population growth in the study areas exert pressure on land use, resulting in the conversion of wetlands and open spaces into built-up areas.

17.4.3 Environmental Challenges in Managing Wetlands

17.4.3.1 Discharge of Effluent

Urban surface water quality in Zimbabwe is significantly compromised by negligent discharge of effluent by local authorities, particularly affecting the Marimba, Mukuvisi and Nyatsime rivers within Harare Metropolitan Province. The Environmental Management Agency (EMA) reported that approximately 336 mega litres of effluent was discharged into the environment, due to the inadequate capacity of sewerage treatment facilities (Muleya 2020). This issue is not isolated to Harare; many local authorities across Zimbabwe struggle with similar waste management failures (CoB 2021). Additionally, industrial activities contribute to the pollution of wetlands, leading to elevated phosphate levels and increased risks of waterborne diseases within local communities (Chumuriwo 2017; Mujuru 2024). The discharge of raw sewerage not only degrades water quality but also undermines the aesthetic and ecological health of urban wetlands, contradicting the principles of the national development strategy (NDS) 1 aimed at sustainable wetland management and the vision for Zimbabwe to become an upper-middle-income economy by 2030. The challenges noted here are similar to those identified by Adeeyo et al. (2022), who found that pollutants from various sites negatively impact sustainable wetland management in South Africa.

17.4.3.2 Wetland Cultivation

Wetland cultivation poses a significant threat to the ecological integrity of Zimbabwe's wetlands (Marambanyika et al. 2016; Musasa and Marambanyika 2022). Many informal residents, particularly those on the urban periphery, engage in unauthorized urban agriculture. Climate change has exacerbated this issue by disrupting seasonal rainfall patterns, leading to increased reliance on wetland cultivation in areas like vei and dambo wetland typologies that support diverse ecosystems (GoZ 2020a, b). Such practices deplete water resources, introduce unsustainable crops and facilitate the spread of invasive species, thereby altering hydrological processes and degrading natural ecosystems (ibid.). While local authorities, like Ruwa Local Board, have attempted to enact by-laws to regulate these activities, enforcement remains challenging due to public reliance on urban agriculture for livelihoods (Siamonga 2014; Shayanewako 2022). The rapid expansion of Harare has not kept pace with the provision of essential services, including water, leading to increased competition for this vital resource (Finn and Bandauko 2024). These findings are consistent with those of Kuusaana et al. (2021), who noted that various wetland typologies in Ghana are being utilized for multiple agricultural purposes by a range of stakeholders.

17.4.3.3 Policy and Governance

The constitutional framework provides a legal basis for environmental rights, particularly sections 28 and 74, emphasizing the right to shelter and the necessity of court orders for property eviction. However, governance challenges, including political polarization and the mismanagement of wetlands, have exacerbated degradation. Many urban local authorities, often led by opposition parties, are required to report to a minister voted and appointed through the Zimbabwe African National Union–Patriotic Front (ZANU-PF), which has seen municipal governance politicized, with land allocation based on political loyalty rather than ecological considerations (Rocco and van Ballegooijen 2018). This dynamic undermines the mandate of local authorities and the Environmental Management Agency (EMA) in effectively managing wetlands. The reluctance of these bodies to prepare and implement Local Environmental Action

Plans (LEAPs) further jeopardizes wetland conservation efforts, leading to the degradation of several wetlands and confrontation with Wetland Trusts (Mhlanga et al. 2014; Harare Wetlands Trust 2020). The function of controlling development is vested within local authorities, leaving a functional gap, as EMA does not have the power to demolish structures built on wetlands. Weak governance has resulted in urban wetlands becoming sites for unconventional housing, urban agriculture and infrastructural development, thereby threatening the resilience of these ecologically sensitive areas (Matenga 2019; De Gobbi 2020).

Wetland invasion under the jurisdiction of Ruwa Local Board is distinct when compared to other local authorities from Harare Metropolitan Province. RLB governs primarily privately owned land. The local authority has power in Ruwa Local, a community developed on state land. With increased demand for land, the local authority subdivided a wetland near Spar, Mavambo, creating both home industry and church stands. The former created lots sought to support local livelihoods, while church leaders were given lease agreements, clearly articulating that they were only permitted to construct temporary structures and were not owning land, but rather, the local authority was leasing the land to them for a specified timeframe. These two clauses create the opportunity for restorative urbanism, provided the local authority feels the form of urban development is becoming unsustainable.

17.4.4 Conservation Opportunities

The Harare City Council has historically struggled to meet the water supply needs of its residents, with daily demand reaching 1200 megalitres while only 300 megalitres are supplied. Wetlands play a crucial role in natural water purification and management. In response, residents, community organizations and Harare City Council have initiated a collaborative community-led wetland restoration program at Monavale Vlei, recognizing the vital role of wetlands in mitigating urban development impacts. The initiative was launched after extensive loss of wetlands due to urbanization, with 3716 hectares—nearly 16% of the wetland area—lost to development by 2021 (OAG 2021).

17.4.4.1 Restoration Projects: Success Stories and Strategies for Wetland Rehabilitation

Monavale Wetland, designated as a Ramsar site and part of Manyame sub-catchment area—Harare's primary water source—has seen successful restoration efforts led by Conservation of Monavale Trust (COSMO). Established in response to inadequate protection for wetlands as 'natural infrastructure', COSMO implements a model of community engagement that includes monitoring and preventing unlawful encroachments on the 34-hectare wetland. Since its inception, the initiative has yielded positive ecological outcomes such as improved water quality and increased biodiversity, including the presence of migratory bird species like the striped crake (*Aenigmatolimnas marginalis*) and the streaky-breasted flufftail (*Sarothrura boehmi*) (Gabay and Chitagu 2024).

However, the wetland has faced threats from proposed developments, including a contentious plan for a 21-cluster housing project. Legal challenges arose when both the Harare City Council and the EMA issued permits for the development, which the courts deemed irregular due to their potential to cause irreversible ecological damage (Munyoro 2021; RWI 2023). These findings highlight a critical gap in the capacity of local authorities to manage wetlands effectively, emphasizing the need for community-led initiatives like COSMO to foster sustainable practices in urban wetland management.

17.4.5 Resilience Strategies

Wetlands are increasingly becoming recognized as nature-based solutions (NBS) that can enhance urban resilience. The national government is pursuing policy initiatives, including the National Development Strategy 1, Presidents call to Action—no compromise to service delivery, and the Zimbabwe National Human Settlement Policy (ZNHSP), to leverage wetlands as critical components in building resilient urban environments. Through following such initiative, its inherent ability in co-creating and experimentation of sustainable solutions, nature allows wetlands to be treated as Urban Living Labs. This approach enables stakeholders to collaborate on real-world problem-solving while developing practical solutions that can be effectively adopted under Harare Metropolitan Province.

17.4.5.1 Ecosystem Services: Valuing Services Provided by Wetlands

In Budiriro West, Harare, the construction of 42 homes on a wetland resulted in significant displacement during the rainy season of 2023–2024, prompting government intervention to relocate affected families to Dzivarasekwa, another high-density suburb in Harare. This incident underscores the vulnerability of wetlands, which the Harare city council has classified as flood-prone areas, leading to efforts for reclamation and alternative housing solutions as part of the Zimbabwe National Human Settlements Policy of 2020 (GoZ 2020c). The reclamation initiative aims to restore the wetlands' stormwater management capabilities, allowing them to mitigate flooding risks while preserving critical habitats for endangered species.

The proximity of wetlands to the Marimba River enhances their role in stormwater management, allowing for effective water infiltration and percolation while minimizing risks to human safety. The natural process helps mitigate flooding and supports urban resilience against climate impacts. Furthermore, wetland reclamation provides vital opportunities for the conservation of endangered flora and fauna, facilitated by reduced human interference in these ecosystems. In the context of global climate action, local wetlands also serve as crucial carbon sinks, contributing to carbon sequestration efforts. As of 2023, Zimbabwe's contribution to global carbon dioxide emissions from fossil fuels and industry was less than 0.5% (Ritchie et al. 2020). The reclamation initiatives undertaken by the city of Harare represent significant progress in recognizing wetlands as essential components of environmental infrastructure. These efforts not only bolster the ecological integrity of the Harare metropolitan province but also play a critical role in addressing the multifaceted challenges posed by climate change. By preserving wetlands in their natural state, integrating wetlands into urban planning and management strategies, Zimbabwe can enhance its capacity to store carbon dioxide, thereby mitigating potential worst-case climate scenarios. This aids in fostering greater resilience, ensuring sustainable development, benefiting both communities and the environment.

17.4.5.2 Adaptive Management: Approaches to Enhance Wetland Resilience Against Urban Pressure

Study findings clearly depicted the ongoing degradation of wetlands through both legal and illegal developments, calling for innovative strategies to bolster their resilience. Embracing wetlands as Urban Living Labs (ULLs), as exemplified by the Monavale Initiative, allows for collaborative experimentation among experts, community members and local governments to address urban challenges. These labs provide a platform for continuous learning and adaptation, demonstrating how integrating wetlands into urban planning can contribute to sustainable city development and resilience against climate change impacts.

17.5 Conclusion

The wetlands within the Harare metropolitan province are undergoing significant degradation, primarily due to encroachment. Notable exceptions include the effective management observed at Monavale Wetland and Chiremba Wetland, where community trusts and local authorities have implemented strategies to protect these vital ecosystems. The reclamation of flood-prone areas, such as those in Budiriro, highlights the importance of recognizing wetlands as essential components of urban infrastructure. The study advocates for the adoption of wetlands as Urban Living Labs, harnessing their potential as nature-based solutions to support resilient urban development in the face of climate change.

Acknowledgements We greatly appreciate the time spent by all research participants in support of this project. Constructive criticism from unanimous reviewers is also appreciated.

Potential Conflicts of Interest On behalf of all authors, the corresponding author states that there is no conflict of interest.

Informed Consent Informed consent was obtained from all individual participants included in the study.

Authors' Contribution All authors have contributed equally to this work.

Data Availability Statement The datasets used during and/or analysed during the current study are available from the corresponding author on reasonable request.

AI Discloser AI-assisted technology was not used in the preparation of this work.

Funding The research was not supported by any funding.

References

Adedeji OH, Adeofun CO, Tope-Ajayi OO, Ogunkola MO (2020) Spatio-temporal analysis of urban sprawl and land use/land cover changes in a suburb of Lagos and Ogun metropolises, Nigeria (1986-2014). IFE J Sci 22(1):10.4314/ijs.v22i1.4

Adeeyo AO, Ndlovu SS, Ngwagwe LM, Mudau M, Alabi MA, Edokpayi JN (2022) Wetland resources in South Africa: threats and metadata study. Resources 11(6):54. https://doi.org/10.3390/resources11060054

Afonin N, Kozlovskaya E, Moisio K, Kokko ER, Okkonen J (2024) Frost quakes in wetlands in northern Finland during extreme winter weather conditions and related hazard to urban infrastructure. Cryosphere 18(2223–2238):2024. https://doi.org/10.5194/tc-18-2223-2024

Ansari AH, Irini A, Jariah I, Ismail I (2011) Ramsar convention and convention on biological diversity as mechanism for promoting biodiversity conservation of Malaysian wetlands. J Appl Sci Res 7:2385–2393

Bernard J, Saunois M, Salmon E, Ciais P, Peng S, Berchet A, Serrano-Ortiz P, Gnanamoorthy P, Joachim J (2024) Satellite-based modeling of wetland methane emissions on a global scale (SatWetCH4 1.0). https://doi.org/10.5194/egusphere-2024-1331

Bulkeley H, Chan S, Fransen A, Landry J, Wagner A, Seddon N, Deprez A, Kok M (2023) Building synergies between climate & biodiversity governance: a primer for COP 28. United Nations, United Arab Emirates

Chumuriwo B (2017) An assessment of water quality along Mukuvisi River, Harare, Zimbabwe, University of South Africa, Available on https://core.ac.uk/outputs/83637591/. Accessed on 2 Aug 2024

City of Bulawayo (2021) Bulawayo local environmental action plan, 2021–2025

CivActs (2021) CIVACTS Bulletin 3: March 2021, www.nayoafrica.org

Cobbinah PB, Korah PI, Bardoe JB, Darkwah RM, Nunbogu AM (2022) Contested urban spaces in unplanned urbanization: wetlands under siege. Cities 121:103489. https://doi.org/10.1016/j.cities.2021.103489

Convention on Wetlands (2021) Global wetland outlook: special edition 2021. Secretariat of the Convention on Wetlands, Gland

Cunliffe R (2020) Current status of Harare wetlands. Harare Wetlands Trust, Harare

De Gobbi M (2020) Environmental integrity and doing business in Zimbabwe: challenges and engagement of sustainable enterprises, ILO working paper 2. ILO, Geneva

Dixon A, Wood AP, Hailu A (2021) Wetlands in Ethiopia: lessons from 20 years of research, policy and practice. Wetlands 41

Donde, O., Otieno, A. & Muia, W. (2023). Wetlands for remediation in Africa: threats and opportunities. In: Lobato de Magalhães, T., Otte, M.L. (eds) Wetlands for remediation in the tropics. Wetlands: ecology, conservation and management, vol 9. Springer, Cham. https://doi.org/10.1007/978-3-031-23665-5_10

Eastman JR (2012) IDRISI Selva manual. Clark University, Worcester, MA

Eyvaz M, Albahnasawi A (2023) Wetlands, new perspective. Environ Sci IntechOpen. https://doi.org/10.5772/intechopen.111469

Ferreira CSS, Kašanin-Grubin M, Solomun MK, Sushkova S, Tatiana Minkina T, Wenwu Zhao W, Kalantari Z (2023) Wetlands as nature-based solutions for water management in different environments. Curr Opin Environ Sci Health 33:100476

Finn BM, Bandauko E (2024) Dwindling funds and increased responsibilities: decentralization, unfunded mandates, and Harare's infrastructure crisis. Habitat Int 48(2024):103087

Fluet-Chouinard E, Stocker B, Zhang Z, Malhotra A, Melton J, Poulter B, Kaplan J, Goldewijk KK, Siebert S, Minayeva T, Hugelius G, Joosten H, Barthelmes A, Prigent C, Aires F, Hoyt A, Davidson N, Finlayson M, Lehner B, McIntyre P

(2023) Extensive global wetland loss over the past three centuries. Nature 614:281–286. https://doi.org/10.1038/s41586-022-05572-6

Gabay A, Chitagu T (2024) Community-led wetland restoration may hold key to Harare's water crisis, *Mongabay*, https://news.mongabay.com/2024/10/community-led-wetland-restoration-may-hold-key-to-harares-water crisis/

Gojendro S, Ch B, Das S, Bharti V, Peetambari N, Bidyasagar S, Meetei W, Sharma P (2023) Wetlands: importance and conservation strategy for the loosing ecosystem. 3:73–79

Govender-Ragubeer Y, Meeuwis J, Mckay TM (2014) A series of unfortunate events: how the Battle to save an urban wetland was both won and lost. J Transdiscipl Res S Afr 10(1):149–168

Government of Zimbabwe (2013) Constitution of Zimbabwe

Government of Zimbabwe (2020a) National development strategy 1: 'Towards a prosperous & empowered upper middle-income society by 2030,' Harare, Zimbabwe

Government of Zimbabwe (2020b) Zimbabwe Wetland Policy (Draft)

Harare Wetlands Trust (2020) Current status of harare wetlands

Iqbal MF, Khan IA (2014) Spatiotemporal land use land cover change analysis and erosion risk mapping of Azad Jammu and Kashmir, Pakistan. Egypt J Remote Sens Space Sci 17:209–229

Kabiri S, Allen M, Okuonzia JT, Akello B, Ssabaganzi R, Mubiru D (2022) Detecting wetland encroachment and urban agriculture land classification in Uganda using hyper-temporal remote sensing. AAS Open Res. https://doi.org/10.12688/aasopenres.13040.2

Kingsford R, Bino G, Finlayson M, Falster D, Fitzsimons J, Gawlik D, Murray N, Grillas P, Gardner R, Regan T, Roux D, Thomas R (2021) Ramsar wetlands of international importance–improving conservation outcomes. Front Environ Sci 9: 643367. https://doi.org/10.3389/fenvs.2021.643367

Krauss KW, Zhu Z, Camille L, Stagg CL (2021) Wetland carbon and environmental management, Geophysical monograph series. Wiley, Hoboken

Kudumba B (2022) Transforming urban policy to combat wetland degradation in Harare. Open J Soc Sci 10:479–505. https://doi.org/10.4236/jss.2022.1011031

Kuusaana ED, Ahmed A, Campion BB, Dongzagla A (2021) Characterisation and typology of urban wetlands in Ghana: implications for the governance of urban commons in secondary cities in Africa. Urban Govern 1(1):2021. https://doi.org/10.1016/j.ugj.2021.09.002

Mantel S, Xoxo S, Mahlaba B, Jane Tanner J, le Maitre D (2021) The role of ecological infrastructure (EI) in mitigating the impacts of droughts. Water Research Commission

Marambanyika T, Beckedahl H, Ngetar NS, Dube T (2016) Assessing the environmental sustainability of cultivation systems in wetlands using the WET-health framework in Zimbabwe. Phys Geogr 38(1):62–82. https://doi.org/10.1080/02723646.2016.1251751

Matenga L (2019) Placing Climate change in wetland conservation and urban agriculture contestations in Harare, Zimbabwe. In: The geography of climate change adaptation in urban Africa. Springer, Cham, pp 155–181. https://doi.org/10.1007/978-3-030-04873-0_6

McInnes RJ, Davidson NC (2018) Convention of migratory species (CMS) and wetland management. In: Finlayson CM et al (eds) The wetland book. Springer, Dordrecht. https://doi.org/10.1007/978-90-481-9659-3_111

Mhlanga B, Maruziva R, Buka L (2014) Mapping wetland characteristics for sustainable development in Harare: the case of Borrowdale west, highlands, National Sports Stadium and Mukuvisi woodlands wetlands. Ethiopian J Environ Stud Manag 7(5):488–498

Mohanty S, Pandey P, Pandey M, Srivastava P, Dwivedi C (2024) Wetlands contribution and linkage to support SDGs, its indicators and targets- a critical review. Sustain Dev 1-45. https://doi.org/10.1002/sd.2948

Mujuru L (2024) Polluted water threaten Zimbabwe's town life. The Daily News, Available at https://dailynews.co.zw/polluted-waters-threaten-zimbabwean-towns-life/#google_vignette. Accessed 2 Aug 2024

Muleya T (2020) EMA raises concern over urban Centers sewerage discharge. Chronicle. https://www.chronicle.co.zw/ema-raises-concern-over-urban-centres-sewage-discharge/. Accessed 1 Aug 2024

Mumuni AS, Mensah H, Asamoah S, Simpeh EK (2025) A systematic review on the causes and effects of urbanization on wetlands in sub-Saharan Africa. Urban Sustain Soc 2(1):153–177. https://doi.org/10.1108/USS-10-2024-0063

Munhende L (2020) Zimbabwe: land barons feast on Chitungwiza wetlands, All Africa. https://allafrica.com/stories/202006100449.html. Accessed 15 Dec 2024

Munyoro F (2021) Top court stops Monavale wetland development, The Herald, https://www.herald.co.zw/top-court-stops-monavale-wetland-development/. Accessed on 15 Dec 2024

Musasa T, Marambanyika T (2020) Threats to sustainable utilization of wetland resources in Zimbabwe: a review. Wetlands Ecol Manag 28(4):681–696. https://doi.org/10.1007/s11273-020-09732-1

Musasa T, Marambanyika T (2022) The nexus between drivers of wetland utilisation patterns, land use/land cover change and ecosystem services in two wetlands, Zimbabwe. Trans R Soc S Afr 77(2):119–131. https://doi.org/10.1080/0035919X.2022.2079156

Musasa T, Muringaniza KCR, Manyati M (2023) The role of stakeholder participation in wetland conservation in urban areas: a case of Monavale Vlei, Harare. Sci Afr 19. https://doi.org/10.1016/j.sciaf.2023.e01574

Office of the Auditor General OAG (2021) Protection of wetlands. Environmental Management Agency, Harare

Olawole MO, Msimanga L, Adegboyega SA, Adesina FA (2011) Monitoring and assessing urban encroachment into agricultural land—a remote sensing and GIS based study of Harare, Zimbabwe. IFE J Sci 13(1):149–160

Oliver E, Marsters L (2024) Nature-based solutions in sub-Saharan Africa for Climate and water resilience: A methodology for evaluating the regional status of Investments in Nature-Based Solutions from a scan of multilateral development Bank portfolios. World Resources Institute, Washington, DC

Ozturk D (2015) Urban growth simulation of Atakum (Samsun, Turkey) using cellular automata-Markov chain and multi-layer perception-Markov chain models. Remote Sens 7:5918–5950. https://doi.org/10.3390/rs70505918

Pachawo V (2023) Land-use impacts on water resources in semi-arid regions: the case of Manyame catchment Zimbabwe. Published Doctoral thesis, Universidade do Minho Escola de Engenharia

Raoul Wallenberg Institute RWI (2023) Final papers of the 2022 National Symposium on ten years of the declaration of rights in the Zimbabwean constitution. Raoul Wallenberg Institute of Human Rights and Humanitarian Law, Lund

Ritchie H, Roser M, Rosado P (2020) CO2 and greenhouse gas emissions. Published online at OurWorldinData.org. Retrieved from: https://ourworldindata.org/co2-and-greenhouse-gas-emissions [online resource]

Rocco R, van Ballegooijen J (eds) (2018) The Routledge handbook on informal urbanization, 1st edn, Routledge. https://doi.org/10.4324/9781315645544

Shayanewako S (2022) Come and get permits, Harare tells urban farmers. The herald. Available at https://www.herald.co.zw/come-and-get-permits-harare-tells-urban-farmers/. Accessed on 2 Aug 2024

Siamonga E (2014) Urban agriculture: authorities found wanting. The Chronicle. Available at https://www.chronicle.co.zw/urban-agric-authorities-found-wanting/. Accessed 2 Aug 2024

Sinthumule NI (2021) An analysis of communities' attitudes towards wetlands and implications for sustainability. Glob Ecol Conserv 27:e01604. https://doi.org/10.1016/j.gecco.2021.e01604

Spataro E, Rojas I (2020) Analysis of the decisions of the convention on biological diversity: a tool for use at the national level. Internal Briefing, Friends of the Earth International

The Environmental Management Act, [20:27] Government of Zimbabwe

Tsiko S (2022) Wetlands: unsung heroes of the climate change crisis, The Herald. Available at https://www.herald.co.zw/wetlands-unsung-heroes-of-the-climate-change-crisis/. Accessed 2 Aug 2024

Waly MM, Mickovski SB, Thomson C, Amadi K (2022) Impact of implementing constructed wetlands on supporting the sustainable development goals. Land 2022(11):1963. https://doi.org/10.3390/land11111963

Wang J (2023) A review on urban wetland as nature-based solutions in achieving SDG 11. https://doi.org/10.13140/RG.2.2.25773.46561

World Bank (2021) Catalogue of nature based solutions for urban resilience. World Bank Group, Washington, DC

Zimbabwe National Statistics Agency (ZimStat) (2002) Census 2002: preliminary report

Zimbabwe National Statistics Agency (ZimStat) (2012) Census 2012: preliminary report

Zimbabwe National Statistics Agency (ZimStat) (2022) Zimbabwe 2022 population and housing census report

Government of Zimbabwe [GoZ] (2020c) Zimbabwe National Human Settlements Policy. https://ucaz.org.zw/documents/zimbabwe-national-human-settlements-policy/

Abdullah HM, Mukti M, Miah MG, Kiram MA, Tanzir T, Hossain S (2024) Development at the cost of unsustainable degradation of wetlands: Unraveling the dynamics (historic and future) of wetlands in the megacity of Dhaka. World Development Sustainability 4. https://doi.org/10.1016/j.wds.2024.100131

Ramsar Convention Secretariat (2016) An introduction to the Ramsa Convention on Wetlands, 7th Edition. (previsously The Ramsar Convention Manual). Gland,
Switzerland. handbook1_5ed_introductiontoconvention_final_e.pdf

Kumar A, Kanauja A (2014) Wetlands: Significance, Threats and their conservation.

Office of the Auditor General [OAG] (2019) Protection of Wetlands by the Environmental Management Agency Under the Ministry of Environment, Tourism and Hospitality Industry. https://www.veritaszim.net/node/5463

Xu T, Weng B, Yan D, Wang K, Xiangnan L, Bi W, Meng L, Cheng X, Liu Y (2019) Wetlands of International Importance: Status, Threats, and Future Protection. International Journal of Environmental Research and Public Health, 16(10), 1818. https://doi.org/10.3390/IJERPH16101818

18 Integrating Wetlands into Urban Planning for Resilience of Cities: A Pathway to the Achievement of Sustainable Development Goals

Chilungu Mwiinde, Thomas Simbule, and Ephraim Kabunda Munshifwa

Abstract

Ecosystems, particularly wetlands, provide services that are key to supporting and enhancing societal well-being. However, rapid urbanisation and its associated population growth have resulted in the allocation of land for urban development and the growth of unplanned settlements in wetlands and flood-prone areas. This situation affects the resilience of cities as it results in increased intensity and frequency of harmful flood events, which exacerbate environmental degradation and associated climate change effects. Integrating wetlands into urban planning for the resilience of cities, thus, becomes central in improving the societal well-being of the growing urban population. Focusing on the three cities of Ndola, Livingstone and Kitwe, this chapter applied the case study and qualitative approaches in exploring the integration of wetlands into urban planning for the resilience of cities. Semi-structured interviews were used for collecting primary data from key informants, including planners and environmentalists in the local authorities of the three cities. GIS and Google Maps were also used, and secondary data were collected through a literature review. Data collection and analysis were based on key aspects including the status, effects and strategies with regard to the integration of wetlands into urban planning. It then links the findings to strategies for the achievement of the SDGs, affirming that the prevailing situation threatens the achievement of such goals. The chapter argues that there is minimal mainstreaming of wetlands in Zambian cities. It thus recommends the need for the Zambian Government, through the local authorities, to think about ways of actualising sustainable integration of wetlands into urban planning for the resilience of cities.

Keywords

Urbanisation · Wetlands · Climate change · Integrated urban planning · Sustainable development goals (SDGs)

18.1 Introduction

The rapid growth of the global urban population has highly pressurised the natural landscapes of

C. Mwiinde (✉)
Planning Unit, The Copperbelt University, Kitwe, Zambia
e-mail: chilungu.mwiinde@cbu.ac.zm

T. Simbule
Planning Department, Livingstone City Council, Livingstone, Zambia
e-mail: Livingstone.Council@grz.gov.zm

E. K. Munshifwa
Lusaka Campus, The Copperbelt University, Lusaka, Zambia
e-mail: Ephraim.munshfwa@cbu.ac.zm

S. K. Chakraborty et al. (eds.), *Wetlands and Urban Resilience*, Sustainable Development Goals Series,
https://doi.org/10.1007/978-3-032-06881-1_18

cities and resulted in substantial loss of natural wetlands, which are key aspects of hydrology, human health, biodiversity and climate change (Rogerson et al. 2021). Despite wetlands covering about 1.2 billion hectares (ha) of the global area, they have been declining at a faster rate. For instance, the available data show about 35% loss of wetlands since 1970 (Convention on Wetlands 2021). The quality of available wetlands has thus been adversely affected by rampant real estate development, pollution, invasive species, drainage, unsustainable utilisation and climate change (Mitra 2024). This has adversely affected the functioning of wetlands and consequently, the social, economic and environmental conditions of cities and people. For instance, most cities currently experience unsustainable utilisation of wetlands and natural resources, resulting in environmental degradation (Rogerson et al. 2021; Amaranath and Subrahmanian 2020). This affects the well-being of people, especially the urban poor (Chiluba et al. 2021). The key concern is how to best ensure sustainable utilisation and management of wetlands, especially in developing countries.

Globally, studies have shown the importance of wetlands in the resilience of cities (Rogerson et al. 2021; Amaranath and Subrahmanian 2020; Davies et al. 2020; FAO 2020; Tang et al. 2020; Ro et al. 2020; WWT Consulting 2018). This has resulted in global strategies including SDGs Target 6.6, which focuses on protecting water ecosystems (Convention on Wetlands 2021) and SDGs Target 11.4, specifically, which focuses on strengthening efforts for protecting and safeguarding cultural and natural heritage (UN-Habitat 2023; Kometa et al. 2018). SDG 11, in totality, focuses on ensuring inclusiveness, safety, resilience and sustainability of cities and communities and thus becomes a key global call for action in dealing with such challenges. Despite such planning initiatives for promoting livable cities, the loss of wetlands has been continuous with rapid urbanisation, resulting in environmental degradation and climate change-associated disaster risks (Rogerson et al. 2021; Amaranath and Subrahmanian 2020). This situation threatens the attainment of SDGs, especially in developing countries (Chen et al. 2022; Convention on Wetlands 2021; Casey and Stazen 2021; Chiluba et al. 2021). Ensuring sustainable utilisation of wetlands and other natural resources (Convention on Wetlands 2021), therefore calls for an urgent need to integrate wetlands into urban planning as a pathway to the achievement of the SDGs, at local, regional, national and global levels.

This chapter thus explores the integration of wetlands into urban planning for the resilience of cities. Focusing on three Zambian cities of Ndola, Livingstone and Kitwe, the specific objectives include exploring the status of wetlands, the status of integrating wetlands into urban planning, the effect of integrating wetlands into urban planning on the resilience of cities and the strategies for sustainable integration of wetlands into urban planning for the resilience of cities. The chapter then links the findings to strategies for achievement of the SDGs. The chapter outcome will be helpful to the Zambian Government through local authorities in understanding and actualising sustainable integration of wetlands into urban planning for the resilience of cities.

After this introduction, the chapter highlights the rationale, which is followed by a section highlighting the study area. The chapter then highlights the materials and methods used that are followed by a section on results and discussions. Limitations of the chapter are then discussed, followed by sections on recommendations and conclusions. The chapter ends with acknowledgments.

18.2 Rationale of the Chapter

Some authors consider wetlands to be the kidneys of the earth, as they provide a key link between land and water. These are dynamic aquatic ecosystems (including floodplains, swamps) and exist everywhere globally (Amaranath and Subrahmanian 2020; Tang et al. 2020). Urban wetlands include those that continued despite the historical development associated with urbanisation and those recently constructed in urban settings. Wetlands are key in absorbing extra rainfall, regulating floods, as well as minimising consequent economic and infrastructure damages resulting from storms. They are also helpful in improving the quality of water and air, as well as moderating city temperatures (WWT Consulting 2018). Furthermore, the excellent benefits of wetlands include ensuring more livable cities and well-being of humans through supply of water, moderation of climate, treatment of wastewater, biodiversity habitation, agriculture and aquaculture, tourism, recreation, education and research (SDG 11: Targets 11.4) (Amaranath and Subrahmanian 2020; Tang et al. 2020). However, unplanned and rapid urbanisation has resulted in rapid population growth, which has led to human settlements and activities in wetlands, especially in the developing world. For instance, over 55% of the global population currently lives in cities. This is estimated to reach about 60% by 2030 and 65% by 2050 (Mwiinde and Munshifwa 2024; Bodo 2019; WWT Consulting 2018). Despite the global speed of urbanisation, it is anticipated that developing countries across Asia and Africa will account for about 90% of the future global population growth (Convention on Wetlands 2021), resulting in serious economic, social and environmental implications (Chen et al. 2022; Bodo 2019).

This situation threatens the functioning of wetlands globally, as human settlements and associated rapid population growth evolve more on land associated with wetlands and natural resources than in uplands. Rapid urbanisation has thus resulted in encroachments and continuous loss of wetlands through conversions aimed at accommodating developmental activities (Amaranath and Subrahmanian 2020; WWT Consulting 2018). Consequently, this has resulted in increased intensity and occurrence of harmful events, including floods and environmental degradation (Mwiinde and Munshifwa 2024; Munshifwa et al. 2021; World Bank 2022; Chiluba et al. 2021; Amaranath and Subrahmanian 2020). For instance, some literature shows that about 40% of wetlands in India have been lost between 1991 and 2001 due to rapid population growth and associated water and food demands. This therefore results in wetlands degradation and adverse climatic conditions (Amaranath and Subrahmanian 2020) that greatly affect the social, economic and environmental conditions for people globally (Akinluyi and Adedokun 2014; Bodo 2019; Chen et al. 2022; Tonne et al. 2021). Since the growth of human settlements evolves around wetlands, ensuring the resilience of cities and sustainable development requires sustainable integration of wetlands into urban planning (Amaranath and Subrahmanian 2020). How then can we ensure sustainable utilisation and management of wetlands for the resilience of cities? Thus, there is need for this chapter.

18.2.1 Wetlands and Urban Planning

Sustainably utilised wetlands are capable of enhancing the economic, social and environmental benefits for cities (Mitra 2024). For instance, Cambodia has been able to provide services worth about US$30.12 million with regard to food, water provisioning and wastewater treatment annually through the Boeng Cheung Aek Wetland in Phnom Penh (Convention on Wetlands 2021; Ro et al. 2020). Despite some studies proving the positive impact of wetlands-ecosystem services on economic, social and environmental activities, wetlands have not been fully integrated into urban development planning (Amaranath and Subrahmanian 2020). This situation makes wetlands susceptible to unsustainable utilisation, consequently, poor governance, as no agency would be fully responsible for their sustainability (WWT Consulting 2018). Considering the value of wetlands, ensuring sustainable development (SDG 11 Target 11. 4) requires conserving, restoring, managing and maintaining their functioning. However, not integrating wetlands into urban planning makes their conservation and sustainable utilisation difficult, thereby resulting in threats which include (WWT Consulting 2018) (i) housing demand and other developments, (ii) biodiversity loss through conversion of wetlands to recreation centres and public parks, (iii) pollution from solid and water waste, (iv) river and stream channelisation, (v) wetlands hydrological disconnection from watercourses, (vi) using hard rather than green infrastructure solutions and (vii) losing native species due to invasive species.

Such threats adversely affect the sustainability of wetland functioning, since cities use about 75% of the natural resources globally and produce over 70% of the total global waste (carbon emissions), despite occupying only about 3% of the land on earth (Convention on Wetlands 2021; WWT Consulting 2018). Considering wetlands sustainability as a solution for cities in mitigating climate change-related risks; supporting food production and generating income through activities including tourism and recreation is key (Mitra 2024; WWT Consulting 2018). Urban planners should thus consider integrating wetlands and other economic, social and environmental natural resources into urban planning (Rogerson et al. 2021; Amaranath and Subrahmanian 2020). Integrated urban planning entails incorporating all efforts for wetland conservation into the present planning framework. This calls for interagency communication, which is key in building collaborative partnerships, maximising areas of expertise, organising multi-objective visions whilst building consensus and bridging diverse agency programmes and geographic boundaries (Tang et al. 2020). Some of the suggested strategies for sustainable integration of wetlands into urban planning have been highlighted in Sect. 18.2.2.

18.2.2 Global Strategies for Sustainable Integration of Wetlands into Urban Planning

Some literature suggests a number of recommendations for ensuring the sustainable integration of wetlands into urban planning for the resilience of cities. Generally, there is need to consider (i) inclusivity and participatory approaches: for engagement of all key stakeholders through establishment of communication, education and public awareness programmes; (ii) incentive systems: for protecting wetlands through encouraging payments for environmental services (within and beyond cities); (iii) comprehensive approaches: for mitigating adverse impacts on wetlands through localised development solutions for green supply chains and energy neutral (low carbon); and biodiversity hotspots: for biodiversity maximisation through understanding native wetland habitats, landforms, hydrology and soil types (Mitra 2024; Amaranath and Subrahmanian 2020).

Urban planners should specifically consider (i) embarking on integrated planning: for promoting wetlands as natural green infrastructure for conserving ecological environment including (water resources and waste water treatment elements); (ii) clearly articulating urban wetlands: for informed processes of decision-making; (iii) developing specific urban wetland management planning: by mapping wetlands features; (iv) mapping, monitoring and conducting baseline surveys: for analysing wetlands ecological features; (v) coordinating planning approaches: with related public and private agencies; (vi) framing appropriate participatory methods: for stakeholders' involvement and empowerment; (vii) selecting certain urban wetlands: as natural wastewater treatment systems for sustainably mitigating urban pollution and sedimentation; and (viii) formulating comprehensive plans: for adapting anticipated urban developments to prevailing economic, socio-cultural physical and ecological traits (Mitra 2024; Bharwani et al. 2023; Amaranath and Subrahmanian 2020).

Policy makers should consider (i) developing holistic urban planning policies: by including development frameworks and spatial planning for protecting wetland ecosystems and associated services; (ii) addressing water management issues: at various and appropriate scales through specific management planning for wetlands; (iii) formulating legislative and regulatory framework: for reactiveness in wetlands protection; (iv) providing inspiring regulatory charters: for promoting public-private partnerships; (v) involving all stakeholders: for promoting participatory planning and management approach; and (vi) incentivising economic activities: through subsidies and payments for ecosystem service (Mitra 2024). Developers should also consider (i) promoting restoration and rehabilitation of degraded wetlands: through local government support; (ii) fostering various positive benefits: resulting from integrating wetlands in developments; (iii) promoting wetlands: as natural green infrastructure in urban development activities; and (iv) promoting local communities involvement: in all planning and management processes (Mitra 2024; UNECR n.d.; Amaranath and Subrahmanian 2020).

In ensuring sustainable utilisation of wetlands for enhanced economic, social and environmental conditions and, consequently, well-being of humans, the integration of wetlands into urban planning should be based on the following resolutions of the Convention on Wetlands (Convention on Wetlands 2021; Ramsar Convention 2018).

(i) **Resolution XI.11:- Principles for planning and management of urban and peri-urban wetlands,** which required decision-makers and urban planners to consider adopting practical principles for avoiding destruction of existing wetlands; incorporating restoration and creation of wetlands in urban infrastructure nature-based solutions; understanding wetlands value and benefits; stakeholders engagements in wetlands decision-making and full integration of wetlands into urban planning.

(ii) **Resolution XIII.16: Sustainable urbanisation, climate change and wetlands,** which discourages activities with negative impacts on wetlands. City initiatives, including sponge cities, city and biodiversity, as well as garden city, have practically adopted such principles and incorporate wetlands into integrated urban waste management, well-being benefits and biodiversity conservation.

(iii) **Resolution XII.10: Wetland city accreditation of the Ramsar Convention,** which inspires cities' application for wetland city accreditation and celebrates key roles of wetlands in livability and sustainability of cities. This has so far resulted in the accreditation of 18 cities in France, China, the Korean Republic, Hungary, Sri Lanka, Tunisia and Madagascar.

(iv) **Existing multilateral environmental agreements (MEAs),** which provide a platform for unique action scope and ambitions, for the effective implementation of wetlands goals through collaborations with stakeholders, including institutions, MEAs and contracting parties. International importance wetlands also help as laboratories for sustainable wetland management.

18.3 Chapter Study Area

The chapter study area covers the three Zambian cities of Ndola, Livingstone and Kitwe, located on the Copperbelt and Southern Provinces of the country. Ndola District is the provincial capital for the Copperbelt Province of Zambia. Ndola is the third-largest district in the country. Livingstone District is the provincial capital of the Southern Province of Zambia. Kitwe is the second-largest district in Zambia. The district is located in the central part of the Copperbelt Province (Fig. 18.1).

18.3.1 Wetlands in Zambian Cities

According to the Ministry of Lands and Natural Resources (GRZ-MLNR, 18), in its National Policy on Wetlands, wetlands are considered to be one of the main global productive systems. Earthwide, wetlands are mainly relevant in water supply and sanitation, food resource, timber and thatch production, recreation and tourism, sediment trap and carbon sink, research and education and beauty value. The desire to promote conservation of wetlands and sustainable management of wetland resources led to the adoption of the Ramsar Convention on Wetlands of International Importance (RCWII) in 1971, to which the Republic of Zambia is a party. The major obligation of the Convention on Wetlands is to ensure

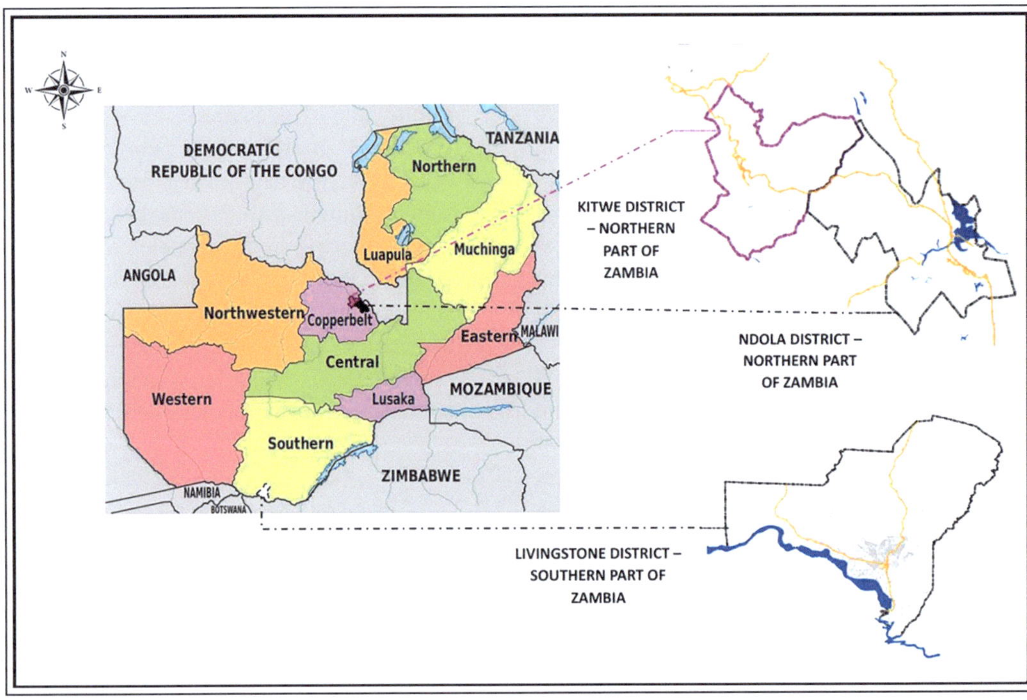

Fig. 18.1 The chapter study area. (Source: Author generation based on Open Street Maps 2025)

the implementation of principles on "wise use" of wetlands worldwide (GRZ-MLNR, 18). Zambia's wetlands cover approximately 14% to 19% of the total area of the country, which supports various economic sectors in the country. According to the MLNR (GRZ-MLNR, 18), tourism, agriculture, fisheries and forestry, among others, depend on wetlands. Further, the report by MLNR indicates that wetlands also support livelihoods and provide ecosystem services such as reducing the impacts of storm damage and flooding. They also help in maintaining good water quality in rivers, recharging groundwater, storage of carbon and nutrient cycling, ultimately helping in controlling pests and stabilising climatic conditions. Further to this, MLNR indicates that wetlands are key for biodiversity and the prevention of soil erosion. Wetlands across Zambia also provide habitat for many species of animals and birds, whilst being a rich source of fish as well as grazing ground for rural pastoralists.

18.3.2 Historical and Current Status on Urban Developments in Wetlands of Zambian Cities

Between the 1890s and 1960s, Zambia experienced urban development with displaced indigenous people. British Government policies controlled locals through the British South Africa Company, promoting development without urban planning due to minimal costs for municipal services and administration (Makasa 2019). The colonial Cecil Rhodes' plan to build a rail line connecting Cape Town to Cairo in Zambia, aimed at linking Africa with Europe, led to the declaration of 50Km of land as Crown land and native reserve order in 1928 (Makasa 2019). Tait (1997) revealed the land where Cecil Rhodes' rail line in Zambia, built to connect Africa with Europe, was declared Crown land and a native reserve in 1928. This led to the removal of indigenous Africans and the loss of customary rights. The land was turned into Crown Land, supporting European settlement and development and allowing informal development as a preserve of African human settlements. The discovery of copper deposits led to the opening of mines, townships and cities, some connected by railways. Colonial rule introduced policies like the hut and poll tax, industrialisation without urbanisation, rural-urban migration, labour migration and urban segregation. Capitalist governments stimulated cash needs among Africans (Makasa 2019). Colonial policies led to the introduction of a poll tax and restricted markets for African produce, prompting African men to seek cash by trading their labour for material goods. This initiated and increased rural-urban migration over time. However, the lack of balance between industrial development and urban planning resulted in poorly organised urban areas, racial segregation and housing shortages, with many squatters living on the outskirts of planned settlements, which grew in wetlands, stimulating climate change and disaster risks (Tait 1997). It can be deduced from this that some wetlands would have already been affected as the development of human settlements was not guided by any plan down to the time when Zambia gained independence.

After Zambia gained independence in 1964, the new government was sensitive to the colonial past policies which saw Africans suffer the injustice of urban planned housing and segregation. The government thus prioritised to deal with housing needs of the Africans and consequently abolished the hut and poll tax; allowed rural-urban migration; built more housing to address industrialisation without urbanisation of the Africans. Furthermore, the government banned migration of labour and gave people the freedom to reside in places of their choice, hence promoting high rural-urban migration. This resulted in an influx of people into towns along the line of railway where they anticipated finding employment with a reasonable income (Jayarajan 1979). During the time of population influx, Zambia had not yet developed spatial development strategies which would have protected wetlands. Consequently, the pattern of settlements disregarding wetlands had continued to increase (Kachali 2007). The changes that came with Zambia gaining independence brought about rapid growth in population. According to a recent census report, 45% of the total country's population now lives in the urban areas (CSO 2022). The report equally pointed out that Lusaka, being the capital city, has been the largest and fastest-growing city. Lusaka's population has more than tripled since independence, which has challenged sustainable urbanisation in relation to the environmentally sensitive areas such as wetlands.

The previous section discussed that wetlands have been associated with human activity due to their limitless supply of water, rich alluvial soils for the production of root crops and cereals, as well as abundant natural resources including vegetation, fish and wildlife (Kachali 2007). In addition to this, Chabwela et al. (2017) equally showed wetlands to be a source of life that provided a source of food in terms of protein to humans. To this day, wetlands are an arena for several human and animal activities.

Human development on natural resources is taken to be the greatest constraint in a number of developing countries. Disturbance of watersheds in the recent past has brought long-term negative effects on the quantity and quality of land and water resources (Kamweneshe and Richard 2017; Darghouth et al. 2008). For instance, during the last few decades, disturbances of watersheds have led to serious problems in connection with the environment and people at all levels. This negative change results from a series of natural and anthropogenic factors, including natural soil erosion, changes in farming systems, overgrazing, deforestation and pollution. Environmental costs, together with socio-economic impacts, have resulted in the development of watershed management approaches. Urban developments, such as land use and cover changes stimulated by demographic dynamics, contribute more than other processes to land cover changes and are the primary factors responsible for wetland degradation. This has transformed such environmentally sensitive areas at all levels, including global, regional and local perspectives over the past few decades (Amare 2015; Begg 1987).

In the Zambia context, wetlands are endangered by various factors including the spread of invasive alien species, shrub disturbance, human settlement generation, pollution contamination, mineral exploitation, dam construction for hydro-power generation, draining, climate change, genetically modified organisms and channellisation (GRZ-MLNR, 18). GRZ (2011) deliberately focused on environmentally sensitive attributes and recognised wetlands as worthy of special protection. According to the National Policy for Wetlands, the Act provides for the declaration of identified wetlands as protected areas, although very little has been achieved since the enactment of the law. It further indicates the absence of integrated policy direction on wetlands management in relation to land uses, thus the manifestation of unsustainable use of wetlands.

18.4 Materials and Methods

Focusing on the three Zambian cities of Ndola, Livingstone and Kitwe, the chapter applied a combination of case study and qualitative approaches in exploring the integration of wetlands into urban planning for the resilience of cities. The approach involved getting inputs from town planners and environmentalists with the aim of deepening the understanding of the status, effects and strategies with regard to the integration of wetlands into urban planning in three cities. The chapter started with the identification of the key objectives and aspects with regard to wetlands, urban planning and resilience of cities based on some reviewed literature and field observations of the cities of focus. Semi-structured interviews were then prepared based on key identified aspects for the collection of primary data. Primary data were then collected from key informants from the town planning and environmental sections of the local authorities in the three cities of focus. Besides capturing council resolutions on these specific subjects, telephone interviews were used. GIS and Google Maps were also used to analyse the spatial relations with regard to wetlands in the cities of focus. Secondary data were collected through a literature review. The understanding of wetlands, urban planning and the resilience of cities in relation to SDGs was based on a combination of peer-reviewed published literature and expert knowledge (Table 18.1). The Google Scholar and Research Gate search engines were used in filtering published research in peer-reviewed journals or authoritative public reports from official organisations. The review process involved searching the literature based on key identified aspects with regard to wetlands, urban planning and the resilience of cities. The arguments and evidence found in the literature and reports, together with the expertise of town planners and environmentalists sections in local authorities of the three cities, guided the discussions with regard to the key aspects of wetlands, urban resilience and SDGs. The findings are helpful to the Zambian Government through local authorities in improving the understanding and actualising the sustainable integration of wetlands into urban planning for resilience of cities.

18.5 Results and Discussion

Three cities were analysed in relation to developments in wetlands. In this case, GIS and Google Maps were used to analyse the spatial relations in these locations. Besides capturing Council Resolutions on these specific subjects, telephone interviews were conducted with Town Planning and Environmental Sections of the local authorities. In this case, Ndola, Livingstone and Kitwe city councils were the target.

(i) **Ndola District**

Ndola District is the provincial capital of the Copperbelt Province of Zambia. Ndola is the third-largest district in the country. It had a total population of 624,579 in 2022 and an annual population growth rate of about 2.7% (CSO 2022). Ndola District is endowed with mineral resources as well as vast natural resources, including forests and wetlands. Spatial analysis for the

Table 18.1 Material and methods

Phase	Involves	Target objectives include	Tools	Mixed
Phase 1	Identification of objectives and key aspects (wetlands, urban planning, resilience of cities and SDGs)	1. Exploring the status of wetlands	Literature review Field observations	Case study and qualitative approaches
		2. Exploring the status of integrating wetlands into urban planning		
		3. Exploring the effect of integrating wetlands into urban planning on the resilience of cities		
		4. Identifying the strategies for sustainable integration of wetlands into urban planning for the resilience of cities		
		5. Exploring the link between integrating wetlands into urban planning, resilience of cities and SDGs		
	Preparation of semi-structured interview questions	All objectives and key aspects		
Phase 2	Data collection and analysis	Based on all the objectives and key aspects	Interviews GIS and Google Maps Literature review Site observations	Case study and qualitative approaches
Output	Strategies for understanding and actualising sustainable integration of wetlands into urban planning for urban resilience			

Source: Authors (2025)

Ndola district revealed increased human settlement development in the city's wetlands. Plans, Works and Development Committee reports pointed to the fact that the Ndola City Council (NCC) conducted an operation between 2018 and 2020, which attempted to demolish what was regarded as illegal residential buildings. According to property ownership documents reviewed for this chapter, 90% of the plans revealed that land was allocated by the local authority and that these properties were on the layout plan created by NCC and numbered by the MLNR. In reference to Statutory Instrument (SI) No. 1 on the protection of public streams and sources of water, it was evident that most of the properties built along the Kafubu River are right within the river buffer of 50 meters. It was also noticed from the spatial relations that even properties built beyond 50 metres were still technically in wetlands. According to the Town Planning section of NCC, it was revealed that human settlement planning in that area was done without strong consideration of the natural characteristics of the land, hence the creation of the residential plots within the wetland.

Ndola City Council's position is that wetlands along the Kafubu River, being in the Central Business District (CBD), attract a lot of interest from developers, putting pressure on the council to allocate within these areas. The Environmental Planning section of the city also asserted that the Water Resources Management Authority (WARMA) had not been vigilant in protecting wetlands, making them porous to illegal constructions in such locations. The recommendation from the Environmental Section was that joint efforts between WARMA and the Local Authority were needed in order to protect these wetlands. As also noted by Munshifwa et al. (2021), the city of Ndola has continued to allocate land for development in environmentally vulnerable areas, including the Kafubu River basin (Figs. 18.2 and 18.3).

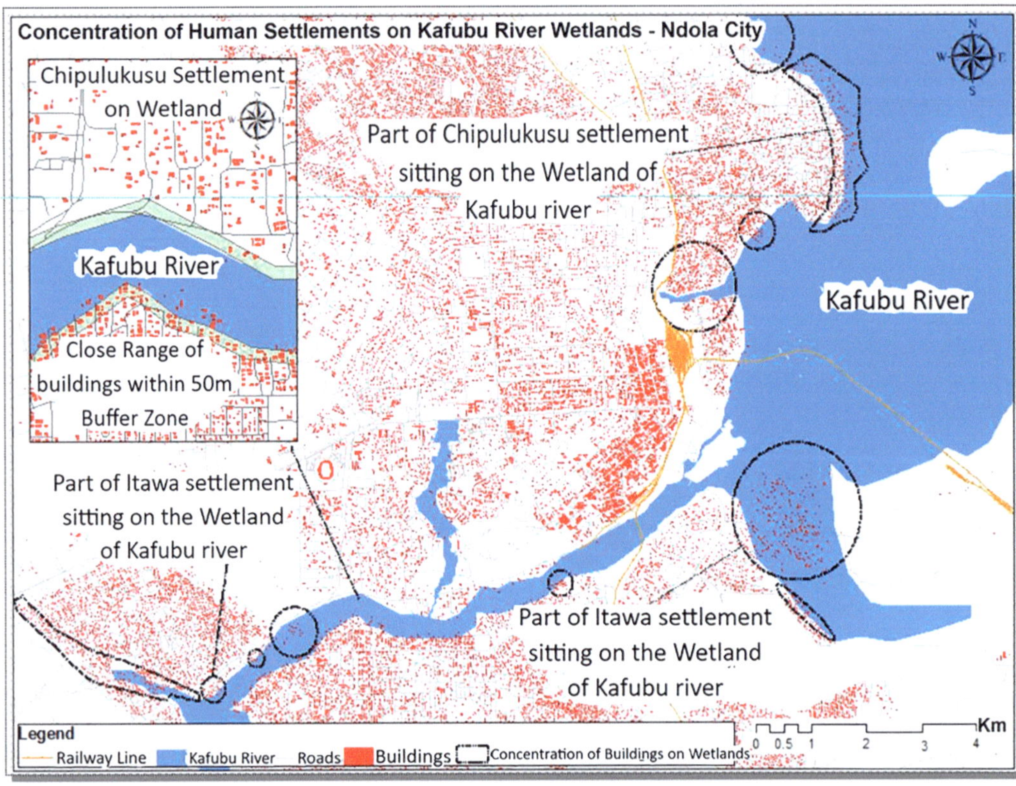

Fig. 18.2 Human settlements on wetlands. (Source: Author generation based on Open Street Maps 2024)

(ii) **Livingstone District**

Livingstone district is the provincial capital of the Southern Province of Zambia. The district had a total population of about 177,393 in 2022 and an annual population growth rate of about 2% (CSO 2022). Livingstone district has special spatial aspects which include harbouring the Victoria Falls (world heritage site), the Zambezi River, an archaeological site, a historical rail bridge and Mosi-o-tunya National Park and wetlands.

The Planning and Environmental Sections of the Livingstone City Council (LCC) were engaged on the sustainable use of wetland within the district. The Environmental Planning section indicated that the city is not heavily endowed with wetlands due to its climatic conditions. However, it was mentioned that Dambwa Central had wetlands which have over time been developed upon. The Town Planning section also confirmed that the wetlands in question were encroached upon illegally, with full involvement of the local authority. However, Plans, Works and Development committee reports provided for the regularisation of the buildings without considering the waterlogged nature of the land. Furthermore, the Environmental section expressed its concern over frequent floods in the areas and the vulnerability of residents to disaster risks. It was equally observed that over time, the nature of disasters has changed with increased severity, indicating that the area is not suitable for such development or human settlement. Figure 18.1 shows a detailed analysis of the spatial planning for Livingstone (Fig. 18.4).

(iii) **Kitwe District**

Kitwe is the second-largest district in Zambia, both spatially and population-wise. The district is located in the central part of the Copperbelt Province. It had a total population of 661,901 in 2022 and an annual population growth rate of about

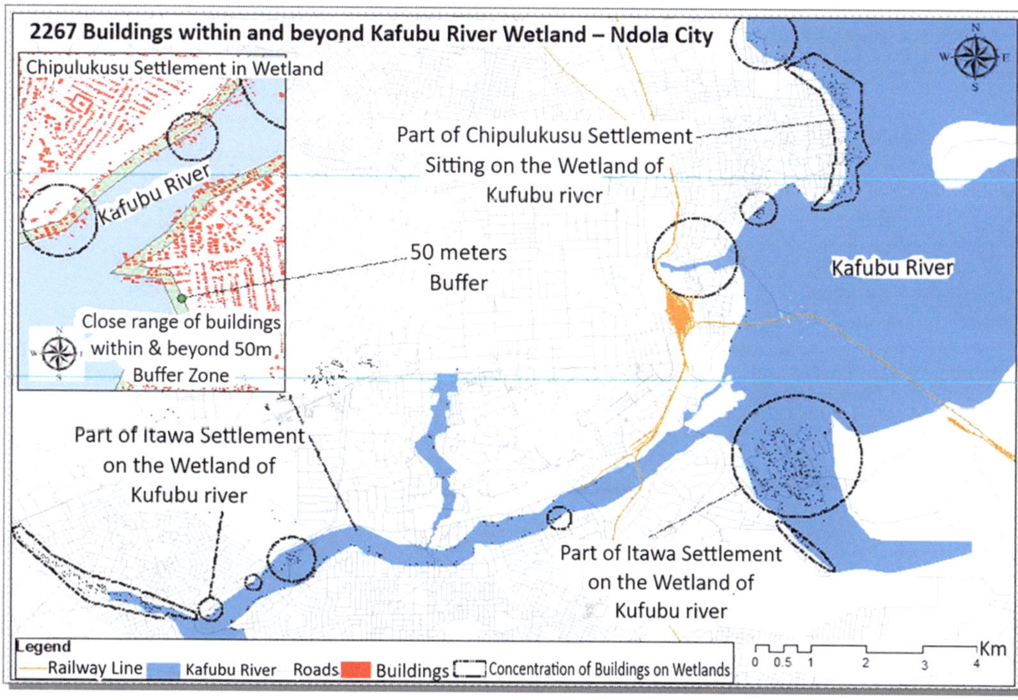

Fig. 18.3 Human settlements on wetlands beyond the river buffer. (Source: Author generation based on Open Street Maps 2024)

2.1% (CSO 2022). Kitwe district is endowed with mineral resources as well as vast natural resources, including forests and wetlands.

Kitwe has young inhabitants, with slightly over 66% of the population below the age of 25 years. The city is the second-largest industrial centre in Copperbelt Province, with the core economic base being mining and other associated mining service industries. Being the mining industry for a long time, it has retrenched residents who are now self-employed in the informal sector (Chileshe et al. 2023). In the 1990s, unemployment rates grew with the decline of the mining sector. These economic activities have attracted large numbers into the city and, as such, put pressure on land to establish settlements and commercial entities (Chileshe et al. 2023).

The Town Planning and Environmental Sections of Kitwe City Council noted that building on the wetland of Buchi stream, which is disaster prone, started illegally and was later regularised by the local authority. The Town Planning section further pointed out that the local authority had no option but to create the plan in accordance with what was already obtained on the ground. Contrary to the observation from the environmental section, portions of Buchi were found to be lying in the disaster-prone areas. The environmentalist also argued that residents are exposed to floods and the likelihood of landslides. They pointed to cracks in buildings attributed to ground movements over time. The spatial analysis below gives more insight on the situation obtained on the ground (Fig. 18.5).

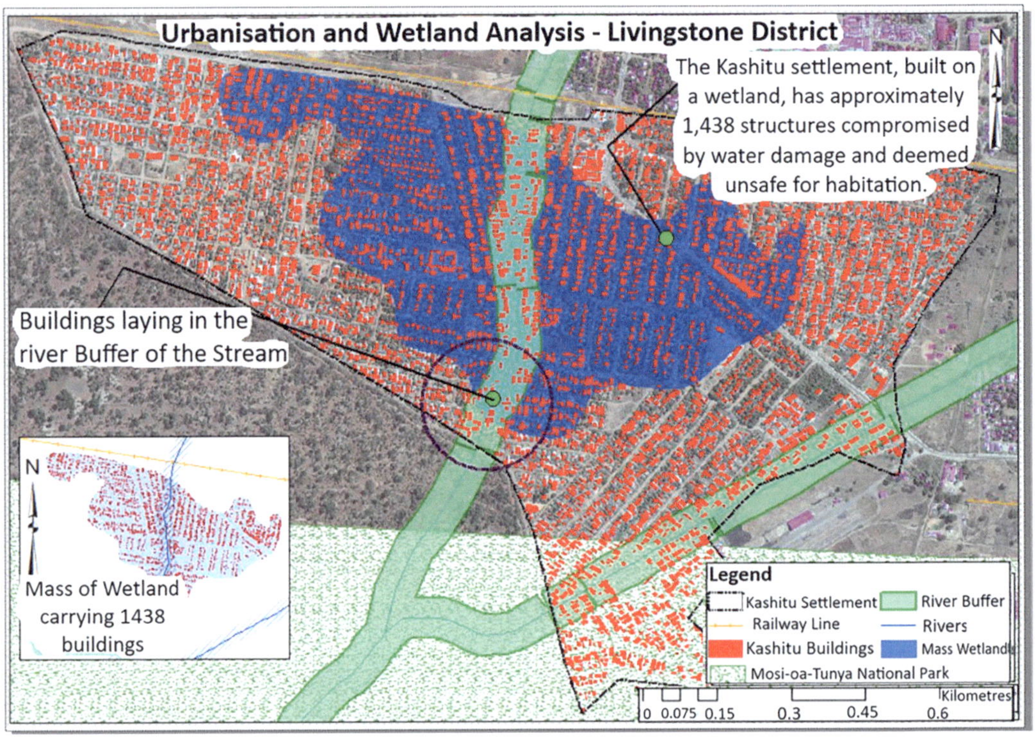

Fig. 18.4 Spatial planning for Livingstone. (Source: Generated by the authors, 2024)

18.5.1 Strategies for Sustainable Integration of Wetlands into Urban Planning for Resilience of Zambian Cities

Zambia has been practicing urban planning for over 100 years. This was initiated by the creation of some towns in the country under the colonial regime, although comprehensive planning and planning as a process started after Zambia's independence. The Town Planning ordinance was enacted in 1929 to serve as an urban planning governing tool and a legal basis for town planning schemes prepared under the colonial era, and regulated development control in the country. Following independence, the Town and Country Planning Act (TCPA), 283 of the Laws of Zambia, was enacted to determine and control spatial matters in the country. Among other things, the Act provided for the appointment of planning authorities; establishment of the town and country planning tribunal; preparation, approval and revocation of development plans, and control of development and subdivisions (GRZ 1965).

The TCPA, as a principal planning tool, then provided for various land uses and standards. It equally took care of public participation in land use planning, indicating that stakeholders were supposed to be consulted in any land use planning decisions. The engagement with the Engineering and Planning departments at Ndola and Kitwe City Council revealed that public consultation never took place, and there were no records at the local authority that indicated public consultations were conducted during spatial planning decisions that would have probably. Further to this, the respondent department revealed that until the recent past, Zambia had an insignificant number of urban planning professionals trained to advise and enforce planning standards as provided for in the Act, and that the Ministry of Water Resources Management Authority (WARMA) of Zambia was

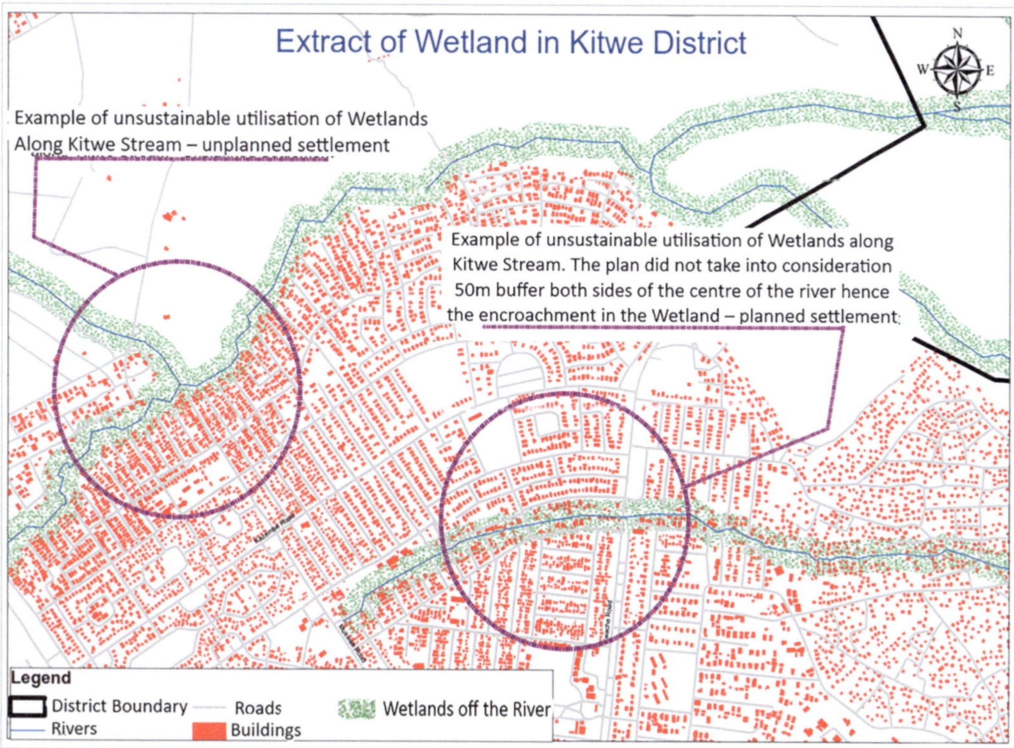

Fig. 18.5 Human settlement on wetlands of Kitwe Streams. (Source: Author generation based on Open Street Maps 2024)

recently established in 2011 to manage and protect water resources and wetlands. According to the Planning Department at Livingstone City Council it was observed that TCPA never recognized squatter settlements and such were regarded as nonexistent, much as they were established in wetlands. Further to this, Livingstone City Council Planning Department highlighted that Institutional and Organizational arrangements were not collectively put in place to address comprehensive land use planning that would have catered for wetlands.

The TCPA was in force up to 2015 and gave way to the Urban and Regional Planning (URP) Act number 3 of 2015. This is because the TCPA ceased to be responsive to a number of current demands, and spatial planning issues in the built environment are not able to respond to the arising circumstances. The URP Act No. 3 of 2015 vests administrative and legal powers for planning in the minister responsible for URP, whose powers have been delegated to the local authorities, who are required to work with the provisions of the law. The establishment of the URP Act brought about new features which were not part of the TCP Act. Such features comprise planning for the entire district, including customary land through planning arrangements; joint integrated development planning across districts; an integrated planning approach that captures all sectors within the specified geographical space; seamless planning for the entire district and a spatial development framework for the locality. The URP Act provides a principal guide on planning processes, including management and administration; planning framework and systems; improvement areas; and planning application and permission. The Integrated Development Plans (IDPs) for Kitwe, Ndola and Livingstone City Councils indicate that spatial development frameworks are being used to enforce, capture and integrate wetlands in land use plans. As earlier mentioned, the URP Act provided for the production of the IDP as a planning instrument to guide spatial development for the cities. Unlike before, spatial planning has increasingly employed technology such as GIS and remote sensing in determining sustainable use of land whilst providing a seamless approach to spatial planning and harmonising different characteristics of land over a given space (GRZ 2018).

Establishment of Water Resources Management Authority (WARMA) of Zambia in 2011 and enactment of URP Act No. 3 of 2015, jointly enforced, would bring about the integration of wetland in land use planning. It is expected that local authorities are expected to mainstream wetlands in urban planning. Integrated Development Plan for Ndola City Council revealed deliberate mainstreaming in wetlands in land use planning for sustainable management of nature to counter climate change and disaster risk reduction. The IDP has not provided any guidance on how to address compromised wetlands, which suggests that wetlands in Ndola are not being utilized in a sustainable manner. The Planning Department for both local authorities suggested that meaningful planning that would integrate wetlands would be possible if a comprehensive land use zoning plan had been in existence. According to these local authorities, there has been a zoning plan after colonial rule, and not the whole land is accounted for in the development plan. The department expressed concern that land, such as wetlands, is subjected to illegal subdivision and unsustainable uses because it is not protected by the approved zoning plan. The WARMA Department, being the custodian of wetland management, was equally interviewed on its strategies to protect wetlands from unsustainable use. WARMA revealed that they have conducted a nationwide wetland mapping as a practical aspect of protecting such resources. It has been observed by WARMA that much of the wetland has been compromised upon, hence the need to adhere to planning standards need to be employed when working with wetlands. Like Livingstone City Council, WARMA suggested that there is need to establish planning arrangements between WARMA and local authorities in respect of land use planning as a way of enhancing sustainable use of wetlands that will contribute to the reduction in climate change and disaster risk.

A brief discussion of the three cases above points to a clear disregard of regulations prohibiting the allocation of land and development in wetlands. Although often emerging as illegal settlements, evidence indicates the routine legalisation of such settlements. The Urban and Regional Planning (URP) Act of 2015 made provision for a paradigm shift in the integration of spatial planning in Zambia by mandating the local authorities to manage urban planning through integrated development plans. Thus, Chileshe et al. (2023) argued that integrated development Planning (IDP) has both the broader municipal strategic planning and the spatial development planning, which is supposed to limit organic generation of settlements. However, contrary to the provisions of the URP Act, spatial planning is often guided by the structure plans, which are static maps with limited forecasting capabilities, leading to the development of towns without a spatial framework. The result is the manifestation of developments in wetlands and other environmentally protected areas (Banda, et al., 2021). In addressing uncoordinated spatial planning, the URP Act further provided for the Local Authorities to initiate preparation or review of IDPs within 6 months after general elections (GRZ 2018).

The process of developing the IDP promotes interlinked workflows, including planning and issues surveys, development framework, spatial development framework and implementation plan, followed by approval and implementation. LCC indicated that members of the public and technical stakeholders (District Development Coordinating Committee) are heavily consulted, as evidenced in the issues survey report. LCC further indicated that technical stakeholders came from across the whole service providers in the district, as expected by the IDP Guideline of 2018. This approach pushed spatial planning to be seamless, encompassing spatial elements in a given geographic area. It is expected that the aspect of the spatial framework gives a general segregation of land uses, mainly based on their natural characteristics, to best suit the proposed land use. In view of this, IDPs are expected to enforce mainstreaming of wetlands (disaster risk reduction) spatial development strategies. It was, therefore, appreciated that wetlands can be protected and sustainably managed once they are captured and incorporated into the spatial development framework of IDPs.

The previous section discussed that WARMA is primarily responsible for wetland management but they are not the Planning Authority. URP Act reveals that stakeholders like WARMA should be part of the planning process as meaning of providing input during the planning process, especially the Local Area Plan, which is a more specific spatial relation considering socio-economic, cultural and environmental benefit analysis. Review of IDPs Spatial Development Plans for Ndola, Kitwe and Livingstone districts pointed out that spatial aspects of IDPs are not translated into Local Area Plans; however, all three local authorities have been producing location plans without considering what was provided for in the IDP. Nonetheless, a sample of layout plans from all three local authorities suggests that buffer zones for wetland were considered in those plans, which shows recognition of wetland in land use planning at a lower level, though this is not sustainable enough because wetlands have been tampered with on the ground.

18.5.2 Integrating Wetlands into Urban Planning, Resilience of Cities and SDGS

As pointed out in various sections of this chapter, there is a strong link between integrating wetlands into urban planning, resilience of cities and the SDGs. Wetlands play a key role in the achievement of SDGs and other global sustainability commitments (Mitra 2024; Convention on Wetlands 2021; Pascual et al. 2017). For instance, wetlands provide water (the bloodstream) that sustains people and ecosystems. Wetlands are thus key in regulating water quantity and maintaining water quality, consequently supporting economic, social and environmental development, which enhances human well-being. Achieving international targets associated with biodiversity, climate change and sustainable development thus requires wise utilisation of wetlands. The role of wetlands in protecting risks associated with water is also helpful for society in focusing beyond the built infrastructure (Barman et al. 2021; UNECR n.d.; Convention on Wetlands 2021).

This chapter argues that integrating wetlands into urban planning forms the basis for the protection and wise utilisation and management of wetlands. Wise utilisation and management of wetlands have a greater effect on the resilience of cities and are consequently considered to be a key pathway in achieving most of the SDGs (SDGs Indicator 6.6.1), including benefits for climate change, biodiversity conservation and disaster risk reduction associated with global targets. This helps in tracking the changes in the extent of water-related ecosystems over time (Mitra 2024). Integrating wetlands into urban planning is thus linked to outcomes of sustainable development in various ways, as their effects on resilience of cities and delivery of SDGs go beyond protecting water ecosystems (SDGs Target 6.6) (Convention on Wetlands 2021). For instance, sustaining wetlands helps in coastal environment protection, thus supporting the delivery of life below water (SDG 14) and sustaining aquaculture productivity, thus supporting the delivery of zero hunger (SDG 2). This is evidenced through the literature showing that fish consumption accounted for 17% of animal protein intake globally in 2017, and two-thirds of the total global fish consumption depends on coastal wetlands (FAO 2020). Integrating wetlands into urban planning also forms the basis for improving the quality of the environment, thus benefiting human health (SDG 3) through recreation activities (Mitra 2024; Convention on Wetlands 2021; UNECR n.d.).

Whilst ecosystems such as wetlands are vulnerable to climate change effects, integrating them into urban planning provides opportunities for climate change mitigation and adaptation as well as resilience of cities (UNECR n.d.). For instance, since wetlands are connectors of climate, allocating land for development in ecologically sensitive areas such as wetlands results in the degradation of ecosystems and climate change, subsequently disaster risks, mainly associated with floods (Mwiinde and Munshifwa 2024; Xu et al. 2022; Munshifwa et al. 2021; Maes et al. 2019). Addressing this requires coordination and partnerships across actions in the reduction of climate change and disaster risk, as well as ensuring sustainable development. Wetlands are also key carbon stores, thus supporting climate action (SDG 13) and peacemaking (SDG 16) (Mitra 2024; Convention on Wetlands 2021). Integrating wetlands into urban planning would thus enhance the key role of wetlands in enhancing spiritual life, aesthetics and culture. For instance, the high altitude, Himalayan Wetland Mansarovar is visited by numerous pilgrims annually for spiritual atonement (Verschuuren 2016). Thus, recognising the rights of nature within urban planning and legal frameworks becomes a key dimension of justice in ensuring sustainable well-being for humans and the planet (Davies et al. 2020). When sustainably managed, wetlands assist in buffering risks associated with water and become a key pathway for ensuring built infrastructure sustainability (Chausson et al. 2020).

This chapter, however, argues that there is minimal mainstreaming of wetlands in Zambian cities. The development trajectory of most cities, as evidenced by the cities of Ndola, Livingstone and Kitwe, is unlikely to result in sustainable utilisation of wetlands and other natural resources towards the achievement of SDGs. Ensuring sustainable development requires considering sustainable integration of wetlands into urban planning and decision-making (Convention on Wetlands 2021) for enhancing the value of functional ecosystems and biodiversity for the resilience of cities (Chen et al. 2022). Integrating wetlands (as a key water management infrastructure element) into urban planning is thus capable of ensuring sustainable utilisation of wetlands and other natural resources, consequently, forming the basis for resilience of cities (Bharwani et al. 2023; Amaranath and Subrahmanian 2020). This, therefore, helps in achieving the SDGs, aimed at providing peace and prosperity for the people and the planet through enhanced green infrastructure (water-sensitive urban design approaches) and environmental sensitivity. Consequently, such a situation promotes sustainable development through ensuring numerous benefits for the following (WWT Consulting 2018): (i) wildlife: including creating biodiversity habitat and habitat connectivity; (ii) residents: including enjoying nature; sense of community; improved health and well-being; (iii) economy: including employment and economically sustainable green infrastructure; and (iv) developers: including improved economic, social-cultural and environmental performance as well as customer satisfaction. The chapter thus recommends the need for the Zambian Government, through the local authorities, to think about ways of actualising sustainable integration of wetlands into urban planning for the resilience of cities.

18.6 Limitations of the Chapter

This chapter was limited to the data collection process and focused largely on literature review and brief primary data on the three case studies. However, this chapter provided insights into the integration of wetlands into urban planning for the resilience of cities based on three major Zambian cities of Ndola, Livingstone and Kitwe as viewed in the light of SDGs. It thus contributes to knowledge by highlighting the practical situation and recommendations for sustainable integration of wetlands into urban planning for the resilience of cities and attainment of SDGs.

18.7 Conclusion

This chapter was aimed at exploring the sustainable utilization of wetlands as a means to relate to climate change and disaster risk reduction, integrated urban planning and resilience of cities as seen in the light of SDGs. The chapter focused on three major cities in Zambia (Ndola,

Livingstone and Kitwe). The study found a strong link between the use of wetlands and urbanisation, integrated urban planning and resilience of cities in relation to climate change and disaster risk on human settlements. It is thus argued that mainstreaming wetlands into comprehensive urban planning helps in achieving sustainable utilisation and management of wetlands, consequently adapting to climate change and achieving disaster risk reduction, hence resilience of cities. This improves economic, city attractiveness, convenience, social and environmental conditions, and living standards for urban dwellers; moreover, it will preserve hard-earned investment and save government money that would otherwise go towards disaster rescue operations. However, rapid urbanisation has resulted in unsustainable utilisation of wetlands, which generates various problems, including vulnerability to climate change-related risks such as flood events and associated environmental degradation. In most cases wetland is a common-pool resource that provides vulnerable communities with food such as fish, wild vegetables and other aquatic products. Once tempered with, as seen on all the case studies, it leaves such dependent communities more vulnerable.

The chapter also highlighted strategies for ensuring the sustainable integration of wetlands into urban planning for the resilience of cities. The general strategies include considering inclusivity and participatory approaches, incentive systems, comprehensive land use planning approaches, production of spatial development strategies and biodiversity hotspots. Urban planners should specifically consider embarking on integrated planning; clearly articulating urban wetlands; developing specific urban wetland management plans; mapping, monitoring and conducting baseline surveys; coordinating planning approaches; framing appropriate participatory methods; selecting certain urban wetlands; formulating comprehensive plans and formulating spatial development strategies targeting wetlands. Policymakers should consider developing holistic urban planning policies, addressing water management issues, formulating legislative and regulatory framework, providing encouraging regulatory charters, involving all stakeholders and incentivising economic activities. Developers should also consider promoting restoration and rehabilitation, various positive benefits and local communities' involvement in all wetland planning and management processes.

With regard to the Zambian context, it has been observed that Zambia, from the onset, has not been deliberate on comprehensive urban planning, and that some useful institutional and organizational arrangements came into existence in the last decade, rendering urban planning vulnerable to unrealistic utilisation of land and other natural resources such as wetlands. Urbanisation initiated lots of changes economically and upward shifts in population and demography, which resulted in the rapid growth of informal human settlements. The Town and Country Planning Act No. 283, being the first guiding principal law on urban development in Zambia, has influenced spatial planning in the country. However, despite the enactment of this law, much of the developments on the ground have remained uncoordinated, with the manifestation of unplanned settlements on environmentally sensitive areas. It appears there has been little enforcement from the mandated organisation to ensure the implementation of spatial development plans meant to sustainably utilise such wetlands. It takes a little bit longer for settlements to grow in wetlands, meaning mandated institutions have been very passive in sustainable utilization of such wetlands in the city. These have no regard for the natural usefulness of these wetlands.

The evolution of planning ushered in the Urban and Regional Planning Act No. 3 of 2015, which provided for every Local Authority to operate on the basis of an Integrated Development Plan. This was aimed at promoting spatial relations for the purpose of attaining sustainable use of land and natural resources. It was, thus, anticipated that municipalities would increasingly manage land within their jurisdictions in a sustainable manner, since enforcement of the spatial development framework is enshrined in them.

Acknowledgements The authors wish to thank the town planners and environmental specialists in the cities of focus for their helpful comments during interviews.

References

Akinluyi ML, Adedokun A (2014) Urbanization, environment and homelessness in the developing world: the sustainable housing development. Mediterr J Soc Sci 5(2):261–271

Amaranath T, Subrahmanian RR (2020) Water urbanism-integrating urban wetlands into urban landscape for sustainable development. PalArch's J Archaeol Egypt/Egyptol 17(9):7421–7436

Amare S (2015) Land use/cover change at Infraz watershed, Northwestern Ethiopia, North Western Ethiopia. J Landsc Ecol 8(1):69–83

Barman A, Jha R, Raja F (2021) Impacts of urbanization and encroachment in urban wetlands, a case study of ganga riparian wetlands in Patna, s.l.: s.n

Begg GW (1987) The wetlands of Natal (part I) an overview of their extent, role and present status, Natal town and regional planning report 70. Pietermaritzburg

Bharwani S et al (2023) Supporting climate-resilient urban planning: 10 lessons from cities in southern Africa. Bettina Koelle, s.l.

Bodo T (2019) Rapid urbanisation: theories, causes, consequences and coping strategies. Geogr Res 2(3): 32–35

Casey L, Stazen L (2021) Seeing homelessness through the sustainable development goals 15. Eur J Homelessness 15(3):147–155

Chabwela H, Chomba C, Thole L (2017) The habitat structure of Lukanga Ramsar site in Central Zambia: an understanding of wetland ecological condition. Open J Ecol 7:406–433

Chausson A et al (2020) Mapping the effectiveness of nature-based solutions for climate change adaptation. Glob Chang Biol 26:6134–6155

Chen M, Chen L, Cheng J, Yu J (2022) Identifying interlinkages between urbanization and sustainable development goals. Geogr Sustain 3(4):339–346

Chileshe RA et al. (2023) Rethinking urban sprawl: moving towards sustainable urban planning practice in Zambia. s.l.:s.n

Chiluba L, Kabwe MS, Chibesa F (2021) Causes of homelessness in Zambia. s.l.:s.n

Convention on Wetlands (2021) Global wetland outlook: special edition 2021. Secretariat of the Convention on Wetlands, Gland

CSO (2022). www.zamstats.gov.zm/census-and-statistics. Central Statistics Report., s.l.: CSO

Darghouth S, Ward C, Gambarelli G, Styger E (2008) Watershed management approaches, policies, and operations: lessons for scaling up. s.n, Washington, DC

Davies GT et al (2020) Towards a universal declaration of the rights of wetlands. Mar Freshw Res 72(5):593–600

FAO (2020) The state of world fisheries and aquaculture 2020. Rome ed. s.l.: FAO

GRZ (1965) The government of Zambia (1965) town and country planning. Government Printers, Lusaka

GRZ (2011) Environmental management act no. 12 of 2011. Government Printers, Lusaka

GRZ (2018) Integrated development plan guideline. Government Printers, Lusaka

GRZ-MLNR, 18. National policy on wetlands. Lusaka: Government Printer

Jayarajan CK (1979) Some aspects of urban planning in Zambia. Department of Town and Country Planning, Research Unit, Lusaka

Kachali RN (2007) Stakeholder interactions in wetlands: implication for social ecological system sustainability. Lund University, Lund

Kamweneshe B, Richard BJ (2017) Hydrology and water resources of tropical Africa. In: Kamweneshe B, Beilfuss R (eds) Zambia Crane and Wetland Conservation Project, Working paper 3. Elsieview Scientific Publishing Company, New York, p 208

Kometa SS, Kimengsi NJ, Petiangma DM (2018) Urban development and its implications on wetland ecosystem Services in Ndop, Cameroon. Environ Manag Sustain Dev 7(1):21–36

Maes MJA, Jones KE, Toledano MB, Milligan B (2019) Mapping synergies and trade-offs between urban ecosystems and the sustainable development goals. Environ Sci Pol 93:181–188

Makasa PLK (2019) Housing economics and policies in Zambia. A&S Printers, Port Elizabeth

Mitra S (2024) Governing Urban Wetlands in India: A Pathway to Sustainable Urbanisation. ORF Issue Brief No. 690, February 2024. Observer Research Foundation, s.l.

Munshifwa EF, Mwenya CM, Mushinge A (2021) Urban development, land use changes and environmental impacts in Zambia's major cities: a case study of Ndola. In: Sustainable real estate in the developing world. Emerald Publishing Limited, s.l

Mwiinde C, Munshifwa EK (2024) Urbanization, poverty, and homelessness in Zambian cities: a threat to the achievement of sustainable development goals. In: Homelessness to hope: research, policy and global perspectives. Elsevier, s.l., pp 37–51

Pascual U et al (2017) Valuing nature's contributions to people: the IPBES approach. Curr Opin Environ Sustain 26:7–16

Ramsar Convention (2018) Wetlands: essential for a sustainable urban future, Fact sheet no. 10. Secretariat of the Convention on Wetlands, Gland, Switzerland

Ro C et al (2020) The economic value of peri-urban wetland ecosystem services in Phnom Penh. In: The international conference of science and applied geography. IOP conference series: earth and environmental science. IOP Publishing, Phnom Penh

Rogerson RJ, Horgan D, Roberts JJ (2021) Integrating artificial urban wetlands into communities: a pathway to carbon zero? Front Built Environ 7:777383

Tait J (1997) From self-help housing to sustainable settlement: capitalist development and urban planning in Lusaka, Zambia. Aldershot, Avebury

Tang Z, LaGrange T, Zhang L, Cyboron J (2020) Integrating wetland conservation into local planning, University of Nebraska-Lincoln, Lincoln, Nebraska, – s.n

Tonne C et al (2021) Defining pathways to healthy sustainable urban development. Environ Int 146(106236)

UNECR (n.d.) United Nations Environment Agency Resolution 73/284: United Nations Decade on Ecosystem Restoration (2021–2030). UNECR, s.l

UN-Habitat (2023) SDG 11 synthesis report 2023: tracking progress towards inclusive, safe, resilient and sustainable cities and human settlements. s.n, Nairobi

Verschuuren B (2016) Religious and spiritual aspects of wetland management. In: The wetland book. Cham, Springer, pp 1405–1415

World Bank (2022) Zambia urbanization review: leveraging cities and towns for Zambia's future—policy note. World Bank, s.l.

WWT Consulting (2018) Good practices handbook for integrating urban development and wetland. John Holmes/FLP, Slimbridge

Xu Z et al (2022) Responses of spatial relationships between ecosystem services and the sustainable development goals to urbanization. Sci Total Environ 850(157868)

An Integrated Approach to the Planning and Management of a Peri-urban Wetland: A Case Study of Chatra Beel, English Bazar, Malda District, West Bengal

19

Sanjoy Dutta and Debjani Roy

Abstract

Wetlands are considered one of the world's most productive biological ecosystems, but anthropogenic abuse has reduced their number by three times faster than forests (Global Wetland Outlook 2018). The chapter explores the critical role of wetlands in enhancing urban resilience amidst growing environmental and climatic challenges. Blue-Green Infrastructure (BGI) is the new approach in urban planning where natural resources like trees, parks, lawns, rivers, and wetlands are prioritized. By integrating wetlands into urban planning and infrastructure, urban towns and cities can strengthen their capacity to adopt remedial measures and recover from environmental crises. The study highlights a case study taking Chatra Beel (a peri-urban wetland located in the periphery of the English Bazar town of Malda district) and policy recommendations to promote sustainable urban development through wetland management and restoration. The spatial peripheral changes of this wetland have been studied through an overlay analysis prepared from imageries that have been downloaded from USGS (of the four mentioned years) and digitized using ArcGIS software. The result shows that 51.84% of the area of the Chatra Beel has been encroached on by the local people for settlement by infilling, and the remaining part is polluted by the disposal of solid-liquid waste from different point and non-point sources of pollution (NPSP).

A comprehensive conservation and management plan is urgently needed to save wetlands from impending disaster. This chapter prepares a plan layout map of Chatra Beel using thematic mapping with ArcGIS software. The proposed plan suggests that 26% of the area can be used for plantation and agriculture along the banks of the wetland. In contrast, the remaining parts can be used for fishing, aquaculture, and aqua park activities.

Keywords

Anthropogenic abuse · BGI · Environmental degradation · NPSP · Productive biological ecosystem · Urban encroachment

19.1 Introduction

India is at the tipping point of transition from a rural to an urban society, so it is crucial to maintain the balance between nature and ever-increasing demand. As urbanization comes with

S. Dutta (✉)
Department of Geography, Pakuahat Degree College, Malda, West Bengal, India

D. Roy
PG Department of Geography, Nirmala College, Ranchi, Jharkhand, India

S. K. Chakraborty et al. (eds.), *Wetlands and Urban Resilience*, Sustainable Development Goals Series,
https://doi.org/10.1007/978-3-032-06881-1_19

many opportunities, it also throws up tough challenges for the urban resilience of a city and its citizens in the form of environmental degradation, extreme weather conditions, water pollution, drinking water crises, groundwater depletion, etc. To survive in this adverse situation, several global and Indian cities have incorporated nature-driven solutions like Blue-Green Infrastructure (BGI) in their urban planning and design. While there is no established definition of BGI, a literature review of most terminologies and outputs can be categorically adjusted under nature-based solution (Udas-Manikikar and Driver 2021), focusing on three pillars—climate change, health, and urban resilience. BGI assimilates blue infrastructure (water bodies like rivers, lakes, wetlands, canals, ponds, and floodplains) and green infrastructure (trees, parks, lawns, fields, and forests) in urban planning for sustainable livelihood. By global standards, the "Rain City Strategy" of Vancouver, Canada (Humes et al. 2024), the "Grey to Green Initiative" of Portland, USA (Bureau of Environmental Services, The City of Portland Oregon 2005), the "Sponge City Programme" of Wuhan, China (Jing 2019), and the "Active, Beautiful and Clean(ABC) waters programme" (Singapore's National Water Agency) of Singapore are vivid examples of BGI. The Active, Beautiful and Clean Waters (ABC Waters) Programme was launched in 2006 by PUB -Singapore's National Water Agency(Goh *et al.* 2017), with two objectives: improving water quality and enhancing liveability. It aims to transform Singapore's canals, rivers, and reservoirs into beautiful recreational spaces where people can be close to water and naturally become stewards of the waterways and water bodies (Centre for Liveable Cities, Singapore 2017). In India, the Blue-Green Master Plan was initiated in 2014 by cities like Bengaluru, Madurai (Kumar and Vasimalai 2016), Bhopal (National Institute of Urban Affairs 2016), and in 2021 by Delhi (Delhi Development Authority, Planning, India 2021). In the 2014 report "Future-proofing Indian Cities: Bengaluru Action Plan for Water and Sanitation Infrastructure" (Wankhade, Kavita, and IIHS 2014), top priority was given to water bodies for flood prevention, water purification, and providing a buffer zone for water recharge. In the early days, civilization grew around the water bodies, and now, with the rapid rate of urbanization, human beings are trying to grab the water bodies. The human disturbance score protocol technique, which takes into account pollution levels (chemical, physical, and biological contamination), hydrological alterations (water abstraction, drainage, and dams), and changes in land use (urbanization, agriculture, and industry), is used to determine the extent of human disturbance to the wetlands. These factors collectively assess the extent of anthropogenic pressure on ecosystems (Krishnaraj and Mathesh 2022). The degradation and loss of wetlands have been related to several variables, including overfishing and overexploitation, global climate change, infrastructure expansion, land conversion, eutrophication and pollution, infrastructure withdrawal, and population growth (Chen et al. 2019). These degraded water bodies hurt human beings and their livelihood. About 54–57% of the world's wetlands have been lost since 1900, and this has commonly been attributed to the rapid increase in agricultural drainage, irrigation schemes, dam construction, and conversion for residential or industrial use that go hand in hand with economic growth-centered interpretations of development (Davidson 2014). Even in some incidents, the state government encroached on the wetland area for tourism or other purposes (De Roy 1992). Under these circumstances, integrating wetland management plans into urban resilience strategies is essential for sustainable urban development. A comprehensive management plan involves assessing the existing wetland ecosystem, identifying areas for restoration or conservation, and aligning this effort with broader urban planning initiatives. The United States first researched rebuilding and restoring damaged wetlands (Zhou et al. 2020).

English Bazar Municipality is struggling with a growing population with infiltration from neighboring Bangladesh. Like other cities, English Bazar also harnesses the Blue and Green Infrastructure to keep balance with development. As a side effect of urbanization, urban wetlands and *peri*-urban wetlands are considered the most vulnerable environmental units (Ramachandra et al. 2012). This vulnerability is due to the gradual conversion of wetlands into urban areas (Mondal et al. 2017; Pal et al. 2017), a terrible example of this is Chatra Beel (colloquially, beel is termed for wetlands), a peri-urban wetland located in the area adjacent to the English Bazar municipality. The Chatra Beel has lost its areal circumference since 2000, about 52.84% of its total area, where 14 new settlement areas have developed. The rest is used for garbage disposal, where approximately 20% of the English Bazar municipality's solid and liquid waste is disposed of daily. As a result, Chatra Beel has turned into a "Contaminated, smelly, closed polluted reservoir." In this situation, Chatra Beel could no longer hold the floodwater, resulting in inundation of the area every rainy season. The problem arising from this situation can be addressed in two ways, i.e., firstly, the Chatra Beel wetland is experiencing significant ecological degradation due to expanding urban activities. Encroachments, pollution (mostly garbage and sewage), and land use changes disrupt natural hydrology and ecosystem balance. Secondly, due to the deterioration of the wetland, critical services like flood regulation (district Malda is a flood-prone area), groundwater recharge, and temperature moderation are being diminished. As a result, urban resilience capacity of Chatra Beel is being compromised. Qualitative and quantitative standards of Chatra Beel have changed drastically from its genesis, which is a serious threat to the environment and ecology of the area.

So, restoration of the Chatra Beel and other district wetlands within and around the urban areas not only revitalizes ecological functions but will also provide natural infrastructure, enhancing the capacity of the region to cope with environmental challenges. Wetlands conservation does not mean setting wetlands aside as protected areas but positively using them as an essential resource (Jerath 1992). To ensure sustainability and decision-making, fruitful planning should incorporate multi-stakeholder engagement, including local communities, urban planners, environmental scientists, and policymakers. Wetland management is an ongoing process or cycle where plans or objectives are regularly updated or adapted based on knowledge or learning about the site. It is considered necessary that a process of continuous monitoring and review is implemented to understand whether wetland management is achieving its objectives or not. When specific management actions do not deliver the desired outcomes, these can be integrated, improved, and adapted within subsequent management planning cycles (Ramsar Convention Secretariat 2010).

Key components may include zoning regulations to protect the beel areas, creating green buffer zones, pollution control measures to stop garbage disposal, and using Nature-based Solutions (NbS) to mitigate flood-prone areas. Wetland conservation is urgently needed by embedding wetland conservation and restoration into the urban policy framework. Hopefully, the Chatra Beel area can improve water quality, economic activities, and recreational and aesthetic benefits to residents. This planning will not only safeguard ecological integrity but also strengthen urban infrastructure to support future environmental challenges.

While the study of *Chatra Beel* highlights the critical role of wetlands in enhancing urban resilience—particularly in terms of flood mitigation, ecological balance, and supporting local livelihoods—there remains a **limited integration of socio-economic dimensions and policy frameworks** in the analysis. The study focuses primarily on ecological functions and urban planning perspectives, but does not sufficiently address (1) how **local communities** perceive and interact with the wetland, (2) the **institutional challenges** in managing the wetland within urban governance frameworks, (3) the **long-term socio-economic trade-offs** involved in wetland conservation versus urban development, (4) and a **lack of real-time data or predictive models** to assess the resilience contribution of Chatra Beel under future climate change and urban expansion scenarios. This gap points to a need for **interdisciplinary research** that integrates ecological, social, technological, and governance perspectives to fully evaluate and enhance the role of wetlands like Chatra Beel in urban resilience planning.

In this study, the key objective is to assess the spatio-temporal changes faced by Chatra Beel in the last 20 years and to find out the role of anthropogenic activities behind it. Based on this finding, a proper plan layout will be framed, balancing the people's aspirations and ecological integrity to ensure the sustainable and wise use of Chatra Beel in the context of urban planning for the English Bazar Municipality. The proposed plan will combine water bodies with farming, forestation, and aqua-park, which generate employment for local people and promote community engagement and governance model by exploring the participatory approaches to wetland management that involve local communities, NGOs, and government bodies. The study's originality lies in highlighting the importance of sustainable management practices of wetlands to protect the environment and socio-economic integrity by restoring urban resilience. The study also provides a spatiotemporal observation for a long period to frame a scientific plan to integrate the ecological perspective with the rapid rate of urbanization.

The significance of this research lies in its potential to illuminate how unchecked urbanization (the rate of urbanization is 13.58% of Malda district in 2011) threatens wetland systems, thereby threatening not just the biodiversity but also the urban communities that depend on this particular wetland (Chatra Beel) for water security, flood control, and controlling local climate.

Study Chatra Beel provides critical insights into the local consequences of global environmental trends. Focusing on Chatra Beel as a specific case study, the chapter underscores the urgent need for policy reforms, inclusive planning, and community-based conservation efforts. The chapter aims to contribute to broader discourses on sustainable urban development and wetland conservation, offering a practical pathway for harmonizing growth with ecological balance.

19.2 Study Area

Malda district of West Bengal covers an area of 3113 Sq.km is situated between Lat. of 24°40′20″ N and 25°33′08″ N and Long 87°45′50″ E and 88°28′10″ E and demarcated by the state boundary of Jharkhand in the west and the international boundary of Bangladesh in the east. The River Ganges forms the southwestern boundary of the district. The district extends to North Dinajpur in the north and Murshidabad in the south, which is marked by the east-west flowing River Ganga. Chatra Beel is located in the peripheral part of the south-central English Bazar Municipality area, with an elevated portion around the beel of about 25–30 meters in height. The average depth of Chatra Beel is 3–3.5 meters (Fig. 19.1 and Table 19.1).

After the construction of NH-34, the population of Mouzas (a type of administrative unit comprising one or more settlements) around Chatra Beel has rapidly increased due to good connectivity and the availability of basic and essential services. The land use patterns of the Chatra Beel area's eastern and western portions are entirely different. The eastern part is the outskirts of the English Bazar Municipality; human interference and its effects are more prominent here. In Chatra Beel, it is observed that in the last 30 years, approx. 360% (from 9.1 to 42.30 hectares) increase in the built-up area in the eastern portion and a proportional decrease in the water body. New colonies grew at a rapid rate, and population pressure increased equally (Table 19.2).

19.3 Materials and Methods

The methodology provides technical guidance for the practical design of the research work. An extensive literature review has been done after defining the problem related to the anthropogenic abuse of wetland property. After that, secondary data were collected from secondary sources, and

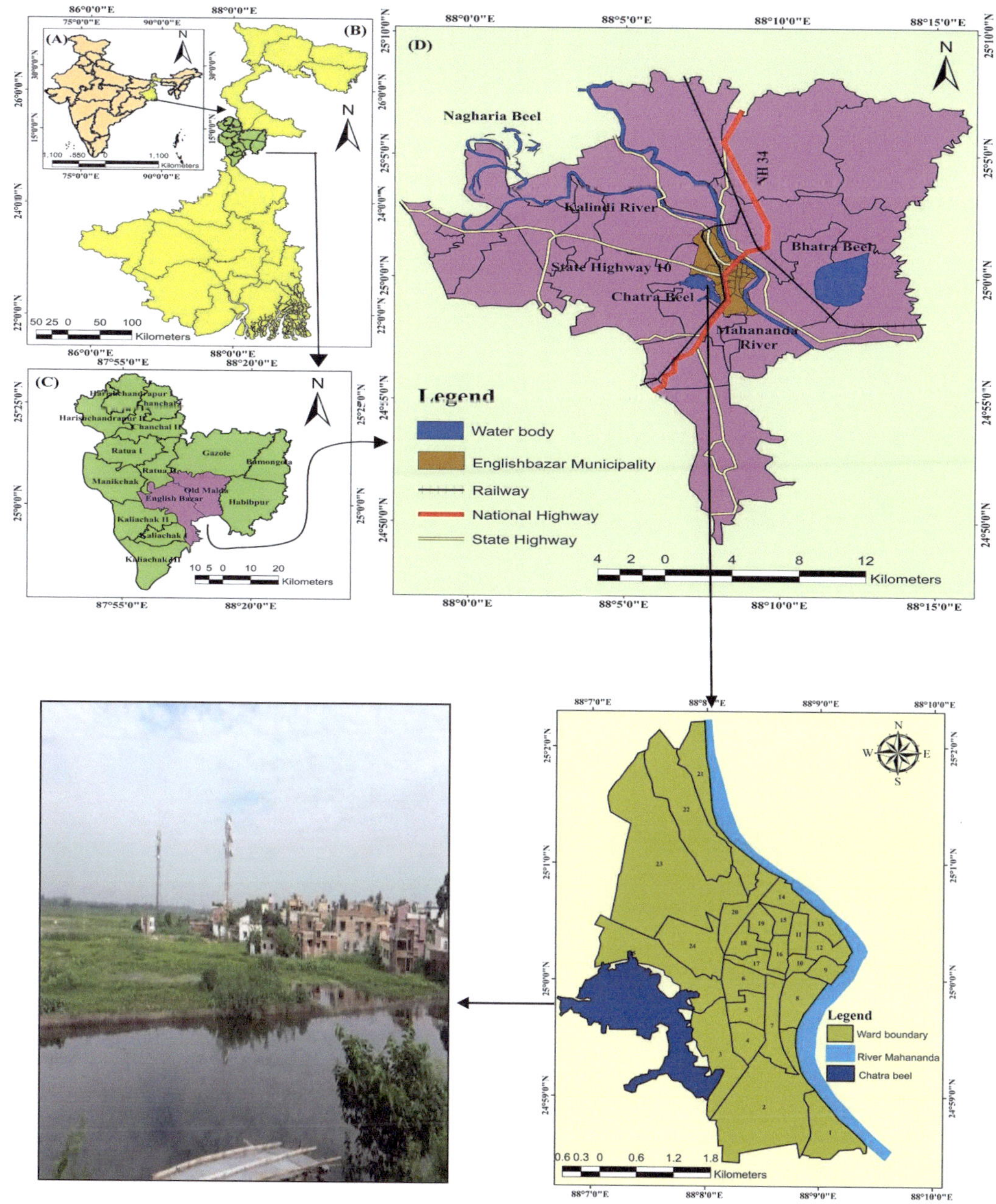

Fig. 19.1 Location map of the study area. (Source: National Atlas and Thematic Mapping Organization)

Table 19.1 Administrative location of Chatra Beel

Name of the wetland	Administrative location	Lat. and long.	Mouza no. and name
Chatra Beel	English Bazar block	24°59′02″N to 25°00′5″N & 88°6′30″E to 88°8′08″E	69-Piroj Pur, 70-Arajidilal pur 71-Uttar Ramchandrapur 88-Uttar Jadupur

Source: District Census Handbook, Malda

Table 19.2 Population growth of the Mouzas around Chatra Beel

			1981			1991			2001			2011		
JL. no	Name of Mouza	Area insq. km	Total population	Density (sq./km)	Growth rate	Total population	Density (sq./km)	Growth rate	Total population	Density (sq./km)	Growth rate	Total population	Density (sq./km)	Growth rate
70	Arazi Dilalpur	0.6	423	705	32.3	615	1025	45	810	1350	32	1061	1768	31
71	Uttar Ramchandrapur	2.37	623	263	39.7	989	417	59	1401	595	42	1934	816	38
88	Uttar Jadupur	5.26	1253	238	46.1	1960	373	52	4221	803	121	5911	1124	40

Source: District Census Hand Book, Malda, Series-20, Part XII-B (2011)

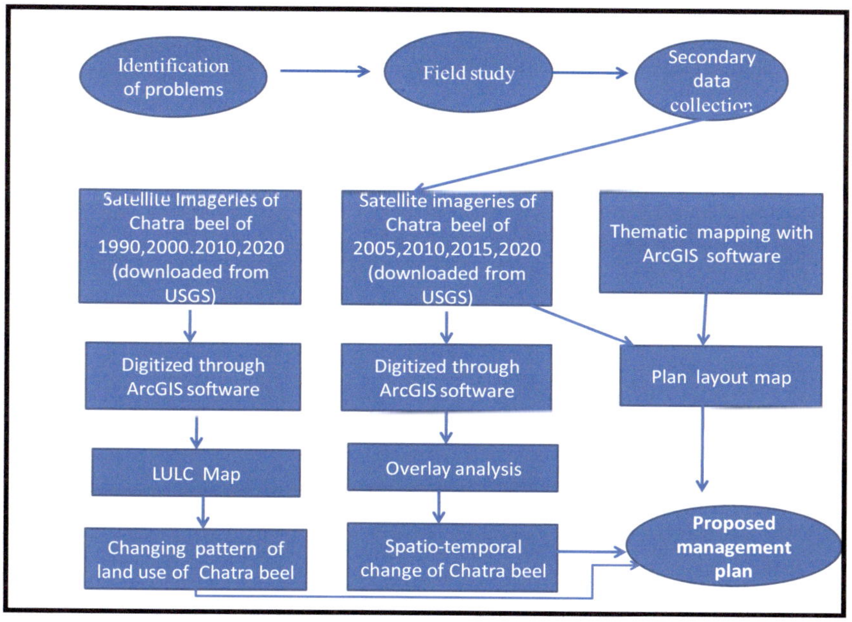

Fig. 19.2 Methodological flowchart. (Source: Compiled by the authors)

an intensive field study was done to observe the current status. This field study around the mouzas of Chatra Beel revealed the inter-relationship between people and wetlands. Next, the data about encroachment were collected and analyzed with the help of GIS software, which maintained the following steps (Fig. 19.2). Finally, a proper management plan was proposed with the help of thematic mapping.

Table 19.3 Areal changes of Chatra Beel (2005–2020)

Chatra Beel		
Year	Area (in hectares)	Areal change (%)
2005	407	–
2010	256	−37.10
2015	208	−18.75
2020	196	−5.76

Source: Authors (compiled from Google Map)

19.4 Results and Discussion

With several phases of transformation, Chatra Beel has gradually changed in quantitative and qualitative dimensions. The changes are mainly observed on two grounds: areal changes, land use and land cover (LULC) changes, and related changes in human activities.

From Table 19.3, it is clear that Chatra Beel has been reduced sequentially. Approximately 51.84% of the area has decreased in only 15 years (from 2005 to 2020). The reduction rate was highest, approximately 37.10%, between 2005 and 2010. After 2015, these trends were controlled slightly, decreasing the percentage of area loss to 5.76%.

The map of Fig. 19.3 shows the spatio-temporal changes faced by Chatra Beel between 2005 and 2020. The "overlay analysis" map was prepared from imageries of 4 years (2005, 2010, 2015, and 2020) downloaded from USGS and digitized using ArcGIS software. The analysis revealed that between 2005 and 2020, the Chatra Beel area went through a considerable change, which negatively impacted both the physical-social environment and ecology of the English Bazar municipal area. The reduction rate is higher due to the rapid extension of settlement (since 1972) in the southwestern part of the Arazi Dilalpur and Uttar Jadupur area, i.e., about 59%

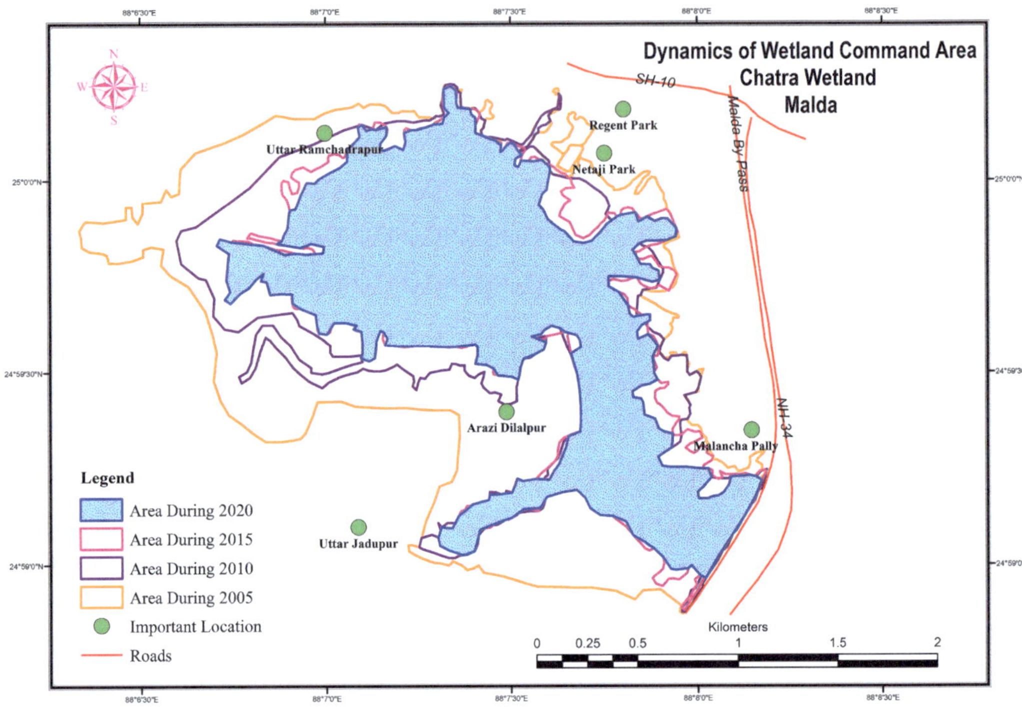

Fig. 19.3 Overlay analysis map of Chatra Beel (2005–2020). (Source: Authors (computed through overlay analysis from Google Map))

disrupting the local ecology, including loss of habitat for native species and pisciculture. Here, population density in the English Bazar block has increased from 899/per sq.km in 2005 to 1100/per sq. Km in 2020, and the wetland area has conveniently been converted into land for settlement and agricultural land.

The anthropogenic encroachment around the beel area has led to the mushrooming of smaller settlement colonies. Any encroachment in the beel area acts as an impervious shield adjacent to wetlands, thereby increasing the rate and volume of runoff, loading sediment and other pollutants into the water body. The Malanchapally colony has gradually extended toward the beel side, reducing 12% of the beel area. There is less encroachment toward the north. Uttar Ramchandrapur is gradually developing, and the town extends toward the north.

19.4.1 Changing Land Use Pattern of Chatra Beel

Accurate spatial information and temporal imagery proved to be effective in analyzing wetland characteristics and dynamics in space and time (Toyra et al. 2005; Xie et al. 2010), making satellite imagery one of the most useful sources of information in wetland monitoring. Land use maps derived from remote sensing imagery thus play a vital role in monitoring human-environmental interactions such as landscape changes, ecological services (conservation), urban planning, and management. In this study, between 1990 and 2020, four maps have been prepared over a gap of 10 years to assess the impact of urbanization on Chatra Beel (Fig. 19.4 and Table 19.4).

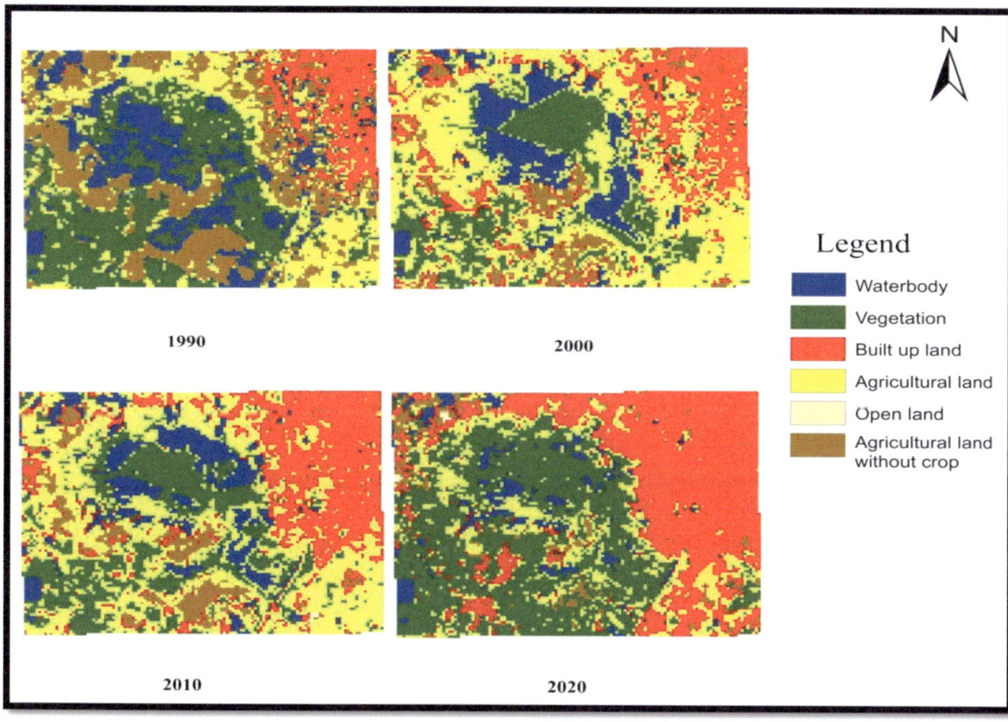

Fig. 19.4 LULC maps of the Chatra Beel area (from1990 to 2020). (Source: USGS)

Table 19.4 Land use categories of the Chatra Beel area (from 1990 to 2020)

Sr No.	Categories of land use (Chatra Beel)	1990		2000		2010		2020	
		%	Change (%)	%	Change (%)	%	Change (%)	%	Change (%)
1	Land under agriculture	20.9	–	36.2	73.0	28.1	−22.37	12.5	−55.51
2	Land under vegetation	8.9	–	13.1	47.19	16.7	27.48	33.1	98.2
3	Land under Agriculture without crop	8.6	–	2.1	−75.58	2.1	0	1.9	9.5
4	Land under water bodies	51.6	–	30.2	−41.47	20.1	−33.44	10.2	−49.25
5	Land under built up areas	9.1	–	17.7	94.5	32.1	81.35	42.3	31.77
6	Land under open land	0.9	–	0.7	−0.22	0.9	0.22	0.9	0

Source: District Land and Land Reforms Office, Malda

The categories under land use in Chatra Beel area are divided into six types (agriculture, vegetation, agriculture without crops, water bodies, built-up areas, and open land). The remarkable changes are noticed under four categories. There has been a significant reduction in the land under water bodies and agriculture (approximately 49% and 55%) and has increased remarkably in the land under settlement and vegetation, mainly water hyacinth (approximately 49% and 55%). One of the primary reasons for such an exponential increase in the built-up area is land acquisition at a much lower price than the prevailing market price in the adjoining areas. The impact of urbanization is responsible for this spatiotemporal characterization of this growth in the built-up areas and wetland encroachment. In this changing pattern of land use map, the percentage of open land has decreased from 0.9% in 1990 to 0.22% in 2020, signifying human encroachment.

19.4.2 Present Position of Chatra Beel

After the construction of NH-34, the population of Mouzas around Chatra Beel has rapidly increased due to the good connectivity and availability of basic and essential services. The LULC of the eastern and western portions of the Chatra Beel area is entirely different. The eastern part is the outskirts of the English Bazar Municipality, and human interference and its effects are more prominent here. New colonies are growing up in the eastern portion at a rapid rate, and population pressure is increasing equally (Fig. 19.5 and Table 19.5).

Between 1980 and 2015, several new settlements were created around Chatra Beel. Along NH-34, the first settlement was formed around 1980, namely Malanchapally in ward no. 3. In early 2000, only three localities, viz. Malanchapally, Krisnapally, and Netaji Park, existed in the eastern part of the Chatra Beel. After 2000, a rapid rate of encroachment began on both sides of the beel. Lake Garden, Regent Park, Bapuji Colony, Sonatala, Kadirpara, Muchipara, and Pakupara were created in the eastern part. After 2005, the western part of the Chatra Beel, mainly Uttar Jadupur and Arazi Dilalpur mouza, was encroached, covering approximately 4.2 sq.km. Between 1991 and 2011, the population of Uttar Jadupur and Arazi Dilalpur increased by 371.74% and 150%, respectively (Table 19.2). To accommodate this population pressure, the beel area was converted into built-up and agricultural land. The area under vegetative cover, mainly invasive species like water hyacinth, has increased from 8.9% in 1990 to 33.1% in 2020.

Uttar Ramchandrapur is the last to come into existence along the Ghorapir-Kazigram road in the northern portion of Chatra Beel. People have built houses in the beel's water and bridges with bamboo to stay connected with the land. New constructions are going on indiscriminately without respecting the sanctity of the beel area and

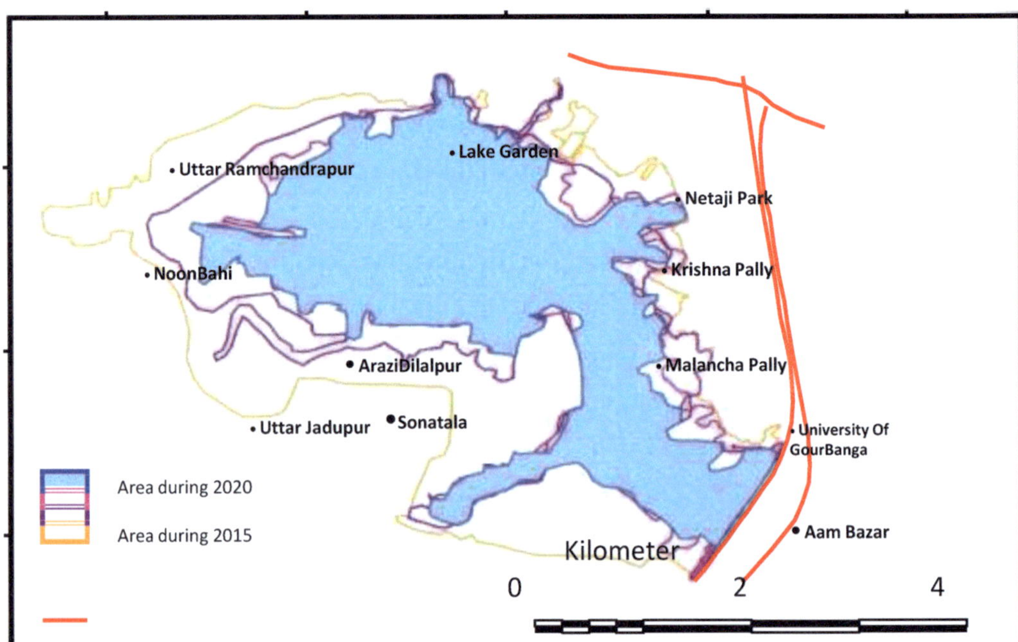

Fig. 19.5 Origin of new colonies around Chatra Beel. (Source: Authors (prepared from the base map obtained from the English Bazar Municipality))

Table 19.5 New colonies around Chatra Beel

Sl.no.	Year of establishment	Locality	Location
1	1980	Malanchapally	Southeast
2	1990	Krisnapally	East
3	1996	Netaji Park	East
4	2003	Lake garden	Northeast
5	2010	Uttar Ramchandrapur	North
6	2010	Noonbahi	Northwest
7	2008	Sonatala	West
8	2008	Kadirpur	West
9	2008	Muchipara	West
10	2008	Paku para	West
13	2010	Arazi Dilalpur	West
14	2008	Uttar Jadupur	Southwest

Source: Field survey

environment. Even nursing homes and water purifying factories are built without considering the effect of effluent and nondegradable products being dumped into the water body. The most terrifying form of human interference with nature can be observed here. Malanchapally and Krisnapally are the oldest settlements that were developed here. Both developed between 1985 and 1990, and the water body shrank in most areas. In the northern part of the Malanchapally area, the culverts, which are the outlet of the beel, are filled with city garbage and sewerage, obstructing the natural flow of the beel into the river, and have turned into a closed, polluted reservoir. The lower part of the Krisnapally area remains permanently waterlogged, and people engage in fishing. *Sol, Telapia, Ruhi, Katla, and Koi* are some fish varieties cultivated here. Netaji Park and Lake Garden are relatively new areas that were developed between 2003 and 2005. The people of these areas are mainly engaged in the tertiary sector, and the affluent are ready to pay any price for the land, which has increased the land prices.

19.4.3 Previously Conceived Master Plan (Laxmipur Drainage Plan) for Chatra Beel

A survey was conducted under Laxmi Roy, formerly Sub Assistant Engineer, 1977. After the initial quantification of the city area, the design evolved as the "Laxmipur Drainage Scheme." According to Nabin Das (former surveyor of this scheme), the accumulated water of the city under this scheme would come out through the Godrail bridge located in the southern part of Chatra Beel, and the water would pass through today's mango market, agricultural farm, and finally toward the River Mahananda. The sluice gate near Mahananda would be the central passage to drain the city's accumulated rainwater. The state government approved the scheme of two crores. Despite an allotment of a grant, this project could not be materialized.

To highlight the reasons for abandoning the proposal, a dispute arose between the irrigation department and the English Bazar Municipality over the Laxmipur Drainage Scheme; the two departments wanted to take ownership of the work. The entire project faced a serious problem when the English Bazar Municipality questioned "why has the irrigation department been given the responsibility of constructing city's drainage?" After this tug of war, the total work was divided into two entities: a) the Laxmipur Drainage Scheme and b) Laxmipur Drainage Prakalpa. It has been decided that under the Laxmipur Drainage Scheme, the irrigation department would take care of the work from Godrail Bridge to Mahananda to discharge the city water. On the other hand, the city's sewage system would be modernized by the English Bazar Municipality, named Laxmipur Drainage Prakalpa. The second obstacle came from the central government, which notified that the city's wastewater could not be dumped into the river. As a result, high drains were constructed in different parts of the city, but the planned drainage scheme from the Godrail Bridge was dumped. The height of the Godrail Bridge continued to increase due to the construction of a second railway line, and the scheme was abandoned.

Next, in 2003, the former MLA of the English Bazar assembly submitted a new plan to the West Bengal Government. The plan was to build a township centered around the Chatra Beel, and the city's accumulated water would be mixed after treatment and, later, flow out toward the Bhatia Beel. This proposed project showed that the expenditure incurred would be recovered from the township development around Chatra Beel. But this project, too, could not materialize. Even today, the sale of this land is still ongoing. Construction of indiscriminate housing projects is continuously growing around the wetland.

19.4.4 Conservation and Management Plan for Chatra Beel

Many research articles, Ph.D. theses, and government reports have been published on the conservation of the Chatra Beel, yet the Chatra Beel is on the verge of extinction. It is high time to make a concerted effort and take action to conserve and restore Chatra Beel. Chatra Beel extends from north to south. In the middle between Krishnapally and Araji Dilalpur, the width is a minimum of around 2 km, and in the extreme south, only a drainage outlet system is located, which connects Chatra Beel with the River Mahananda. Researchers have prepared and proposed a restoration plan for the local Public Health & Engineering Department, keeping the existing layout system (Figs. 19.6 and 19.7; Table 19.6).

19.4.4.1 Preliminary Phase

At the initial stage, the area of the Chatra Beel should be demarcated by well-established boundaries or fencing. Stringent measures should be taken to stop the conversion and encroachment in any form within the demarcated area of the Chatra wetland. The English Bazar Municipality should refrain from sanctioning building plans and approval letters for any construction within the periphery of the Chatra Beel area.

Desilting, digging, or excavating the lake bed will increase Chatra Beel's storage capacity. Tree cover helps prevent siltation, so that afforestation will be initiated along the boundary beel area.

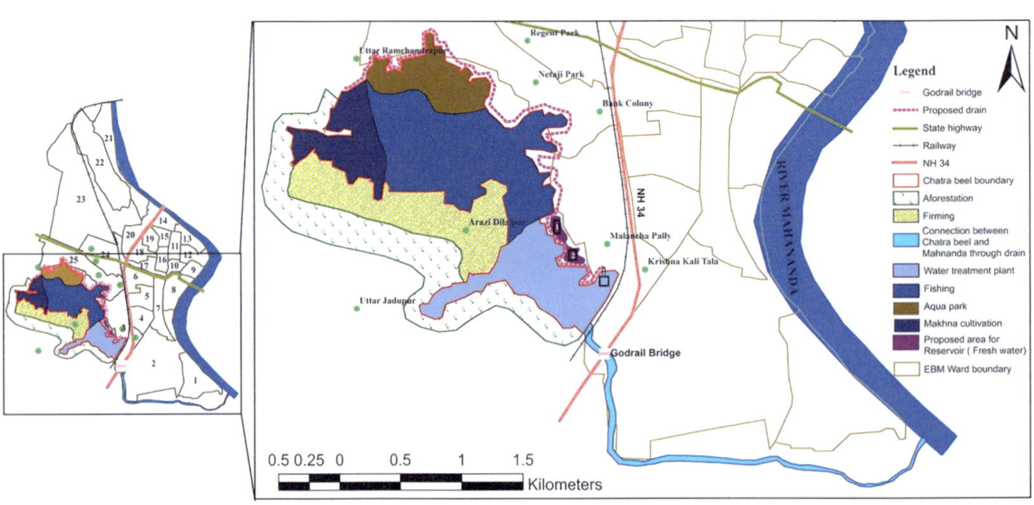

Fig. 19.6 Proposed plan lay out map of Chatra Beel. (Source: Authors (prepared from the base map obtained from the English Bazar Municipality))

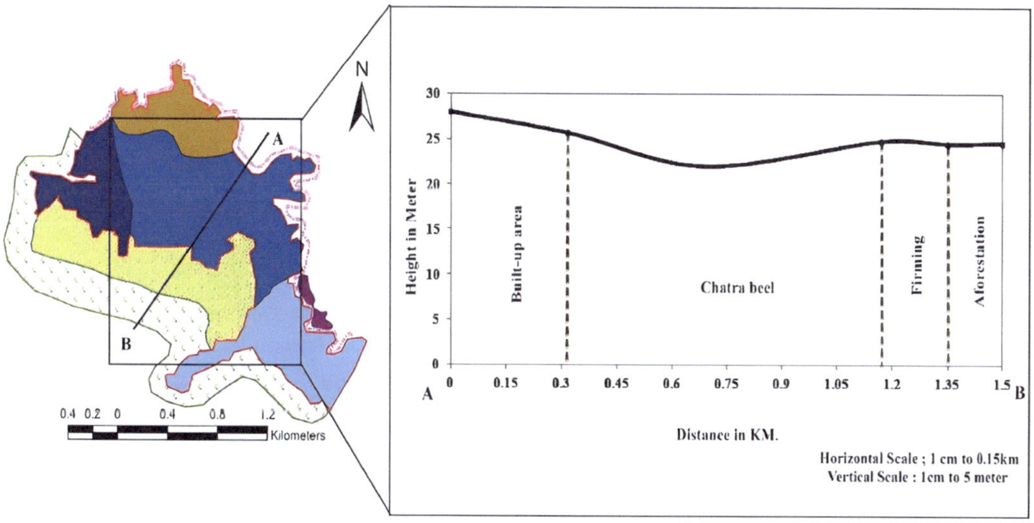

Fig. 19.7 Cross section of the Chatra Beel area (along baseline AB). (Source: Authors (prepared from the base map obtained from the English Bazar Municipality))

Table 19.6 Distribution of land in the proposed plan lay out map

Sl. No.	Activities	Area (Hectares)	%
1	Floriculture and fishing	50.96	26
2	Aqua park	21.56	11
3	Makhna (gorgon nut) cultivation	17.64	9
4	Water treatment plant	25.48	13
5	Firming	39.2	20
6	Plantation	41.16	21

Source: Authors (compiled from the plan lay out map)

After that, the Chatra Beel area must remove the water hyacinth, which causes eutrophication. Physical (cutting and removing the plant using machines or manual labor) or chemical (herbicides and pesticides) methods must be applied to eradicate or control water hyacinth.

Permanent solid waste dumping ground area needs to be restricted outside city limits, and all solid waste should be transported from across the city and dumped in the demarcated land to stop the environmental pollution in and around the Chatra Beel area. A solid waste management project is also to be initiated to manage a vast quantity of waste, almost 200 metric tons generated per day by the English Bazar Municipality, and around 20% of it is nondegradable plastic. Implementing integrated solid waste management as per the national solid waste management (SWM) guidelines is needed to keep the wetland litter and pollution-free. Liquid waste must be treated before it enters Chatra Beel. In this blueprint, a waste treatment plant is to be constructed in this area, and the reuse of treated water must be emphasized.

Developing a strategy to control nonpoint source pollution (NSP), such as brick kilns, scrap factories, and plastic factories, and conducting regular monitoring should be done to stop nonpoint sources of pollution (NSP).

Regular monitoring of water quality standards is essential to maintaining the chemical parameters of the wetland.

19.4.4.2 The Final Phase

In the final phase, a drainage plan will be implemented in which a high drain will be constructed between Uttar Ramchandrapur and Malanchapally area along the eastern side of Chatra Beel, which falls in Chatra Beel in the Malanchapally area. All existing drains from Uttar Ramchandrapur to Malanchapally (Municipal ward number 3) must be linked to this high drain. The southern portion must be left for accumulated water storage in the English Bazar Municipality. A water treatment plan has to be constructed in the Malanchapally area, which treats the accumulated water disposed of in the beel through a high drain. After treatment, fresh water was stored in a constructed reservoir in the Malanchapally area to supply the English Bazar Block area, mitigating the water crisis. The remaining residual water is transported to the River Mahananda through Godrail Bridge for disposal. A sluice gate will regulate the wastewater before it is disposed of. During the monsoon season, excess rainwater also passes through this outlet to the River Mahananda, which relieves the city from water stagnation in flood situations.

An aquatic park is proposed in the northern part and will be restricted to economic activities. After the project, the aqua park will be leased to a third party for revenue generation. Local people will get jobs in the proposed aqua park, benefiting thousands of families.

The water of Chatra Beel is being seriously polluted due to the overuse of pesticides, insecticides, fungicides, and other chemicals in makhana (Gorgon nut) cultivation. Cultivators of this region, for better yield, apply several chemicals. Particularly after a few years of Gorgon culture, when natural nutrients are exhausted, high doses of urea are applied (Khatun and Anwaruzzaman 2012). Furthermore, the big-sized leaves of mature plants prevent sunlight and oxygen from entering the water and make it difficult for fish to survive. As a result, the fish population started declining due to lower oxygen levels and chemical toxicity. Considering this adverse effect on the environment and water quality of Chatra Beel, the practice of makhana cultivation should be gradually decreased.

The central portion of the beel will be restricted to fishing, floriculture, and aquaculture, such as lotus, Paniphal, Thankuni, Brahmi, and Kulekhara cultivation. Rainwater harvesting is also proposed in this portion, and it can be used for irrigation under the "Jal Dharo and Jal Bharo" project of GoWB.

19.4.5 Initiative of National Green Tribunal (NGT)

In March 2023, the Eastern Zone bench of the National Green Tribunal (NGT) took a crucial step by constituting a committee to inspect and provide urgent remedial measures to restore the Chatra Beel in response to a petition filed against the encroachment in Chatra Beel (Aiman and Chishti 2023). This action clearly indicated the pressing need to address the situation. On April 24, 2023, a six-member committee, under the leadership of the principal secretary, environment department, GoWB, visited the Chatra Beel area and submitted a comprehensive report to NGT. In an interim order, NGT directed the district administration to halt any land conversion in five mouzas surrounding Chatra Beel. This was a significant move to prevent further damage. In its final verdict, NGT ordered the district administration that

Detailed mapping of Chatra Beel needs to be done with the help of GIS.

Defining the area and evaluating the quality of water of Chatra Beel.

Stop the conversion of land under the area of Chatra Beel with immediate effect.

To ensure that no further encroachment occurs, strict legal action will be taken against those engaged in it.

19.5 Challenges of the Study

Restoration of wetlands is complicated by several issues that are more social, cultural, economic, and stakeholder than ecological (Root-Bernstein and Frascaroli 2016). The principal attributes of the restoration mechanism are wetland policies, anthropogenic threats, and conservation strategies. Chatra Beel has been exploited and neglected for an extended period. The main

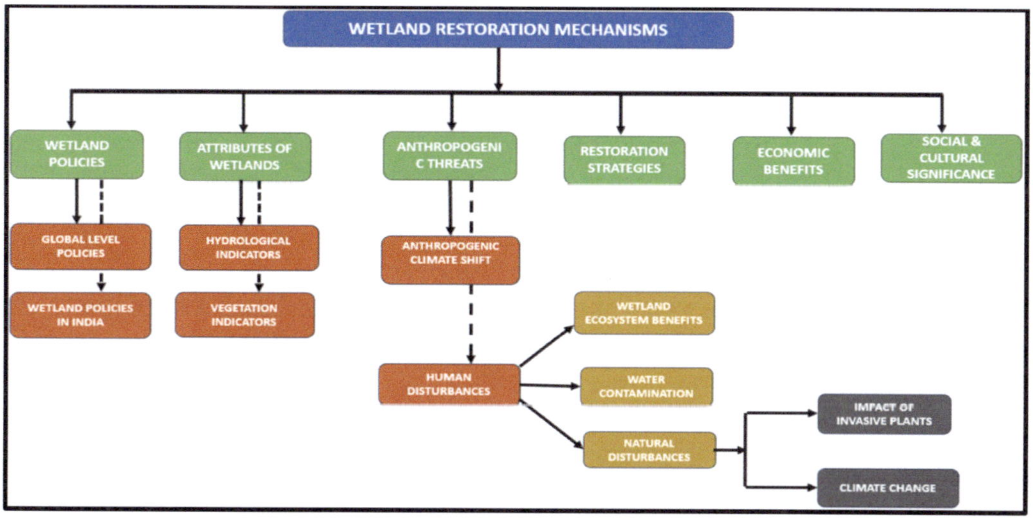

Fig. 19.8 Flowchart on the wetland restoration mechanism. (Source: Prasanya et al. (2024): A review of the wetlands restoration mechanisms and its economic and social benefits)

challenges of the restoration process are (1) weak administrative policies for restoring wetlands, (2) inadequate long-term monitoring mechanism, (3) unavailability of real-time data for decision-making, (4) lack of monetary support for policy implementation, and (5) community participation (Fig. 19.8).

19.6 Conclusions

Reckless urbanization and population growth in the English Bazar Block and Malda district have obliterated the hydrology, ecology, and environment, bringing along unhygienic living conditions and socio-economic misery. Both life and livelihood stand threatened due to the changing nature of the economy and environment of the wetlands in the study area. Simultaneously, the wetlands of Malda district are gradually losing their existence, and almost 50% of the total wetlands are no longer on the map. A spurt in population till 2011, as emphasized in Table 19.2, is responsible for the gradually shrinking boundary of this peri-urban water body. With the decrease in area and increased anthropogenic activities around the beel, Malda town's "life saver" has now been transformed into a wastewater body. Thus, community participation and awareness programs are the need of the hour. Local government authorities, NGOs, and different environmental organizations should come forward to protect the natural environment and ecology of Chatra Beel. The author has provided a comprehensive management plan to properly conserve and manage this natural water body resource and ensure uninterrupted wetland services. It is further recommended that participatory processes are used when setting objectives and ensuring that local communities are fully engaged with the objectives (McInnes et al. 2023), especially as it is increasingly recognized that wetlands are a social-ecological system, not just an ecological system (Kumar et al. 2023). If this restoration plan is approved and put into action, the Chatra Beel will help address several issues in the English Bazar Block, including the freshwater crisis, flooding, and rising unemployment. However, many technical challenges need to be overcome, and success will depend on support from the government, legal entities, and expert organizations. Unless the wetlands of the Malda district become free from reckless urbanization and anthropogenic activities that impede the sustenance of human life and wetlands, the life of flora and fauna will be threatened. As they say, "Well begun is half done," so this minor effort of restoring and saving this peri-urban wetland will also be a significant achievement in the future.

Acknowledgments Our sincere thanks go to all the staff of SD&DLRO, PHE, PWD, Fisheries Department, and District Administration. We would also like to thank the people living around the Chatra Beel area for cooperating during the primary survey.

AI Disclosure The authors declare that no part of this chapter was taken from the AI sources.

References

Aiman J, Chishti JA (2023) NGT Forms Committee To Provide Remedial Measure To Restore North Bengal's Wetland Chatra Beel. https://www.livelaw.in/news-updates/national-green-tribunal-chatra-beel-natural-wetland-encroachment-225093. Accessed 20 Sept 2023

Bureau of Environmental Services (2005) Actions for watershed health, 2005 Portland watershed management plan. The City of Portland Oregon, Portland Oregon. https://doi.org/10.15760/mem.35

Centre for Liveable Cities Singapore (2017) The active, beautiful, clean waters programme. Water as an Environmental Asset, Urban system studies, 1

Chen W, Cao CLD, Tian R, Wu C, Wang Y, Qian Y, Ma G, Bao D (2019) An evaluating system for wetland ecological health: Case study on nineteen major wetlands in Beijing-Tianjin-Hebei region, China. Sci Total Environ 666:1080–1088

Davidson CN (2014) How much wetland has the world lost? Long-term and recent trends in global wetland area. CSIRO Publishing, pp 934–941. https://doi.org/10.1071/mf14173

Delhi Development Authority. Planning, India. https://dda.org.in/planning.aspx. Accessed on 22 Mar 2021

District Census Hand Book, Malda (2011) Census of India, Kolkata, Series-20, and Part XII-B

Humes C, Spraakman S, Lukes R (2024) City of Vancouver. Engineering Department. Green Infrastructure Implementation Branch

Jing L (2019) Inside China's leading 'sponge city': Wuhan's war with water city'. The Guardian International Edition. https://www.theguardian.com/cities/2019/jan/23/inside-chinas-leading-sponge-city-wuhans-war-with-water. https://doi.org/10.1007/978-3-030-01488-9_2

Khatun J, Anwaruzzaman AKM (2012) Gorgon cultivation - a serious threat to wetland bio-diversity. Golden Res Thoughts 1(2)

Krishnaraj V, Mathesh S (2022) Association of Wetland Size Versus Ecological Scenarios and Ecosystem Services within the Agroclimatic zone (Namakkal District), Tamilnadu. Res Square. https://doi.org/10.21203/rs.3.rs-1863880/v1

Kumar AM, Vasimalai MP (2016) Future proofing of an Indian city: Lessons from Madurai the website of United Nations Office for Disaster Risk Reduction Prevention Web. https://doi.org/10.18356/9789211067545

McInnes, R.J., Gell, P.A., Finlayson, C.M. (2023), 14 – Management effectiveness of wetland-protected areas, Gell, P.A., Davidson, N.C. , Finlayson, C.M., Ramsar wetlands, Elsevier, 357-376 https://doi.org/10.1016/B978-0-12-817803-4.00002-4

Mondal B et al (2017) Urban expansion and wetland shrinkage estimation using a GIS-based model in East Kolkata Wetland, India. Ecol Indic 83:62–73. https://doi.org/10.1016/j.ecolind.2017.07.037

National Institute of Urban Affairs (2016) RFP for appointment of planning consultant for green and blue master plan for Bhopal Smart City. Bhopal Smart City Development Corporation Limited

Pal S et al (2017) Detection of land use and land cover change and land surface temperature in English Bazar urban centre. Egypt J Remote Sens Space Sci 20:125–145. https://doi.org/10.1016/j.ejrs.2016.11.003

Prasanya J, Kanmani S, Senthil Kumar P (2024) A review of the wetland's restoration mechanisms and its economic and social benefits. Water Pract Technol 19(11): 4355–4377. https://doi.org/10.2166/wpt.2024.241

Ramachandra TV et al (2012) Insights to urban dynamics through landscape spatial pattern analysis. Int J Appl Earth Obs Geoinf 18:329–343. https://doi.org/10.1016/j.jag.2012.03.005

Ramsar Convention Secretariat (2010) National Wetland Policies: developing and implementing, Ramsar hand books for the wise use of wetlands, vol 2, 4th edn. Ramsar Convention Secretariat, Gland, p 41. https://doi.org/10.1017/cbo9780511975301.015

Root-Bernstein M, Frascaroli F (2016) Where the fish swim above the birds: configurations and challenges of wetland restoration in the Po Delta, Italy. Wiley, Restor Ecol. https://doi.org/10.1111/rec.12369

Roy RD (1992) Case Study of Loktak Lake of Manipur. Wetlands of India, South Asia Books, New Delhi, pp 37–69.

Toyra J et al (2005). Towards operational monitoring of a northern wetland using geomatics-based techniques) Remote Sens Environ 97(2). https://doi.org/10.1016/j.rse.2005.03.012

Udas-Manikikar S, Driver B (2021) Blue-green infrastructure: an opportunity for Indian cities, ORF occasional paper no. 317. Observer Research Foundation. https://doi.org/10.1109/scsp.2016.7501030

Wankhade K, IIHS (2014) Future proofing Indian cities: Bangalore action plan for green-blue infrastructure. https://doi.org/10.24943/9789351568377

Xie Z et al (2010) Analyzing qualitative and quantitative changes in coastal wetland associated to the effects of natural and anthropogenic factors in a part of Tianjin.

China. Estuar Coast Shelf Sci. https://doi.org/10.1016/j.ecss.2009.03.040

Goh X P , Radhakrishnan M, Zevenbergen C, Pathirana A (2017) Effective Control Legislation and Active beautiful, Clean (ABC) Waters Design Features in Singapore ,Water,9(8),627. https://doi.org/10.3390/w9080627

Zhou D, Yu J, Guan B et al (2020) A comparison of the development of wetland restoration techniques in China and other nations. Wetlands 40:2755–2764. https://doi.org/10.1007/s13157-020-01305-5

Global Wetland Outlook (2018) Wetlands-disappearing three times faster than forests,warns new report-The Convention on Wetlands-Ramsar. (www.gloabalwetlandoutlook.ramsar.org)

Jerath N (1992) The Environs and Problems of Harike Wetland : Some Strategies for Conservation. Wetlands of India, Ashish Pub. House,South Asia Books, New Delhi, pp 103–124

Kumar R, Pierre Horwitz P, Finlayson MC (2023) 19-Wetlands as social-ecologial systems: Bridging nature and society, Editors(s): Peter A, Gell Nick C, Davidson C, Max Finlayson, Ramsar Wetlands, Elsevier 525–553. https://doi.org/10.1016/B978-0-12-817803-4.00021-8

Synergistic Impact of Urbanization of Kolkata Metropolitan and East Kolkata Wetlands on the Mangrove-Estuarine Complex of Sundarbans, India

20

Arundhati Ganguly and Susanta Kumar Chakraborty

Abstract

The East Kolkata Wetland, designated as a wetland of global importance under the United Nations Ramsar Convention, located on the eastern flank of the Kolkata metropolitan city, renders valuable ecosystem services on one hand and releases a substantial amount of wastes, both solid and liquid, to the Sundarbans Biosphere Reserve by a network of estuaries, creeks, and tidal canals. The release of such a huge amount of waste material ultimately results in drastic alteration of the total ecology of the mangrove ecosystem in general and physicochemical parameters of soil and water in particular. In addition, undesirable changes in the land-use pattern, imposition of modern technology-based aquacultural and agricultural practices, development of brick kilns, modification of the natural drainage system, increased urban settlement, enhancement of the concentration of toxic city waste, etc., ultimately lead to decline of natural vegetation cover along with their associated faunal components in this region. All of these environmental stresses have resulted in massive destruction and mortality of mangrove assemblages and faunal diversity, putting millions of regional humans at risk. This chapter has dealt with all of these issues with an integrated and holistic approach in order to chalk out proper eco-management strategies in a sustainable manner.

Keywords

East Kolkata wetland · Urbanization · Toxic city waste · Potential ecological risk · Sundarbans mangrove ecosystem

20.1 Introduction

The Sundarbans, the world's largest mangrove forest and a UNESCO World Heritage Site, has a significant influence and connection with the Kolkata Metropolitan Area (KMA) across ecological, economic, socio-cultural, and climatic dimensions. It acts as a natural buffer against cyclones and tidal surges from the Bay of Bengal, shielding Kolkata and its surrounding areas from the brunt of many extreme weather events, protecting lives and infrastructure in the metropolitan area. The vast mangrove forest acts as a carbon sink, gives regional climatic stability, and helps in pollution mitigation and climate change effects for Kolkata.

The linkages between Kolkata Metropolitan City, the East Kolkata Wetlands (Ramsar Site),

A. Ganguly (✉)
Department of Environmental Science, Asutosh College, Kolkata, West Bengal, India
e-mail: arundhati.ganguly@asutoshcollege.in

S. K. Chakraborty
Department of Zoology, Vidyasagar University, Midnapore, West Bengal, India

S. K. Chakraborty et al. (eds.), *Wetlands and Urban Resilience*, Sustainable Development Goals Series,
https://doi.org/10.1007/978-3-032-06881-1_20

and the Sundarbans form a complex, interdependent system that connects urban life, natural ecosystems, and rural livelihoods. The hydrological continuum connects the three zones from dense urban development to peri-urban wetlands to coastal mangrove forested deltas. The EKW plays a role in recharging groundwater and regulating salinity levels in soil and water, affecting both Kolkata's water security and crucial for the ecological health of the delta. The Kolkata metropolitan economic centre generates waste and relies on EKW for natural sewage treatment. The EKW, biodiversity corridors, and the Sundarbans, biodiversity hotspots, are a hub of wetland-dependent birds (including migratory) and fish, crab, and molluscs species, supporting the livelihood of peri-urban people through sewage-fed fish farming, agricultural practices, and ecotourism, benefiting the economic growth of the state. Climate change, rising sea levels, and frequent cyclones are pushing residents of the Sundarbans to migrate to Kolkata in search of safer living conditions and better livelihood opportunities. This migration is reshaping the city's demographics and placing increasing strain on its infrastructure. The Sundarbans mangrove ecosystems are increasingly affected by upstream urban activities, land-use changes, and the degradation of the East Kolkata Wetlands—both of which are vital for the sustainability of Kolkata. This chapter explores the impact of urbanization on the East Kolkata Wetlands, the riverine connection of Kolkata city and EKW to the Indian Sundarbans, and the heavy metal pollution extends to the Sundarbans Biosphere Reserve of India.

Sundarbans, located in the Bengal Basin of the Ganga-Brahmaputra delta (GBD), is drained by many small-to-medium-size rivers, narrow channels, and tidal creeks, which are developed by a series of tectonic movements, paleoclimatic changes, and eustasy. The geomorphology of the Sundarbans mangrove ecosystem is determined by morphological variation, erosion, and deposition of sediment by the rivers (Mandal et al. 2019). The Indian portion of the Sundarbans covers an area of 9630 km^2 and comprises 102 islands, of which 54 are inhabited by approximately 4.2 million people. UNESCO designated the Sundarbans as a World Heritage Site in 1987, and it has been recognized as a Biosphere Reserve by the IUCN in 1989. However, nearly 55% of this unique ecosystem has been reclaimed to date (Das 2015). The landscape has been reclaimed and modified for human settlements (nearly 55%) mainly by the constructions of polders, which results in significant alteration in channel hydrodynamics, siltation process, amplification of tidal ranges, anthropogenic stress factors at a large spectrum, while the climate change constraint in Sundarbans is alarming (Higgins et al. 2018; Pethick and Orford 2013).

The East Kolkata Wetlands (EKW), being an amalgamation of innumerable small to shallow water channels, used for sewage-fed fisheries and agricultural production, are located in the eastern part of Kolkata City. These water bodies, formerly part of the Sundarbans, once served as the spill basin for the mighty Bidyadhari river. Initially saline in nature, they earned the name 'Salt Lake'. The region lies between the Hooghly river in the West and the Bidyadhari river in the East, and is generated by intricate drainage channels and physiologically related to the moribund delta building process. The East Calcutta Wetland lies between 22°25′N & 22°30′N latitude and 88°24′E & 88°35′E longitude. The East Calcutta wetland region is under threat due to increasing population pressure of Kolkata metropolitan city, which is expanding towards the east on the marshy lands at a higher rate over the last 50 years (Everard et al. 2019). In 2002, the Government of India designated the East Kolkata Wetlands (EKW) as a Ramsar Site (Wetland of International Importance under the Ramsar Convention), recognizing its ecological, economic, and socio-cultural significance (Chakraborty et al. 2023). However, this region has been identified as a waste recycling region as wastewater is being treated and used in fish farming, while solid waste is reused for vegetable farming.

The Indian Sundarbans, lying between 21°13′ N & 22°40′N latitude and 88°03′E & 89°07′E longitude, is demarcated by an imaginary Dampier-Hodges line (prepared between 1829 and 1830), marking the northernmost limit of the Sundarbans eco-region having tidal influence. The line runs from the Bay of Bengal in the south, Kulpi in the northwest, to Basirhat in the north, and the Ichhamati-Raimangal in the northeast, which separates the Indian part of Sundarbans from Bangladesh (Banerjee et al. 2017). Hooghli, Muriganga, Saptamukhi, Thakuran, Matla, Bidyadhari, and Raimangal-Harinbhanga are the dominant estuarine rivers of the Indian Sundarbans. The Ichhamati river originates in Nadia and bifurcates into the Sahebkhali and Kalindi rivers. River Bidyadhari also originates from Nadia at Haringhata, West Bengal, and flows through North 24 Parganas. Later, the Bidyadhari river joined with Kalindi (downstream flow channel of Ichhamati) at Nazat, Sandeshkhali II block, and the combined stream flows as the Raimangal river southwards through South 24 Parganas. The Raimangal flows in the Sundarbans Reserve Forest as Harinbhanga and meets the Bay of Bengal. This Ichhamati-Bidyadhari-Rainmangal serves as the international boundary between India and Bangladesh (Das 2015). The Bidyadhari river flows through both North and South 24 Parganas and is named as Haroa river, Chaumuha river, Malancha river, and Kolagachia river, respectively, from north to south.

The Bidyadhari river occurs upstream of the Matla river, another wide estuarine river after the Hugli river, and the geographical separation of settlement islands with reserve forest (Gour 2012; Das 2015). The Matla river flows in the central part, and the Bidyadhari river in the eastern and northern sectors of the Sundarbans, India. The Matla river originates at the confluence of the Bidyadhari, Karatiya, and Rampur Khal near Canning II, and joins with the Kuriabhanga stream, and flows southwards to join with the Bidya river at Jharkhali (demarcation between Herobhanga reserve forest and inhabited Basanti Block) and finally meets the Bay of Bengal. Another notable river Thakuran flows through the buffer zone of the Sundarbans Biosphere Reserve, connecting with the Bidhyadhari-Matla, and the Saptamukhi River. The Matla river is now fully dominated by tidal water and highly saline, receives a little amount of freshwater mainly in the rainy season, and through the interconnecting flows from Icchamati-Raimangal and Bidya rivers.

Rivers of the deltaic region of South West Bengal provide valuable ecological services, the hydrological processes maintain the quality of water and sediments, support floral and faunal population dynamics in the coastal region, and, moreover, irrigation systems, domestic water supply, and a lifeline for industrial activities. The freshwater rivers from the lower Bengal basin, loaded with urban waste on joining with the saline estuarine network of the Sundarbans mangrove ecosystem of India, constitute a well-mixed aquatic ecosystem with a macro-tidal setting (>4 m), where the physicochemical parameters of water are governed by seasonal climatic conditions (Ghosh and Mukhopadhyay 2016). The transitional intertidal zone of the estuarine environment of Sundarbans, India, is very dynamic, where the exchange of materials, transport of solutes, water flux, etc., regulates the functioning of the ecozones (Song et al. 2020). Non-tidal sediment transportation during the winter months tends to deposit southward of the estuary, raising the riverbeds of Matla, Herobhanga, Bidyadhari, and Raimangal rivers. Therefore, silt dredging from the riverbeds is very common for embankment constructions and brick production (Barua et al. 1994). Fine materials tend to deposit at the river channels rather than at the river mouth (Bhattacharya and Giosan 2003). River Bidyadhari was regarded as a dying estuary due to riverbed siltation from geogenic reasons and carries floodwater to a limited extent, resulting from discontinuities in the flow path, disconnections with the parent river Ganges, and with the other interlinking tributaries due to continuous heavy siltation (Bhadra et al. 2017). Now, rainfall in the catchment area is the only freshwater source while the area is under the influence of semidiurnal tides. The Bidyadhari river has undergone significant ecological and morphodynamic changes due to heavy siltation, disconnection from the Hooghly river and several other freshwater tributaries, land-use changes, and sewage discharge. This river was once tidally active in this region, but began to decline from the 1930s onward. As a result, the East Kolkata Wetlands (EKW) region was gradually altered. Later, a stormwater canal was also excavated, linking the area to Kultigong and eventually channelling flow through the Bidyadhari and Raimangal rivers into the Sundarbans mangrove ecosystem. Additionally, filling up the channels, cutting off the connection between rivers and wetlands for developing fish farms, or protecting the agricultural land from tidal wash by constructing dams further affect the natural drainage process in the long term. Upstream land-use changes and waste disposal from the adjacent Kolkata metropolitan city aggravated the concentration of pollutants or contaminants within the riverbed of the Sundarbans estuarine complex of India because of a drastic reduction in water volume.

Among all these factors, sediment attracts special attention in the river environment as the quality of sediments is connected with vegetation, productivity, and land-use processes, as it acts as a reservoir of pollutants, especially heavy metals. Assessing estuarine sediment that may have contamination by Potential Toxic Elements (PTEs) is necessary to ascertain the fate of the dynamic region. Potential Ecological Risk Index (PERI) is a simple and fast practical protocol for environmental assessment, exclusive for contaminated soil, as stability and concentration of contaminants present in soil sediment is higher than the water column; easy to sample and represents an integrated value in time-space. The risk index increases with increasing sediment contamination; the number of pollutants increases, so does the risk index, although different pollutants show different risk levels (Hakanson 1980).

Most of the heavy metals are predominantly deposited in sediment by hydrolysis, adsorption, and coprecipitation processes (Suresh et al. 2015). Heavy metal contamination is chronic, magnifies, gets accumulated in the biota, and enters the trophic level. Urban sewage, agricultural run-off, and industrial effluents from the upstream and surrounding areas are the prime sources of metals in estuarine rivers of tropical regions (Rinklebe et al. 2019).

Resilience studies include assessing changes in water quality, fish diversity, land-use patterns, and ecosystem services, focusing on evaluating how the wetlands respond to environmental and human-induced stressors, exploring the adaptive capacity of local communities that depend on these wetlands, particularly how they cope with shrinking resources and changing livelihood patterns. Soil sediment and surface water quality assessment to understand the influence of natural and manmade changes, as well as variable alteration in physicochemical and biological parameters in accordance with seasonal and temporal variations, are vital in aiming for the environmental protection and safety of public health (Singh and Prasad 2015).

The present study focused on the interfluves of Ichhamati and Bidyadhari, extended to Raimangal and Malta rivers of mature and active deltaic parts of the Sundarbans Biosphere Reserve, to identify the ecological risk of the Sundarbans mangrove ecosystem from the perspective of environmental degradation. So far, most of the monitoring and conservation activities are restricted to the core protected area of Sundarbans. Since nature's ecological services of Sundarbans extend beyond human-defined boundaries, the pollution seeping into the water and soil of this precious ecosystem cannot be adequately managed merely by marking core or buffer zones. Integrating Sundarbans and EKW into the city's urban planning and policy frameworks is a key to enhancing resilience. Protecting and restoring the Sundarbans and wetlands can help to maintain ecological functions, reduce flood risks, ensure water security, and support sustainable urban development. Thus, resilience in EKW and the Indian Sundarbans is not just ecological, but also socio-economic, linking environmental health with community well-being.

20.2 Description of the Study Area

Rivers Ichhamati and Bidyadhari are flowing towards the south-eastern side following the general slope of the Sundarbans delta. The entire stretch of the Matla-Bidya-Raimangal river channels showed various environmental dimensions. The north-western part of this interfluve region experienced the natural land built-up process, and in many places, the human interventions are responsible for the changing behavioural pattern of the rivers, and later, the abandoned river channels are converted into fish ponds (Mondal and Satpati 2020). In order to understand the seasonal dynamics of heavy metal concentration in different stretches of the estuarine network, different study sites were selected, which are as follows:

(I) **Malancha (S-I)**—This sampling site is located in the extended part of East Calcutta wetlands at the bank of the Bidyadhari river, North 24 Parganas (22°30′29″ N; 88°46′30″ E), West Bengal, India. The river banks are protected by the concrete embankment structures with lots of human activities such as aquaculture and fishing, brickfields, leather or tannery processing units, profuse horticulture, and thriving against the onset of a continuous flow of sewage coming down from the Kolkata metropolitan area (WB, India) loaded with toxic industrial effluents and solid municipal waste. The area is connected with the Ichhamati river and still connected with some rivulets, sewage canal (locally known as Nallah), and expanded swallow wetland networks. Scanty mangrove-associated plants and grasses are still in existence. A very minimal variety of mangrove and mangrove-associated vegetation is present here. One of the largest fishing harbours and fish retail markets of South West Bengal is present, where different fish varieties from wetlands, rivers, and marine fish are available.

(II) Dhamakhali (S-II)—This study site is located in Sandeshkhali II (22°21′35″ N; 88°51′58″ E), North 24 Parganas, West Bengal, at the downstream part of the Ichhamati–Bidyadhari interfluve region, locally known as Kolagachia river and Dhansa river, meet and jointly flow as Raimangal river, which demarcates the international boundary between India and Bangladesh. It is a gateway to the Sundarbans Biosphere Reserve in the eastern sector and a popular tourist site. Unlike Malancha, this part is characterized by more tidal influence, carrying a higher volume of water. The environmental flow of this sector of the estuarine network has been hindered and modified by several human activities, such as diversion of water flow for the purpose of aquaculture, creation of brickfields, construction of concretized roads, jetty ghats, as well as concrete and earthen embankment with bamboo fencing, ensuring human settlement. The presence of mangrove vegetation like *Avicennia* sp., *Exocaria* sp., *Acanthus* sp., *Derris* sp., grass salt marsh like *Sueda* sp., *Salicornia* sp., alongside some inclusion of mesophytic species, was observed. Mangrove afforestation aimed to protect the land but failed due to the recurrent impact of monsoon spring tides and cyclonic storm surges (Fig. 20.1).

(III) **Jharkhali (S-III)**—This study site is situated in Basanti Block (22°01′08″N; 88° 40′56″E), South 24 Parganas, West Bengal, between two mighty rivers, the Matla (on the west) and the Bidya (on the east), located just opposite to the Herobhanga Reserve Forest. The area is a sheltered and protected zone under the surveillance of the Forest Department, and a tiger rehabilitation centre was established on the site. As an outcome of the eco-restoration programme, enhanced diverse mangrove vegetation diversity and biological productivity in the area. Although aquaculture, fishing activities, and agriculture practices are the main livelihoods of the people, the progressively increasing Sundarbans mangrove eco-tourism is inviting rapid construction of road

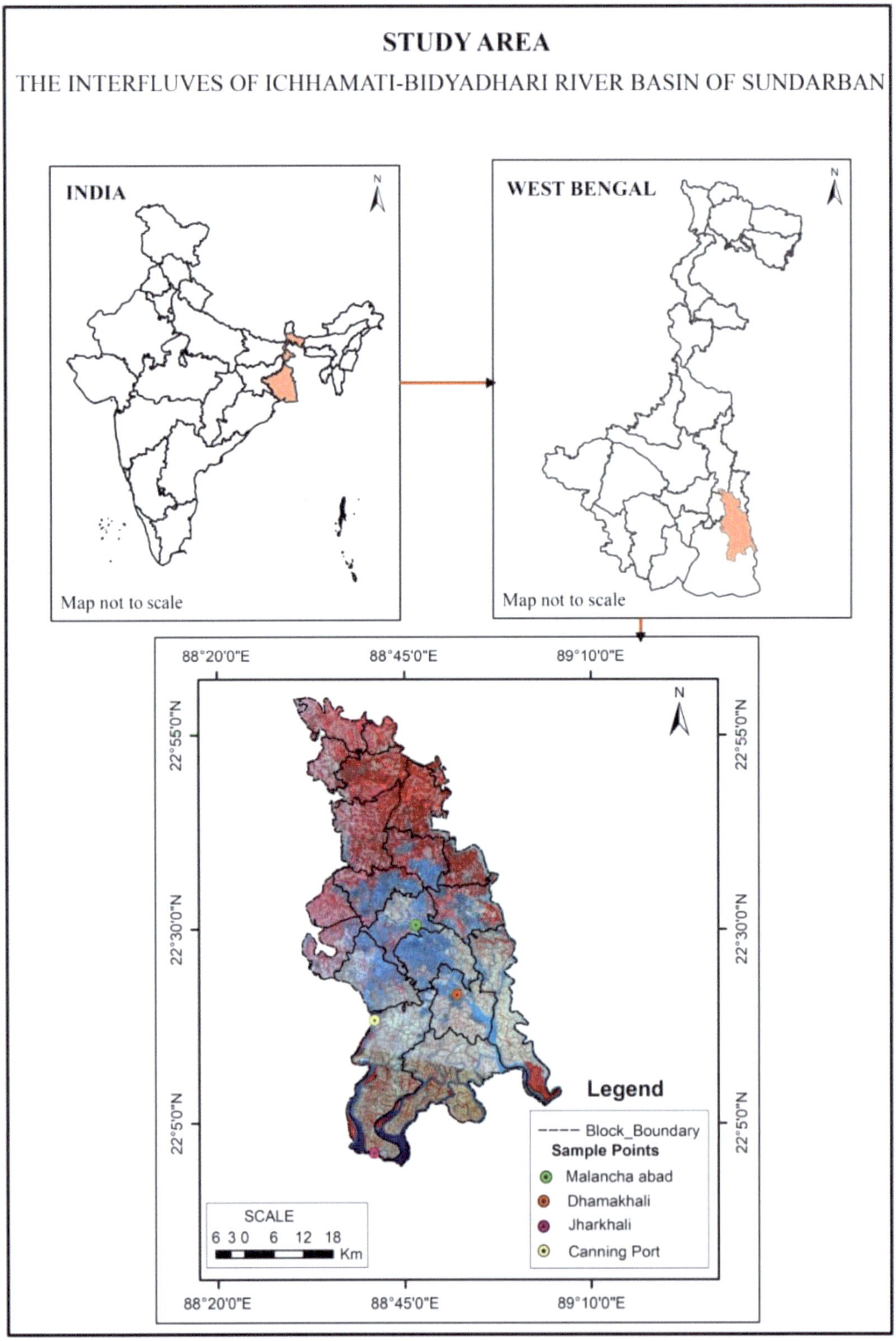

Fig. 20.1 Study location map

networks, hotels, by clearing off mangrove vegetated areas, parking zones, ferry services, etc.

(IV) **Canning Port (S-IV)**—This study site is also located in Basanti Block, South 24 Parganas, West Bengal, (22°18′13″N; 88°40′59″E), on the bank of the Matla river. From the British period, it was a past riverine port of Kolkata. It is a very important site for the transportation of goods, communication of the people of Sundarbans with the mainland, and other services. Intensive fishing activities on the river, along with agriculture, aquaculture, and commercial activities, are predominant. Collection of prawn seeds and crab potting affects the natural succession of mangroves on the tidal mudflats. Although afforestation programmes have been conducted by the Sundarbans Development Board to revive the mangroves and maintain the natural beauty of the landscape.

20.3 Materials and Methods

A. **Analysis of Physicochemical Parameters of Soil and Water**

Soil organic carbon was analysed by the Walkey and Black method (Trivedi and Goel 1984), and available phosphorus content of soil was determined by using Olsen's sodium bicarbonate extractable method due to the alkaline nature of the soils in the Indian Sundarbans (Olsen and Sommers 1982).

B. **Heavy Metal Analysis**

A modified wet digestion procedure was used for the determination of selected metals (lead, mercury, copper, cadmium, chromium, nickel, and zinc) in soil and water in seasonal variation from the study sites in AAS (Chernoff 1975).

All the data were calculated based on standard deviation, maximum, and minimum values by using the XLSTAT, PAST, and SPSS 20.5 versions.

C. **Potential Ecological Risk Index Estimation**

The following factors were calculated for PERI, an established method by Hankinson (1980).

- **Contamination Factor** *(CF):* CF is a useful tool to monitor contamination in sediments over time. It is the ratio of every metal in the present sample to the background values of the same metal.

$$\mathrm{CF} = C/C0,$$

where C is the mean concentration of the given metal in sediment and *C*0 is the background concentration of the metal in the sediment [Background value for Pb-30, Hg-0.061, Cu-23.8, Cd-0.34, Cr-89.6, Ni-40.2, Zn-84]

- **Degree of Contamination** (DC): The degree of contamination is determined by the sum of all contamination factors (CFs):

$$\mathrm{DC} = \Sigma\mathrm{CF}$$

- **Toxic Response Factor (TRF)**: TRF depends on the sedimentological sensitivity factor (SSF). SSF influences the bioavailability of metals, which depends on the trophic status. To derive the SSF value, the relations between total phosphate (TP), organic matter (OM), and total organic carbon (TOC) were tested. The bio-production ratio (BPR) was calculated by the following method:

$$\mathrm{SSF} = \mathrm{BPR} = (\mathrm{TP\ in\ mg/g})/(\mathrm{TOC\ in\%}) \times 100$$

$$\mathrm{Hg} = 40.5/\mathrm{BPR}; \mathrm{Cd} = 30.51/2/\mathrm{BPR}1/2; \mathrm{Pb} = 5.51/2/\mathrm{BPR}1/2;$$

$$Cu = 5.51/2/BPR1/2; Cr = 2.51/2/BPR1/2; Zn = 1.51/2/BPR1/2$$

- Potential ecological risk index (PERI): PERI is the sum of all PER calculated for each metal in one site: PERI = Σ PER.

D. The interfluves of the Ichhamati-Bidyadhari river are extended over (latitude 22°10′0″ N to 23°0′0″ N and longitude 88° 30′0″ E to 90° 0′0″ E), covering 16 blocks in North 24 Parganas and 8 blocks in South 24 Parganas of West Bengal, the reclaimed islands of Indian Sundarbans.
  - **Land-Use-Land-Cover (LULC) Map**: Satellite images Landsat 7 ETM+ (2009, 2010) and Landsat 8 OLI (2019, 2020) from the United States Geological Service (USGS) were used to prepare the land-use land-cover map by onscreen digitization using Arc-GIS 10.3.
  - **Normalized Difference Vegetation Index (NDVI):** This is a commonly used vegetation index that quantifies the difference between near-infrared (0.7–1.4 μm, which gets reflected by green vegetation) and red band (0.6–0.7 μm, which gets absorbed by chlorophyll of green vegetation) (Lillesand et al. 2015).

$$NDVI = (NIR - RED)/(NIR + RED)$$

- **Normalized Difference Salinity Index (NDSI):** This index is used to assess the salinity-affected area, which also deals with the NIR band (Khan et al. 2005).

$$NDSI = (RED–NIR)/(RED + NIR).$$

NDVI and NDSI were prepared by ERDAS-14. Landsat7 ETM+ image with a total of 8 bands was processed by applying the formula [(Band 4−Band 3)/Band 4 + Band 3] to prepare NDVI (2010) and NDSI (2010). Landsat 8 OLI, with a total of 11 bands, was processed by applying the formula [(Band 5−Band 4)/Band 5 + Band 4] for NDVI (2020) and NDSI (2020) maps.

- **Land Surface Temperature (LST):** These maps were prepared from the data of Landsat 8 by ERDAS 14. To estimate the brightness temperature, the TIR band 10 was used, while for calculating NDVI, band 4 and band 5 were used. The input of band 10 was used to retrieve the Top of Atmospheric Spectral Radiance (Barsi et al. 2014). After the conversion of digital numbers (DNs) to reflection, the spectral radiance of TIRS band data was converted to brightness temperature (Kelvin) by using thermal constants from the metadata file, and finally converted to Celsius. Calculation of NDVI and proportion of vegetation was important as they are highly related to the land surface emissivity to get the Land Surface Temperature (Jiménez-Muñoz and Sobrino 2006).

20.4 Results and Discussion

The Sundarbans mangroves, along with their adjoining East Kolkata wetland ecosystem, provide numerous ecological and economic benefits, which invite rapid population growth in this sensitive and fragile estuarine region. Extreme exploitation of the biological resources and the livelihood options, like the collection of finfish, shellfish, prawn seed, crab potting, intensive aquaculture, and agro-farming practices, leads to a reduction in mangrove vegetation cover, degradation of ecosystem health, natural succession, and regeneration of mangrove swamp to the overall biological diversity loss of the ecosystem (Chowdhury et al. 2017).

The level of heavy metals in different components of the ecosystem, like water, sediment, and biota, is vastly dominated by many hydrological, physical, and chemical changes with seasonal variability (Saha et al. 2001). The estuarine sediments are major reservoirs of different metals, and seasonal variability regulates the distribution and immobilization of toxic substances regulated by several physicochemical factors. Many similar studies on heavy metals have been conducted in various parts of the world (China, Chile, United States, Turkey, Bangladesh, and India) their widespread sources, long residue time, and easy accumulation (Wang et al. 2021).

Malancha (S-I), Dhamakhali (S-II), Jharkhali (S-III), and Canning (S-IV) are situated in the Ichhamati-Bidyadhari river basin of Sundarbans, and are differentially affected by industrial waste and agricultural works, which in turn results in the pollution of the estuarine system. The monsoon rain reduces the contents of Pb, Hg, Cd, Ni, Cu, Zn, and Cr in water and sediments, followed by an increase in the post-monsoon period. These differences in metal content in water and sediment during monsoon and post-monsoon months indicate sediment-geomorphic processes that regulate the mobilization of metals in the environment. Freshly deposited sediment can accumulate a greater quantity of heavy metals (Mukherjee et al. 2020). Irregular estuarine geometry, variable tidal levels, and seasonal variation strongly governed the metal deposition in sediments of the Hugli river (Mukherjee and Kumar 2012). In this study, heavy metals were found in higher amounts in water than in soil, except for Pb throughout the 3 years of study for all the study sites. Therefore, it can be said that rivers are the major source or carriers for bringing down the pollutants to the Sundarbans. River Bidyadhari (S-I), as it receives Kolkata city waste, and its downstream part at Dansa-Raimangal (S-II), was found to be more polluted than the river Matla (S-IV) and river Herobhanga (S-III). Chromium (Cr) concentration was maximum, followed by Pb>Cu> Hg>Ni> Zn> Cd, both in water and soil. Cr is largely used in the Bantala leather complex, and the tannery waste contributes huge Cr to the Bidyadhari river. Hg, Cd, Cu, and Zn were negatively correlated with the abundance of species in Canning (S-IV); Pb, Ni, and Cd in Dhamakhali (S-II); Pb and Zn in Jharkhali (S-III) (Figs. 20.2, 20.3, 20.4, and 20.5). Cr, Ni, Cu, Zn, Cd, Pb, and Hg are found to be the most common metalloids, and their hazardous impact depends upon the availability and duration of exposure (Zakaria et al. 2021). Mangroves can

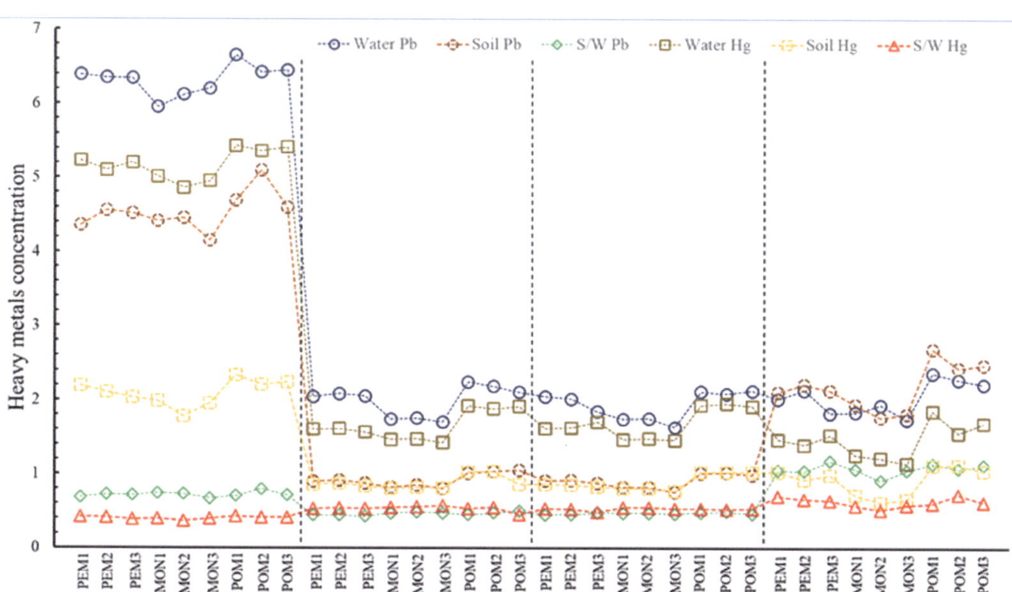

Fig. 20.2 The concentration (in ppm) of lead and mercury in water and soil in the study sites [(Malancha (S-I), Dhamakhali (S-II), Jharkhali (S-III), and Canning (S-IV)]. (Source: authors' field study and laboratory analysis)

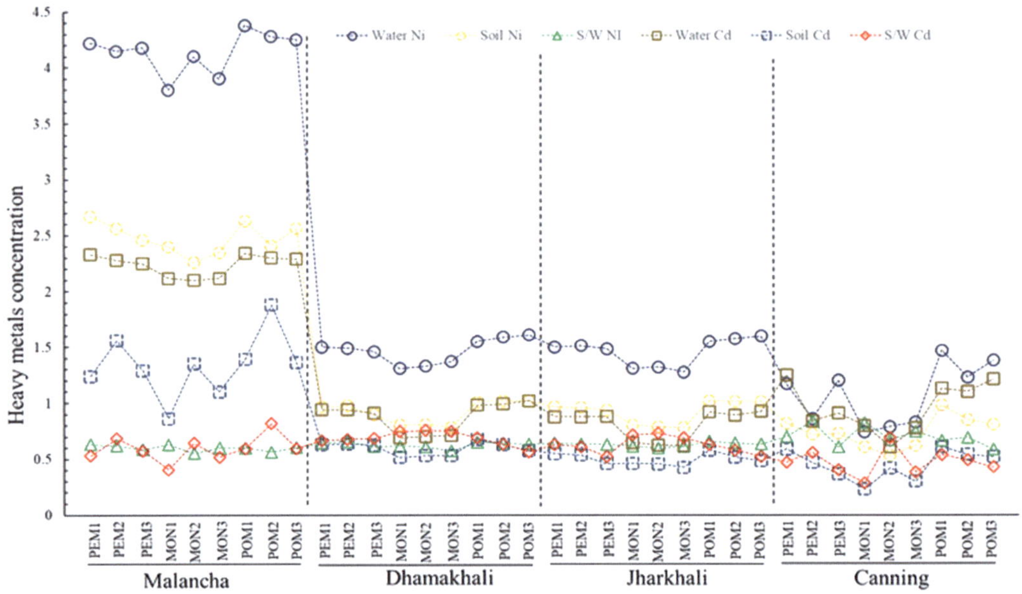

Fig. 20.3 The concentration (in ppm) of nickel and cadmium in water and soil in the study sites [(Malancha (S-I), Dhamakhali (S-II), Jharkhali (S-III), and Canning (S-IV)]

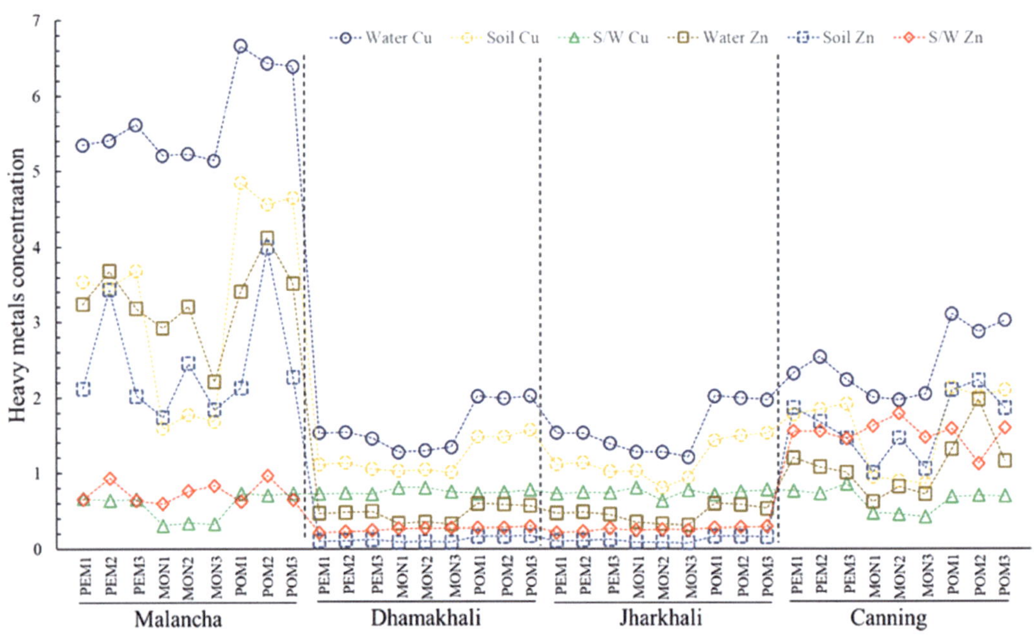

Fig. 20.4 The concentration (in ppm) of copper and zinc in water and soil [(Malancha (S-I), Dhamakhali (S-II), Jharkhali (S-III), and Canning (S-IV)]

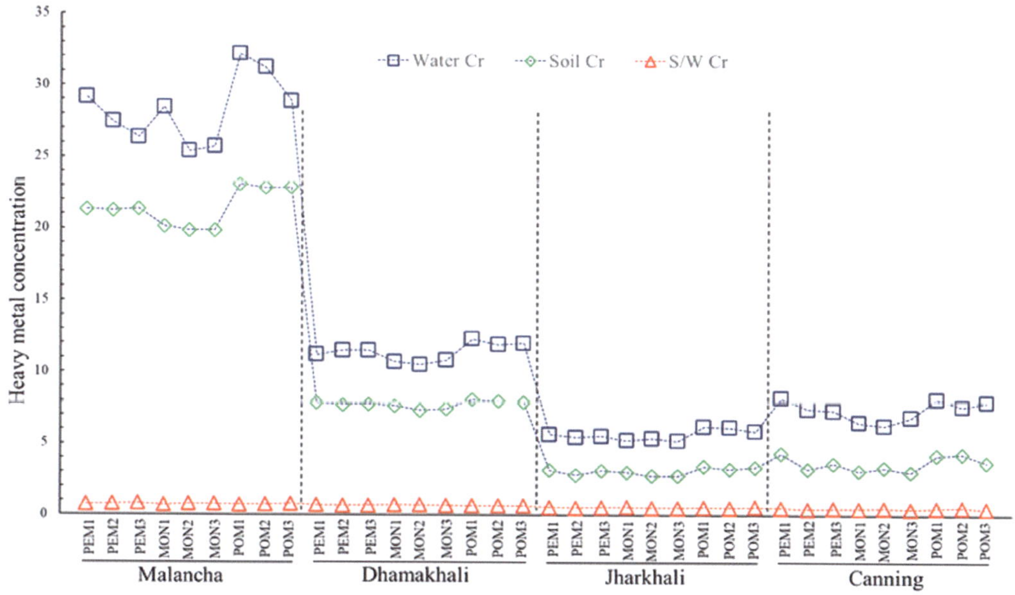

Fig. 20.5 The concentration of chromium (in ppm) in water and soil [(Malancha (S-I), Dhamakhali (S-II), Jharkhali (S-III), and Canning (S-IV)]. (Source: authors' field study and laboratory analysis)

potentially trap heavy metals and retain them in their body that in the long term affect plant growth (Nguyen et al. 2020).

Fluctuations in tidal flooding of the region control the availability of heavy metals for plant uptake or phytoremediation (Yang et al. 2021). The rate of litterfall increased in metal-contaminated soil, whereas the rate of litter decomposition decreased in the polluted soil (Lucisine et al. 2015). Still, there is a huge research gap regarding the magnitude of uptaking metal, the translocation process in different organs, the release by litterfall, and the total reflux of metals in the mangrove system (Ray et al. 2021).

Concentration factors (CF): The concentration factor (CF) of mercury (Hg) was very high in all the study sites. It was maximum in Malancha S-I (37.05), and for the other study sites, S-II (14.14), S-III (14.62), and S-IV (14.50) values were very close. CF value for cadmium was classified as 'Considerable' ($3 \leq CF < 6$) for S-I (4.8) and 'Moderate' ($1 \leq CF < 3$) in S-II (1.75), S-III (1.72), and S-IV (1.2). For the rest of the heavy metals, the CF value was found at a low level (Figs. 20.6 and 20.7).

Degree of contamination (DC): The DC values of all the sites were found to be 'Very high' ($DC \geq 6$), and the value was maximum in S-I (42.50), followed by S-III (16.49), S-II (16.08), and S-IV (15.94) (Fig. 20.9).

Potential Ecological Risk: Potential Ecological Risk (PER) value for Pb, Cu, Cr, and Zn was low (<1) (Figure: 9.3.4.1). The PER value for Cd (<11) was also at a low level. The PER value for Hg in S-I (110.48) was at 'Considerable level' ($80 \leq PER < 160$) and in S-II at 'Moderate level' ($40 \leq PER < 80$) (61.67), while it was low in S-III (21.22) and S-IV (30.22) (Figs. 20.7 and 20.8).

Potential Ecological Risk Index (PERI): Finally, the Potential Ecological Risk Index (PERI) for S-I (121.27), S-II (67.4), S-III (23.22), and S-IV (32.12) was at 'low level' (150). In the comparative account of the study, it was observed that the DC value was high in S-I and S-II, and so was the PERI value. But in S-III and S-IV, the PERI value was less in comparison to the DC value of the study sites (Fig. 20.9).

High levels of mercury in the coastal environment of the Indian Sundarbans were already established by several research findings (Bhattacharya et al. 2015). The degrees of

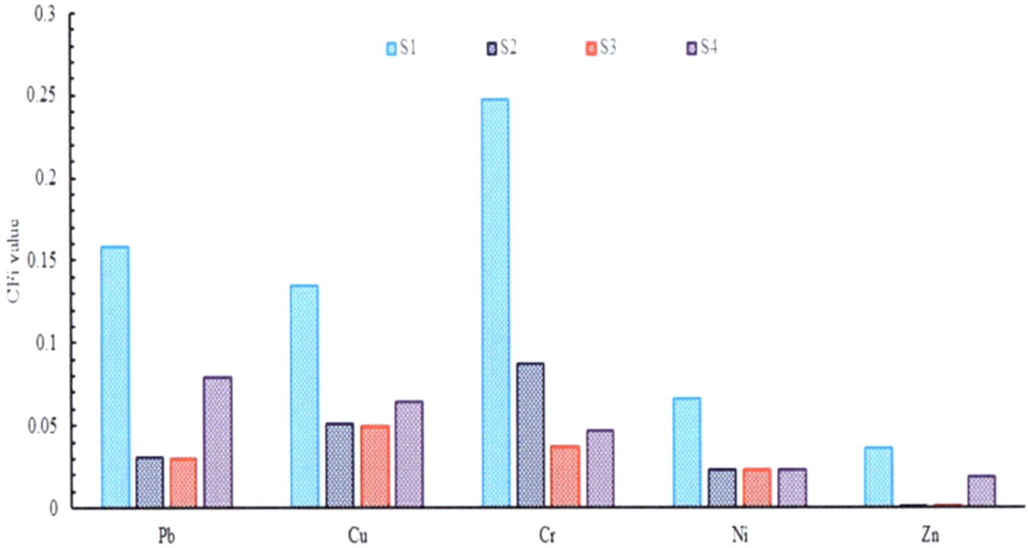

Fig. 20.6 The concentration factor (CF) of heavy metals (Pb, Cu, Cr, Ni, Zn) in the soil [S-I-Malancha (S1), S-II Dhamakhali (S2), S-III Jharkhali (S3), and S-IV Canning (S4)]. (Source: authors)

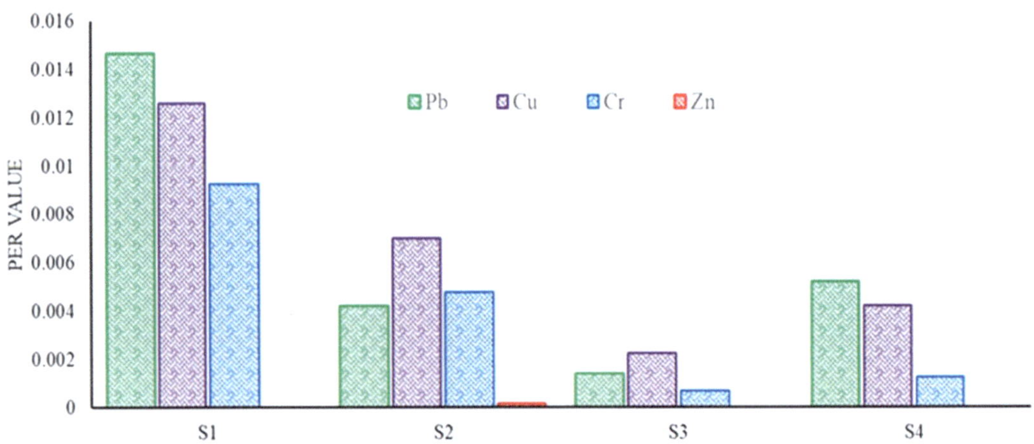

Fig. 20.7 Potential ecological risk from heavy metals Pb, Cu, Cr, and Zn in study sites. [S-I-Malancha (S1), S-II Dhamakhali (S2), S-III Jharkhali (S3), and S-IV Canning (S4)]. (Source: authors)

contamination ranking showed that the Malancha S-I was highly threatened (DC rank 1) by heavy metal contamination, as Bidyadhari receives all the metropolitan chemical wastes, leather complex waste throughout the year, and brings down pollution load along with it to the downstream part.

Metals have a phytotoxic effect on plants, while mangroves show a very significant role in the accumulation and remediation of metals from sediment. In the Sundarbans, Pb and Cd were abundant as potential toxicants to the mangroves, while *Avicennia* species, *B. gymnorrhiza, and Acanthus ilicifolius* have the maximum potential for metal accumulation, as the metal translocation factor was >1. The accumulation pattern in the below-ground biomass of mangroves from sediment increased threefold (Ray et al. 2021). A study of PERI in the Sri Lankan mangrove estuary revealed that, despite their phytoremediation

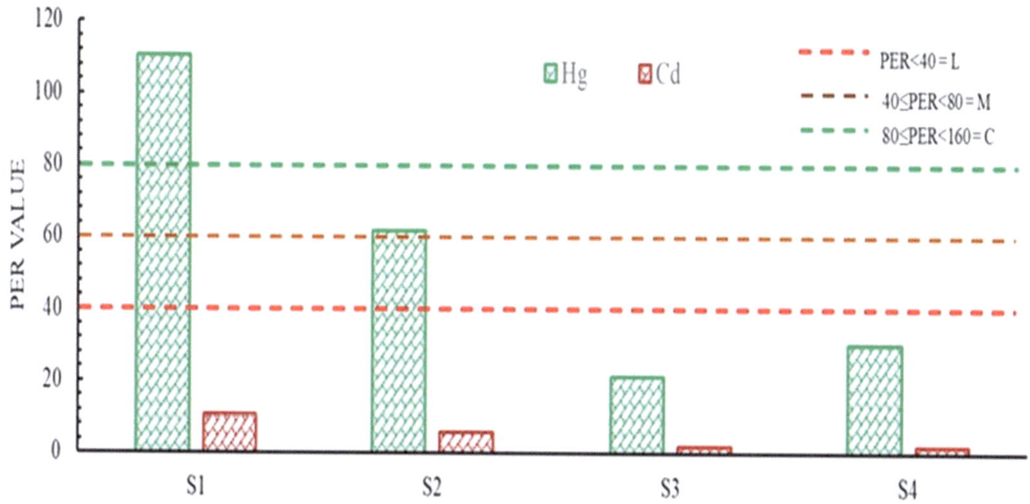

Fig. 20.8 Potential ecological risk from Hg and Cd at different study sites. [S-I-Malancha (S1), S-II Dhamakhali (S2), S-III Jharkhali (S3), and S-IV Canning (S4)]. (Source: authors)

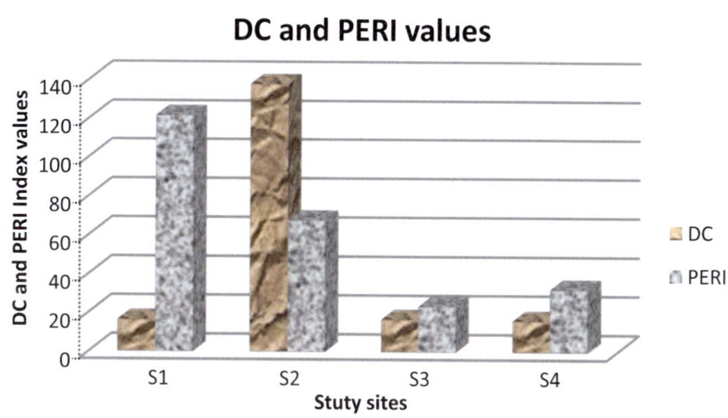

Fig. 20.9 Bar diagram showing degree of contamination (DC) and potential ecological risk index (PERI) for all the study sites [DC < 1-low, 1 ≤ DC < 3-moderate, 3 ≤ DC < 6-considerable, DC ≥ 6-very high; PERI<150-low, 150 ≤ PERI<300-moderate, 300 ≤ PERI<600 considerable, PERI ≥600-very high]

potential, mangroves should not be considered hyperaccumulators. The capacity of accumulation is a maximum of 5 gm per kg of biomass for mangrove trees (Abeywardhana et al. 2022). Metal oxidization, uptake, and exudation processes, and symbiosis of microbes in the rhizosphere affect the metal mobilization in a flood plain (Li et al. 2013). S-II and S-IV are the gateways to enter the Protected Area Network of Sundarbans, and Ichhamati-Bidyadhari river carries fresh water to a large extent to the Sundarbans region, along with their toxic substance contribution. The PER value individually from Hg was at a considerable level in S-I and S-II, but as per the PERI value, the potential risk factor of the ecosystem is still low. The PERI value was not too far from a considerable risk limit. From this present study, Hg and Cd as PTEs may impose the maximum threats to the Sundarbans ecosystem. Gradual degradation of the coastal soil of the Hooghly estuary from heavy metals was observed, but the concentration factor in biota indicated the phytoremediation potential of the mangroves, as the concentration of metals in mangrove roots was proportional to sediment concentration (Ghosh et al. 2022). Analysis of

PERI in Sundarbans, Bangladesh, has also highlighted the significance of mangroves and mangrove-associated plants as phytoextractors and phytoprotectors (Rahman and Rahman 2021). Various indices on sediment contaminants revealed progressive degradation of surface sediment at moderate to high levels, instead of having a low ecotoxicological risk. Mangrove vegetation density observed to be high at S-III and S-IV, playing a crucial role in reducing threats from metals. Toxic elements of the sediment were observed to be correlated with antimicrobial activities of the mangrove leaves of *Avicennia alba* and *Acanthus* sp. (Bakshi et al. 2018). A significantly higher amount of mercury was found in the scalp hair of the residents of the Indian Sundarbans region who consumed local fish on a regular basis (Gibb et al. 2016). A less ecological risk was found in the sediment based on a study of the Effective Range Low (ERL) value, but the level of mercury in drinking water was often above the desirable limit by WHO (Bhattacharya et al. 2015). In the present ecotoxicological study, the existence of heavy metals in soil and the bioaccumulation pattern were established to the relevant impacts on the resident species of the Sundarbans mangrove ecosystem, which in turn imposes threats to the people who are largely dependent on the living resources. Various conservation strategies are already implied still the Sundarbans estuarine environment of India is under constant threat from human perturbation (Figs. 20.10, 20.11, 20.12 and 20.13).

Ecological resilience against urbanization and decadal transformation of the Ichhamati-Bidyadhari river basin with the application of satellite images and interpretation in view of demographic changes

- **Changes in the land-use pattern**—Estuarine geomorphology indicates the biophysical vulnerability. A considerable change in the last decade (2009–2019) was observed from the LULC maps (Fig. 20.14). A decrease in natural vegetation cover on a large scale at 4.7197 ha/year was observed. The open forest-land was cleared off and converted to

Fig. 20.10 Wastewater discharge loaded with heavy metals to the Bidyadhari river at Kultigong

Fig. 20.11 Scanty mangrove vegetation on the riverbank of the Bidyadhari river

Fig. 20.12 Fishing activities at the Raimangal river

Fig. 20.13 Wetland fisheries. (Source: authors' field observation)

agricultural land and settlement structures. An increase in the canal or river area (444.12 ha) and water body (285.60 ha) indicates the submergence of low-lying land resulting from gradual sedimentation and failure of the interconnecting drainage network to dissipate rainwater or tidal overflow to the downstream part.

A large-scale decrease in wetland area (1064.58 ha) and an increase in aquaculture area (3728 ha) were observed within 10 years, where the probable cause is the conversion of wetlands into aquaculture practices or intensive fish farming for greater economic benefit. A decline in agricultural land (2905.9 ha) also indicates that the general conversion of agricultural land into aquaculture is gaining popularity among marginal farmers in the entire coastal belt of West Bengal, also for fast economic growth in a short period. The rural settlement was increased as expected with growing population pressure (Table 20.1).

- **Changes in the vegetation cover:** Human activities have significantly reduced the vegetation cover of the Earth's surface. NDVI maps help to visually interpret or detect vegetation dynamics or degradation due to natural or manmade causes. NDVI indicates the health of vegetation cover, green biomass, leaf area index, and production pattern (Thenkabail et al. 2004). NDVI is classified into three major classes—water body, shrub land, and healthy vegetation. The NDVI value varies from -1 to $+1$, where water bodies are represented by the low value (-1 to 0), areas of shrubs and meadows are represented by a moderate value (0.2–0.35), and the large value (0.6–0.8) represents dense vegetation and tropical forest. In this present study, a decrease in healthy vegetation cover, mostly towards the central and southern part of the interfluves, was observed in parallel with increasing human settlements, agricultural, and aquaculture activities' healthy vegetation structures decreased. A decline in healthy mangroves by 20.545 ha covering Gosaba, Basanti, Canning 2, part of Bangar of South 24 Parganas, Sandeshkhali 1 and Sandeshkhali 2, Hasnabad, and the major part of Hingalganj of North 24 Parganas was observed between 2010 and 2020. Still, healthy mangrove vegetation cover was observed at Hingalganj along the Raimangal river in the south of the interfluve. The degradation of the healthy vegetation covers followed by an increase in shrubland (open forest or less dense vegetation cover) along the riverbank of Bidyadhari, Dansa, Gosaba, Bidya, and Matla in North and South 24 Parganas of West Bengal, within the study period (2010–2020), was observed. This indicated the emerging threats of salinity and associated changes in the edaphic factors, as a sequel of reduced freshwater flow and nutrients load by surface runoff from the agro-farming and urban sectors containing heavy metals and other toxicants (Fig. 20.15; Table 20.2).

- **Changes in the salinity:** Increasing soil and water salinity in coastal areas all over the

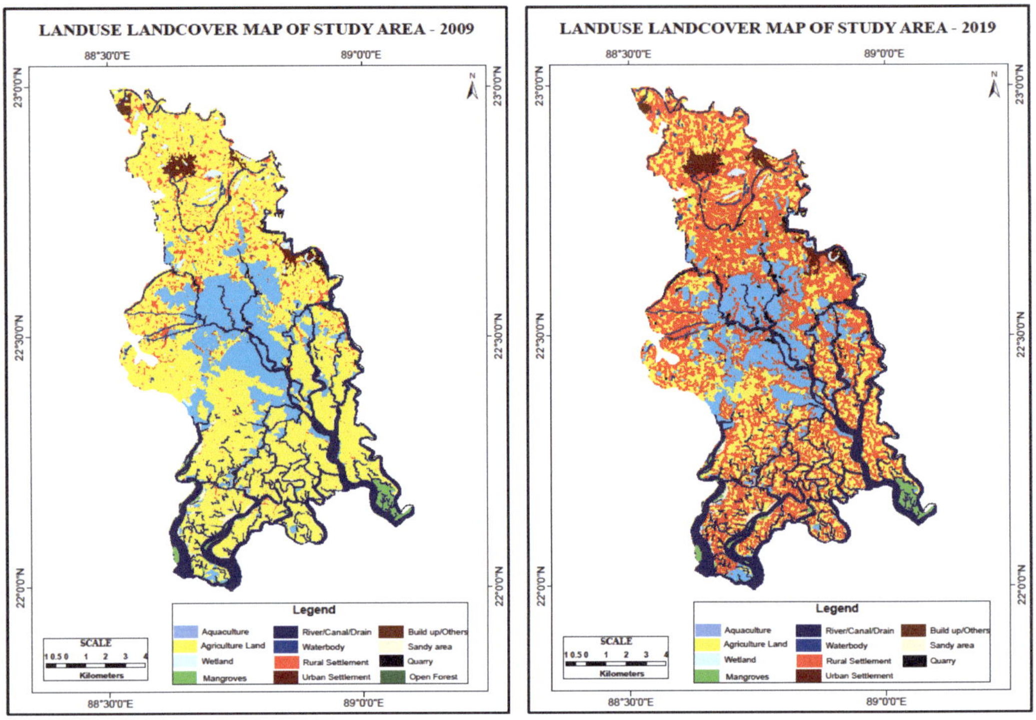

Fig. 20.14 Land-use land-cover changes from 2009 to 2019 of the Ichhamati–Bidyadhari interfluves region

Table 20.1 Different classes of land-use patterns and comparative changes from 2009 to 2019

2009			2019		
Land-use/land-cover classes	Area in hectares	Area in percentage (%)	Land-use/land-cover classes	Area in hectares	Area in percentage (%)
River/canal	31806.880	9.501	River/canal	32251.009	9.633
Waterbody	399.312	0.119	Water body	684.906	0.205
Open forest	2.061	0.001	Mangrove	5361.850	1.602
Mangrove	5409.047	1.616			
Sandy area	3049.370	0.911	Sandy area	2689.715	0.803
Wetland	1773.882	0.530	Wetland	709.321	0.212
Agricultural land	155144.884	46.341	Agricultural land	152239.33	45.474
Aquaculture	49506.027	14.787	Aquaculture	53233.953	15.901
Urban settlement	6804.660	2.033	Urban settlement	6653.783 1.987	1.987
Rural settlement	77086.434	23.026	Rural settlement	77244.576	23.073

Source: Data analysed by authors

world is the most damaging environmental problem. In the NDSI map of 2010, low saline zones were observed mostly in the northern sector, i.e. in North 24 Parganas, though many blocks of South 24 Parganas, excluding the river channels, water bodies having tidal influence were observed to have very high salinity. Soil surface areas of moderate salinity were centered around Basanti, Gosaba, Hingalganje, Sandeshkhali-1, and Sandeshkhali-2 in a scattered manner. The moderately saline soil is suitable for agriculture and aquaculture practices. In the NDSI map of 2020, the total land cover with high

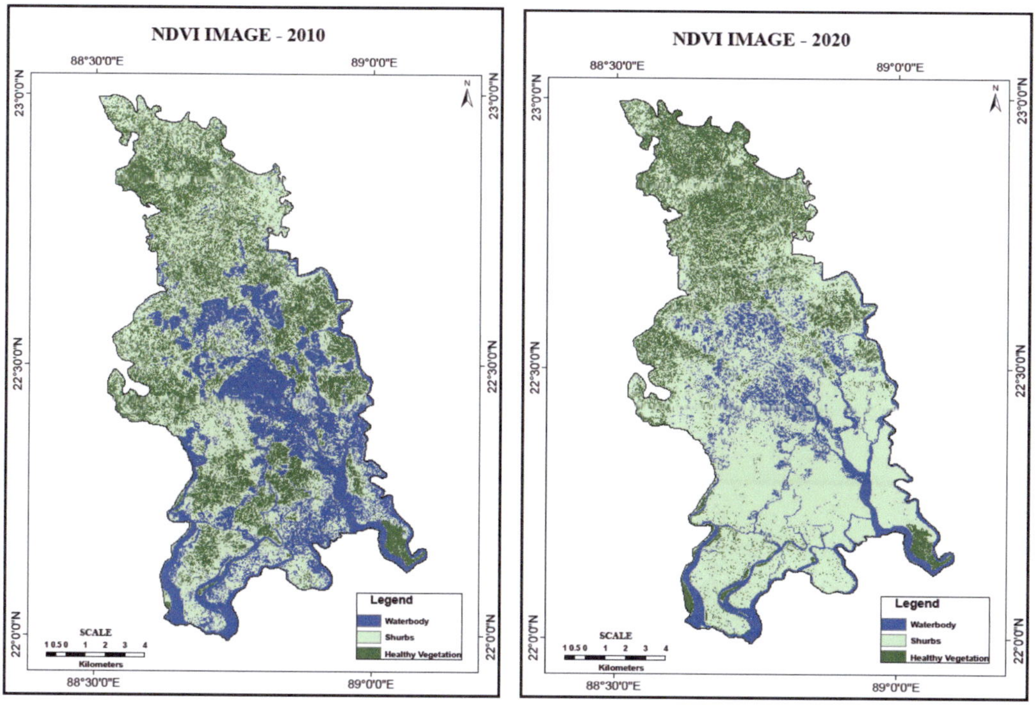

Fig. 20.15 Normalized difference vegetation index map of the Ichhamati–Bidyadhari interfluves region of 2010 and 2020

Table 20.2 Changes in the vegetation cover in the Ichhamati-Bidyadhari interfluve region between 2010 and 2020

CLASS	2010 (Area in ha.)	2020 (Area in ha.)
Water body	97749.5	51557.7
Shrubs (open forest, degraded forest land)	147,041	213,777
Healthy vegetation	91045.1	70500.3

Source: Data analysed by authors

to very high soil salinity encroachment was increased mostly in the southern blocks of the interfluves, indicating the salinity encroachment in the interfluves area of Sundarbans (Fig. 20.16).

Changes in land surface temperature: A significant increase in the average temperature of the land surface of the present interfluves from 2009 to 2019 was observed in the area of aquaculture (0.84 °C), rural (0.69 °C), and urban settlements (0.92 °C), and also in the zones with mangrove cover (0.62 °C) (Fig. 20.17).

The observed changes or the difference in max temperature were highest for the water body (27 ° C to 30.13 °C), followed by the sandy (27.02 °C to 29.14 °C) and the quarry (28.60 °C to 31.40 ° C) (Table 20.3).

The Sundarbans coastal ecosystem has become susceptible to both climate change hazards and social changes. Based on the Rennel Map I and XX, it was noted that the inhabitant of Sundarbans extended their land area by polderization along the low tide line during the nineteenth century (extensively in Hasnabad, Canning, Haroa, Kultali, Bhangar, Patharpratima) until few areas of Sundarbans declared to be protected in the year 1878 (Chatterjee 1990; Ghosh and Mistri 2020). As a consequence,

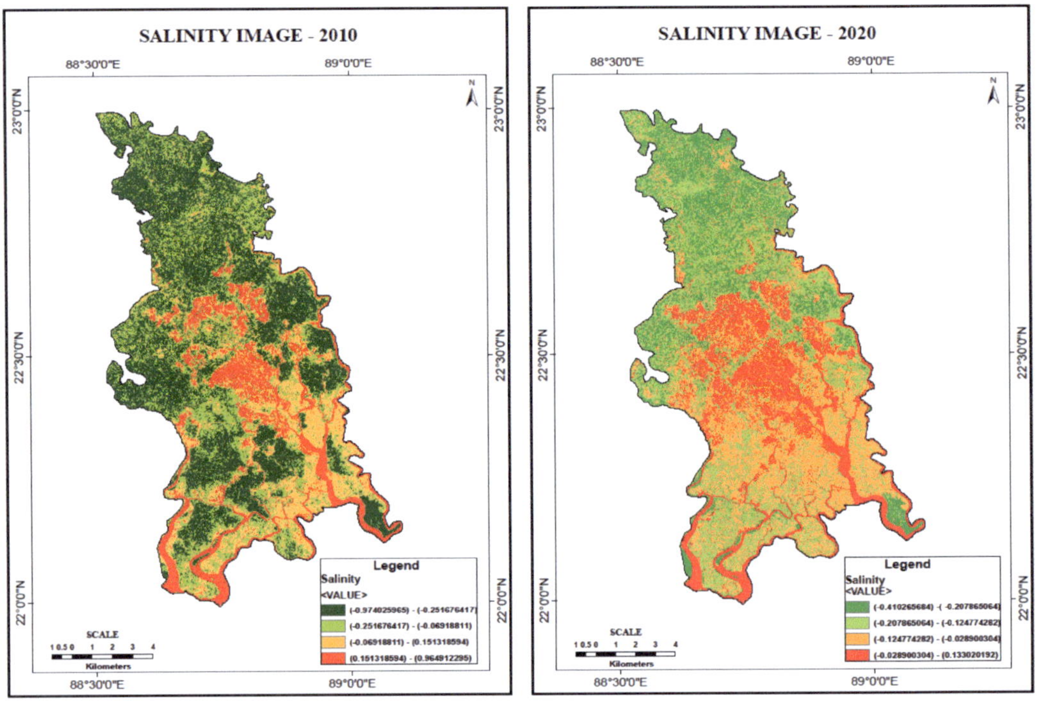

Fig. 20.16 Normalized difference salinity index map of the Ichhamati-Bidyadhari interfluve region, 2010. (Source: Data analysed by authors)

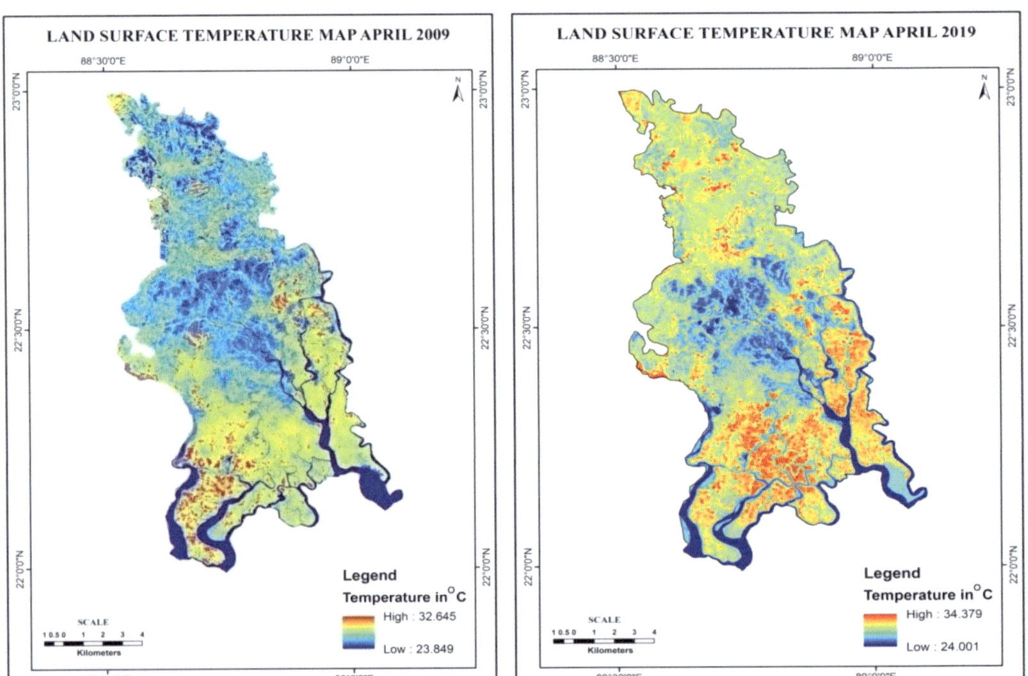

Fig. 20.17 Land surface temperature map of 2009 and 2019 of the Ichhamati-Bidyadhari interfluve region

Table 20.3 Land surface temperature changes between 2009 and 2019

	Land surface temperature (°C)						
	2009			2019			Decadal increase in average temperature (°C)
Land use/land cover class	Min	Max	Mean	Min	Max	Mean	
Aquaculture	24.12	28.50	26.31	25.01	29.30	27.15	0.84
Agriculture land	25.50	29.70	27.60	25.00	30.30	27.65	0.05
Water body	26.01	27.00	26.50	24.02	30.13	27.07	0.57
River/canal/drain	23.84	26.91	25.37	24.00	27.21	25.60	0.23
Wetland	24.82	27.55	26.18	25.00	28.21	26.60	0.42
Rural settlement	28.00	31.01	29.50	28.09	32.29	30.19	0.69
Urban settlement	28.36	32.64	30.50	28.50	34.34	31.42	0.92
Built-up (others)	27.87	32.00	29.93	27.03	33.29	30.16	0.23
Sandy area	25.27	27.02	26.14	24.00	29.14	26.57	0.43
Quarry	26.74	28.60	27.67	25.00	31.40	28.20	0.53
Mangroves	24.21	28.32	26.26	24.70	29.06	26.88	0.62

Source: Data analyzed by authors

huge silt carried by the tidewater or rivers failed to distribute on the tidal mudflats, raised the channel bed, and potentially influenced the river hydrodynamics. In addition, the conditions of coastal living and non-living environmental components are changing drastically in the Sundarbans following the intense climate change impact. The interfluves of the Ichhamati-Bidyadhari river basin, situated at the lower deltaic part of the Ganges-Brahmaputra delta at around 6 m elevation, have gone through remarkable land-use changes where the livelihood and economy of the densely growing population largely depend on natural resources (Ranjan et al. 2017). Embankments in the Indian Sundarbans imposed a wide range of direct and indirect impacts; they also have some positive and many negative effects on the environment and the people (Dhara and Paul 2016). The premature reclamation of land collapses river linkages and natural drainage systems, driving saltwater far upstream during high tide. Earthen embankments are built to prevent tidal inundation during cyclonic events and bank erosion. During seasonal cyclonic events or in the monsoon, embankment breaching, coastal floods, and overtopping of rivers in the low-lying areas failed to drain off due to the reverse slope of the embankments, creating waterlogged pockets. People who converted the many marshy wetlands to fish farms face economic loss due to such ecological change. However, many lock gates have been built along the polders as artificial drainage systems but do not function well. The embankment or coastal engineering structure whether concrete or muddy is vulnerable to extreme climatic events in Sundarbans. The Amphan cyclone pushed the water level to around 5.45 m, which was above the general height of the embankment and destroyed around 87 km out of 3500 km long coastal engineering structures, including dikes, earthen dams, bamboo fencing with sandbag, or laterite boulders in Sundarbans. Embankment breaching along the Bidyadhari riverbank at Minakhan in Amphan 2020 and Yash 2021 devastated all the fisheries and agricultural land surrounding the area (Report from Down to Earth on 23rd May, 2020, and The Telegraph, 16th June, 2021). Increasing soil and water salinity in the coastal islands of the Indian Sundarbans for the period 1975–2000 was well documented in many studies, and the situation became alarming in the aftermath of Aila, 2009. Basanti, Gosaba, and Sandeshkhali blocks in Sundarbans, India, were already reported to have a high level of soil salinity in the presence of hypersaline channels in the surroundings (Das et al. 2016). Since the sixteenth century, the drying of the Bidyadhari river, which used to serve as a conveyor belt for supplying freshwater to the central sector and eastern sector of the Indian Sundarbans, is a significant reason of ingressive salinity in the central part of the region (Mitra et al. 2009). A gradual increase in the salinity profile is the cumulative impact of river runoff, changes in the precipitation pattern, and evaporation, which subsequently reduce the biomass of *Heritiera fomes*, the glory of Sundarbans (Banerjee et al. 2017). A large mangrove vegetated land along the Matla-Bidya riverbank areas had slumped off. Decreasing canopy cover with increasing human pressure allows direct heating of the land surface, followed by a higher rate of evaporation, except in monsoon months, which is found to be an important driver of increasing surface soil salinity.

According to the Millennium Ecosystem Assessment (2005), human population growth and economic development are major drivers of wetland degradation, primarily through land-use changes, infrastructure development, pollution, resource overuse, and invasive species. Wetlands vary in size, catchment area, population, and development stage, but all face direct pressures from surrounding economic activities. Key sectors influencing these wetlands include agriculture, aquaculture, fisheries, tourism, urbanization, shipping, industry, and mining. Human activities such as land reclamation, damming, water extraction, port construction, pollution, and logging (especially in mangroves) lead to habitat loss, disrupted hydrology, and pollution. These pressures result in environmental degradation—like erosion, subsidence, and hypoxia—and ecological damage, including the loss of salt marshes, seagrasses, and mangroves. This, in turn, diminishes ecosystem services such as climate regulation, carbon sequestration, and nutrient cycling, negatively affecting human welfare. The situation is worsened by indirect threats like sea-level rise. Despite their value, coastal wetlands and mangrove ecosystems are undervalued, leading to unsustainable land-use changes. All coastal wetlands located in or near urban areas around the world have been found to be impacted in various ways, regardless of their conservation status. These include Mackenzie-Beaufort Coastal Wetland, near urban settlements in northern Canada; Coastal Wetlands in North America, including the Mississippi River Delta Coastal Wetlands, adjacent to major urban centres in the United States; Bahia Blanca Wetlands, near the city of Bahía Blanca, Argentina; Venice Lagoon Wetlands, within the urban area of Venice, Italy; Watamu, Mangrove Wetland of Mida Creek, close to urbanized coastal areas in Kenya; Vembanad Wetlands, surrounding urban centres such as Kochi in India; Huang He (Yellow River) Delta Wetland, adjacent to industrial and urban zones in the People's Republic of China; and Mangrove Forests of the Gulf of Papua, influenced by nearby urban development in Papua New Guinea (Newton et al. 2020). These wetlands are increasingly under pressure from urban expansion, industrialization, and associated human activities, which contribute to ecological degradation despite conservation efforts. Conserving and managing these ecosystems is crucial for achieving several UN Sustainable Development Goals (especially SDGs 2, 3, 6, 12, 13, 14, and 15).

India has implemented several laws and policies aimed at wetland conservation and is a signatory to key international agreements such as the Ramsar Convention and the Convention on Biological Diversity (CBD). Despite this, the enforcement of these regulations remains weak, leaving coastal wetlands increasingly vulnerable to threats like land-use changes, rapid and unregulated development, and the growing impacts of climate change. Most conservation efforts so far have focused on designating Marine and Coastal Protected Areas (MCPAs). While such designations may help curb specific threats like timber extraction, they fall short in addressing other complex and widespread drivers of degradation. Challenges such as inadequate infrastructure and limited resources have further hindered the effectiveness of MCPAs in achieving their intended conservation goals.

The National Environmental Policy of 2006 led to the introduction of the Wetlands (Conservation and Management) Rules in 2010, which were revised by the Ministry of Environment, Forest and Climate Change (MoEF&CC) in 2017. Additionally, the Coastal Regulation Zone (CRZ) Notification of 2018 marked a significant step by classifying coastal wetlands as ecologically sensitive areas, providing them with a stronger legal safeguard.

Coastal wetlands possess an inherent capacity for vertical and horizontal adjustment, allowing them to adapt to both environmental and human-induced changes. However, preserving this adaptive potential requires urgent action. Strengthening governance, enhancing enforcement, and adopting site-specific management approaches are essential for the effective and long-term conservation of coastal wetlands in India (Vincent and Owens 2021).

The major population of the Indian Sundarbans region is shifting their economy from agriculture to fishery, exerting pressure on the ecosystem by their livelihood dependency on the living resources like shrimp, shell collection, and crab potting, which affect the natural regeneration of mangrove sapling and the land accretion process. Many rural families were engaged in supplying Tiger shrimp seed to fish farms of local and neighbouring states. A field-based survey in post-disaster of Aila super cyclone explained the embankment failure as an aggregated impact of long-term erosion of the river channels, steepness of embankment structures, and seepage water exerted pore pressure in clay particles (Bhattacharya et al. 2014). Therefore, mangrove afforestation and conservation as a natural coastal engineering structure became foremost important to protect the hinterland of the Sundarbans.

20.5 Conclusion

In the history of human civilization, coastal environments have undergone rapid transformation, as half of the world population lived within 200 km of the sea (Cosby et al. 2024). As coastal megacities grow, land reclamation has become a common strategy to create space and combat erosion, with major projects seen in cities like Mumbai, Tokyo, Hong Kong, and the Netherlands. Over half (55%) of the 2303 Wetlands of International Importance are affected by multiple impact factors, mainly pollution (54%), resource use (53%), natural system changes (53%), and agriculture/aquaculture (42%) (Xu et al. 2019). Rising heavy metal levels in coastal waters and wetlands were reported in Mumbai, Goa, and Singapore. However, this urban expansion poses challenges for environmental sustainability and disaster management. Kolkata's urban pollution and river dumping affect the health of the delta, and in turn, the Sundarbans. Rapid urbanization and climate change amplify the need for integrated urban-rural-environmental planning between the KMA and the Sundarbans region. The East Kolkata Wetlands (EKW) with an area of 125 km^2 recognized as 'a rare example of environmental protection and development management,' and considered as an ecozone of global importance because of its potential for natural recycling of urban waste through wastewater-fed fisheries and agricultural activities. The Bidyadhari river was once vital for supplying freshwater to the central and eastern Indian Sundarbans. The interfluves of the Ichhamati-Bidyadhari region located in the reclaimed part of Indian Sundarbans have been facing challenges of greater siltation in the riverbed, receiving less amount of freshwater from the upstream, introduction of exotic species, and alteration of traditional land-use pattern, while being brought to bear city wastes and agricultural-aquaculture waste from surroundings over the last three decades.

Over the past few decades, land reclamation pressures have influenced the topography and hydrology of the EKW and the Sundarbans. Unplanned urbanization, the construction of embankments in water channels, fragmentation of the wetlands to smaller water bodies (EKW) to expand the Kolkata Metropolitan Area, urban agriculture, and fisheries, altered meteorological conditions such as changing rainfall patterns and urban heat island effects, reduced freshwater flow, and increased pollution load. Additionally, infrastructure development like the EM Bypass, unregulated industrial activities (e.g. Bantala leather complex), and the spread of brick kilns further disrupt the ecosystem. Together, these drivers hinder the settlement, regeneration, and natural propagation of mangroves and the faunal diversity they support. Multifaceted problems in this mangrove estuarine belt such as climate change-induced sea-level rise, global warming, cyclonic storm surges, progressive salinity intrusion into water and soil, clearing of mangroves, habitat fragmentation, and the construction of concrete embankments hamper the formation of tidal mudflats, impose varying pressures on the ecosystem's productivity, and threaten forest-based livelihoods and the survival of the poor inhabitants of this coastal region. The ecological studies and data-driven analysis identified how the urban wastewater extends and degrades the Sundarbans mangrove ecosystem in the long term. The Sundarbans once had a remarkable natural bioremediation capacity, with its mangrove forests acting as natural filters that trapped pollutants, absorbed excess nutrients, and supported microbial processes to break down harmful substances. This self-cleansing ability played a crucial role in maintaining water quality, sustaining biodiversity, and strengthening the ecosystem's resilience. However, this capacity has significantly deteriorated due to increasing pollution and heavy metal accumulation in flora and fauna. Native microfloras have been disrupted, fish diversity has declined, and pathogenic, drug-resistant microbes are on the rise, leading to health risks and economic losses. Beneficial probiotic species have diminished, making way for harmful bacteria such as Aeromonads, Pseudomonads, and Coliforms. Additionally, key water-purifying species like clams, oysters, and molluscs are also rapidly disappearing. Integrating wetlands into urban planning is essential for fostering long-term ecological and societal sustainability, ensuring cities are better equipped to adapt to environmental stresses while preserving the well-being of their communities. Wastewater quantity and quality, particularly heavy metal content and the presence of pathogenic organisms, should be maintained within safe limits. Efforts are needed to conserve wetland habitats and the biodiversity of migratory water birds, control invasive fish species, and address livelihood vulnerabilities by providing alternative sources of income. Additionally, the promotion of salt-tolerant and climate-resilient traditional rice varieties, supported by AI tools for disaster navigation, response, and recovery, is essential.

Wetland management remains inadequate, particularly in Africa and Asia. Effective global wetland protection and restoration should focus on five key areas: management and policy, monitoring, restoration, knowledge, and funding. Analyzing landscape and land-use changes can inform policymaking, guiding sustainable development and the management of natural resources. Assessments are needed on people who collect resources and depend on wetlands and coastal ecosystems and on the eco-management strategies of the stakeholders, including NGOs, forest protection committee, eco-development committee, to ensure sustainable resource utilization. Eco-monitoring (indicator-based assessment, vulnerability indices to episodic and chronic factors) research activities linked with restoring biodiversity conservation efforts to protect, restore the Sundarbans estuarine complex of India should extend beyond the protected landmass of the biosphere region to surrounding landscapes like watersheds, forests, flood plains, and river basins that are ecologically interconnected. Violations of the Wetlands (Conservation and Management) Rules, 2017, and the East Kolkata Wetlands (Conservation and Management) Act, 2006, must be curbed.

Potential Conflicts of Interest On behalf of all authors, the corresponding author states that there is no conflict of interest.

Informed Consent Informed consent was obtained from all individual participants included in the study.

Funding The research was not supported by any funding.

References

Abeywardhana DC, Adikaram NM, Kularatne RKA (2022) Are mangrove forests reliable sinks of heavy metals due to phytoremediation and other mechanisms? A Sri Lankan perspective. Mar Pollut Bull 177:113453. https://doi.org/10.1016/j.marpolbul.2022.113453

Bakshi M, Ghosh S, Chakraborty D, Hazra S, Chaudhuri P (2018) Assessment of potentially toxic metal (PTM) pollution in mangrove habitats using biochemical markers: a case study on Avicennia officinalis L. in

and around Sundarban, India. Mar Pollut Bull 133: 157–172. https://doi.org/10.1016/j.marpolbul.2018.05.030
Banerjee K, Gatti RC, Mitra A (2017) Climate change-induced salinity variation impacts on a stenoecious mangrove species in the Indian Sundarbans. Ambio 46(4):492–499. https://doi.org/10.1007/s13280-016-0839-9
Barsi JA, Schott JR, Hook SJ, Raqueno NG, Markham BL, Radocinski RG (2014) Landsat-8 thermal infrared sensor (TIRS) vicarious radiometric calibration. Remote Sens 6(11):11607–11626. https://www.mdpi.com/2072-4292/6/11/11607
Barua DK, Kuehl SA, Miller RL, Moore WS (1994) Suspended sediment distribution and residual transport in the coastal ocean off the Ganges-Brahmaputra river mouth. Mar Geol 120(1–2):41–61. https://doi.org/10.1016/0025-3227(94)90076-0
Bhadra T, Mukhopadhyay A, Hazra S (2017) Identification of river discontinuity using geo-informatics to improve freshwater flow and ecosystem services in Indian Sundarban Delta. In: Environment and earth observation: case studies in India. Springer, Cham. https://doi.org/10.1007/978-3-319-46010-9_10
Bhattacharya JP, Giosan L (2003) Wave-influenced deltas: geomorphological implications for facies reconstruction. Sedimentology 50(1):187–210. https://doi.org/10.1046/j.1365-3091.2003.00545.x
Bhattacharya S, Dash JR, Patra PH, Dubey SK, Das AK, Mandal TK, Bandyopadhyay SK (2014) Spatio-temporal variation of mercury in Bidyadhari River of Sundarban delta, India. Explor Anim Med Res 4(1): 19–32
Bhattacharya BD, Nayak DC, Sarkar SK, Biswas SN, Rakshit D, Ahmed MK (2015) Distribution of dissolved trace metals in coastal regions of Indian Sundarban mangrove wetland: a multivariate approach. J Clean Prod 96:233–243. https://doi.org/10.1016/j.jclepro.2014.04.030
Chakraborty SK, Sanyal P, Ray R (2023) Eco-biological uniqueness of East Kolkata wetland, a Ramsar site in India. In: Wetlands ecology: eco-biological uniqueness of a Ramsar site (East Kolkata wetlands, India). Springer, Cham, pp 257–301. https://doi.org/10.1007/978-3-031-09253-4_5
Chatterjee S (1990) Land reclamations in the Sundarbans-an overview. In: Proceedings of the Indian history congress, vol 51. Indian History Congress, pp 440–446. https://www.jstor.org/stable/44148257
Chernoff B (1975) A method for wet digestion of fish tissue for heavy metal analyses. Trans Am Fish Soc 104(4):803–804
Chowdhury GR, Pal N, Zaman S, Saha A, Mitra A (2017) Shrimp seed collection in Indian Sundarban estuaries: a threat to overall estuarine ecosystem services. J Environ Soc Sci 4(1):127
Cosby AG, Lebakula V, Smith CN, Wanik DW, Bergene K, Rose AN et al (2024) Accelerating growth of human coastal populations at the global and continent levels: 2000–2018. Sci Rep 14(1):22489
Das GK (2015) River systems and tidal courses. In: Estuarine Morphodynamics of the Sunderbans. Springer, Cham, pp 23–45. https://doi.org/10.1007/978-3-319-11343-2_2
Das S, Choudhury MR, Nagarajan M (2016) Earth observation and geospatial techniques for soil salinity and land capability assessment over Sundarban Bay of Bengal coast, India. Geodesy Cartogr 65(2)
Dhara S, Paul AK (2016) Embankment breaching and its impact on local community in Indian Sundarban: a case study of some blocks of South West Sundarban. IJISET 3(2):23–32
Everard M, Kangabam R, Tiwari MK, McInnes R, Kumar R, Talukdar GH, Das L (2019) Ecosystem service assessment of selected wetlands of Kolkata and the Indian Gangetic Delta: multi-beneficial systems under differentiated management stress. Wetl Ecol Manag 27:405–426
Ghosh S, Mistri B (2020) Geo-historical appraisal of embankment breaching and its management on active tidal land of Sundarban: a case study in Gosaba Island, south 24 Parganas, West Bengal. Space Cult India 7(4):166–180. https://doi.org/10.20896/saci.v7i4.587
Ghosh A, Mukhopadhyay S (2016) Quantitative study on shoreline changes and Erosion Hazard assessment: case study in Muriganga–Saptamukhi interfluve, Sundarban, India. Model Earth Syst Environ 2(2):75. https://doi.org/10.1007/s40808-016-0130-x
Ghosh S, Bakshi M, Mahanty S, Chaudhuri P (2022) Assessment of role of rhizosphere process in bioaccumulation of heavy metals in fine nutritive roots of riparian mangrove species in river Hooghly: implications to global anthropogenic environmental changes. Mar Pollut Bull 174:113157. https://doi.org/10.1016/j.marpolbul.2021.113157
Gibb H, O'Leary KG, Sarkar SK, Wang J, Liguori L, Rainis H et al (2016) Hair mercury concentrations in residents of Sundarban and Calcutta, India. Environ Res 150:616–621. https://doi.org/10.1016/j.envres.2016.03.028
Gour J (2012) Changing fluvio-geomorphological environment in the Matla-Bidyadhari interfluve- a model unit of active & mature Indian Sundarbans. Geo Anal 2:No.1
Hakanson L (1980) An ecological risk index for aquatic pollution control. A sedimentological approach. Water Res 14(8):975–1001
Higgins SA, Overeem I, Rogers KG, Kalina EA, Chadwick O, Passalacqua P (2018) River linking in India: downstream impacts on water discharge and suspended sediment transport to deltas. Elem Sci Anth 6. https://doi.org/10.1525/elementa.269
Jiménez-Muñoz JC, Sobrino JA (2006) Error sources on the land surface temperature retrieved from thermal infrared single channel remote sensing data. Int J Remote Sens 27(05):999–1014. https://doi.org/10.1080/01431160500075907
Khan NM, Rastoskuev VV, Sato Y, Shiozawa S (2005) Assessment of hydrosaline land degradation by using a simple approach of remote sensing indicators. Agric

Water Manag 77(1–3):96–109. https://doi.org/10.1016/j.agwat.2004.09.038
Li T, Liang C, Han X, Yang X (2013) Mobilization of cadmium by dissolved organic matter in the rhizosphere of hyperaccumulator sedum alfredii. Chemosphere 91(7):970–976. https://doi.org/10.1016/j.chemosphere.2013.01.100
Lillesand T, Kiefer RW, Chipman J (2015) Remote sensing and image interpretation. Wiley
Lucisine P, Lecerf A, Danger M, Felten V, Aran D, Auclerc A et al (2015) Litter chemistry prevails over litter consumers in mediating effects of past steel industry activities on leaf litter decomposition. Sci Total Environ 537:213–224
Mandal U, Nayak D, Mullick S, Samui A, Mahanta K, Raut S, Roy S, Kumar A, Burman D (2019) Trend analysis of weather parameters over Indian Sundarbans. J Agrometeorol 21:10.54386/jam.v21i3.253
Mitra A, Gangopadhyay A, Dube A, Schmidt ACK, Banerjee K (2009) Observed changes in water mass properties in the Indian Sundarbans (northwestern Bay of Bengal) during 1980–2007. Curr Sci 97(10):1445–1452. http://www.jstor.org/stable/24107336
Mondal M, Satpati L (2020) Human intervention on river system: a control system—a case study in Ichamati River, India. Environ Dev Sustain 22:5245–5271
Mukherjee DP, Kumar B (2012) Evaluation of metal contamination in freshly deposited sediment of Hugli estuary, India. Arch Appl Sci Res 4(2):1155–1168
Mukherjee I, Singh UK, Singh RP, Kumari D, Jha PK, Mehta P (2020) Characterization of heavy metal pollution in an anthropogenically and geologically influenced semi-arid region of East India and assessment of ecological and human health risks. Sci Total Environ 705:135801. https://doi.org/10.1016/j.scitotenv.2019.135801
Newton A, Icely J, Cristina S, Perillo GM, Turner RE, Ashan D et al (2020) Anthropogenic, direct pressures on coastal wetlands. Front Ecol Evol 8:144
Nguyen TH, Hoang HNT, Bien NQ, Tuyen LH, Kim KW (2020) Contamination of heavy metals in paddy soil in the vicinity of Nui Phao multi-metal mine, North Vietnam. Environ Geochem Health 42:4141–4158
Olsen SR, Sommers LE (1982) Phosphorus. In: Page AL (ed) Methods of soil analysis part 2 chemical and microbiological properties. American Society of Agronomy, Soil Science Society of America, Madison, pp 403–430
Pethick J, Orford JD (2013) Rapid rise in effective sea-level in Southwest Bangladesh: its causes and contemporary rates. Glob Planet Chang 111:237–245. https://doi.org/10.1016/j.gloplacha.2013.09.019
Rahman MS, Rahman MS (2021) Effects of elevated temperature on prooxidant-antioxidant homeostasis and redox status in the American oyster: Signaling pathways of cellular apoptosis during heat stress. Environ Res 196:110428. https://doi.org/10.1016/j.envres.2020.110428
Ranjan AK, Sivathanu V, Verma SK, Murmu L, Kumar PBS (2017) Spatio-temporal variation in Indian part of Sundarban Delta over the years 1990–2016 using geospatial technology. Int J Geomat Geosci 7(3):275–292
Ray R, Mandal SK, González AG, Pokrovsky OS, Jana TK (2021) Storage and recycling of major and trace element in mangroves. Sci Total Environ 780:146379. https://doi.org/10.1016/j.scitotenv.2021.146379
Rinklebe J, Antoniadis V, Shaheen SM, Rosche O, Altermann M (2019) Health risk assessment of potentially toxic elements in soils along the central Elbe River, Germany. Environ Int 126:76–88. https://doi.org/10.1016/j.envint.2019.02.011
Saha SB, Mitra A, Bhattacharyya SB, Choudhury A (2001) Status of sediment with special reference to heavy metal pollution of a brackishwater tidal ecosystem in northern Sundarbans of West Bengal. Trop Ecol 42(1):127–132
Singh A, Prasad SM (2015) Remediation of heavy metal contaminated ecosystem: an overview on technology advancement. Int J Environ Sci Technol 12:353–366
Song C, Rohr RP, Vasseur D, Saavedra S (2020) Disentangling the effects of external perturbations on coexistence and priority effects. J Ecol 108(4):1677–1689. https://doi.org/10.1111/1365-2745.13349
Suresh G, Ramasamy V, Sundarrajan M, Paramasivam K (2015) Spatial and vertical distributions of heavy metals and their potential toxicity levels in various beach sediments from high-background-radiation area, Kerala. Mar Pollut Bull 91(1):389–400. https://doi.org/10.1016/j.marpolbul.2014.11.007
Thenkabail PS, Enclona EA, Ashton MS, Van Der Meer B (2004) Accuracy assessments of hyperspectral waveband performance for vegetation analysis applications. Remote Sens Environ 91(3–4):354–376. https://doi.org/10.1016/j.rse.2004.03.013
Trivedi RK, Goel PK (1984) Chemistry and biological methods of water pollution studies. Environmental Publication, Karad
Vincent SGT, Owens KA (2021) Coastal wetlands of India: threats and solutions. Wetl Ecol Manag 29(5):633–639
Wang W, Lin C, Wang L, Jiang R, Liu Y, Lin H, Chen J (2021) Effects of human activities on the spatial distribution, ecological risk and sources of PTEs in coastal sediments. Int J Environ Res Public Health 18(23):12476
Xu T, Weng B, Yan D, Wang K, Li X, Bi W et al (2019) Wetlands of international importance: status, threats, and future protection. Int J Environ Res Public Health 16(10):1818
Yang W, Liu D, Wang Y, Hussain B, Zhao F, Ding Z, Dawood M (2021) Variations in phytoremediation potential and phytoavailability of heavy metals in different Salix genotypes subjected to seasonal flooding. J Environ Manag 299:113632
Zakaria Z, Zulkafflee NS, Mohd Redzuan NA, Selamat J, Ismail MR, Praveena SM et al (2021) Understanding potential heavy metal contamination, absorption, translocation and accumulation in rice and human health risks. Plants 10(6):1070. https://www.mdpi.com/2223-7747/10/6/1070

Integrating Science, Policy, and Stakeholder Engagement in Wetland Conservation: Strategies for Effective Management and Sustainable Outcomes

21

Azamat Ali and Shohrat Ali

Abstract

Wetland conservation plays three vital roles such as protecting biodiversity, managing water cycles, and offering ecosystem services. Effective management of wetlands necessitates the integration of scientific research, policy frameworks, and stakeholder engagement. The chapter examines the cooperative use of these elements to sustainably conserve vital areas. The study focused on wetland ecology and modern scientific achievements demonstrates how strong scientific study supports sound policy development. The Ramsar Convention, together with national and regional policies, undergoes assessment for its implementation consequences while addressing their obstacles. The chapter stresses that stakeholder engagement stands as an essential element, so it offers methods to involve local people together with state agencies and nonprofit organizations in protection work. The chapter uses successful integration cases to show readers how innovative and best practice approaches operate for wetland management. This chapter discusses prevailing obstacles when merging science with policy and stakeholder participation while presenting methods to advance their cooperative practices and effectiveness results. The study gives direction to current and future wetland conservation practices through the discussion of upcoming trends. The method includes all necessary parts to establish enduring, sustainable wetland management techniques that sustain the ecological health of wetlands in the long term.

Keywords

Wetland conservation · Stakeholder engagement · Policy integration · Ecosystem management · Sustainable outcomes

21.1 Introduction

Wetlands hold fundamental importance throughout the world because they generate various essential ecological and economic functions and cultural benefits. Under their natural filtering mechanism, pollutants are removed, which leads to improved water quality for maintaining ecosystem health (Zedler and Kercher 2005). A prime feature of wetlands is their ability to support thriving biodiversity since they serve as habitats for both common and endangered species. Freshwater wetlands act as habitats which support 40%

A. Ali (✉)
School of Liberal and Creative Arts (Journalism and Mass Communication), Lovely Professional University, Phagwara, Punjab, India

S. Ali
Department of Civil Engineering, Central University of Jharkhand, Ranchi, India
e-mail: shohrat.ali@cuj.ac.in

S. K. Chakraborty et al. (eds.), *Wetlands and Urban Resilience*, Sustainable Development Goals Series,
https://doi.org/10.1007/978-3-032-06881-1_21

of all species across the world (Xu et al. 2019). Peatlands, among other wetlands, contain quantities of carbon that exceed those found in tropical rainforests; thus, they play a critical role in climate change mitigation. Wetlands represent major components of significant economic value. Wetlands generate essential resources such as irrigation water and fisheries aid, as well as an ecotourism boost. Local communities use these activities to ensure their food security while earning income in numerous geographic regions (Barbier 2011). Wetlands support environmental sustainability through their cost-saving wastewater treatment process known as constructed wetlands (Vymazal 2011). The instances underline how wetlands produce economic advancement because of their dual contribution to financial development.

Various cultures within societies strongly recognize the worth of wetland areas. The Sundarbans and the Okavango Delta serve multiple purposes because these famous wetlands support many plant and animal species while forming a foundation for traditional local beliefs and spiritual practices. The cultural value of wetlands extends to accurate portrayal through art as well as literature and traditional rituals, where Baral et al. (2016) documented this type of human cultural enrichment. The cultural importance that surrounds wetlands increases their total value dimension. The rising number of threats endangers the importance of wetlands. The destruction of habitat presents an enormous threat to ecosystems because of human actions involving urban development, together with farming and industry. Drainage and land reclamation during the early twentieth century resulted in the excessive loss of 64% of global wetland territories (Ballut-Dajud et al. 2022). Decreased wetlands result in decreased biodiversity and reduced ecological benefits provided by these ecosystems. The process of pollution makes wetlands worse off in their overall condition. The combination of untreated sewage waste from agriculture and industry produces pollution that causes water contamination, which leads to eutrophication that decreases oxygen levels, thus killing aquatic animals. Industrial chemical emissions poison water supply systems, thus causing harm to wildlife, together with human communities (Mushtaq et al. 2020).

The environmental threat currently affecting wetlands comes from climate change. Changes in rainfall patterns lead to drying up of interior wetlands at the same time, while rising ocean water erodes coastal wetlands. The temperature increase of peatlands poses an urgent problem because it reduces biodiversity levels while enabling carbon storage loss, which intensifies global warming. The presence of invasive species damages wetlands when these species outcompete local flora and fauna, which results in ecosystem degradation together with decreased biodiversity (Zedler and Kercher 2005). Weak policies together with poor governance create additional obstacles for effective wetland conservation. Implementation problems, along with insufficient policy enforcement, reduce the success rate of Ramsar Convention guidelines (Bridgewater and Kim 2021). Wetland threats become more difficult to manage due to the absence of comprehensive management plans. Wetlands security demands continuous collaboration between scientists together with policymakers and local residents. Multiple efforts, such as stronger laws and enhanced international partnerships and investments in mangrove planting and peatland restoration, have been identified as essential by Xu et al. (2019) for wetland conservation success. The implementation of wetlands in climate strategies becomes essential because they serve as fluid carbon reservoirs and water management systems. The conservation of natural resources receives greater public backing when awareness programs are implemented.

This chapter introduces an innovative approach to wetland conservation through a model where scientific research enables policy development supported by community participation. This chapter unites previous isolated studies by analyzing their supportive interactions rather than conducting separate evaluations. The study primarily demonstrates policy formation with scientific research alongside a review of practical implementation across international and national domains, together with strategies for community outreach in wetland conservation. Real-life examples serve as a basis for the chapter to demonstrate more effective, sustainable strategies for protecting wetlands.

21.2 Ecological Significance of Wetlands

21.2.1 Role of Wetlands in Biodiversity Conservation

A wide range of plant and animal species find their habitat in wetlands alongside multiple species considered threatened and endangered. Wetlands provide three vital survival services to animals by functioning as fish nurseries and providing shelters for amphibians and serving as breeding habitats for birds, mammals, and insects. The Sundarbans in India holds both the Bengal tiger with many aquatic species, as well as being a UNESCO World Heritage Site (Mistri et al. 2020). The African Okavango Delta maintains several large mammals, including elephants, lions, and hippos, in addition to various bird species. During their seasonal movements, the migratory Siberian crane depends on the Indian wetland Keoladeo National Park for sanctuary (Baral et al. 2016). Genetic diversity in ecosystems remains possible because of the importance of wetlands. Wetlands provide organisms with the ability to adapt their genetics to respond to environmental changes. The ecosystems function as links between land and water to support the nutrient cycles and energy flow while providing complex habitats. The maintenance of both local and global biodiversity depends on wetlands (Mitsch and Gosselink 2015).

21.2.2 Ecosystem Services Provided by Wetlands

21.2.2.1 Water Purification

Natural water purification occurs through wetlands by removing sediment, together with heavy metals and pollutants from water. These water habitats act as nutrient scarcity reserves that remove excessive nitrogen and phosphorus elements to block downstream algal blooms. Scientists have proven that wetlands effectively lower nitrate content by 90% (Vymazal 2011). The wetlands located in Kolkata, India, remove wastewater from the region while providing benefits to biodiversity in addition to sustaining community employment. Constructing wetlands has become a common method for wastewater treatment across municipal and industrial facilities because they successfully improve water quality. Wetlands achieve two critical processes through water recharge and flow regulation, which simultaneously reduce floods while coping with drought conditions. Wetlands act as shoreline defenders to minimize both coastal storms and floods. Water purification, hydrological regulation, biodiversity support, and coastal protection represent only some of the valuable services that wetlands provide to foster global socioeconomic significance (Omondi and Navalia 2020).

21.2.2.2 Flood Control

The proper functioning of wetlands serves as natural flood protection since they absorb rainfall to retain floodwaters. Both inland and coastal wetlands serve different functions within the environment, where inland wetlands work as sponges while coastal wetlands, including mangroves and salt marshes, provide defense against storm surges and hurricanes. According to Costanza et al. (2008), the flood damage reduction across the Mississippi River Basin wetlands reaches billions of dollars annually. Strategic water management in wetlands produces substantial flood protection enhancements (50% improvement), which yield economic profits exceeding their expenses through 1.9 monetary benefits. Wetlands lower flood dangers through their capacity to store and hold water, along with water purification. Wetlands in a watershed context deliver major flood control services, foster habitats, enhance water quality, and enable recreational activities (Leon et al. 2018).

21.2.2.3 Carbon Sequestration

Wetlands serve as climate change mitigation hubs by keeping considerable amounts of carbon. The small area of land covered by peatlands represents only 3% but these lands contain 30% of global soil carbon stocks. The breakdown of peatlands leads to the emission of greenhouse gases that escalate global temperatures. Peatlands in Indonesia become more fire resistant, while their carbon release decreases following restoration activities (Harenda et al. 2018). Mangroves simultaneously absorb substantial amounts of carbon, which supports coastal human settlements. The conservation of wetlands remains imperative because they operate as key carbon-trapping sites to help fight climate change. Conversion or drainage of wetlands releases trapped carbon substance, which turns into carbon dioxide gas (Moomaw et al. 2018). Active management alongside inland migration helps protect wetlands from releasing their stored carbon dioxide.

21.2.3 Case Studies: Success Stories in Wetland Restoration

21.2.3.1 The Everglades, United States

The Florida Everglades received major rehabilitation efforts to solve problems related to contamination and declining habitats. The Comprehensive Everglades Restoration Plan (CERP) has dedicated efforts to manage water movement, while fighting contamination in addition to enabling bird species restoration (Xin 2014). The project demonstrates restoration of wetlands through the successful combination of scientific understanding along with technology, and collaboration between different entities. Multiple restoration techniques have been implemented for many years in the Everglades to study how hydrologic restoration affects wetland ecology. Floridian and federal governance bodies, along with other stakeholders, have developed a restoration plan by constructing around 17,000 hectares of Stormwater Treatment Areas to address both nutrient excess and water flow restoration needs. Surface runoff enters treatment wetlands that purify the water before the Everglades obtains it (Chimney and Goforth 2001).

21.2.3.2 Chilika Lake, India

The largest brackish water lagoon in Asia, known as Chilika Lake, faced an issue with sediment buildup and ecological system damage. Dredging activities, along with biodiversity protection techniques in restoration projects, generated better fish stocks, which drew more migratory birds and bettered local fishing communities' economic conditions (Nazneen et al. 2021). Various endangered and nonendangered plants and animals thrive in the ecological habitat of Chilika Lake, including the Irrawaddy dolphin. The bird sanctuary on Nalaban Island within the lake functions as an essential living space for resident birds, together with seasonal bird migrants. Chilika Lake obtained Ramsar wetland status in 1981, and restoration programs successfully removed it from the Montreux Record. Research conducted about Chilika Lake hydrology and salinity, along with circulation dynamics, demonstrated seasonal changes that sustain the lagoon's productivity and biodiversity (Jayaraman et al. 2007).

21.2.3.3 Danube Delta, Europe

Agricultural pollution, together with habitat degradation, has affected the Danube Delta in Europe. The restoration project has achieved improved water flow along with pollution reduction that resulted in the fish population rebound and recovery of pelicans and eagles among bird species. The restoration measures implemented by the region have made the Delta better prepared to handle droughts and floods. In the Mekong Delta of Vietnam, restoration work includes floodplain connection restoration and improving nutrient efficiency while maintaining biodiversity (Tran et al. 2019). The Mekong Delta deals with continual threats stemming from climate change, along with upstream dam development that hurts its natural water flow. A total of over 500 million people make their homes in river deltas, which also hold the status of high-productivity ecosystems. The destruction of deltas faces multiple threats due to human activity, combined with climate change and development pressure. Different challenges, including repetitive flooding and increased saltwater intrusion as well as sediment loss, impair the region. The restoration process for both hydrology and ecosystem systems enhances delta resilience but normally entails significant financial expenses (Tran et al. 2019).

21.3 Role of Scientific Research in Wetland Conservation

21.3.1 Understanding Wetland Ecology Through Research

Science research acts as the foundation for interpreting wetlands, along with building protective measures for these environments. Researchers now study the connections between water products and species which exist in these ecosystems. Scientists who study wetland plants have proved their capability to transform according to changes in water depth levels alongside changes in ecological nutrients. Climatic adjustments enable wetland vegetation to develop stronger resistance against global warming (Keddy 2010). Scientists have recognized that microorganisms within wetlands have substantial functions that maintain ecosystem stability. Microorganisms regulate methane production together with nitrogen fixation as fundamental processes for preserving ecological equilibrium (Reddy and DeLaune 2003). Scientists use DNA barcoding and environmental DNA (eDNA) as new technologies for speeding up the identification of rare and invasive species. The conservation of biodiversity, together with effective harm species management, benefits from these novel methods (Taberlet et al. 2018).

21.3.2 Using Geospatial Tools for Wetland Monitoring

The monitoring of wetlands becomes effective through present-day Geographic Information Systems (GIS), together with remote sensing technology. Through satellite images, scientists can detect modifications that occur in both wetland territories and vegetation and water quality over time periods. Landsat satellite data serve as a tool for scientists who have used it to identify the Sundarbans and the Mississippi Delta wetland reduction. The special tool known as Synthetic Aperture Radar (SAR) plays an important role in seasonal flooding mapping. Such data hold particular importance regarding bird species which rely on wetlands for their survival. The effectiveness of Synthetic Aperture Radar remains constant irrespective of what weather conditions prevail (Klemas 2013). Detailed mapping of small wetlands becomes possible with drone technology, which uses the name unmanned aerial vehicles (UAVs). The mapping system displays information related to vegetation health in addition to soil moisture and erosion. Wetland restoration planning receives valuable support from these collected data sources (Anderson and Gaston 2013). Strategic predictions regarding the influence of climate change and human interaction on wetlands result from analyzing geospatial data supported by artificial intelligence methods. The combination of these systems into a unified approach provides better conservation and management planning tools.

21.3.3 Benefits of Ecological Modeling and Impact Studies

Scientists depend on ecological models to better understand wetlands, together with their ability to anticipate conservation results. The models demonstrate the interlinkages between water components, together with nutrients and species. The models help identify areas worth protective measures. The models of peatlands demonstrate that these crucial wetlands function as carbon stores while fighting against climate change (Jellinek et al. 2014). The process of Environmental Impact Assessment evaluates the potential wetland damages that result from development activities, including roads and dams. The assessment process provides recommendations for lowering potential damage. The implementation of EIAs in the Mekong Delta resulted in better practices for water resource management in that region. Studies determine the economic worth of wetland advantages, which include flood prevention and sustainable water systems. Research indicates that wetland preservation costs less money and achieves better results than artificial construction methods (Barbier 2011).

21.4 Policy Frameworks for Wetland Conservation

21.4.1 International Agreements and Conventions

21.4.1.1 Ramsar Convention

Through the Ramsar Convention established in 1971, members work together to protect wetlands worldwide. The Convention promotes wetland sustainability by showing its dual ecological and economic contribution to the world. Countries participating in the Ramsar Convention establish Ramsar Sites through the identification of important wetlands which require conservation planning (Finlayson et al. 2011). The number of designated Ramsar Sites around the world exceeded 2400 sites with an area of approximately 250 million hectares of wetlands at the beginning of 2024. However, threats such as urbanization, agricultural expansion, and climate change challenge many Ramsar Sites. Lack of responsible monitoring and insufficient budget resources create major obstacles for proper management of these regions (Finlayson and van Dam 2023). The establishment of effective legislation, together with the Ramsar principle integration in local development strategies, represents the solution to tackle these problems.

21.4.1.2 Sustainable Development Goals (SDGs)

The UN Sustainable Development Goals benefit greatly from wetlands because they support their targets regarding clean water, as well as climate action and life on land. Wetlands deliver three fundamental functions through water purification, together with carbon storage and biodiversity protection (Kerton 2023). The SDGs motivate countries to practice sustainable approaches for wetland restoration and protection efforts. The challenge exists when implementing wetland protection strategies within broader development initiatives because different interests battle for agricultural expansion, together with infrastructure development and urban growth opportunities (Zedler and Kercher 2005). Wetlands will contribute to SDG goals through successful monitoring processes, along with reporting systems and international bilateral partnerships.

21.4.1.3 National and Regional Policies for Wetland Conservation

Countries worldwide have implemented separate protection policies which suit their individual circumstances. The 2017 Wetlands Rules of India offer leadership on wetland identification together with operational guidance for management alongside prohibitions against damaging activities (Kumar et al. 2021). The US Clean Water Act, through its protective measures, both confronts wetland pollution problems and supports wetland restoration initiatives (Mitsch and Gosselink 2015). International partnerships remain essential for the proper management of boundaries which exist between multiple countries. The International Commission for the Protection of the Danube River leads Danube River Basin countries through collaborative efforts to handle shared water reserves together with wetlands (Sommerwerk et al. 2022). Although successful operations have been achieved, the program faces ongoing obstacles resulting from limited financing support, alongside expert shortages and disagreements between key stakeholders. Better policy execution requires governments to establish stronger institutions that work closely with NGOs and local communities for better cooperation.

21.4.2 Policy-Driven Success Stories

21.4.2.1 China's Poyang Lake

Successful wetland protection occurs at Poyang Lake, which represents China's largest freshwater lake, through implemented policy measures. Millions of people depend on Poyang Lake water, while the habitat sustains large bird migration populations. Through protective measures developed by the Chinese government, Poyang Lake received protection from banned sand mining activities and was designated a protected area (Yan et al. 2018). Scientific research acted as a guide for developing water management tactics. Local residents participated through education initiatives and beneficiary programs for environmentally friendly practices. A combination of strategies improved water quality and restored ecosystem functions, as well as created better biodiversity in the area. The outcome demonstrates how merging specific policies with scientific knowledge plus community-based interventions creates successful results.

21.4.2.2 Natura 2000 in Europe

The European Union established Natura 2000 under the Birds and Habitats Directives for protecting important natural habitats such as wetlands. Natura 2000 achieves its protection goals through stringent measures that also prioritize land sustainability. Nature 2000 protected wetlands confirm the effectiveness of strong policy frameworks since biodiversity has increased while ecosystems have improved (Donald et al. 2007). The time commitment to nature protection in Denmark exceeds that of China because Danish environmental laws emerged during the early part of the twentieth century. The EU Directives expanded Danish nature conservation through the establishment of the Natura 2000 network, which protects important habitats in all areas of the country (Su et al. 2010). China has intensified its focus on environmental protection by creating nature reserves throughout 18% of its territory, which exceeded 11,000 designations starting from 2019. The National Wetland Conservation Program seeks to restore wetlands while capturing carbon, which generates substantial effects on biodiversity preservation and water resources preservation worldwide.

21.5 Stakeholder Engagement in Wetland Management

21.5.1 Defining Stakeholders: Local Communities, Governments, NGOs, and Private Sector

Wetland management needs an active partnership between four essential groups, which include local communities together with governments and NGOs, and the private sector. Wetlands supply local people with three fundamental resources: their economic basis, traditional cultural activities, and water and food supply. Their involvement in conservation practices enables strategies to match their needs and values, which creates sustained commitment according to Díaz et al. (2015). Such communities share their traditional ecological familiarity, which strengthens scientific methods used in management frameworks (Berkes 2017). Governments take primary responsibility to create regulatory frameworks while delivering fundamental and specialized backing for wetland protection (Vélez et al. 2018). The work of NGOs supports conservation by advocacy as well as public awareness campaigns and technical expertise, which helps them oversee projects with community involvement alongside governmental authorities (Azis 2022). Agriculture, tourism, and energy industry businesses participate actively in the conservation effort alongside the public sector. The private sector should integrate sustainable practices while participating in corporate social responsibility initiatives with partnership programs through conservation groups to achieve both environmental protection and wetland conservation targets.

21.5.2 Importance of Participatory Approaches and Community-Based Conservation

Wetland management requires participatory approaches like community-based conservation for achieving its effective implementation. Through these strategies, the stakeholders maintain decision-making power to cooperatively design and implement, and check the success of conservation projects within their communities. Scientific information works together with indigenous methods through CBC. The implementations guarantee environmental and societal respectability through proper planning (Zafar et al. 2023). The people living in local communities maintain a deep understanding of their geographic region. Local communities maintain the ideal position to detect important transformations in their wetland systems. The involvement of local communities during conservation projects increases the probability of achieving successful results. Community members tend to keep conservation efforts going if they rcccivc clcar advantagcs from such measures. The use of participatory approaches leads to the development of social capital through improved community connections. Through their programs, the exchange of knowledge happens between various stakeholder groups. The implementation of management systems which local communities guide promotes enduring sustainability. Community programs in Africa generate enhanced water quality together with greater species diversity while improving adaptive measures to climate change effects (Angmor et al. 2016). People who have invested interest in conservation develop into dedicated protectors of their natural resources.

21.5.3 Challenges in Achieving Effective Stakeholder Collaboration

Wetland management faces difficulties in building strong stakeholder relationships because stakeholders possess varying interests, together with different power positions. The immediate interests of community fishing and agricultural activities compete with the sustainability planning efforts of governments and nongovernmental organizations. The process of reaching consensus needs open conversations between parties with matching objectives alongside a willingness to compromise. Large NGOs together with governments maintain control over resources and decision-making despite the fact that they restrict the authority of local and indigenous communities. Failure to achieve balance in power dynamics creates conditions that cause people to feel frustrated until they disconnect (Larson et al. 2022). The success of conservation depends on having complete indigenous input during every stage of conservation activities. The implementation of organizational partnerships necessitates programs to increase capacities among participants. In order to properly manage wetlands, local communities demand training programs, and both government agencies and nongovernmental organizations need technical and enforcement support (Vélez et al. 2018). Insufficient financial backing, together with limited availability of resources, obstructs ongoing conservation campaigns. Wetlands exist across numerous administrative areas, which demand coordinated action between different governmental levels and branches. To achieve successful management, it is necessary for public entities and private organizations from the agriculture and water sectors to unite with conservation agencies in strong collaborations across administrative boundaries (Berkes 2017).

21.6 Integration of Science, Policy, and Stakeholder Engagement

21.6.1 Benefits of Interdisciplinary Collaboration in Wetland Management

Successful wetland management depends upon integrated collaboration between experts of natural sciences and social sciences alongside policy specialists and holders of local knowledge. Wetlands function as complex ecosystems which require a whole-system management approach because they combine multiple environmental elements alongside economic, financial resources, and social communities, as well as cultural attributes. The successful implementation of conservation arises when scientists unite with policymakers and stakeholders through their collective support. The management of wetlands requires environmental experts who understand hydrology and biology, and ecology to work alongside social scientists who handle governance and policy rules, alongside the study of human conduct (Ali 2024; Vélez et al. 2018). Interdisciplinary teamwork guarantees that biodiversity defense matches both the financial and social requirements of community residents. The integration of multiple areas of expertise leads to improved scientific integrity, together with more practical outcome possibilities. The approach helps achieve national development objectives through its ability to link conservation procedures with existing policy measures. Additionally, this approach fosters adaptive management and innovation. Experts from various backgrounds collaborate to develop adaptable science-based strategies which integrate local circumstances, so wetland approaches can effectively handle altering environmental and social situations (Delaney et al. 2022).

21.6.2 Strategies for Harmonizing Scientific Research, Policy Implementation, and Stakeholder Input

Wetland management works best when researchers collaborate with policymakers and stakeholders using transparent communication methods based on mutual trust. Adaptive management supports effective management through its emphasis on ongoing monitoring along with strategic adjustment development, which depends on new evidence and stakeholder input. The approach allows scientific information, along with community voices, to shape all stages of decision-making procedures. Inclusive governance platforms unite local communities with policymakers and NGOs, and the private sector to initiate collaboration through joint work (Díaz et al. 2015). Native species conservation and water management tasks are usually executed by local stakeholders because these programs help scientific findings adapt to community requirements. Working together in workshops and conducting joint research allows stakeholders to exchange ideas and build trust while developing mutual understanding (Berkes 2017). The implementation of policy instruments which display ecosystem service values allows stakeholders to harmonize their objectives through showing the economic sustainability of conservation efforts (Barbier 2013).

21.6.3 Examples of Integrated Approaches to Wetland Conservation

There are a number of case studies that show that integrated strategies of wetland conservation are successful. The example of the Danube Delta restoration in Eastern Europe, which is a UNESCO World Heritage site, proves the importance of the cooperation between local communities, governments, non-governmental organizations, and scientists. The management of water flow, species protection, and ecosystems, as well as the scientific research directed restoration in communities, led to better biodiversity, water quality, and livelihoods (Andrews et al. 2022). The Everglades restoration project in the United States demonstrated the potential to work together between federal, state, and local governments, NGOs, scientists and indigenous peoples. Integrating hydrology information, environmental study, and folk knowledge resulted in habitat recovery, decreased pollution, and endangered species protection, which increased climate resilience. Internationally, the Ramsar Convention brings together the stakeholders to harmonize conservation and development; this is witnessed in designation of Yellow Sea wetlands in China and South Korea. In the same vein, there are mutual efforts of the Great Barrier Reef wetlands in Australia, which combine the knowledge of science and cultural appropriateness in the management of the wetlands (Bridge et al. 2019).

21.7 Challenges and Barriers to Integration in Wetland Conservation

21.7.1 Conflicts Between Development and Conservation Goals

The main obstacle in safeguarding wetlands stems from the opposing development demands versus environmental protection ambitions. Most wetlands exist in regions showing high potential for economic growth since they are situated near both coastal zones and urban fringes. Wetlands located in these areas encounter permanent risks from industrial development and urbanization, alongside infrastructure expansion. Economic development leads to the draining and conversion of wetlands into agricultural areas, because it receives priority above ecological protection (Hu et al. 2017). Governments occasionally choose development profits in the present moment instead of protecting the natural value of wetlands. Wetland habitat destruction, together with habitat deterioration, results from these human activities, which subsequently damage their sustainability capabilities and ecological functions. Central to the successful management of conservation areas and development is a proper land-use planning system. The preservation of both economic interests along ecological preservation needs equal consideration. It becomes demanding to strike this balance because regions tend to prioritize industrial and agricultural projects in their wetland territories (McCartney and Houghton-Carr 2009). Wetland conservation should become a policy priority, while development projects need authorization to perform development activities without damaging habitats.

21.7.2 Lack of Awareness and Education Among Stakeholders

A significant barrier in preserving wetlands stems from insufficient education among three key groups of stakeholders, which include policymakers and developers, as well as neighboring community members. The decision-making processes exclude wetlands as valuable resources because stakeholders lack understanding about the importance of wetlands in protecting biodiversity and water purification, and their roles in flood control and climate regulation (Mitsch et al. 2015). The insufficient knowledge about watersheds, as well as short-term financial gains from fishing and farming, causes decision-makers to fail in protecting both watersheds and ecology during land-use planning. The absence of technical expertise among policymakers creates difficulties when they aim to include scientific and local knowledge in policy creation (Vélez et al. 2018). To improve this knowledge deficit, all stakeholder groups need to receive extensive environmental education alongside awareness promotion programs. Community-based conservation demonstrates effective capabilities for improving ecological understanding while strengthening sustainable practices. Wetland preservation receives extended support from stakeholders through educational programs which explain the ecological balance and direct human outcomes of these ecosystems, according to Berkes (2017).

21.7.3 Resource Constraints and Funding Limitations

Wetland conservation projects struggle to overcome challenges because they have limited resources and insufficient financial backing. The necessary funding requirements for restoration work, along with monitoring operations and policy execution, exceed the limited financial support which environmental projects obtain specifically within developing countries since healthcare and poverty elimination and infrastructure establishment take precedence (Rohr et al. 2018). Sustainable management of wetlands requires significant financial investments because of their complex ecosystems, yet insufficient financial rewards make ecosystem conservation less attractive. The acquisition of public and private funds remains challenging due to inadequate policy frameworks which do not accurately represent wetland ecosystem service worth (Turner et al. 2000). Subscription to better funding methods, including ecosystem service markets and payment for ecosystem services programs and public-private partnerships, helps determine the economic worth of key ecological services, which include water purification and flood control alongside carbon sequestration. Organizations, including the Global Environment Facility, along with international aid agencies, deliver support through funding projects to financially vulnerable areas (Xu et al. 2019).

21.7.4 Potential Solutions to Address Barriers

The conservation of wetlands in land-use planning demands various unified solutions to break down operational restrictions. The implementation of inclusive governance platforms stands as a crucial element which combines scientists and policymakers with local communities and industry representatives and their representatives to find and execute solutions (Ali 2024; Vélez et al. 2018). Stakeholders who use dialogue-based frameworks can exchange knowledge while developing united agreements. Sustainable support requires policies to fully understand the complete ecological, economic, and social benefits wetlands provide through their water purification, flood mitigation, and biodiversity proactive roles. Through the Ramsar Convention, along with other instruments, nations establish conservation practices through constant support for appropriate wetland use and management (Finlayson 2014). The resolution of information deficiencies requires equal attention because purposeful educational programs increase both official decision-making and public understanding regarding conservation goals. Payment for ecosystem services (PES), together with carbon credits and wetland restoration bonds, represents innovative funding mechanisms that local communities can use to seek sustainable funding for conservation. This enables local communities to support their conservation efforts with economic incentives (Canning et al. 2021).

21.8 Future Directions in Wetland Conservation

21.8.1 Emerging Trends and Innovations in Wetland Conservation

21.8.1.1 Nature-Based Solutions

Wetland conservation, along with restoration, now shows itself as an effective solution based on natural principles. Through their natural ecosystem functions, wetlands act as filters for water and flood regulators, and carbon storage units, to resolve environmental problems as well as promote biodiversity and human wellness. The strategy includes three parts: preserving existing wetlands, rebuilding damaged ones, and establishing new wetland areas. Wetland restoration initiatives involving mangroves, salt marshes, and seagrasses represent crucial NbS practices since these measures enhance carbon sequestration while lowering greenhouse gas emissions (Iram et al. 2022). Wetland restoration programs create stronger systems to resist both droughts and floods as they simultaneously enhance biodiversity preservation (Vélez et al. 2018). The integration of NbS reduces economic uncertainties that local communities face because of climate impacts by protecting their water resources while safeguarding agricultural production. An increasing number of authorities, together with worldwide organizations and local population groups, integrate NbS solutions within their strategies for wetland conservation. As stated by the Ramsar Convention on Wetlands, nature-based solutions serve as a foremost approach to reach its objectives and achieve optimal ecological health of wetlands (Seddon 2022).

21.8.1.2 Technological Advancements: AI and Machine Learning in Ecosystem Monitoring

The combination of modern technological advances, particularly AI and machine learning strategies, provides numerous advantages for wetland preservation through improved ecological monitoring techniques and better evaluation of biological system health, as well as identification of natural shifts and predictive analytics. AI analyzes vast remote sensing information from satellites and drones that enable researchers to observe wetland changes and monitor vegetation patterns together with water quality assessment (Guo et al. 2017). Machine learning models use the analysis of datasets to find patterns which lead to forecasts about climate change effects, land-use transformations, and conservation results. The analysis of wildlife populations together with biodiversity measurement in wetlands benefits from automated systems that process data collected by camera traps and bird calls (Duhart et al. 2018). Among machine learning technologies, the models identify areas prone to degradation, so conservation efforts can be targeted to vulnerable areas (Rohde et al. 2021). These technological solutions enable large-scale monitoring of wetlands by delivering contemporary data, which helps decision makers set proactive conservation goals.

21.8.2 Recommendations for Fostering Global and Local Collaboration

21.8.2.1 Strengthening International and Cross-Border Cooperation

Large wetlands, including river deltas and coastal wetlands in addition to migratory bird habitats, span across numerous countries. The successful protection of such areas needs worldwide collaboration. Through the Ramsar Convention, political entities gain access to international protocols for ensuring the protection of their important wetland areas. Countries sharing wetland resources must strengthen their bilateral cooperation, because this cooperation enables coordinated management across present political borders (Bridgewater and Kim 2021). The challenge for global wetland conservation is to synchronize national policies and priorities. Public authorities need to advance from exclusive national policies toward making inter-state agreements that address scientific research with environmental, economic, and social dimensions of wetland management (Vélez et al. 2018). Better diplomatic ties, along with enhanced knowledge transfer and collective wetland health responsibility, need to be established between neighboring nations.

21.8.2.2 Engaging Local Communities

The conservation of wetlands requires local communities as key actors, specifically when these wetlands provide economic opportunities such as fishing, agriculture, and tourism. To succeed in conservation activities, the programs need to match local requirements and community understanding. Communities must take part in making decisions through participatory management techniques to achieve sustainable conservation results. Wetland region conservation has benefited through community-based programs that merge both indigenous knowledge systems with collaborative resource management approaches (Pyke et al. 2018). Local stakeholders can become better empowered by governments to considerably improve both wetland conservation outcomes and socioeconomic conditions of dependent communities.

21.8.3 Potential Policy Reforms and Funding Mechanisms

Accomplished wetland protection requires national legislation that governments need to establish as their top priority. The implementation of conservation within current institutional and development frameworks proves difficult to achieve. The present policies typically endorse economic expansion above wetland conservation, and this practice continues to destroy natural environments. Wetlands need a complete environmental assessment and economic evaluation, since authorities must implement protection methods which should form part of national strategies about climate adaptation, agricultural planning, and sustainable development goals. Wetlands must receive proper recognition of their disaster-resilience ability to protect against floods and storms (Xu et al. 2019). The implementation of environmentally friendly agricultural practices combined with responsible urban project development strategies minimizes the strain placed on these natural water ecosystems. Financial sustainability needs attention, so new funding pathways through PES programs, coupled with environmental impact bonds and carbon credit markets, serve as important initiatives (Zhou et al. 2023). The PES programs have set systems to give financial returns to wetland protectors and local communities, along with funding from the Global Environment Facility and private partners for large-scale restoration initiatives. The long-term achievement of restoration programs requires reliable funding sources combined with systematic monitoring and institutional capacity development, according to Rohr et al. (2018).

21.9 Conclusion

Sustainable development needs the protection of wetland areas. Wetlands deliver multiple essential ecosystem services to the environment, including water filtration systems and flood regulation, and support the maintenance of biodiversity. The discussion establishes the necessary role of science, together with policy and stakeholder involvement, to achieve conservation outcomes. Advances in nature-based solutions and technological innovations like AI offer new opportunities for wetland restoration and protection. The implementation of management systems based on community settings delivers encouraging results. However, significant challenges remain. The ongoing struggle exists between the objectives of development and conservation. Limited funding is another obstacle. Cross-sector organizations need to work more closely to solve the existing challenges. A call to action is required. The future success of wetland preservation depends on the improvement of joint operational links between scientific studies and government policies, with additional backing from regional and worldwide stakeholder networks. Scientists and governments, as well as communities, need to join forces to synchronize their policies and exchange information. Wetland conservation needs to take precedence in all regional and international initiatives. Successful wetland protection in the future will result from well-organized proactive efforts. We need to make innovation a priority because it stands alongside collaboration and long-term ecological health. We need science-policy-local knowledge synchronization to protect and rehabilitate wetlands, which will safeguard them for future generations.

References

Ali A (2024) Navigating road for net zero policies to aligning policy, politics, and governance for environmental sustainability. In: Singh R, Crowther D (eds) Transition towards a sustainable future. Approaches to global sustainability, markets, and governance. Springer, Singapore. https://doi.org/10.1007/978-981-97-5756-5_10

Anderson K, Gaston KJ (2013) Lightweight unmanned aerial vehicles will revolutionize spatial ecology. Front Ecol Environ 11(3):138–146. https://doi.org/10.1890/120150

Andrews L, Mocorro Powell AD, Rottle N, Engelke J (2022) Advancing equity and justice through community science programming in design, construction, and research of a nature-based solution: the Duwamish floating wetlands project. Soc Ecol Pract Res 4(4): 377–391. https://doi.org/10.1007/s42532-022-00123-9

Angmor EN, Tettey C, Amoah T (2016) Management of community water projects: assessing the challenges of indigenous arrangements. J Geogr Reg Plann 9(6): 122–127

Azis AA (2022) The role of non-governmental organizations (ngo's) in shaping environmental policies. In: Universitas Lampung international conference on social sciences (ULICoSS 2021). Atlantis Press, pp 490–496. https://doi.org/10.2991/assehr.k.220102.064

Ballut-Dajud GA, Sandoval Herazo LC, Fernández-Lambert G, Marín-Muñiz JL, López Méndez MC, Betanzo-Torres EA (2022) Factors affecting wetland loss: a review. Land 11(3):434. https://doi.org/10.3390/land11030434

Baral S, Basnyat B, Khanal R, Gauli K (2016) A total economic valuation of wetland ecosystem services: an evidence from Jagadishpur Ramsar site, Nepal. Sci World J 2016(1):2605609. https://doi.org/10.1155/2016/2605609

Barbier EB (2011) Wetlands as natural assets. Hydrol Sci J 56(8):1360–1373. https://doi.org/10.1080/02626667.2011.629787

Barbier EB (2013) Valuing ecosystem services for coastal wetland protection and restoration: Progress and challenges. Resources 2(3):213–230. https://doi.org/10.3390/resources2030213

Berkes F (2017) Sacred ecology. Routledge

Berkes F (2018) Sacred ecology, 4th edn. Taylor & Francis

Bridge TC, Beaman RJ, Bongaerts P, Muir PR, Ekins M, Sih T (2019) Correction to: the great barrier reef and Coral Sea. Mesophotic Coral Ecosyst:C1–C1. https://doi.org/10.1007/978-3-319-92735-0_20

Bridgewater P, Kim RE (2021) The Ramsar convention on wetlands at 50. Nat Ecol Evol 5(3):268–270. https://doi.org/10.1038/s41559-021-01392-5

Canning AD, Jarvis D, Costanza R, Hasan S, Smart JC, Finisdore J et al (2021) Financial incentives for large-scale wetland restoration: beyond markets to common asset trusts. One Earth 4(7):937–950

Chimney MJ, Goforth G (2001) Environmental impacts to the Everglades ecosystem: a historical perspective and restoration strategies. Water Sci Technol 44(11–12):93–100. https://doi.org/10.2166/wst.2001.0814

Costanza R, Pérez-Maqueo O, Martinez ML, Sutton P, Anderson SJ, Mulder K (2008) The value of coastal wetlands for hurricane protection. Ambio:241–248. http://www.jstor.org/stable/25547893

Delaney JT, Bouska KL, Eash JD (2022) Climate change adaptation thinking for managed wetlands. https://doi.org/10.3133/ofr20211103

Díaz S, Demissew S, Carabias J, Joly C, Lonsdale M, Ash N et al (2015) The IPBES conceptual framework—connecting nature and people. Curr Opin Environ Sustain 14:1–16. https://doi.org/10.1016/j.cosust.2014.11.002

Donald PF, Sanderson FJ, Burfield IJ, Bierman SM, Gregory RD, Waliczky Z (2007) International conservation policy delivers benefits for birds in Europe. Science 317(5839):810–813. https://doi.org/10.1126/science.1146002

Duhart C, Dublon G, Mayton B, Davenport G, Paradiso JA (2018) Deep learning for wildlife conservation and restoration efforts. Int Conf Mach Learn. https://hal.science/hal-04591082/document

Finlayson CM (2014) Linking science to international wetland policy–the Ramsar convention on wetlands. Mar Freshw Res 65(7):573–574. https://doi.org/10.1071/MF14103

Finlayson CM, Davidson N, Pritchard D, Milton GR, MacKay H (2011) The Ramsar convention and ecosystem-based approaches to the wise use and sustainable development of wetlands. J Int Wildl Law Policy 14(3–4):176–198. https://doi.org/10.1080/13880292.2011.626704

Finlayson CM, van Dam AA (2023) Climate change and wetlands: Vulnerability, adaptation, mitigation, resolutions, and scientific societies. In Ramsar Wetlands (pp. 495-524). Elsevier. https://doi.org/10.1016/B978-0-12-817803-4.00016-4

Guo M, Li J, Sheng C, Xu J, Wu L (2017) A review of wetland remote sensing. Sensors 17(4):777. https://doi.org/10.3390/s17040777

Harenda KM, Lamentowicz M, Samson M, Chojnicki BH (2018) The role of peatlands and their carbon storage function in the context of climate change. In: Interdisciplinary approaches for sustainable development goals: economic growth, social inclusion and environmental protection. Springer, Cham, pp 169–187. https://doi.org/10.1007/978-3-319-71788-3_12

Hu S, Niu Z, Chen Y, Li L, Zhang H (2017) Global wetlands: potential distribution, wetland loss, and status. Sci Total Environ 586:319–327. https://doi.org/10.1016/j.scitotenv.2017.02.001

Iram N, Maher DT, Lovelock CE, Baker T, Cadier C, Adame MF (2022) Climate change mitigation and improvement of water quality from the restoration of a subtropical coastal wetland. Ecol Appl 32(5):e2620. https://doi.org/10.1002/eap.2620

Jayaraman G, Rao AD, Dube A, Mohanty PK (2007) Numerical simulation of circulation and salinity structure in Chilika lagoon. J Coast Res 23(4):861–877. https://doi.org/10.2112/04-0225R.1

Jellinek S, Rumpff L, Driscoll DA, Parris KM, Wintle BA (2014) Modelling the benefits of habitat restoration in socio-ecological systems. Biol Conserv 169:60–67. https://doi.org/10.1016/j.biocon.2013.10.023

Keddy PA (2010) Wetland ecology principles and conservation, vol 614. Cambridge University Press

Kerton FM (2023) UN sustainable development goals 14 and 15–life below water, life on land. RSC Sustain 1(3):401–403. https://doi.org/10.1039/D3SU90010J

Klemas V (2013) Using remote sensing to select and monitor wetland restoration sites: an overview. J Coastal Res 289:958. https://doi.org/10.2112/JCOASTRES-D-12-00170.1

Kumar R, Ganapathi H, Palmate S (2021) Wetlands and water management: finding a common ground. In: Water governance and management in India: issues and perspectives, vol 2. Springer, Singapore, pp 105–129. https://doi.org/10.1007/978-981-16-1472-9_5

Larson AM, Barletti JPS, Vigil NH (2022) A place at the table is not enough: accountability for indigenous peoples and local communities in multi-stakeholder platforms. World Dev 155:105907. https://doi.org/10.1016/j.worlddev.2022.105907

Leon AS, Tang Y, Chen D, Yolcu A, Glennie C, Pennings SC (2018) Dynamic management of water storage for

flood control in a wetland system: a case study in Texas. Water 10(3):325. https://doi.org/10.3390/w10030325
McCartney MP, Houghton-Carr HA (2009, May) Working wetland potential: an index to guide the sustainable development of African wetlands. In: Natural resources forum, vol 33(2). Blackwell Publishing Ltd, Oxford, pp 99–110
Mistri A, Das B, Mistri A, Das B (2020) The Sundarban and its environment. In: Environmental change, livelihood issues and migration: Sundarban biosphere reserve, India. Springer, Singapore, pp 21–40. https://doi.org/10.1007/978-981-13-8735-7_2
Mitsch WJ, Gosselink JG (2015) Wetlands. Wiley
Mitsch WJ, Bernal B, Hernandez ME (2015) Ecosystem services of wetlands. Int J Biodivers Sci Ecosyst Serv Manag 11(1):1–4. https://doi.org/10.1080/21513732.2015.1006250
Moomaw WR, Chmura GL, Davies GT, Finlayson CM, Middleton BA, Natali SM et al (2018) Wetlands in a changing climate: science, policy and management. Wetlands 38(2):183–205. https://doi.org/10.1007/s13157-018-1023-8
Mushtaq N, Singh DV, Bhat RA, Dervash MA, Hameed OB (2020) Freshwater contamination: sources and hazards to aquatic biota. In: Fresh water pollution dynamics and remediation. Springer, Singapore, pp 27–50. https://doi.org/10.1007/978-981-13-8277-2_3
Nazneen S, Mahmood G, Jafar Z, Madhav S (2021) Ecosystem services of Lagoon wetlands system in India. In: Wetlands conservation: current challenges and future strategies. Wiley, pp 111–128. https://doi.org/10.1002/9781119692621.ch6
Omondi DO, Navalia AC (2020) Constructed wetlands in wastewater treatment and challenges of emerging resistant genes filtration and reloading. In: Inland waters-dynamics and ecology. IntechOpen. https://doi.org/10.5772/intechopen.93293
Pyke ML, Toussaint S, Close PG, Dobbs RJ, Davey I, George KJ et al (2018) Wetlands need people: a framework for understanding and promoting Australian indigenous wetland management. Ecol Soc 23(3):43. https://doi.org/10.5751/ES-10283-230343
Reddy K, DeLaune RD (2003) Biogeochemistry of wetlands science and applications. CRC Press, Boca Raton
Rohde MM, Biswas T, Housman IW, Campbell LS, Klausmeyer KR, Howard JK (2021) A machine learning approach to predict groundwater levels in California reveals ecosystems at risk. Front Earth Sci 9:784499. https://doi.org/10.3389/feart.2021.784499
Rohr JR, Bernhardt ES, Cadotte MW, Clements WH (2018) The ecology and economics of restoration. Ecol Soc 23(2). https://doi.org/10.5751/ES-09876-230215
Seddon N (2022) Harnessing the potential of nature-based solutions for mitigating and adapting to climate change. Science 376(6600):1410–1416. https://doi.org/10.1126/science.abn9668
Sommerwerk N, Bloesch J, Baumgartner C, Bittl T, Čerba D, Csányi B et al (2022) The Danube river basin. In: Rivers of Europe. Elsevier, pp 81–180. https://doi.org/10.1016/B978-0-08-102612-0.00003-1
Su L, Liu J, Christensen P (2010) Comparative study of water resource management policies between China and Denmark. Procedia Environ Sci 2:1775–1798. https://doi.org/10.1016/j.proenv.2010.10.190
Taberlet P, Bonin A, Zinger L, Coissac E (2018) Environmental DNA: for biodiversity research and monitoring. Oxford University Press
Tran DD, van Halsema G, Hellegers PJ, Hoang LP, Ludwig F (2019) Long-term sustainability of the Vietnamese Mekong Delta in question: an economic assessment of water management alternatives. Agric Water Manag 223:105703. https://doi.org/10.1016/j.agwat.2019.105703
Turner RK, Van Den Bergh JC, Söderqvist T, Barendregt A, Van Der Straaten J, Maltby E, Van Ierland EC (2000) Ecological-economic analysis of wetlands: scientific integration for management and policy. Ecol Econ 35(1):7–23. https://doi.org/10.1016/S0921-8009(00)00164-6
Vélez JMM, García SB, Tenorio AE (2018) Policies in coastal wetlands: key challenges. Environ Sci Pol 88:72–82. https://doi.org/10.1016/j.envsci.2018.06.016
Vymazal J (2011) Constructed wetlands for wastewater treatment: five decades of experience. Environ Sci Technol 45(1):61–69. https://doi.org/10.1021/es101403q
Xin T (2014) Enlightenment of everglade wetland restoration in Florida, USA to Chinese Dongting Lake restoration. Adv Mater Res 955:2139–2144. https://doi.org/10.4028/www.scientific.net/AMR.955-959.2139
Xu T, Weng B, Yan D, Wang K, Li X, Bi W et al (2019) Wetlands of international importance: status, threats, and future protection. Int J Environ Res Public Health 16(10):1818. https://doi.org/10.3390/ijerph16101818
Yan J, Shi W, Li F (2018) Evaluation and countermeasures of the implementation of the lake protection and governance system in Wuhan City, Middle China. Sustainability 10(10):3379. https://doi.org/10.3390/su10103379
Zafar S, Zafar Khan M, Mehmood T, Begum F, Sadiq M (2023) Role of community-based conservation and natural resource management in building climate resilience among vulnerable mountain societies. Clim Dev 15(7):608–621. https://doi.org/10.1080/17565529.2022.2135365
Zedler JB, Kercher S (2005) Wetland resources: status, trends, ecosystem services, and restorability. Annu Rev Environ Resour 30(1):39–74. https://doi.org/10.1146/annurev.energy.30.050504.144248
Zhou Y, Zhao L, Li Z (2023) Wetland ecological restoration and payment for ecosystem service standard: a case study of Ganjiangyuan National Wetland Park. Wetlands 43(3):22. https://doi.org/10.1007/s13157-023-01657-8

22 Urban Resilience and Water Security: A Synthesis of Wetland Strategies

M. C. Nitha Shahul and Ayon Kumar Tarafdar

Abstract

Urban wetlands are vital to natural ecosystems, offering diverse ecosystem services essential for biodiversity and human well-being. With water security emerging as a global concern due to population growth, climate change, and urbanization, this study highlights the critical role of urban wetland conservation in enhancing urban water security, focusing on the Kottooli wetlands in Calicut, Kerala. The study critically examines the urban water demand, supply systems, and the wetland's limited but significant link to water security. By assessing the wetlands' quantitative and qualitative contributions, the research explores socioeconomic impacts of wetland transformation through perception surveys and analyzes factors degrading water quality and quantity. It underscores the implications of these changes for urban water security. Through an extensive literature review and empirical analysis, the study evaluates ecosystem service losses and gains, emphasizing the need for integrated conservation strategies. The findings reveal the wetlands' potential to contribute to urban water needs, including flood control and enhanced water retention, critical for urban resilience. By protecting, restoring, and strategically managing wetlands, Calicut can address environmental challenges, improve urban sustainability, and contribute to global sustainability goals, showcasing the value of wetlands beyond aesthetics and cultural significance.

Keywords

Wetlands · Urban water security · Ecosystem services · Water quality · Conservation & management

22.1 Introduction

Water security is a comprehensive concept that encompasses ensuring consistent and affordable access to an adequate quantity and quality of water for various purposes, including meeting basic human needs, supporting livelihoods, and sustaining local ecosystems. At its core, water security involves guaranteeing that individuals and communities have reliable access to water services, free from interruptions or shortages, and at a cost that is affordable for all members of society, regardless of their socioeconomic status or geographic location. (UNESCO 2012) This aspect of water security underscores the

M. C. Nitha Shahul
Department of Planning, Environmental Planning Programme, School of Planning and Architecture, Vijayawada, India

A. K. Tarafdar (✉)
Department of Planning, School of Planning and Architecture, Vijayawada, India

S. K. Chakraborty et al. (eds.), *Wetlands and Urban Resilience*, Sustainable Development Goals Series,
https://doi.org/10.1007/978-3-032-06881-1_22

importance of equitable distribution and accessibility of water resources to ensure that no one is left behind or marginalized due to a lack of access to water. Water security entails ensuring that the available water is of suitable quality for its intended use. This involves protecting water resources from contamination and pollution, monitoring water quality parameters, and implementing treatment measures to meet health and safety standards for drinking water, irrigation, and ecosystem health. The quality of water is paramount for safeguarding public health, promoting sustainable agriculture, and maintaining the ecological balance of freshwater ecosystems, highlighting the interconnectedness between human well-being and environmental health. It recognizes the fundamental role of water in supporting livelihoods and sustaining local ecosystems. Access to water is essential for fulfilling basic human needs such as drinking, cooking, hygiene, and sanitation, as well as for supporting economic activities such as agriculture, industry, and tourism. Additionally, water plays a vital role in maintaining the health and productivity of ecosystems, including rivers, wetlands, and fisheries, which provide essential ecosystem services such as water purification, flood regulation, and habitat provision. Wetlands are one such resource which has the potential to unlock urban water security. Wetlands are a critical part of natural environments that offer a wide range of ecosystem services. In urban areas, wetlands contribute to the liveability of cities by improving the water quality, carbon sequestration, providing habitats for wildlife species, reducing the effects of urban heat islands, and creating recreational opportunities (Alikhani 2021). One of the key contributions of urban wetlands is their ability to improve water quality. Acting as natural filtration systems, wetlands efficiently remove pollutants and contaminants from stormwater runoff and wastewater. The vegetation and soil within wetlands act as effective filters, absorbing and breaking down pollutants, thus enhancing the overall water quality of urban water bodies such as rivers, lakes, and reservoirs. This improvement in water quality not only benefits aquatic ecosystems but also provides cleaner water for human consumption and recreational activities. Urban wetlands serve as significant carbon sinks, capable of sequestering large amounts of carbon dioxide from the atmosphere. They provide essential habitat for a diverse array of wildlife species. These ecosystems serve as critical breeding, nesting, and foraging grounds for urban wildlife, supporting biodiversity and ecological resilience within cities. By preserving and restoring urban wetlands, cities can enhance habitat connectivity and promote the conservation of native species, contributing to the overall health and sustainability of urban ecosystems.

Water security has risen to the forefront as a critical and urgent global issue, garnering increasing attention from policymakers, researchers, and communities worldwide. At its core, water security encompasses the availability, accessibility, and sustainable use of water resources, with profound implications for human well-being, economic development, and environmental sustainability (Cept 2017). The cognition of patterns of urbanisation and characterisitcs of local stakeholders becomes essential for a territorial understanding of the environmental implications. The paradigm in which an existing urban planning setup influences the development pattern in a city is crucial to the way local urban environment is treated (Tarafdar & Bjønnes 2010). In recent years, growing population pressures, rapid urbanization, climate change impacts, and competing water demands have underscored the need for concerted efforts to ensure water security for present and future generations. The global scenario of water security presents a complex and multifaceted challenge, with diverse geographic, social, economic, and environmental dimensions. At the heart of the issue is the increasing strain on water resources due to a combination of population growth, urbanization, climate change, and unsustainable water management practices. Many regions around the world are experiencing water scarcity, where the demand for water exceeds the available supply. This scarcity is exacerbated by factors such as uneven distribution of water resources, competing water uses, and over-extraction of groundwater. According to the United Nations, by 2025, an estimated 1.8 billion people will be living in regions with absolute water scarcity, while two-thirds of the world's population could be under stress conditions. The UN-Water Facts of 2016 project a concerning future for India's water security, predicting that the country will transition from being "water stressed" to "water scarce" in the coming decades. This projection is based on the anticipated increase in water demand due to population growth, urbanization, and economic development, coupled with the impact of climate change on water availability and distribution. According to the World Resources Institute (WRI), 54% of India faces high to extremely high levels of water stress, where water withdrawals exceed 40% of the available supply. This situation reflects the growing imbalance between water demand and availability, particularly in densely populated and agriculturally intensive regions (Tien Shiao 2015). India's current water challenges, including overexploitation of groundwater, pollution of surface water sources, inefficient irrigation practices, and inadequate water management policies, exacerbate the risk of water scarcity. To address this looming crisis, India must prioritize sustainable water management practices, such as promoting water conservation, improving water use efficiency, implementing rainwater harvesting and groundwater recharge measures, and enhancing water governance frameworks to ensure equitable and efficient allocation of water resources. Waterbodies are an important part of urban ecosystems. They perform significant environmental, social, and economic functions—recharging groundwater to act as sponges, supporting biodiversity, and providing livelihoods (Mahreen Matto 2017). Wetlands play a crucial role in unlocking water security by serving as natural reservoirs and regulators of water flow, thereby mitigating the impacts of water scarcity, floods, and droughts. These ecosystems act as sponges, absorbing and storing excess rainwater during periods of heavy rainfall and gradually releasing it during dry spells, thereby replenishing groundwater supplies and maintaining streamflow (UNEP 2017). Additionally, wetlands help improve water quality by filtering pollutants, sediment, and nutrients from surface water, thus enhancing the availability of clean water for drinking, agriculture, and industry. Moreover, wetlands support biodiversity and provide essential habitats for wildlife, contributing to ecosystem resilience and ecological balance. Recognizing the value of wetlands in water security, it is imperative to prioritize their conservation, restoration, and sustainable management as integral components of water resource management strategies.

This research is grounded in the growing recognition of urban wetlands as pivotal assets in enhancing water security and building urban resilience. The study seeks to analyze the complex dynamics of urban water demand, supply, and distribution, while simultaneously evaluating how wetlands quantitatively and qualitatively contribute to securing urban water resources. It investigates the degradation of water quantity and quality in urban wetlands, uncovering the underlying socio-ecological and governance factors. Additionally, the research delves into the socio-economic consequences of wetland transformation, particularly for vulnerable populations who rely on these ecosystems for livelihood and ecological services. By positioning wetland conservation as a cornerstone of adaptive urban planning, this study proposes a comprehensive urban water security framework integrating wetland conservation.

22.2 Study Area

The study area selected for investigation encompasses the Calicut Corporation area in Kerala, comprising 75 wards, with the focal point being the Kottooli wetlands occupying 214 acres in the heart of the city. These wetlands hold significant importance, not only locally but nationally, as they are recognized as one of the 27 nationally important wetlands listed by the Ministry of Environment and Forests. The choice of Calicut as the study area is strategic, driven by compelling factors that underscore its relevance in the context of urban water security and environmental management. Notably, Calicut has garnered attention for its rapid urbanization and exponential population growth, ranking fourth among the fastest-growing urban areas globally according to the Economist Intelligence Unit (EIU). With a staggering rate of change recorded at 34.5% from 2015 to 2020, the city's expansion presents both opportunities and challenges in terms of water resource management and environmental sustainability. Moreover, the water stress map of India (IWT 2.0) highlights Calicut as a highly stressed zone, further emphasizing the critical need to study and address water-related issues within the region (Tien Shiao 2015). Against this backdrop, the study aims to investigate the role of the Kottooli wetlands in mitigating water stress, improving water quality, and enhancing urban resilience in the face of rapid urbanization and increasing environmental pressures (Figs. 22.1 and 22.2).

The western boundary of Kottooli, the "Canoli" canal was built in 1848 by the then district collector and later known after him. It was built for the inland water transport connecting the Korapuzha River in the north and the Kallaypuzha River in the south. It was used for transporting wood from the renowned business center of Nilambur teak plantation. According to the master plan for the Kozhikode urban area 2035, 64.2% of the population in the planning area depend on well water. Shortage of water is commonly felt in the coastal belt, elevated areas, and some of the peripheral areas. It was found that about 5% of the population of the planning area faces water scarcity for more than 6 months in a year, whereas 14% of the population faces water scarcity up to 6 months (Town and Country Planning Department 2017). There is a huge supply–demand gap in water supply in the planning area; the supply is less than 50% of the demand (Fig. 22.3).

Watershed delineation involves identifying the boundaries of a watershed, also known as a catchment, drainage basin, or river basin. In the study area, the delineation reveals that the fourth-order stream within the corporation limits corresponds to the Kallayi River, extending 5.48 kilometers.

According to the watershed map (Fig. 22.4), the region is primarily characterized by two major watersheds: the Kallayi River and the Korapuzha River, along with two sub-watersheds of the Chaliyar River within the corporation's jurisdiction. This hydrological framework plays a vital role in understanding and managing the area's water resources effectively (Fig. 22.5).

22.3 Material and Methods

The methodology of the study is mainly carried out as four parts, in which the first part focuses on analyzing the urban water demand, supply characteristics, and resource system of the city, in which the overall demand gap assessment for the city is carried out in the sub-basin level. Here, the entire city is divided into five sub-basins based on the watershed delineation (Fig. 22.4). The second part focuses on assessing the quantitative and qualitative contributions of urban wetland toward urban water security, where wetland level quantitative and qualitative aspects are analyzed in detail. Under this part, detailed land use and land cover analyses have been carried out in order to understand the change in area and the groundwater recharge capacity. Subsequently, the quantitative aspect and the groundwater quality for two different years are analyzed, which gives a clear understanding regarding the qualitative aspect. The third part of the study focuses on investigating the socioeconomic consequences of urban wetland transformation. In this part, the overall dependency of the people residing in the nearby wards on the wetland is analyzed based on the primary survey. The fourth part of the study deals with the factors contributing to water quantity and quality degradation in urban wetlands, focusing on water security. Here in this part, the different drivers of wetland degradation are analyzed based on the priority given by the stakeholders to the particular issue. Here, the major drivers considered in this part were encroachment, canal pollution, and urbanization as the top priority drivers of wetland degradation. Finally, a framework is developed integrating both urban water security and wetland conservation. Overall, the study gives the significance of conservation and management of urban wetlands in enhancing water security (Fig. 22.6).

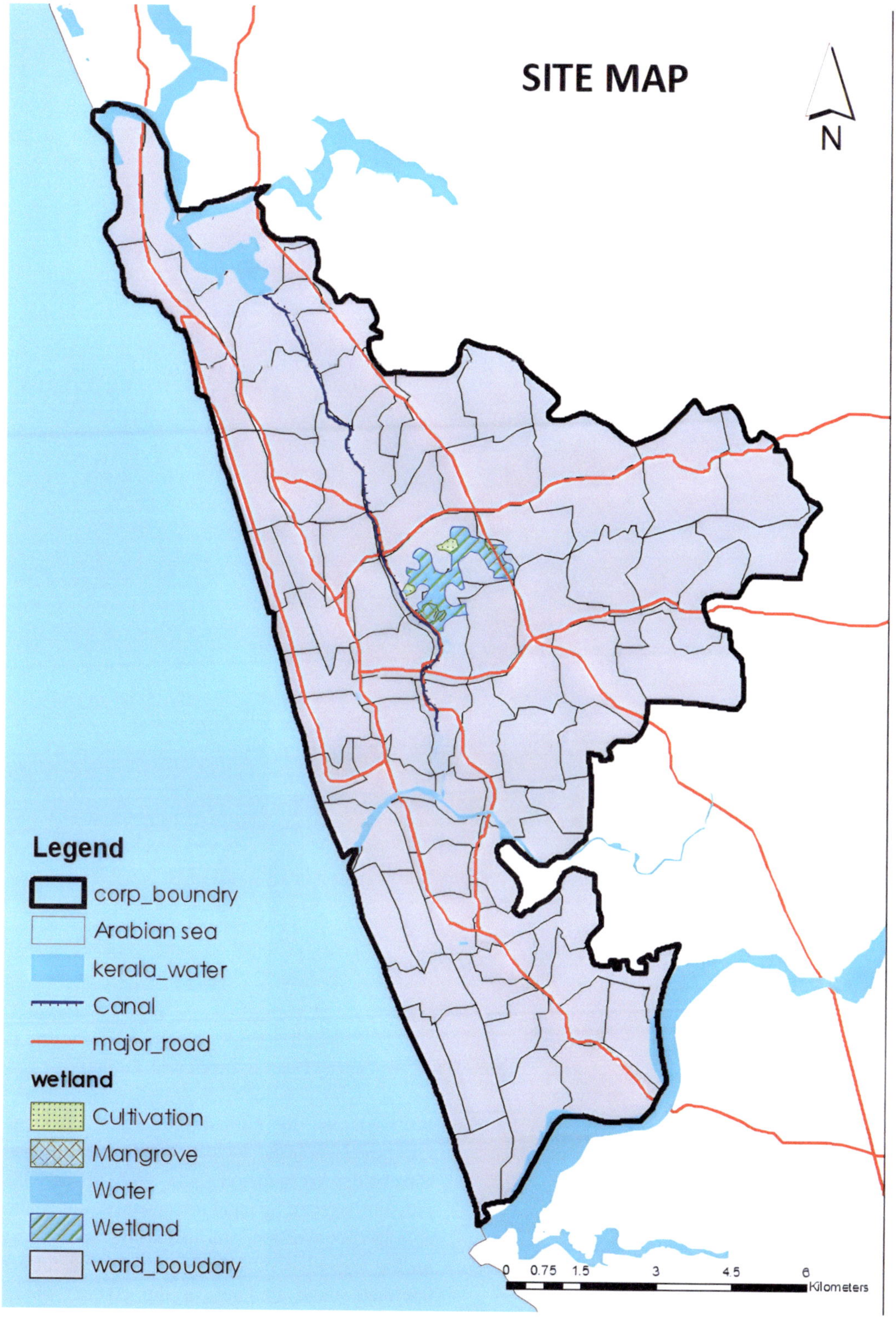

Fig. 22.1 Study area: Calicut Corporation. (Source: Author generated)

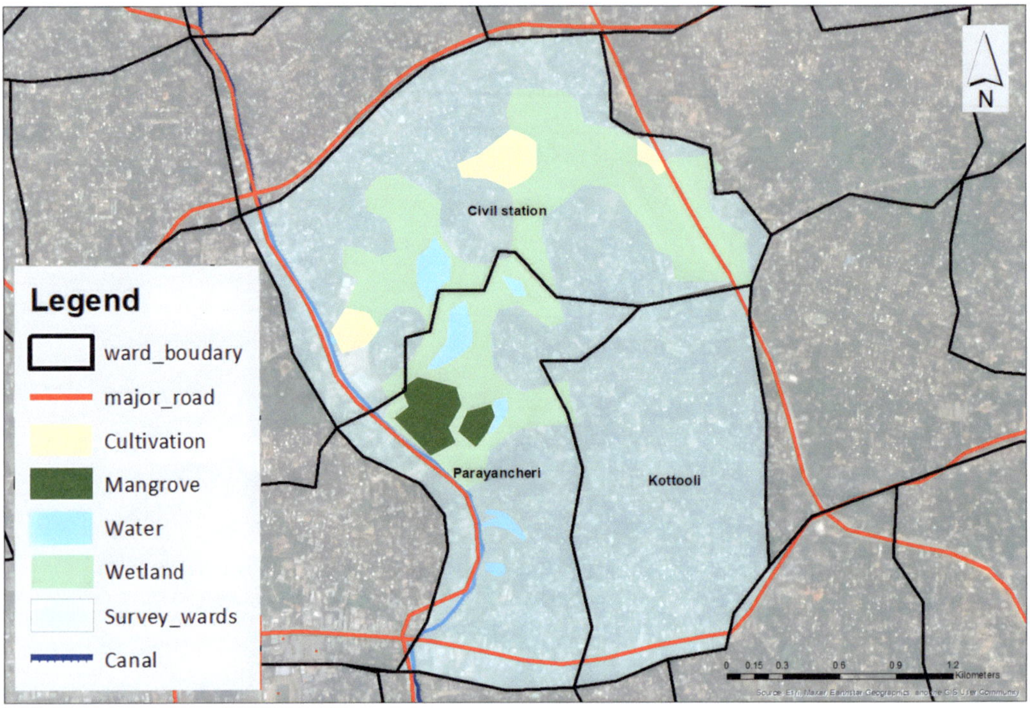

Fig. 22.2 Wetland boundary and the neighboring wards. (Source: Author generated)

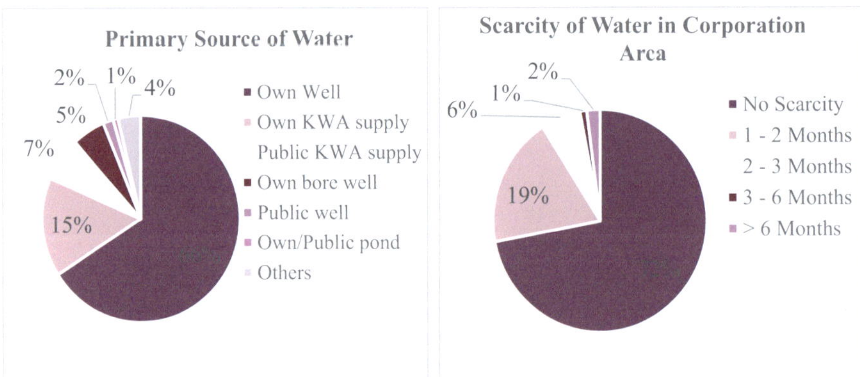

Fig. 22.3 Water statistics: Calicut Corporation. (Source: Town and Country Planning Department, Calicut)

22.4 Results and Discussion

This section presents the key findings derived from an integrated analysis of water demand-supply dynamics, land use and land cover (LULC) changes, wetland transformation, and ecosystem service valuation within the sub-basins. The results reflect both spatial and temporal patterns, highlighting the significant ecological and hydrological shifts that have occurred over the past three decades. Through GIS-based analysis, socioeconomic surveys, and stakeholder consultations, the study reveals the pressures exerted by rapid urbanization, unsustainable land conversion, and increasing water stress on the region's wetlands. The discussion further interprets these findings in the context of an integrated framework of urban water security and wetland conservation and the need for a sustainable, conservation-oriented planning framework.

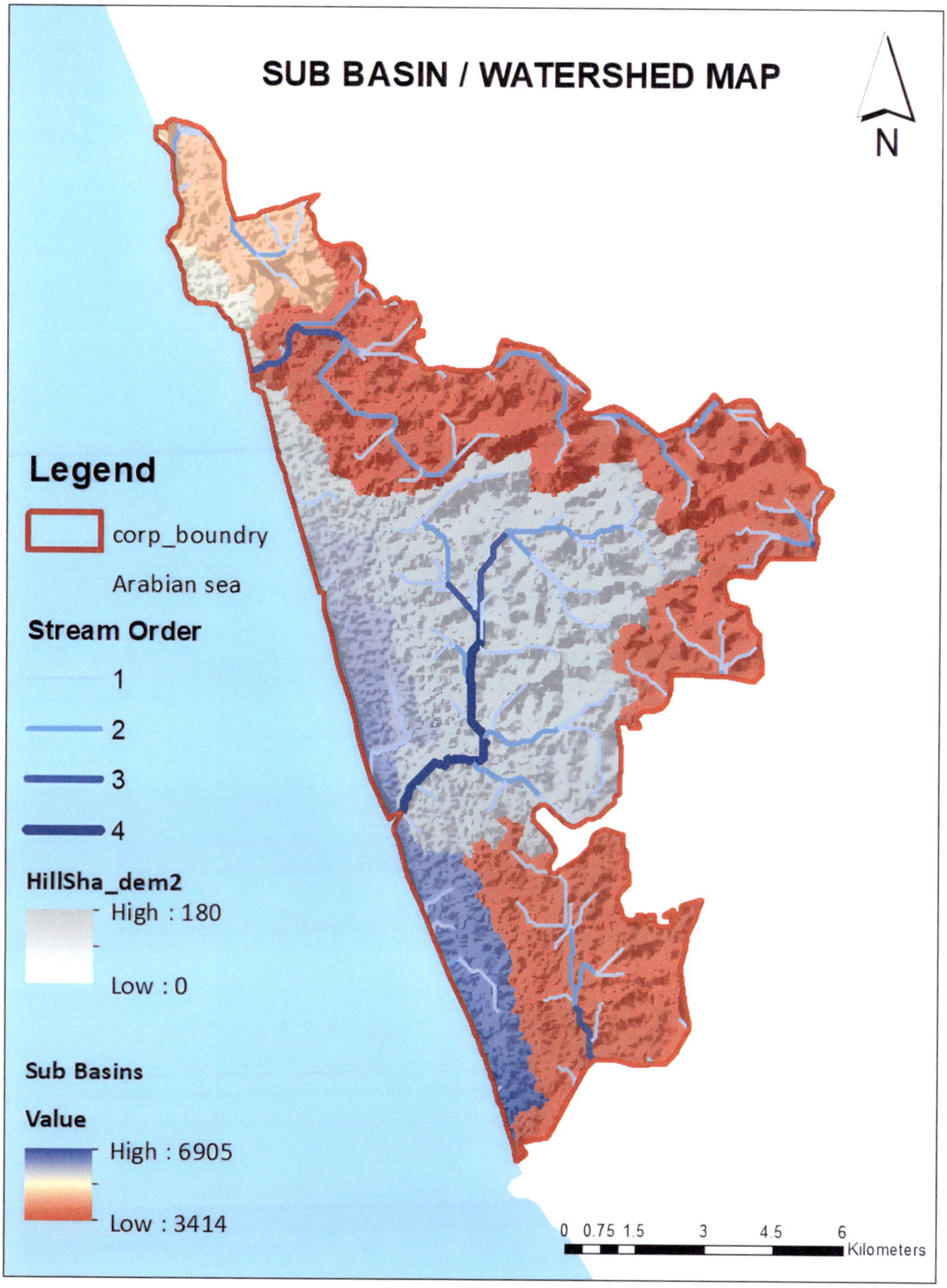

Fig. 22.4 Watershed map. (Source: Author generated)

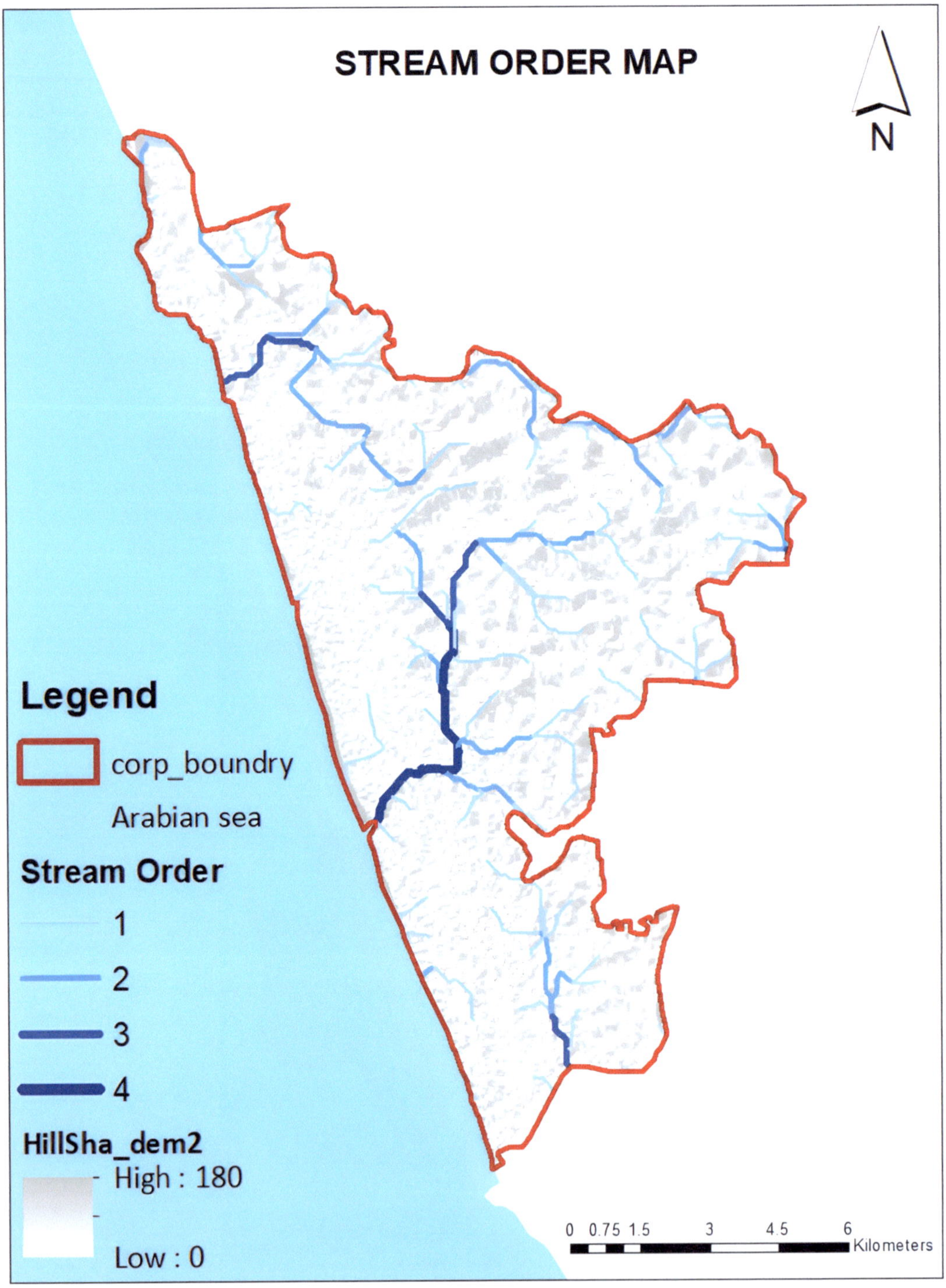

Fig. 22.5 Stream order map. (Source: Author generated)

Fig. 22.6 Methodology flowchart. (Source: Author generated)

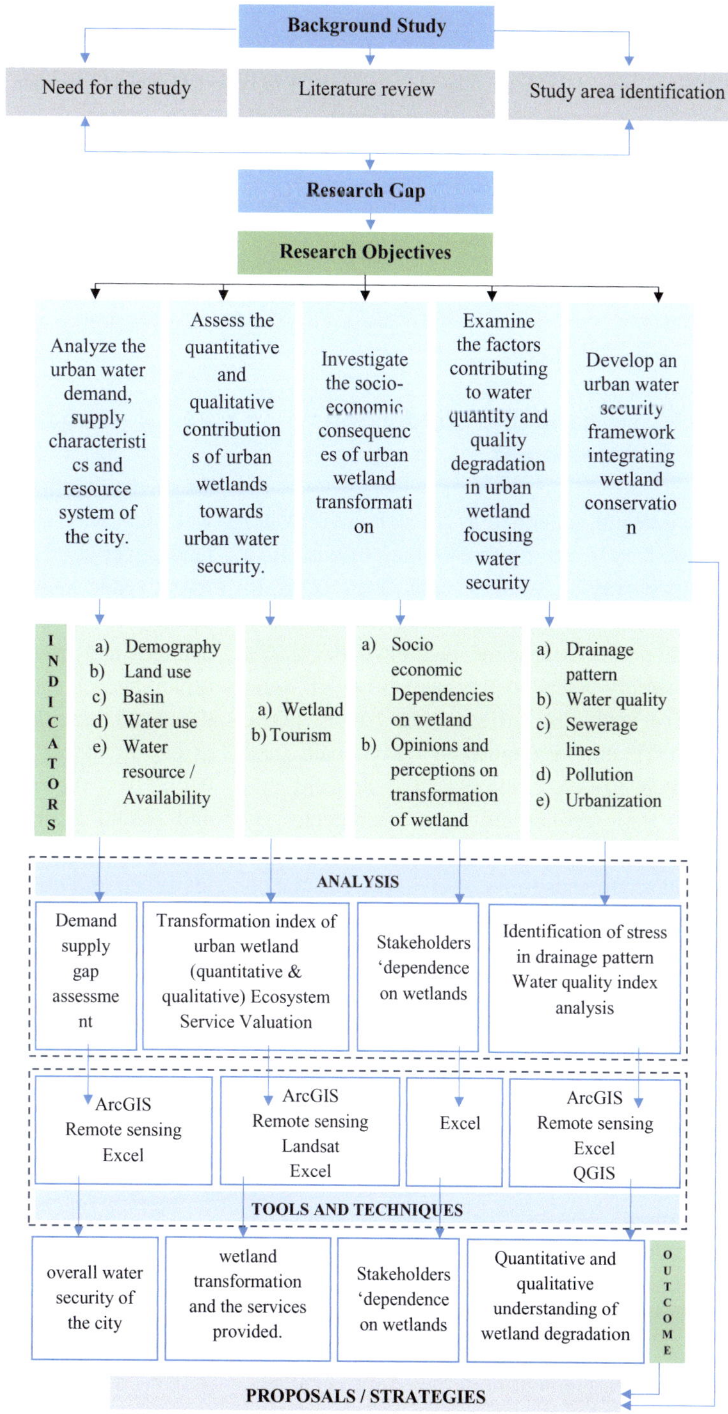

Table 22.1 Demand supply gap assessment 2023

	Water availability		Water demand	Demand supply gap		Inference	
Sub basin	Total (with corp tanks) (ML/year)	Total (without corp tanks) (ML/year)	Total demand (ML/year)	With corp tanks	Without corp tanks	With corp tanks	Without corp tanks
Korapuzha/ Kunoorpuzha	35098.05	23469.15	17679.96	20956.59	9327.6896	Surplus	Surplus
Kallayi	12603.21	3536.606	22866.57	−6979.167	−16045.77	Deficit	Deficit
Chaliyar (a)	16448.40	14842.4	11145.07	8682.9347	7076.9347	Surplus	Surplus
Chaliyar (b)	5192.585	3312.835	6573.94	−558.6549	−2438.405	Deficit	Deficit

Source: Author calculations

22.4.1 Demand and Supply Gap Assessment

Water demand is defined as the volume of water requested by users to satisfy their needs. In a simplified way, it is often considered equal to water abstraction, although conceptually the two terms do not have the same meaning (EEA 1999). Water supply refers to the availability of water resources, including surface water (rivers, lakes, reservoirs) and groundwater (Mynepalli and Sridhar 2020). A comprehensive demand and supply gap assessment was conducted by analyzing the domestic, agricultural, and industrial water demands in the study area. This analysis was complemented by evaluating the existing water supply sources, including groundwater recharge, tap water, and surface water bodies such as streams and ponds. The results indicate a significant water supply deficit in the sub-basins of the Chaliyar River (Table 22.1).

The analysis highlights that the current water supply in these sub-basins is insufficient to meet the prevailing water demand, resulting in a significant supply-demand gap. This shortfall has implications for the domestic, agricultural, and industrial sectors, potentially affecting livelihoods, food security, and industrial productivity in the region. The findings emphasize the urgency of implementing targeted interventions, such as enhancing groundwater recharge, optimizing the use of surface water resources, and adopting water conservation measures.

22.4.2 Urban Wetland Transformation

The quantitative transformation of urban wetlands was analyzed through land use and land cover (LULC) analysis spanning three different years: 1998, 2013, and 2023. This analysis aimed to comprehensively understand the changes and trends in wetland areas within the urban landscape over time. By comparing LULC data from these three distinct time points, one can assess the extent of wetland loss, gain, or degradation over the specified period. Through spatial and statistical analyses, shifts in land use and cover categories, including wetlands, were identified and quantified (Fig. 22.7).

The land use and land cover analyses conducted over the years reveal a concerning trend of decreasing wetlands and expanding built-up areas. This transformation indicates a significant alteration in the urban landscape, with wetland loss attributed to various factors such as urbanization, infrastructure development, and land conversion for human activities (Figs. 22.8, 22.9 and 22.10).

The land use land cover change analysis from 1998 to 2023 (Table 22.2) indicates a substantial reduction in wetland area, declining from 3.12 square kilometers in 1998 to 1.33 square kilometers in 2023. This represents a significant 42.6% decrease in wetland extent over the specified time period. Such a decline highlights the alarming rate of wetland loss within the urban landscape, which can have far-reaching

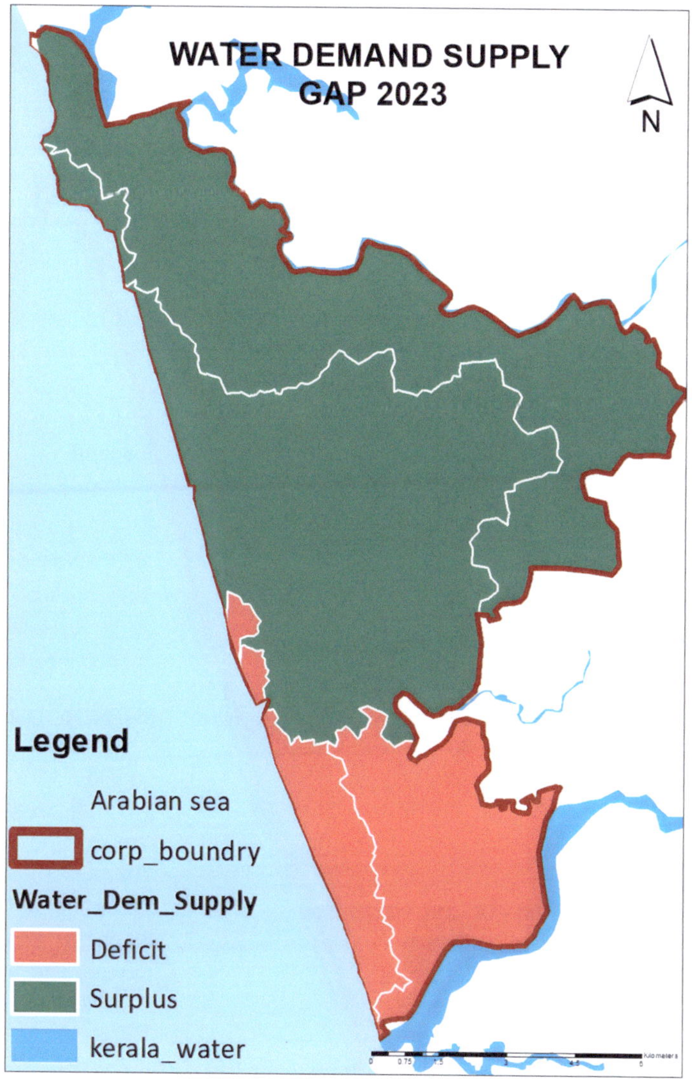

Fig. 22.7 Water demand supply gap. (Source: Author generated)

ecological consequences, including habitat fragmentation, loss of biodiversity, and diminished ecosystem services.

The runoff and groundwater recharge (Fig. 22.11) have been calculated for the three years in order to understand the change in groundwater recharge over the years. The observed decrease in wetland area over the years has indeed contributed to a reduction in groundwater recharge, as wetlands play a crucial role in replenishing groundwater supplies through infiltration and storage of rainwater.

The Normalized Difference Vegetation Index (NDVI) assessment maps of 1998, 2013, and 2023 which are developed after processing of satellite imageries of the region, indicate a drastic decrease in waterbodies and a conversion of water bodies into vegetated land. This remote sensing analysis also shows encroachment happening on the inlet streams of the wetland and increased urbanization in and around the wetland (Figs. 22.12, 22.13, and 22.14).

To understand the qualitative transformation of urban wetland, the Water Quality Index analysis has been carried out for 2017 and 2021. The 2021 readings show the results during COVID. The Water Quality Index (WQI) is a tool used to describe water quality by condensing various

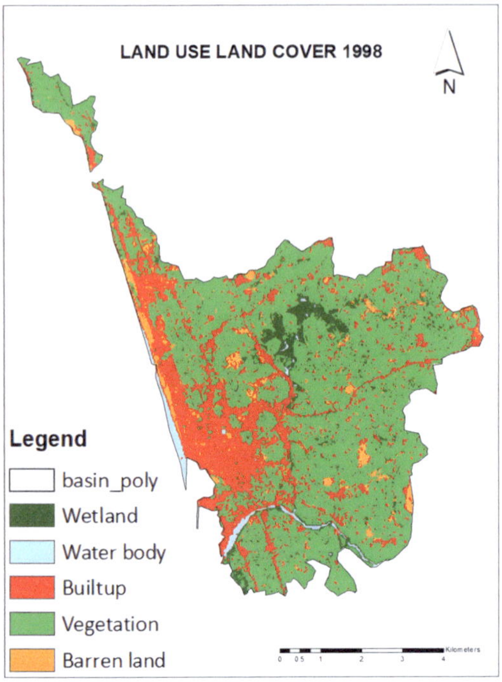

Fig. 22.8 LULC map 1998. (Source: Author generated)

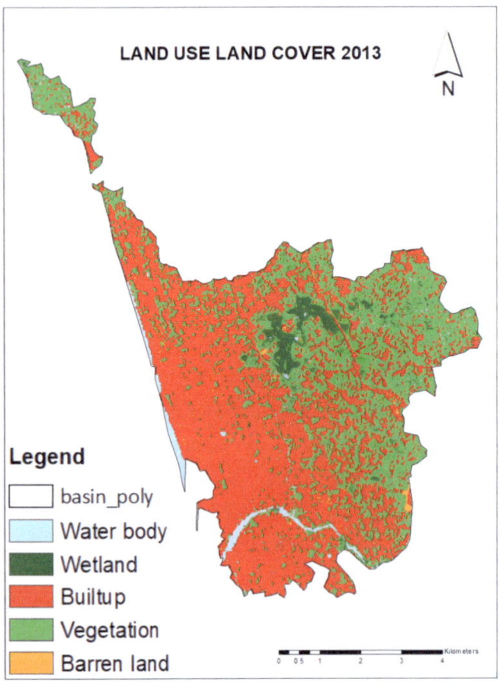

Fig. 22.9 LULC map 2013. (Source: Author generated)

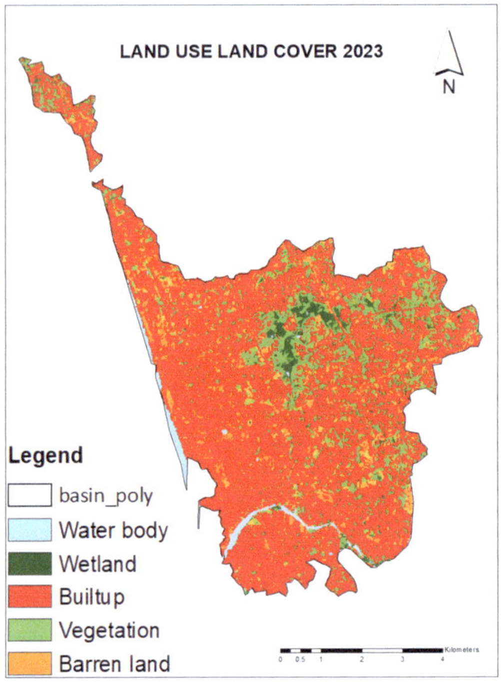

Fig. 22.10 LULC map 2023. (Source: Author generated)

physical, chemical, and biological factors into a single value ranging from 0 to 100 (Chidiac et al. 2023). Here, the water quality parameters considered are pH, turbidity, total dissolved solids, total hardness, calcium, chloride, magnesium, and nitrate. The formula used for the calculation of the Water Quality Index is

$$\mathrm{WQI} = \sum \mathrm{Qi}Wi / \sum wi,$$

where Wi (unit weight) = k/Si

Qi is a quality rating.

Where k is a constant of proportionality.

Qi = Vi / Si * 100, where Vi is the observed value and Si is the standard value (Ahmad et al. 2012). The comparison of the Water Quality Index (WQI) between 2017 and 2021 in and around the wetland reveals a significant improvement in water quality conditions by 2021, which is attributed to the impact of the COVID-19 pandemic. The 2021 WQI indicates better water quality compared to 2017, likely due to reduced anthropogenic activities and pollution during the lockdown periods associated with COVID-19

Table 22.2 LULC change over years

LULC area in sq. km					
Year	Wetland	Water body	BUILTUP	Vegetation	Barren land
1998	3.12	1.3	10.93	30.05	3.5
2013	2.08	1.16	29	16.33	0.39
2023	1.33	0.89	36.88	5.7	4.17

Source: Author calculations

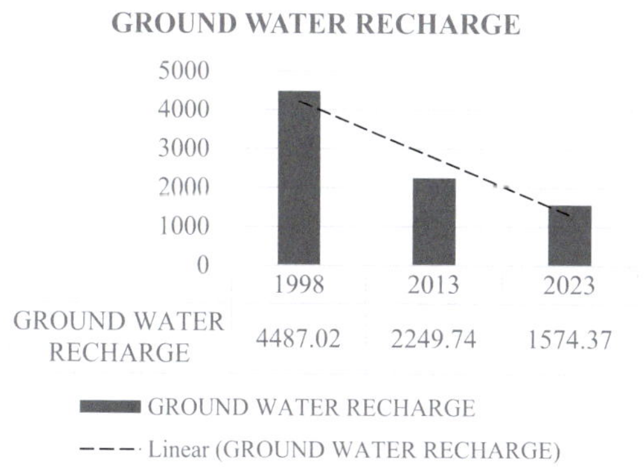

Fig. 22.11 Groundwater recharge changes over the years

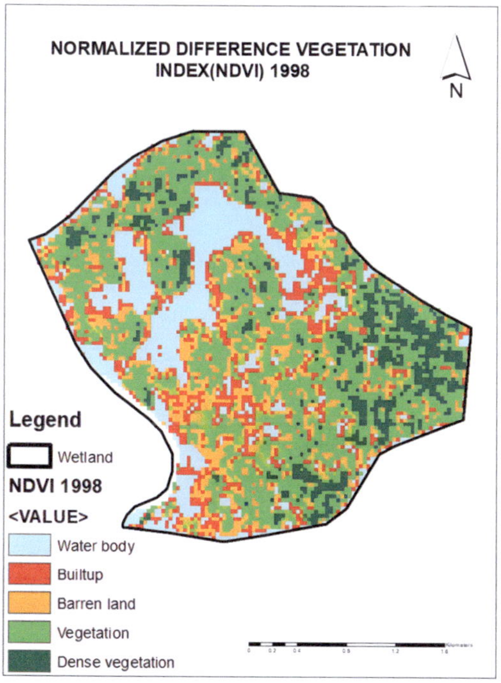

Fig. 22.12 NDVI Map 2013. (Source: Author generated)

restrictions. The decrease in industrial operations, vehicular traffic, and human activities during the pandemic likely resulted in reduced pollution inputs such as industrial discharge, urban runoff, and agricultural runoff into the wetland and surrounding water bodies (Figs. 22.15, 22.16, and 22.17; Table 22.3).

The cost of water treatment is calculated considering Rs 12 as the unit cost per liter according to the Ministry of Urban Development's presentation on water treatment. The cost of treating water annually is estimated to be Rs 4.72 lakhs.

22.4.3 Ecosystem Service Valuation

Ecosystem service valuation for wetlands involves assessing the economic, social, and environmental benefits provided by these ecosystems to society. Through methods such as market-based approaches, nonmarket valuation techniques, and ecological production function analysis. By collecting and analyzing relevant data on wetland characteristics and ecosystem functions, decision-makers can make informed choices regarding land use planning, natural resource management, and conservation strategies.

Valuing the cultural services provided by wetlands involves assessing the nonmaterial benefits that these ecosystems offer to society, including recreation, aesthetics, cultural heritage, and spiritual significance. Cultural ecosystem service valuation methods aim to quantify the subjective perceptions and preferences of individuals or communities regarding the cultural value of wetlands. The contingent valuation method or travel cost methods analyze the costs people

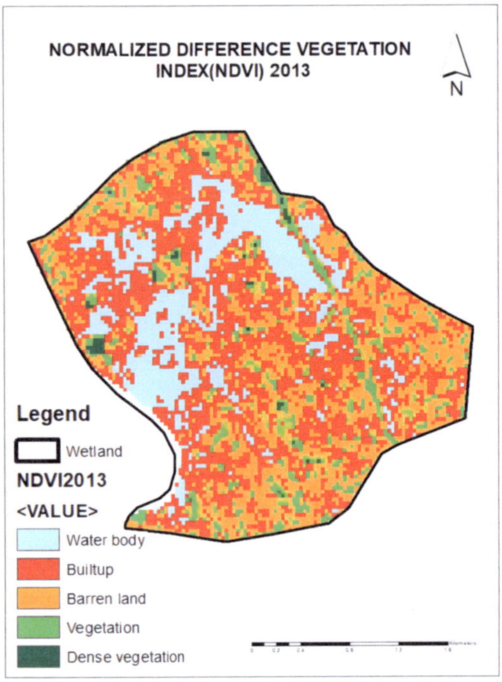

Fig. 22.13 NDVI Map 1998. (Source: Author generated)

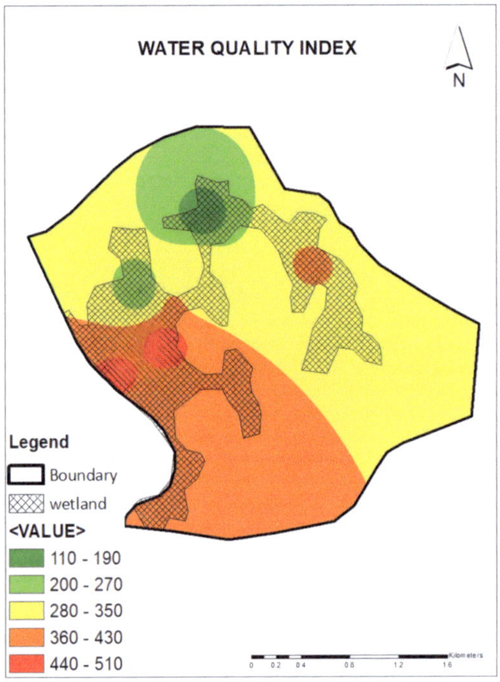

Fig. 22.15 WQI Map 2021. (Source: Author generated)

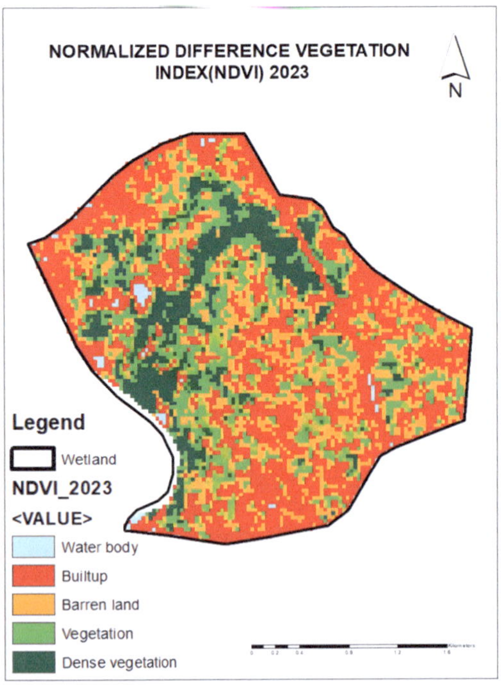

Fig. 22.14 NDVI Map 2023. (Source: Author generated)

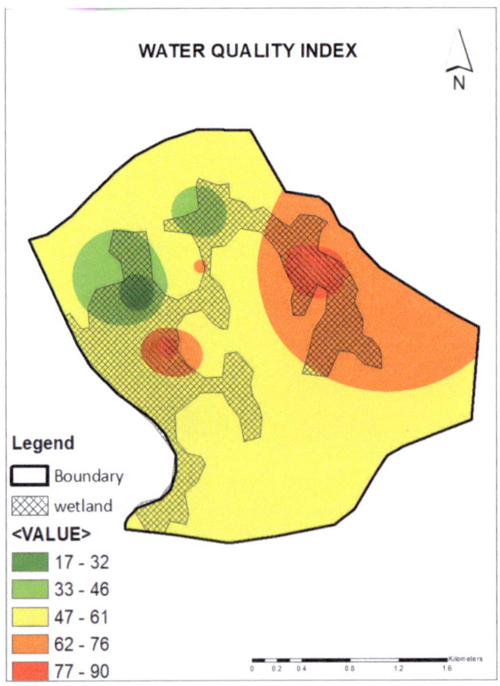

Fig. 22.16 WQI Map 2017. (Source: Author generated)

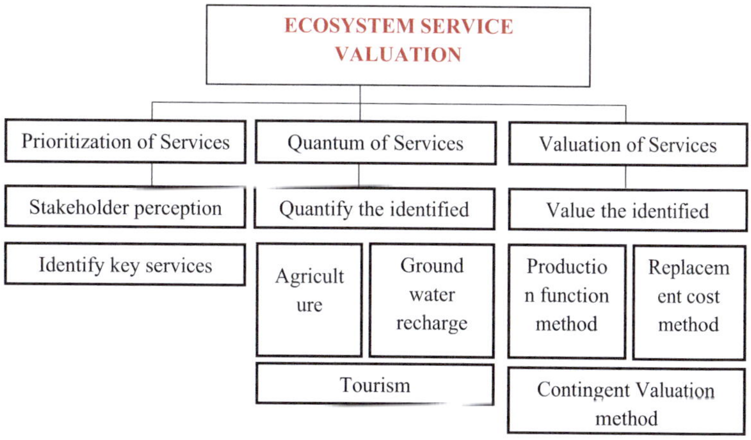

Fig. 22.17 Prioritization of ecosystem service valuation

Table 22.3 Cost of water treatment

Cost of water treatment			
Groundwater recharge	Extractable (25%)	Unit cost (Rs per liter)	Total cost
1574.375 m^3	393.5	12	4,72,200

incur to visit wetland sites for recreational purposes, providing insights into the value of access to these areas (Anna Bronzes 2025).

The existing average number of visitors per day in the Sarovaram biopark located in the Kottooli wetland ranges from 700 to 1200 persons. Additionally, the current entry ticket price is Rs 30 per adult.

To calculate the daily revenue generated from ticket sales, the following formula was used:

$$\text{Daily Revenue} = \text{Number of Visitors} \times \text{Ticket Price per Adult}$$

Using the given ranges:

For 700 visitors: Daily Revenue (700 visitors) = 700 × Rs 30 = Rs 21,000

$$\text{Daily Revenue (1200 visitors)} = 1200 \times \text{Rs } 30 = \text{Rs } 36{,}000$$

So, the Sarovaram biopark generates a daily revenue ranging from Rs 21,000 to Rs 36,000 from ticket sales, depending on the number of visitors.

Hence, the average revenue generated per day currently is Rs 28,500.

According to the survey conducted, taking 20 samples as a sample size using the contingent valuation method, the average willingness to pay was calculated.

Average total willingness to pay = Average willingness to pay entry fee + Average willingness to spend travel cost + Average willingness to pay additional cost.

Using this formula, the average total willingness to pay calculated is Rs 432.

Based on the existing ticket fare and the average visitors per day, the existing revenue generated per day and annually is calculated, which shows that the annual revenue generated is Rs 1.65 crore (Table 22.4).

Physical carrying capacity and rotational carrying capacity are concepts used in the management of natural areas, including parks, reserves, and other recreational areas, to ensure sustainable use while minimizing negative impacts on ecosystems. Physical carrying capacity refers to the maximum number of visitors or users that a natural area can accommodate without causing significant degradation to its ecological integrity, cultural resources, or visitor experience (Donaire et al. 2024). It is determined by factors such as habitat sensitivity, available infrastructure, trail networks, and resource availability. Rotational

Table 22.4 Existing revenue generated from cultural service

Existing revenue generated from cultural service					
Activity	Average visitors/day	Fee per visitor	Revenue generated annually (Rs)	Maintenance cost (Rs)	Net revenue annually (Rs)
Entry	950	30	1,04,02,500	3,25,500	1,00,77,000
Boating	150	120	64,80,000	10,720	64,39,280
Total					1,65,16,280

Table 22.5 Physical carrying capacity and rotation carrying capacity of wetlands

Physical carrying capacityPCC = A × V/a × Rf	
Total area (A)	14,821 sq.m
Area required per user (V/a)	0.5 sq.m
Open period (h)	9 h
Average time per visit (h)	2 h
Rotational factor (Rf) = open period/Avg. time per visit	4.5
PCC (persons/day)	33,347
RCC (persons/day)	10,986

carrying capacity involves managing visitor use through temporal or spatial rotation to distribute impacts more evenly across the landscape and reduce localized degradation. Here, both physical carrying capacity and rotational carrying capacity are calculated to understand the maximum visitors which the wetland can accommodate (Table 22.5).

The physical carrying capacity of the park per day is 33,347 persons, which becomes 1.21 crore persons annually and the real carrying capacity is 10,986 persons per day, which becomes 40.09 lakh persons annually (Table 22.6).

The potential revenue estimation is carried out considering the average willingness to pay and the rotational carrying capacity of the wetland, which gives the potential number of visitors annually. This shows that the wetland has the potential to generate a revenue of Rs 173.19 crore annually (Table.22.5).

Regulatory services provided by ecosystems, such as groundwater recharge, flood regulation, water purification, and climate regulation, play a critical role in maintaining environmental quality and supporting human well-being. The replacement cost method is one approach used to assess the economic value of regulatory services by estimating the cost of artificial alternatives or human interventions that would be required to replicate or replace the services provided by ecosystems.

22.4.4 Socioeconomic Dependency on Wetlands

The socioeconomic dependency on wetlands was assessed through a primary survey conducted in the neighboring wards of the wetland using the simple random sampling method. This approach involved randomly selecting households or individuals within the selected wards to gather data on their reliance on wetland resources and services for livelihoods, income generation, food security, and cultural practices. Survey questions may have focused on activities such as fishing, agriculture, water provision, tourism, and handicrafts, as well as perceptions of wetland importance and threats. By employing a random sampling technique, the survey aimed to ensure representative data collection and provide insights into the diverse ways in which communities depend on wetlands for their socioeconomic well-being. Based on the survey results from a sample size of 30 households, the analysis of ancestral occupation, preceding occupation, and current occupation provides valuable insights into the socioeconomic dynamics and historical

Table 22.6 Potential revenue (cultural services)

Potential revenue generated from cultural service					
Activity	Average visitors/day	WTP per visitor	Revenue generated annually	Maintenance cost	Net revenue annually
Entry	10,986	Rs 432	1,73,22,72,480	3,25,500	1,73,19,46,980

Fig. 22.18 Change in occupation

evolution of livelihoods in the neighboring wards of the wetland. Ancestral occupations reflect traditional livelihood activities passed down through generations, which may include farming, fishing, livestock rearing, or artisanal crafts associated with wetland ecosystems. Preceding occupations represent the transitional phase of livelihoods, indicating shifts or adaptations in response to socioeconomic changes, environmental degradation, or technological advancements. Current occupations reflect the present-day livelihood strategies pursued by households, which may involve a combination of traditional, alternative, or nonresource-dependent activities. By comparing and analyzing ancestral, preceding, and current occupations, the survey aims to understand the changing socioeconomic dynamics, identify drivers of livelihood transitions, and inform targeted interventions or support mechanisms to enhance community resilience, promote sustainable livelihoods, and conserve wetland ecosystems (Fig. 22.18).

The socioeconomic dependency on wetlands has witnessed a significant decline over the years, particularly due to the rapid urbanization and shifting livelihood patterns among the local population. This trend has resulted in the degradation and improper management of wetlands, as traditional reliance on wetland resources diminishes. The drastic reduction in dependency on agricultural activities, from 71% to 2% in the current scenario, underscores the magnitude of this change. As urbanization accelerates, residents increasingly transition away from traditional agricultural livelihoods toward nonagrarian occupations, such as industrial employment, service-sector jobs, and urban-based economic activities. Consequently, wetlands, which were historically central to agricultural practices and rural livelihoods, face neglect, encroachment, pollution, and degradation.

22.4.5 Drivers of Wetland Degradation

As per the stakeholders' perception survey, the principal causes of wetland degradation were pinpointed, with encroachment, pollution of the Canoli canal, and urbanization emerging as the primary drivers. These factors were given the highest priority in the survey analysis, signifying their profound impact on wetland health and sustainability. Encroachment denotes unauthorized occupation and development activities encroaching upon wetland areas, leading to habitat loss and fragmentation. Pollution of the Canoli canal exacerbates water quality degradation, posing threats to aquatic life and ecosystem balance. Urbanization, characterized by the expansion of urban areas into wetland fringes, results in habitat destruction and alters natural hydrological patterns. Detailed analysis of these drivers is essential to comprehend their underlying causes, spatial distribution, and ecological ramifications, facilitating the formulation of targeted interventions and management strategies to mitigate wetland degradation effectively and foster sustainable conservation practices.

Analyzing stress in drainage patterns involves overlaying built-up areas from land use land cover (LULC) data of different years (in this case, 1998, 2013, and 2023) with a stream order map. By doing so, it becomes possible to identify areas where human development, represented by built-up areas, encroaches upon natural waterways, potentially leading to alterations in drainage patterns and reduced water levels in wetlands. The overlay process allows for the visualization of where urbanization or other forms of development have occurred in proximity to streams of varying orders. This analysis aids in understanding the extent of encroachment and its potential impacts on hydrological processes, such as surface runoff, groundwater recharge, and wetland hydrology. By identifying areas of stress in drainage patterns which has been identified through the NDVI analysis around the stream orders, the stakeholders can implement appropriate management strategies to mitigate the adverse effects of development on wetland ecosystems and water resources (Fig. 22.19).

The analysis indicates a significant encroachment on the inlet streams of the wetland by 2023, resulting in a reduced water level within the wetland. This finding shed light on the observed changes in the Normalized Difference Vegetation Index (NDVI) map between 2013 and 2023, where vegetation appears in place of water in the latter. The encroachment of inlet streams, coupled with the associated reduction in water levels, likely led to changes in the wetland's hydrological regime, influencing vegetation dynamics as reflected in the NDVI values. The decrease in water availability may have resulted

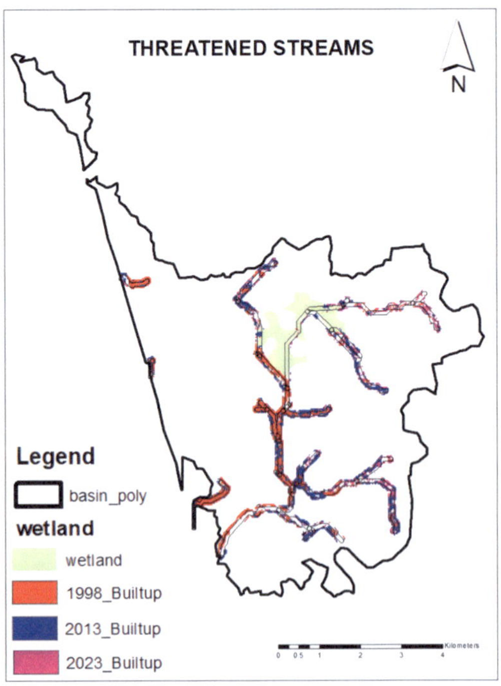

Fig. 22.19 Stream encroachment map. (Source: Author generated)

in the expansion of vegetated areas within the wetland as vegetation adapted to changing moisture conditions or colonized areas left dry due to reduced water levels.

22.4.6 Urban Water Security Framework Integrating Wetland Conservation

The conceptual framework depicted in the diagram underscores the critical integration of urban water security and wetland conservation as a unified approach to addressing contemporary urban environmental challenges. It presents wetlands not merely as ecological assets but as functional components of urban infrastructure capable of alleviating water deficits, regulating hydrological flows, and enhancing the resilience of cities. By positioning wetlands as both providers and regulators of water resources, the framework highlights their potential to bridge the urban supply-demand gap through ecosystem-based solutions.

Fig. 22.20 illustrates how urban water security is intricately connected to the condition of wetland ecosystems. When wetlands are intact, they contribute positively by adding value through their ecosystem services—such as water purification, flood regulation, and groundwater recharge. However, when these systems are degraded, their ability to support water security diminishes significantly, leading to reduced water availability and compromised water quality. This deterioration is often the result of unchecked urban expansion, pollution, and inadequate policy interventions. Thus, the health of wetlands directly influences the sustainability of urban water systems.

The framework proposes a series of interventions to mitigate these issues and restore the functionality of wetlands within urban contexts. These include the conservation of ecosystem services, sustainable extraction of water from wetlands, protection and restoration of riparian zones, perimeter control to prevent encroachment, treatment of polluted inflows, and the overall enhancement of ecosystem functions. These strategies are not only meant to recover lost ecological value but also to meet existing and projected urban water deficits in a sustainable manner. By integrating these measures, the framework offers a pathway for cities to meet water security goals while promoting environmental stewardship.

22.5 Conclusions

The wetland's groundwater recharge capacity, estimated for the year 2023 is 566.75 million liters per year, representing a valuable ecosystem service provided by the wetland within the Kallayi River basin. The wetland's contribution of 566.75 million liters per year plays a significant role in supporting local water availability and ecosystem health. It serves as a natural reservoir, helping to maintain groundwater levels, sustain stream flow, and mitigate the impacts of droughts and water scarcity.

Furthermore, it is noteworthy that this recharge capacity accounts for approximately 2.47% of the total water requirement of the Kallayi River basin. This indicates the importance of the wetland's ecosystem services in meeting the water needs of the surrounding region. By providing a substantial portion of the water required for various purposes such as irrigation, domestic use, and industrial activities, the wetland enhances the resilience and sustainability of water resources within the basin. The wetland's groundwater recharge capacity contributes to the ecosystem service valuation by providing a quantifiable value in terms of water volume replenished annually. This value underscores the importance of preserving and conserving wetland ecosystems for their critical role in maintaining hydrological cycles, supporting biodiversity, and providing essential services to human communities.

The analysis reveals alarming trends in both the quantity and quality of the wetland ecosystem. Over the span of 25 years, from 1998 to 2023, the wetland area has shrunk significantly from 3.12 sq.km to 1.33 sq.km, representing a substantial 42.6% decrease. This reduction signifies a loss in

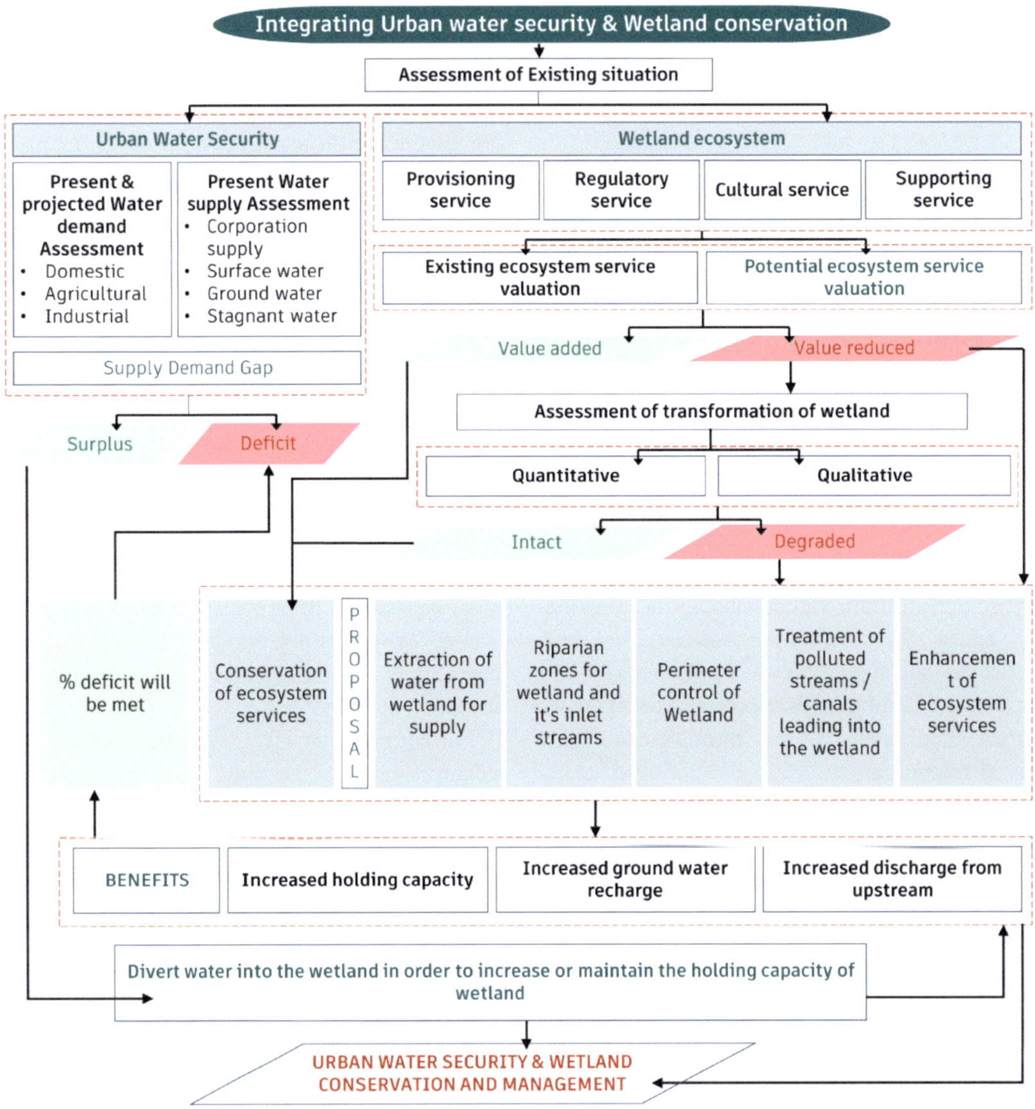

Fig. 22.20 Framework integrating urban water security and wetland conservation. (Source: Author generated)

ecosystem service value (ESV), indicating the diminished capacity of the wetland to provide crucial environmental benefits. Moreover, the quality of the wetland is deteriorating due to external pressures such as urbanization and pollution.

In addition to the costs associated with treating and mitigating the impacts of wetland degradation, it is crucial to consider the value of regulatory services provided by the wetland ecosystem. One such service is the groundwater recharge facilitated by the wetland. The replacement cost of the groundwater recharge pit represents the ecosystem service value (ESV) gained from this regulatory function. This underscores the importance of preserving and restoring wetland ecosystems not only for their intrinsic ecological value but also for the tangible benefits they provide to society, such as groundwater recharge. Incorporating the ESV gained from regulatory services into decision-making processes can help prioritize conservation efforts and ensure the sustainable management of wetland resources for the well-being of both present and future generations.

The analysis reveals a significant disparity between the current revenue generated from the cultural services provided by the wetland and its potential to generate revenue, highlighting the untapped economic value of these services. Currently, the wetland generates revenue amounting to Rs 1.65 crore annually. However, its full potential in terms of cultural service provision suggests a staggering potential revenue of Rs 173.19 crore annually. This represents the potential gain in the ecosystem service value (ESV) that could be realized through the optimal utilization of the wetland's cultural services, such as recreational activities, tourism, and cultural events. Harnessing this potential requires strategic investments in promoting sustainable tourism, enhancing recreational infrastructure, and fostering cultural engagement within the wetland ecosystem. By unlocking the economic value of cultural services, stakeholders can not only bolster local economies but also foster greater appreciation and stewardship of the wetland's natural and cultural heritage.

The framework integrating Urban Water security and Wetland conservation aligns with the book's overarching objective to position wetlands at the center of urban resilience strategies. It synthesizes the core elements explored throughout the study: analyzing urban water systems, assessing wetland contributions, understanding degradation dynamics, exploring socioeconomic impacts, and proposing an integrated framework for urban water security. In doing so, it echoes the broader discourse on *Wetlands and Urban Resilience: Ecological Studies—Analysis and Synthesis*, reaffirming that the future of urban sustainability depends on our ability to harmonize built and natural environments through informed, inclusive, and ecologically grounded approaches.

Acknowledgments We are grateful to the local communities, the wetland authority, and the water supply department of the Calicut District, Kerala, for providing us with relevant data and information on the water supply network and scenario in the study area. The paper also acknowledges the Department of Planning, School of Planning and Architecture Vijayawada for providing the academic and administrative support for undertaking this research.

Potential Conflicts of Interest On behalf of all authors, the corresponding author states that there is no conflict of interest.

Informed Consent Informed consent was obtained from all individual participants included in the study.

Authors' Contribution All authors have contributed equally to this work.

Data Availability Statement The datasets during and/or analyzed during the current study are available from the corresponding author on reasonable request.

Funding The research was not supported by any funding.

References

Ahmad IK, Salih NM, Khadi TR, Nzar YHC (2012) Determination of water quality index (WQI) for qalyasan stream in Sulaimani City/Kurdistan region of Iraq. Int J Plant Anim Environ Sci 2(4):31–50

Alikhani S (2021) Urban wetlands: a review on ecological and cultural values. Water:13–22

Anna Bronzes LH (2025) A comparison of valuation methods for cultural ecosystem services in support of ecosystem accounting. One Ecosyst 10:e108556

Cept C-W (2017) Urban water security planning toolkit. CEPT University

Chidiac S, El Najjar P, Ouaini N, El Rayess Y, El Azzi D (2023) A comprehensive review of water quality indices (WQIs): history, models, attempts and perspectives. Rev Environ Sci Biotechnol 22(2): 349–395

Donaire JA, Galí N, Coromina L (2024) Evaluating tourism scenarios within the limit of acceptable change framework in Barcelona. Ann Tour Res Empir Insights 5(2):100145

EEA (1999) *eea.europa*. Retrieved from https://www.eea.europa.eu/help/glossary/eea-glossary/water-demand

Mahreen Matto SJ (2017) Water efficiency and conservation in urban India. Centre for Science and Environment

Mynepalli KC, Sridhar MA (2020) Chapter 7: Water, sanitation and hygiene (WASH) disease prevention and control in low resource countries. In: Sustainable water engineering. Elsevier, pp 99–120

Tarafdar AK, Bjønness HC (2010) Environmental premises in planning for sustainability at the local level in large southern cities. Case study: Calcutta, India—suggesting the PRETAB planning process model. Int J Sustain Dev World Ecol 17(01):24–38

Tien Shiao AM (2015, February). Retrieved from World Resources Institute: https://www.wri.org/insights/3-maps-explain-indias-growing-water-risks

Town and Country Planning Department, G. o (2017) Master plan for Kozhikode urban area—2035. Town and Country Planning Department, Government of Kerala, Kozhikode

UNEP (2017, January 31). Retrieved from https://www.unep.org/news-and-stories/story/wetlands-limit-impact-floods-drought-cyclones

UNESCO (2012) International hydrological programme (IHP) eighth phase: water security: responses to local, regional and global challenges, strategic plan, IHP-VIII (2014–2021)

MIX
Papier aus verantwortungsvollen Quellen
Paper from responsible sources
FSC® C105338

If you have any concerns about our products,
you can contact us on
ProductSafety@springernature.com

In case Publisher is established outside the EU,
the EU authorized representative is:
Springer Nature Customer Service Center GmbH
Europaplatz 3, 69115 Heidelberg, Germany

Printed by Libri Plureos GmbH
in Hamburg, Germany